QUINOLINES

Part III

This is the thirty-second volume in the series

THE CHEMISTRY OF HETEROCYCLIC COMPOUNDS

THE CHEMISTRY OF HETEROCYCLIC COMPOUNDS

A SERIES OF MONOGRAPHS

EDWARD C. TAYLOR

Editor

QUINOLINES

Part III

Edited by

Gurnos Jones

DEPARTMENT OF CHEMISTRY
UNIVERSITY OF KEELE
STAFFORDSHIRE

Author of this volume

John V. Greenhill

DEPARTMENT OF CHEMISTRY
UNIVERSITY OF FLORIDA
GAINESVILLE
FLORIDA 32611
USA

AN INTERSCIENCE® PUBLICATION

JOHN WILEY & SONS
CHICHESTER · NEW YORK · BRISBANE · TORONTO · SINGAPORE

Other Wiley Editorial Offices

John Wiley & Sons, Inc., 605 Third Avenue,
New York, NY 10158-0012, USA

Jacaranda Wiley Ltd., G.P.O. Box 859, Brisbane,
Queensland 4001, Australia

John Wiley & Sons (Canada) Ltd, 22 Worcester Road,
Rexdale, Ontario M9W 1L1, Canada

John Wiley & Sons (SEA) Pte. Ltd, 37 Jalan Pemimpin   05-04,
Block B, Union Industrial Building, Singapore 2057

Library of Congress Cataloging-in-Publication Data:
(Revised for volume 3)

Quinolines.

   (The Chemistry of heterocyclic compounds; V. 32)
   'An Interscience publication.'
   Includes bibliographical references and indexes.
   1. Quinoline.   I. Jones, Gurnos.   II. Series.
QD401.Q56   547'.596   76-26941
ISBN 0 471 99437 5

British Library Cataloguing in Publication Data:
Quinolines
   Pt. 3
   1. Quinolines
   I. Jones, Gurnos II. Series
   547'.596

   ISBN 0 471 92644 2

Typeset at Thomson Press (India) Ltd., New Delhi
Printed and bound in Great Britain by Biddles Ltd., Guildford, Surrey

Introduction to the Series

The series *The Chemistry of Heterocyclic Compounds*, published since 1950 under the initial editorship of Arnold Weissberger, and later, until Dr. Weissberger's death in 1984, under our joint editorship, was organized according to compound classes. Each volume dealt with syntheses, reactions, properties, structure, physical chemistry, and utility of compounds belonging to a specific ring system or class (e.g. pyridines, thiophenes, pyrimidines, three-membered ring systems). This series, which has attempted to make the extraordinarily complex and diverse field of heterocyclic chemistry as readily accessible and organized as possible, has become the basic reference collection for information on heterocyclic compounds.

However, many broader aspects of heterocyclic chemistry are now recognized as disciplines of general significance which impinge on almost all aspects of modern organic and medicinal chemistry. For this reason we initiated several years ago a parallel series entitled *General Heterocyclic Chemistry* which treated such topics as nuclear magnetic resonance of heterocyclic compounds, mass spectra of heterocyclic compounds, photochemistry of heterocyclic compounds, the utility of heterocyclic compounds in organic synthesis, and the synthesis of heterocyclic compounds by means of 1,3-dipolar cycloaddition reactions. These volumes were intended to be of interest to all organic chemists, as well as to those whose particular concern is heterocyclic chemistry.

It has become increasingly clear that this rather arbitrary distinction between the two series creates more problems than it solves. We have therefore elected to discontinue the more recently initiated series *General Heterocyclic Chemistry*, and to publish all forthcoming volumes in the general area of heterocyclic chemistry in *The Chemistry of Heterocyclic Compounds* series.

Edward C. Taylor

Department of Chemistry
Princeton University
Princeton, New Jersey 08544

Preface

The third Part of Volume 32 deals with two major groups of quinoline compounds, the aldehydes and the ketones. The treatment differs from that adopted in previous Parts, as the material is arranged entirely in the form of information on synthesis. Reactions were found to be largely trivial, and inclusion of many trivial reactions was felt to be unnecessary in a book designed for research workers. Such a treatment necessarily also includes much information on properties, but the chapters make no pretence of recording all information on the chemical and physical properties of quinoline aldehydes or ketones. Nevertheless, in common with Parts I and II, all references to quinoline aldehydes and ketones are included, as far as is humanly possible. It is our hope that this third Part will prove to be as indispensable as a reference work as are those already published.

Contents

CHAPTER 1

Quinoline Aldehydes

JOHN V. GREENHILL[*]

*Pharmaceutical Chemistry, School of Pharmacy,
University of Bradford
Bradford BD7 1DP, England*

[*] Present address: Department of Chemistry, University of Florida, Gainesville, Florida 32611, USA.

List of Tables

I. Introduction

In the preparation of this Chapter the literature has been reviewed to the end of *Chemical Abstracts* Volume 104.

Much of the chemistry involved will be familiar to even the most junior practising chemist, so most of the compounds are listed in Tables with brief indications of the methods of preparation. Where the aldehyde group occurs on a side chain it is impossible to be comprehensive using a manual search and I have relied on CAS online searching for these compounds; see Chapter 2, Section I. Many of these compounds do not lend themselves to tabulation, so their chemistry is discussed in the appropriate sections.

II. Preparation of Quinoline Carboxaldehydes

1. Selenium Dioxide Oxidations

All the unsubstituted quinoline carboxaldehydes have been prepared by selenium dioxide oxidation of the appropriate methylquinolines. Thus 2-methylquinoline (quinaldine) and 4-methylquinoline (lepidine) are readily oxidized in solvent at moderate temperatures. Detailed investigations of the conditions have been reported.[1] In a second study[2] it was shown that 2-methylquinoline was oxidized by selenium dioxide in ethanol or dioxane to the aldehyde contaminated by a trace of quinaldoin. It was stated that the conditions could be changed to make quinaldoin the major product; see Chapter 2, Section III. The 2-methyl- and 4-methyl- but not the 3-methyl-quinolines were oxidized by selenium dioxide in pyridine or dioxane to the corresponding carboxylic acids.[3] The 3-, 5-, 6-, 7- and 8-methylquinolines have been oxidized to aldehydes at high temperatures without solvent. A typical report[4] says that 8-methylquinoline and selenium dioxide were heated to 145–150 °C and a vigorous reaction then began. The temperature rose rapidly to 220 °C. Heating was continued to 250 °C to give a 70% yield of the aldehyde. Clearly a method to be treated with respect! The preparation of the selenium dioxide reagent has been investigated.[5] Best yields were obtained with freshly prepared selenium dioxide, but when the reagent was 12 months old it gave only traces of the aldehydes. The main product from 2-methylquinoline was quinaldoin and that from 4-methylquinoline, 1,2-di(4-quinolyl)ethene. Although one author[6] has claimed better yields with unsublimed material, this is not the experience of most users.

Selenium dioxide has been used to oxidize hydroxymethyl groups to the aldehydes,[7] but most workers prefer manganese dioxide or sodium *m*-nitrobenzenesulphonate for this.[8-10] Dibenzyl ether and selenium dioxide at 180–190 °C gave dibenzoyloxyselenium oxide, which converted 2-methylquinoline to the aldehyde at room temperature.[11] Benzene seleninic anhydride has recently been used to convert 4-methylquinoline to the aldehyde in high yield.[12]

Thiourea and 4-chloroquinaldine gave compound **1**, R = Me which was oxidized (SeO$_2$) to the aldehyde **1**, R = CHO.[13]

1, R = CHO; 9%, m.p. 171–172 °C

Selenium dioxide oxidation of dimethylquinolines has given only low yields of dialdehydes.[14] For example, oxidation of 2,4-dimethylquinoline gave the dialdehyde (4%) and 4-methylquinoline-2-carboxaldehyde (0.5%).[15] Better yields were achieved by ester reduction; see Section II.5.

2. Catalytic Methods

In a gas phase oxidation a mixture of 2-methylquinoline, water vapour and air was passed over 10% molybdenum trioxide on silica gel at 410 °C to give the aldehyde in a yield (62%) which compared favourably with that of other methods. Unchanged methylquinoline could be recovered. If the condensate from the reaction was kept at 80 °C for several hours it gave quinaldoin, but this reaction time could be reduced by the addition of potassium cyanide.[16,17] A trace of quinoline-2-carboxaldehyde was obtained when 2-methylquinoline was passed over a mixture of vanadium pentoxide, molybdenum trioxide and tungsten trioxide at 450–480 °C, but the major product was quinaldoin.[2] In another investigation several different catalysts [V_2O_5, V_2O_5/MoO_3, $Ca(VO_3)_2$ and $Cd(VO_3)_2$] were compared at temperatures from 350–600 °C. The main product was always the acid (11 to 60%) with quinoline-2-carboxaldehyde as a by-product.[18]

3. Ring Closure Methods

The Friedländer synthesis has been applied to the keto-acetal **2** with either acetone or cyclopentanone. The quinolines **3** were not isolated, but treated *in situ* with 4-chlorobenzaldehyde and then hydrolysed (aq. acid) to aldehydes **4**.[19] Compound **2** and the appropriate acetophenone gave the aldehyde **5**, Scheme 1.[20]

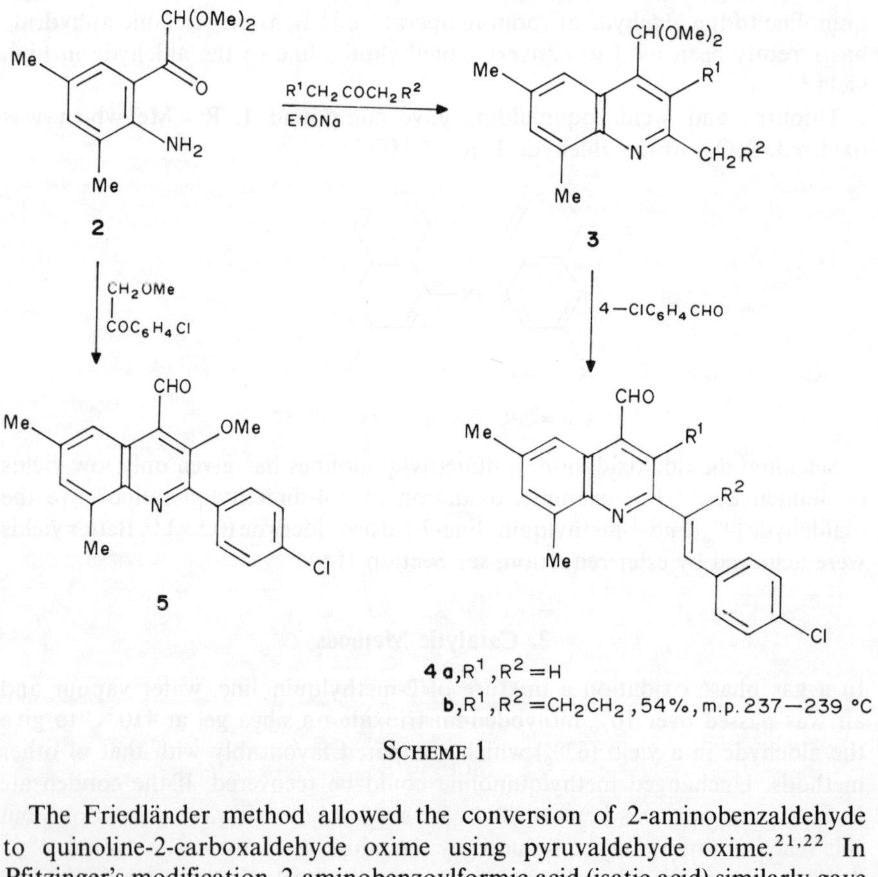

4 a,R^1,R^2 =H

b,R^1,R^2 =CH_2CH_2 ,54%,m.p. 237—239 °C

SCHEME 1

The Friedländer method allowed the conversion of 2-aminobenzaldehyde to quinoline-2-carboxaldehyde oxime using pyruvaldehyde oxime.[21,22] In Pfitzinger's modification, 2-aminobenzoylformic acid (isatic acid) similarly gave the oxime of 2-formylquinoline-4-carboxylic acid.[22] Pyruvaldehyde oxime also reacted with Schiff bases (from 2-aminobenzaldehydes and *p*-toluidine) under basic conditions to give quinoline-2-carboxaldehyde oximes.[23] Although none of these oximes has been hydrolysed, one was converted directly to the phenylhydrazone by phenylhydrazine in acidic alcohol.[23] Another author reported that quinoline-2-carboxaldehyde could be released from its phenylhydrazone by an acid catalysed exchange with 2,4-dinitrobenzaldehyde. The by-product, 2,4-dinitrobenzaldehyde phenylhydrazone, was highly insoluble and easily separated.[24]

The cinnamaldehyde derivative **6** reacted with malonic ester and sodium ethoxide to give a high yield of the thione aldehyde **7**. The suggested mechanism is shown in Scheme 2. The thione **7** could be alkylated with methyl chloroacetate or oxidized with iodine to the disulphide **8**. With ammonium hydroxide and sodium hypochlorite the thione aldehyde **7** gave the new heterocycle isothiazolo[5,4-*b*]quinoline.[25]

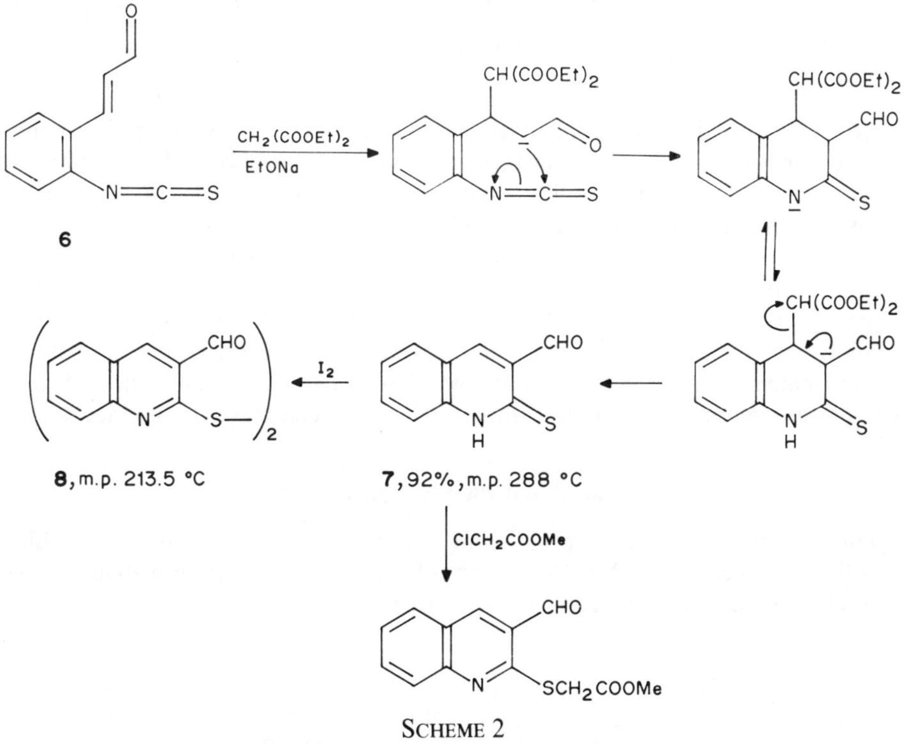

SCHEME 2

An ingenious new route to quinoline-3-carboxaldehydes has been developed.[26-31] Acetanilide or a substituted acetanilide **9**, dimethylformamide and phosphorus oxychloride are refluxed or heated in a sealed tube to give, in

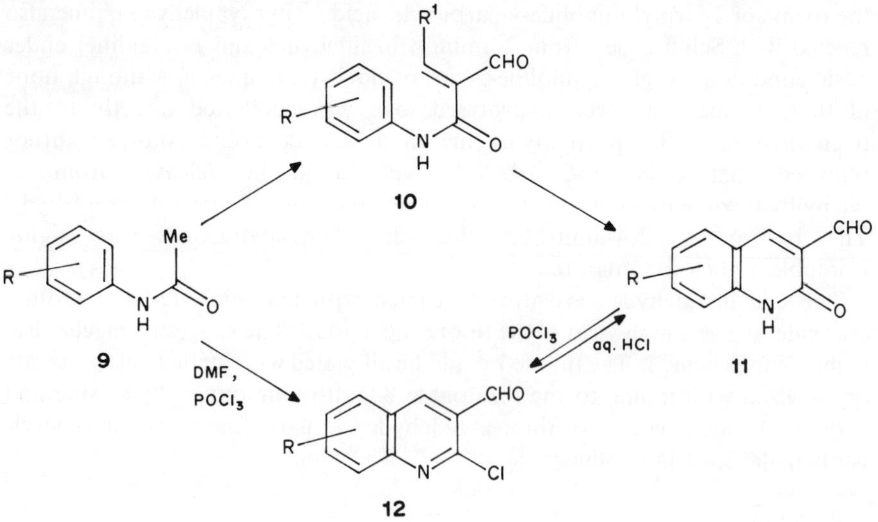

SCHEME 3

most cases, good yields of the aldehydes **12** directly. In a few examples where
the group R was deactivating, the product was the enaminedione **10**, $R^1 = NMe_2$.
This was hydrolysed by sodium hydroxide to the dialdehyde **10**, $R^1 = OH$,
which cyclized in polyphosphoric acid at 150 °C to the quinoline **11**.[29] This
sequence, Scheme 3, presumably indicated the route of the direct reaction. The
chlorine atom could be removed from compounds of type **12** by conversion to
the acetal with ethylene glycol followed by reduction (Zn, NaOH). Better, the
chlorine was exchanged for iodine (NaI, HI, MeCN) and the iodo derivative
was treated with butyllithium and hydrolysed. The chloro-aldehydes **12** are
important intermediates for the synthesis of several hetero fused quinolines.[30,31]

4. Radical and Photochemical Methods

Trioxane with hydrogen peroxide and ferrous ions gave the radical **13**. This
radical attacked vacant 2- or 4-positions of quinoline to give intermediates such
as **14** which could be hydrolysed to aldehydes, e.g. **15**.[32]

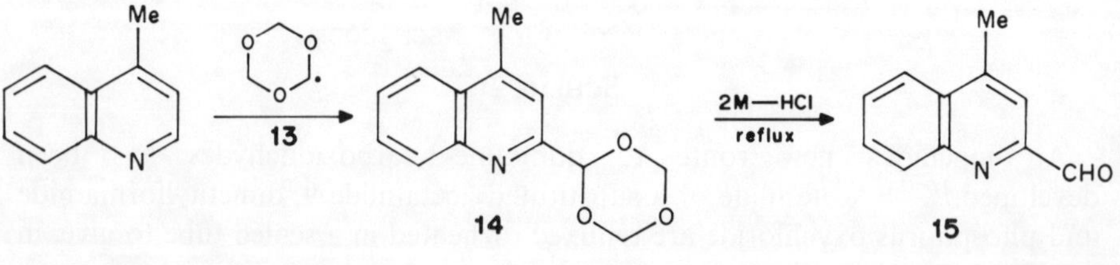

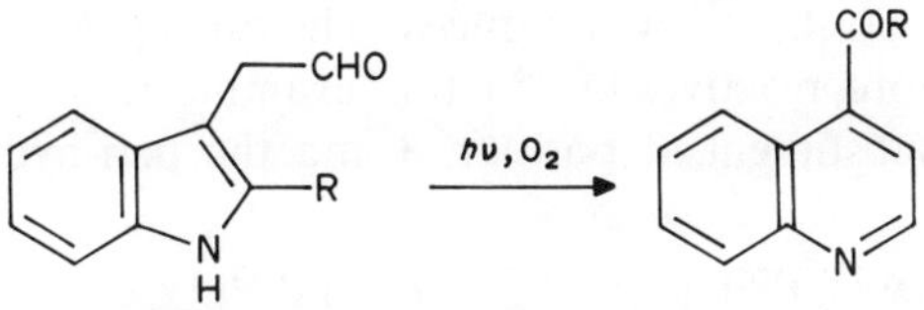

16   **17**

Photochemical oxidation of the indolylacetaldehyde **16**, R = H gave quinoline-4-carboxaldehyde **17**, R = H. The ketone **17**, R = Me was prepared similarly.[33]

5. Ester Reduction

This method has been employed for the conversions—or attempted conversions—of quinoline diesters to the dialdehydes. The most successful techniques[34-36] used diisobutylaluminium hydride (DIBAH) or lithium aluminium hydride (LAH) at −65 to −70 °C. For example, reduction of the dimethyl ester of quinoline-2, 3-dicarboxylic acid with LAH in toluene at −70 °C gave the dialdehyde in 50% yield, but the corresponding diethyl ester with DIBAH in toluene at −70 °C gave 60% of a mixture of the dialdehyde and ethyl 2-formylquinoline-3-carboxylate, which were separated by preparative gas–liquid chromatography (g.l.c.).[34] On the other hand, diethyl quinoline--2,4-dicarboxylate was reduced to the dialdehyde (80%) with DIBAH in toluene at −65 °C, but in THF at the same temperature only the ethoxycarbonyl group in position 2 was reduced to give 95% of the aldehydo-ester.[37] Similarly, when dimethyl quinoline-3, 4-dicarboxylate was treated with LAH in THF at −65 °C

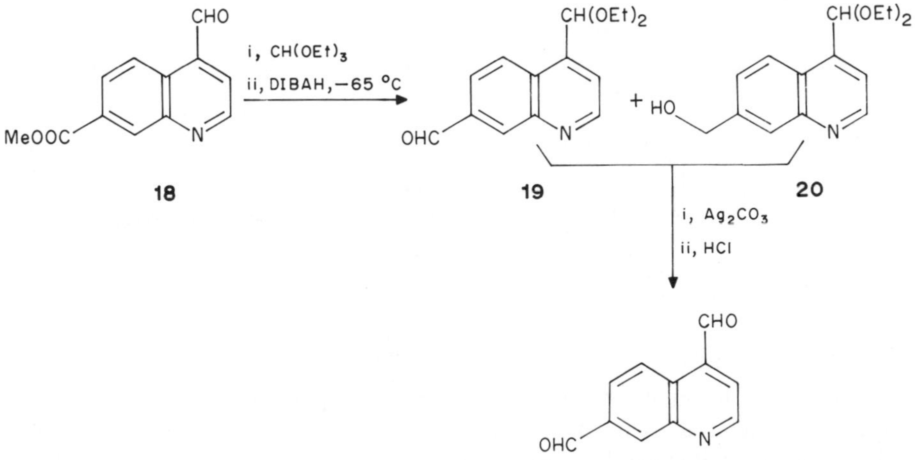

18   **19**   **20**

21

only the group at position 3 was reduced. The ester group in position 4 was expected to be more reactive, but in this example it was considered to be hindered by the substituent at position 3 and the peri-hydrogen at position 5.[38,39]

Compound **18** was prepared from the appropriate methylquinoline and selenium dioxide. The aldehyde group was protected as an acetal before ester reduction to a 1:4 mixture of the aldehyde **19** and the alcohol **20**. Oxidation of the crude mixture followed by acetal hydrolysis gave the 4,7-dialdehyde **21** in 50% overall yield. The 4,6-dialdehyde was prepared similarly.[37]

Some quinoline-2,3-dicarboxamides have also been reduced to dialdehydes with LAH in good yields.[40]

6. Hydrolysis of Fused Quinolines

Nitrous acid reacted with 3-amino-4-methylquinoline to give the triazinoquinoline **23**, presumably via the oxime **22**. Hydrolysis then gave the azido-aldehyde **24**.[41]

22	**23**	**24**

Hydrolysis of the pyrimidinoquinolinium salt **25** gave the amino-aldehyde **26**.[42]

25   **26**

7. Other Methods

Diazotized *p*-nitroaniline reacted with 1,2-dimethylquinolinium iodide in alkali with loss of the *N*-methyl group to give quinoline-2-carboxaldehyde *p*-nitrophenylhydrazone. The 1-ethyl- and 1-isoamyl-2-methylquinolinium salts gave the same product.[43]

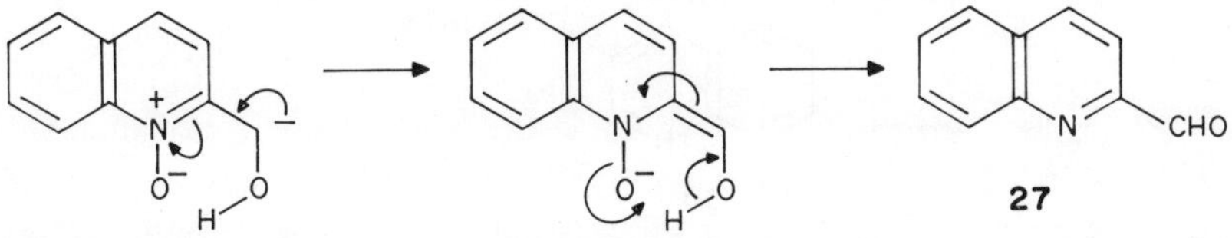

SCHEME 4

When 2-hydroxymethylquinoline *N*-oxide (1 mol) was treated with 1 equivalent of sodium hydroxide at 80 °C it gave quinoline-2-carboxaldehyde **27** in high yield. The suggested mechanism[44] is shown in Scheme 4.

The anion from 2-nitropropane can give both *C*- and *O*-alkylated products; 2-chloromethylquinoline reacted by *O*-alkylation to give an intermediate which was hydrolysed to quinoline-2-carboxaldehyde in poor yield. The 5-nitro derivative gave a mixture of the *C*-alkylated compound (24%) and 5-nitroquinoline-2-carboxaldehyde (21%). The 6-nitro- and 8-nitro-2-chloromethylquinolines gave only *C*-alkylated products.[45] The acidic alkyl groups of 2-methyl- and 4-methyl-quinolines can be deprotonated with sodamide in liquid ammonia and the anions treated with pentyl nitrite to give the corresponding aldehyde oximes in 56% and 67% yields respectively.[46]

A cycloaddition to anthranil has been used to prepare an aldehyde; see Scheme 5.[47]

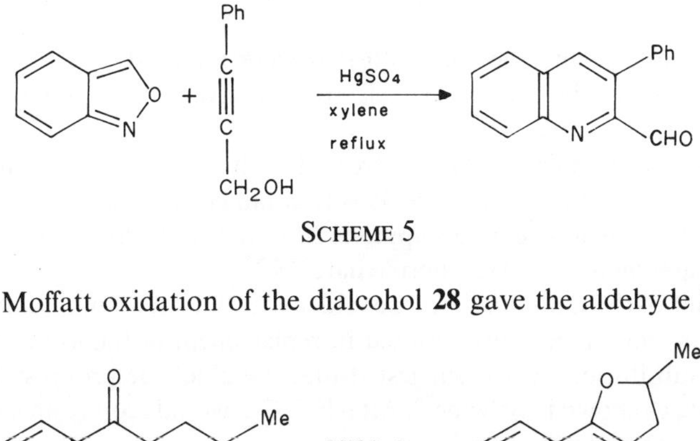

SCHEME 5

Pfitzer–Moffatt oxidation of the dialcohol **28** gave the aldehyde **29**.[48]

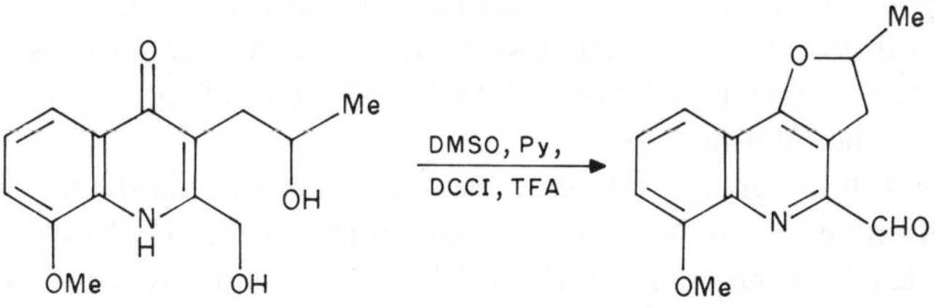

Heating the toluenesulphonylhydrazide **30** with sodium carbonate in ethylene glycol at 160 °C gave quinoline-3-carboxaldehyde in 32.5% yield. The 5-, 6- and 8-aldehydes were prepared similarly.[49]

30

Hydrogenation of 3-cyanoquinoline in the presence of semicarbazide gave the corresponding aldehyde semicarbazone, which was hydrolysed with aqueous acid to the free aldehyde.[50] Several other unsubstituted and substituted quinoline aldehydes have been prepared in this way.[51,52]

Basic hydrolysis of the dichloromethyl group of compound **31** gave the expected aldehyde, but the new group further activated positions 2 and 4 to

31   **32**   **33**

nucleophilic attack and the ring chlorines were replaced as well.[53] Acidic or basic hydrolysis of the product **32** selectively cleaved the 4-methoxy group to give the 4(1H)-quinolone **33**.[53,54]

When the benzazepinone **34** was treated with lithium chloride in DMF at 160 °C for 15 minutes, the aldehyde **36** was obtained. The suggested mechanism is shown in Scheme 6. The benzazepinone **37** gave the aldehyde **36** on oxidation, a reaction assumed to go via intermediate **35**.[55]

Treatment of the quinoline-3-carboxaldehyde **38** with ammonium hydroxide in THF at room temperature resulted in replacement of the 4-methoxy group without 2-subsitution. It was suggested that the aldehyde group stabilized the intermediate as shown in Scheme 7. Attack at C-2 would require an o-quinonoid structure for the intermediate.[56]

The oxazoline group of compound **39** was alkylated by methyl fluorosulphonate and reduced by sodium borohydride. Hydrolysis then gave the aldehyde **40**. In the final step the aldehyde **41** was produced with essentially complete transfer of chirality from C-4 to the biaryl.[57]

Reaction between 4-bromomethylquinoline and 4,6-diphenyl-2-pyridone N-oxide gave compound **42**. When compound **42** was pyrolysed at 160–170 °C it gave quinoline-4-carboxaldehyde. The 2-aldehyde was prepared similarly.[58]

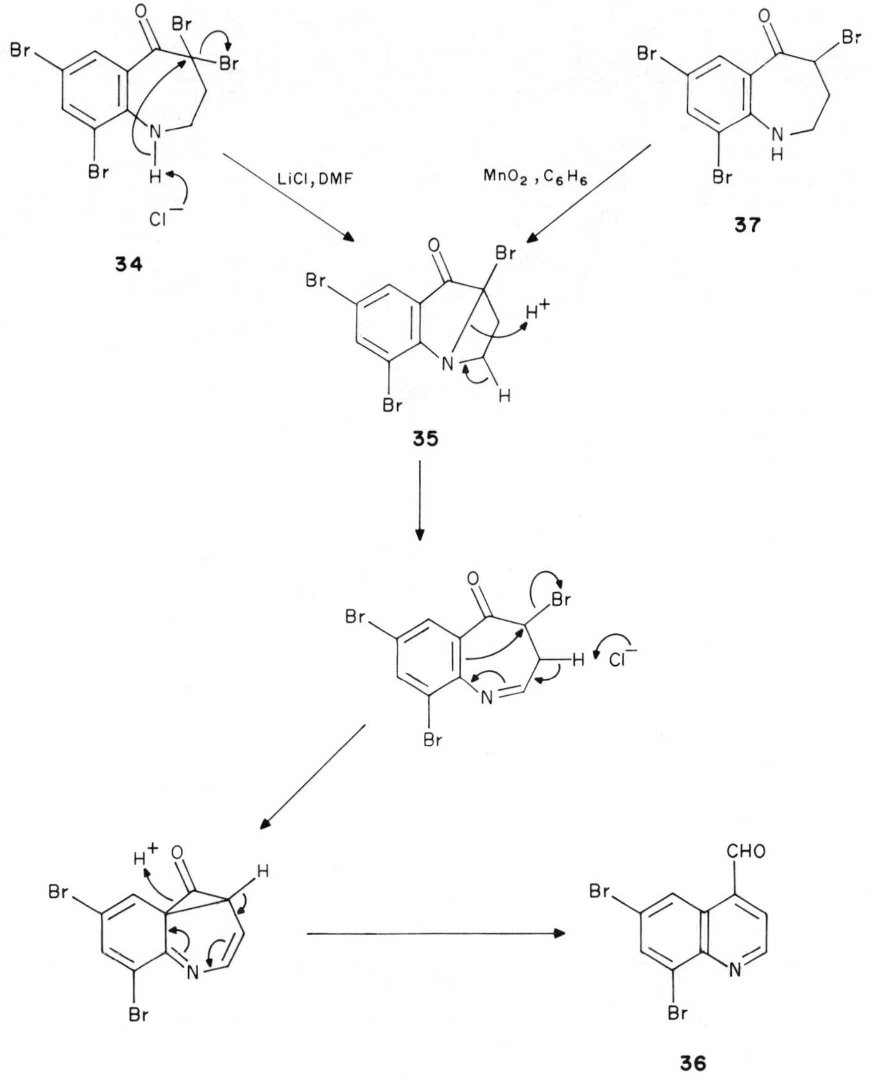

SCHEME 6

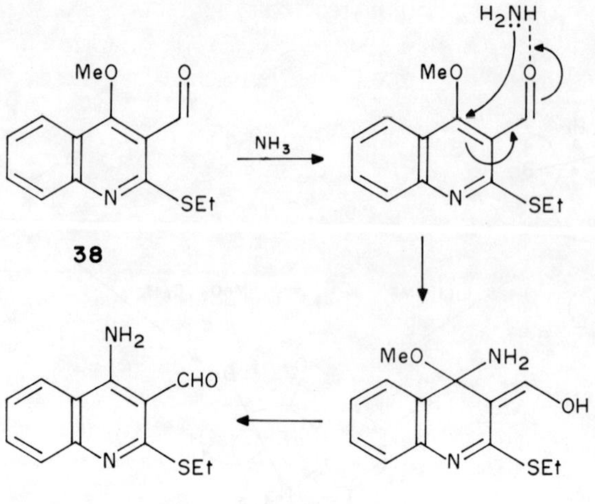

SCHEME 7

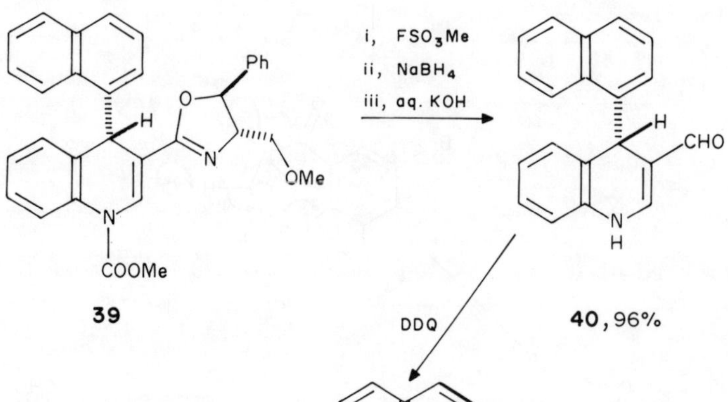

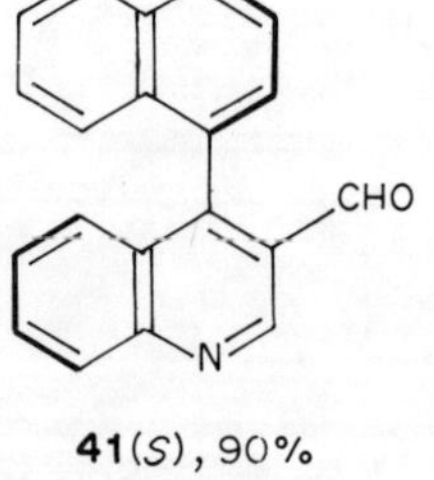

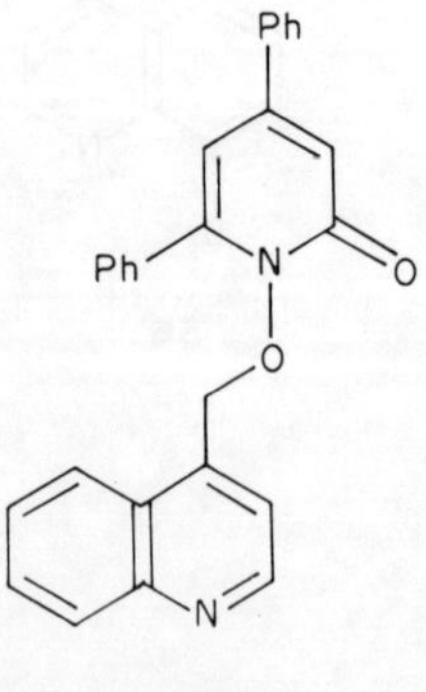

Chloral and 8-hydroxyquinoline reacted in acid to give the alcohol **43** in high yield. When treated with sodium methoxide, sodium acetonate or potassium hydroxide the alcohol **43** was converted into the aldehyde **44**. The suggested route is shown in Scheme 8.[59–61]

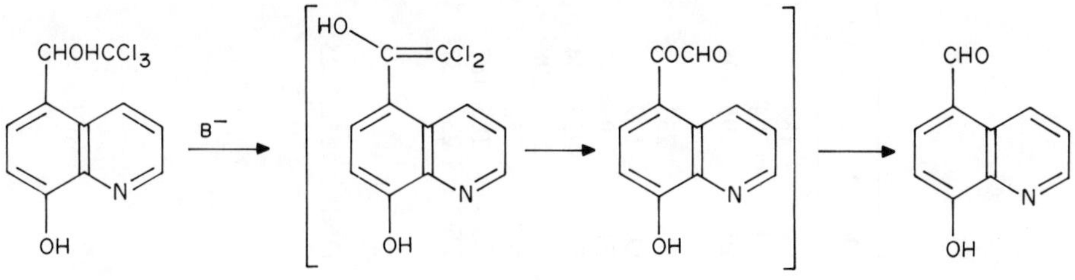

43 **44**

SCHEME 8

Oxidation of 1,2,3,4-tetrahydroquinoline-6-carboxaldehyde to the fully unsaturated aldehyde was achieved with copper (II) chloride in refluxing pyridine or with chloranil in refluxing xylene.[62]

A compound described only as 'mononitro monomethyl quinoline carboxaldehyde' (m.p. 115 °C) was obtained from a crude coal tar base.[63]

III. Quinoline Side-chain Aldehydes

From the hydroxyacid **45**, R = H, 2-quinolylacetaldehyde **46** has been prepared by oxidation with acid permanganate.[257] Alternatively, the ester **45**, R = Me was pyrolysed to give the aldehyde **46**.[258] The aldehyde **46** gave the bromo derivative **47**.[259] Early reports of the preparation of 2-quinolylacetaldehyde[70,260–262] have been shown to be in error. The starting material, thought

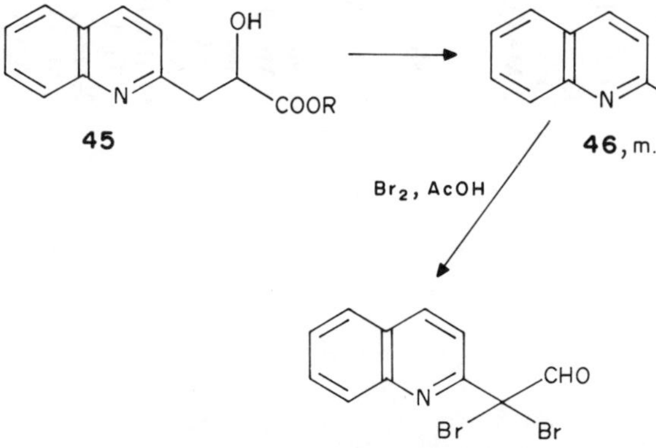

45 **46**, m.p. 98–100 °C

47, m.p. 180 °C

(*continued on p. 38*)

Table 1. Quinoline-2-carboxaldehydes

Substituent(s)	Preparation	Yield (%)	M.p. (°C)	Deriv.*	M.p. (°C)	References
None	QCHBr$_2$, AgNO$_3$	100	71	Ox PH	184 204	64,65
	Q-2-CH=CHCOOH, KMnO$_4$	80		PH	195–198	66–70
	Q-2-CH$_2$OH N-oxide, NaOH, 80 °C	77				44
	Q-2-CH$_2$OH, MnO$_2$	70		SC	233	71
	2-Me-Q, SeO$_2$	66	70–71	Ox PH SC	189 196–197 232–234	5,72,73 74,75,76
	2-Me-Q, SeO$_2$, dioxane	41	67–69	Hy NPH	195–198 250	6,77,78
	2-Me-Q, SeO$_2$, dioxane or ethanol			Ox TSC	185–187 234–237	79
	2-Me-Q, (PhSeO)$_2$O, PhCl	59				80
	2-Me-Q, C$_{14}$H$_{10}$O$_5$Se; see Section II.1					11
	Q-2-CH$_2$Br, 4,6-diPh-pyridone N-oxide	58	68–69	DNP	252.5–255	58
	Q, (HCHO)$_3$, FeSO$_4$, tBuOOH, TFA	47[a]				81
	2-Me-Q, gas phase oxidation	32	70			29,30,31,82
	2-Me-Q N-oxide, Tl(OAc)$_3$, Ac$_2$O, AcOH	20				83
	Q-2-CH$_2$Cl, iPrNO$_2$, EtONa	11	68			45
	Quinaldil, H$_2$, Pt	Trace				2
	Quinaldil, NaOH					84
	Q-2-CH(OMe)$_2$, H$_2$O					85

Position	Preparation	Yield (%)	mp (°C)	Deriv.	mp of deriv. (°C)	References
None	Q-2-CH=NNHPh, 2,4-diNO$_2$C$_6$H$_3$CHO, HCl, 120–130 °C		70–71	Ox	188–190	24,86
				DNP	251–253	
	Q-2-CHOHCHOH-Q, HIO$_4$	87		DNP	247–248	87
	2-NH$_2$C$_6$H$_4$CH=NC$_6$H$_4$-4-Me, AcCH=NOH			Ox	188–189	23
	Q-2-Me, AmNO$_2$, NaNH$_2$	55.5		Ox	189	46
	2-NH$_2$C$_6$H$_4$CHO, AcCH=NOH			Ox	189	21,22
	1-Me-2-PhNHN=CH-Q$^+$I$^-$, 250–280 °C/0.001 mm	53		PH	203–204	88
3-Me	Q-2-CHO, NPH			NPH	244–245	43
	2,3-diMe-Q, SeO$_2$, xylene, 90 °C	50	115–116	NPH	257	88–90
				DNP	274	
	3-Me-Q-2-CH=CHCOOH, KMnO$_4$	42.5	114–115	DNP	261–262	91
3,8-di-Me	2,3,8-triMe-Q, SeO$_2$, EtOH	82	107–108	SC	190–192	92
3,8-di-Me-5-NO$_2$	3,8-diMe-Q-2-CHO, HNO$_3$	85	167	Ox	180–181	92
4-Me	4-Me-Q, (HCHO)$_3$, FeSO$_4$, tBuOOH, TFA	89				81
	4-Me-Q, (HCHO)$_3$, FeSO$_4$, H$_2$O; HCl		77[b]			32
	2,4-diMe-Q, SeO$_2$, dioxane	70	76–78[b]	NPH	340	93
	2,4-diMe-Q, SeO$_2$	0.5	152–154[b]	DNP	282–284	15,90
5-Me	2,5-diMe-Q, SeO$_2$	Trace	78–79	NPH	247–248	90
5,7-di-Me-8-OH	2,5,7-triMe-8-AcO-Q, SeO$_2$; KOH	80	115.5–116	DNP	279–280	94,95
				Ox	204–204.5	
6-Me	2,6-diMe-Q, SeO$_2$, dioxane	40	108–109	PH	198–199	90,96,97
				NPH	269–270	98
	2,6-diMe-Q, gas phase oxidation		106	DNP	252–253	82
				Ox	185	
				PH	196	
				Pic.	164	
	6-Me-Q-2-CHBr$_2$, AgNO$_3$	24.4	105–106			91

Table 1. (*Contd.*)

Substituent(s)	Preparation	Yield (%)	M.p. (°C)	Deriv.*	M.p. (°C)	References
6,8-di-Me	6,8-diMe-Q-2-CH=CHCOOH, KMnO$_4$	10	107			99
6,8-di-Me-7-Cl	2,6,8-triMe-7-Cl-Q, SeO$_2$, dioxane, H$_2$O		147–150			100
6-Me-7-F						101
6-Me-7-Cl	2,6-diMe-7-Cl-Q, SeO$_2$		128–129			102
7-Me	2,7-diMe-Q, SeO$_2$, dioxane	32.4	69–70	PH	202–203	90,96,103
				NPH	231–232	
				DNP	269–270	
	2,7-diMe-Q, gas phase oxidation		87–88			104
8-Me	8-Me-Q-2-Li, DMF	63	83–84			105
	2,8-diMe-Q, SeO$_2$	47	81	NPH	251–252	90
				DNP	256–257	
	2,8-diMe-Q, gas phase oxidation		83.5	Ox	152	82
				PH	104–106	
				Pic.	146	
3-CH$_2$CH$_2$Cl						106
3-CH$_2$CH$_2$Cl-6-Me						107
3-CH$_2$CH$_2$Cl-4-Cl						106
3-CH$_2$CH$_2$Cl-4-Cl-6-Me						106
3-CH$_2$CH$_2$Cl-4,6-diCl						106
3-CH$_2$CH$_2$Cl-4,8-diCl						106
3-CH$_2$CH$_2$Cl-4-Cl-8-OAc						106
3-CH$_2$CH$_2$Cl-4-Cl-5,8-diOMe						106
3-CH$_2$CH$_2$Cl-4-Cl-6-OMe						106
3-CH$_2$CH$_2$Cl-4-Cl-8-OMe						106
3-CH$_2$CH$_2$Cl-4-Cl-6,7-diOBu						106
3-CH$_2$CH$_2$Cl-4-Cl-6-NO$_2$						106
3-CH$_2$CH$_2$Cl-4-Cl-8-NO$_2$						106
3-CH$_2$CH$_2$Cl-5,8-diOMe						106
3-CH$_2$CH$_2$Cl-6-OMe						106

Compound	Preparation	Yield (%)	mp	Derivative	Deriv. mp	References
3-CH$_2$CH$_2$Cl-8-OMe						106
3-CH$_2$CH$_2$Cl-6,7-diOBu						106
3-CH$_2$CH$_2$Cl-6-NHAc						106
3-CH$_2$CH$_2$Cl-8-NHAc						106
8-Et	2-Me-8-Et-Q, SeO$_2$	90		SC	189–190	105, 108
8-iPr	2-Me-8-iPr-Q, SeO$_2$	40				105
8-tBu	8-tBu-Q-2-CH(OMe)$_2$, H$_2$O					85
3-Ph	Anthranil, PhC≡CCH$_2$OH	15		Pic.	214	47
3,4-di-Ph	2-Me-3,4-diPh-Q, SeO$_2$, dioxane	56	97			109
4-Ph			142–143			110
4-Ph-6,7-diOMe	2-Me-4-Ph-6,7-diMeO-Q, SeO$_2$, dioxane, reflux	26[c]	179–180			111
4-(3,4,5-triMeOC$_6$H$_2$)-6,7-diOMe	2-Me-4-(3,4,5-triMeOC$_6$H$_2$)-6,7-diMeO-Q, SeO$_2$, dioxane, reflux	40[d]	185–187			111
4-(5-nitro-2-furanyl)	2-Me-4-(5-NO$_2$-2-furanyl)-Q, SeO$_2$, dioxane			Ox, SC	208–210, 247–248	112, 113
4-(5-NO$_2$-2-furanyl)-7-Me	2-Me-4-(5-NO$_2$-2-furanyl)-7-Me-Q, SeO$_2$, dioxane					113
4-(5-NO$_2$-2-furanyl)-6-Et	2-Me-4-(5-NO$_2$-2-furanyl)-6-Et-Q, SeO$_2$, dioxane					113
4-(5-NO$_2$-2-furanyl)-7-Cl	2-Me-4-(5-NO$_2$-2-furanyl)-7-Cl-Q, SeO$_2$, dioxane					113
4-(5-NO$_2$-2-furanyl)-6-OMe	2-Me-4-(5-NO$_2$-2-furanyl)-6-MeO-Q, SeO$_2$, dioxane					113
4-(5-NO$_2$-2-furanyl)-7-OEt	2-Me-4-(5-NO$_2$-2-furanyl)-7-EtO-Q, SeO$_2$, dioxane					113
4-Cl	2-Me-4-Cl-Q, SeO$_2$, dioxane	58	140	Ox, TSC, DNP, PH	167, 200, 265–266, 213.5–214.5	13, 73, 114; 115
4,7-diCl	2-Me-4,7-diCl-Q, SeO$_2$, xylene	41.4	145–146			116
4,8-diCl						106
4-Cl-6-OMe	2-Me-4-Cl-6-MeO-Q, SeO$_2$, dioxane	82	145–146			114
5-Cl	2-Me-5-Cl-Q, SeO$_2$		134–135			117
5-Cl-6-Me	2,6-diMe-5-Cl-Q, SeO$_2$, aq. dioxane	73	172–173			100, 102

Table 1. (*Contd.*)

Substituent(s)	Preparation	Yield (%)	M.p. (°C)	Deriv.*	M.p. (°C)	References
5,7-diCl-8-OH	5,7-diCl-8-PhCH$_2$O-Q-2-CHO, HCl		211	Hy SC TSC	198–199 300 265	118
5,7-diCl-8-OCH$_2$Ph	2-Me-5,7-diCl-8-PhCH$_2$O-Q, SeO$_2$, dioxane	80	124–125			118, 119
5-Cl-7-I-8-OMe	2-Me-5-Cl-7-I-8-MeO-Q, SeO$_2$, dioxane	46	129–131	Ox	211–213	120
6-Cl	2-Me-6-Cl-Q, gas phase oxidation		140			104
6,8-diCl	6,8-diCl-Q-2-CHBr$_2$, AgNO$_3$					121
7-Cl	2-Me-7-Cl-Q, gas phase oxidation		159			104
8-Cl	2-Me-8-Cl-Q, SeO$_2$, dioxane	70	144–145			122
	2-Me-8-Cl-Q, gas phase oxidation		146			104
4-Br	2-Me-4-Br-Q, SeO$_2$	25	142	Ox TSC DNP	166 200 264	13, 73, 115
6-Br	2-Me-6-Br-Q, SeO$_2$ or gas phase oxidation		164			104
3-OH	2-Me-3-HO-Q, SeO$_2$	90	161–167	Ox	244	76
8-OH	8-AcO-Q-2-CHO, KOH	83	98			123, 124
	8-AcO-Q-2-CHO, HCl	82	98.5–99.5	TSC	243–244	125
	2-Me-8-AcO-Q, SeO$_2$; H$_2$O	54	98.5–99	Ox	167.5–168	95
	2-Me-8-HO-Q, SeO$_2$			Ox	172–173	76
8-OAc	2-Me-8-AcO-Q, SeO$_2$	74	94–95			123
	2-Me-8-AcO-Q, SeO$_2$, dioxane	40	97–97.5	TSC	242–243(d)	125
4-OMe	2-Me-4-MeO-Q, SeO$_2$	12	89	Ox TSC DNP	236 212 267	13, 73, 115
5,8-diOMe	2-Me-5,8-diMeO-Q, SeO$_2$, dioxane	75	113	NPH	268	126, 127
5,8-diOMe-6-NO$_2$	2-Me-5,8-diMeO-6-NO$_2$-Q, SeO$_2$	51	213–215			127
5,8-diOMe-6-NHAc	2-Me-5,8-diMeO-6-AcNH-Q, SeO$_2$	20	203–205			127

Substituent	Method	Yield (%)	mp (°C)	Derivative	Derivative mp (°C)	References
6-OMe	2-Me-6-MeO-Q, SeO₂, toluene	76	105–107			104, 128
6,7-diOMe₂	2-NH₂-4,5-diMeOC₆H₂CH=NC₆H₄-4-Me, AcCH=NOH			Ox	243	23
8-OMe	2-Me-8-MeO-Q, SeO₂, xylene		102	DNP	260(d)	104, 129
4-OCH₂Ph	2-Me-4-PhCH₂O-Q, SeO₂, dioxane	66		TSC	216–220	1
8-OCH₂Ph	2-Me-8-PhCH₂O-Q, SeO₂, dioxane	43	93–94	TSC	176–178.5	125
6-OEt	2-Me-6-EtO-Q, SeO₂		106–107	PH / NPH / DNP	155–156 / 231–232 / 317–318	130, 131
4-OPh	2-Me-4-PhO-Q, SeO₂	31	97	Ox / TSC / DNP	154 / 214 / 277	13, 73, 115
4-OC₆H₄-4-Me	2-Me-4-(4-MeC₆H₄O)-Q, SeO₂	9	111–113	Ox	252–253	13
6-OCH₂O-7	2-NH₂,-3-OCH₂O-4-C₆H₂CH=NC₆H₄-4-Me, AcCH=NOH					23
4-SC₆H₄-4-Me	2-Me-4-(4-MeC₆H₄S)-Q, SeO₂	27	156–157	DiDNP	296	13
3-CHO	Q-2,3-diCON(Me)Ph, LAH	59		diOx	221–222	40
	Q-2,3-diCOOMe, LAH, −70°C	50	163–164	diDNP	318–319	34
3-CHO-7-Me	Q-2,3-diCOOMe, DIBAH	40	161–162			35, 36
	2-I-3-CH(OCH₂)₂-7-Me, BuLi; DMF; HCl	94	162–163			30
3-CH(OCH₂)₂-7-Me	2-I-3-CH(OCH₂)₂-7-Me-Q, BuLi; DMF	69	95–95.6			30
4-CHO	Q-2,4-diCOOEt, DIBAH, −65°C	80	170ᵇ	diNPh	282	14, 37
	2,4-diMe-Q, SeO₂	4	200ᵇ	diDNP	288–290	15, 90
3-COPh	3-PhCO-Q-2-CH₂OH, KMnO₄		132–133			132
3,4-diCOOMe	2-Me-Q-3,4-diCOOMe, SeO₂, dioxane		92			133
3-COOEt	Q-2,3-diCOOEt, DIBAH	e	82–83	Ox	246–249	34, 134
4-COOH	2-NH₂C₆H₄COCOOH, AcCH=NOH	90	220–221	TSC	268–270	22, 135
4-COOH-7-Cl		90	220–221	TSC / Ox	268–270 / 246–249	135

Table 1. (*Contd.*)

Substituent(s)	Preparation	Yield (%)	M.p. (°C)	Deriv.*	M.p. (°C)	References
4-COOEt	Q-2,4-diCOOEt, DIBAH, THF	95	110			37
4-CONH$_2$	2-Me-Q-4-CONH$_2$, SeO$_2$	15	217–219			13
3-SO$_3$H	2-PhCH=CH-Q-3-SO$_3$H, KMnO$_4$			PH	270	136
3-NO$_2$	2-Me-3-NO$_2$-Q, SeO$_2$	60	141			137
4-NO$_2$	2-Me-4-NO$_2$-Q, SeO$_2$	30	135	Ox TSC DNP	195 235 237	13, 73, 115
5-NO$_2$	2-Me-5-NO$_2$-Q, SeO$_2$, dioxane	50	168[b]			138
	2-Me-5-NO$_2$-Q, SeO$_2$	49.5	180–181[b]			117
	2-CH$_2$Cl-5-NO$_2$-Q, iPrNO$_2$, EtONa	21[f]	164[b]			45
6-NO$_2$	2-Me-6-NO$_2$-Q, SeO$_2$		220–221[b]			117
	2-Me-6-NO$_2$-Q, SeO$_2$, xylene		220–204[b]	Ox	248–249	139, 140
8-NO$_2$	2-Me-8-NO$_2$-Q, SeO$_2$, dioxane	55	149[b]			138
	2-Me-8-NO$_2$-Q, SeO$_2$		211–212[b]			117
	2-CHBr$_2$-8-NO$_2$-Q, AgNO$_3$	50	152[b]			64
3-NH$_2$	2-CH$_2$OH-3-NH$_2$-Q, MnO$_2$	50				137, 141
3-NH$_2$-4-PO$_3$Me$_2$	2-CHCl$_2$-3-NO$_2$-4-Cl-Q, P(OMe)$_3$; H$_2$O, EtOH		175–176			142
3-NHCHO	2-Me-3-NHCHO-Q, SeO$_2$	60	182	Ox	215	137
3-NHAc	2-Me-3-AcNH-Q, SeO$_2$, dioxane	41	171	PH	241–242	143
4-N$_3$	2-Me-4-N$_3$-Q, SeO$_2$	60	137–138			13
4-CN	2-Me-4-CN-Q, SeO$_2$	55	184–186			13

*Ox = oxime, Hy = hydrazone, PH = phenylhydrazone, NPH = 4-nitrophenylhydrazone, DNP = 2,4-dinitrophenylhydrazone, SC = semicarbazone, TSC = thiosemicarbazone, Pic. = picrate.

[a] Product mixed with quinoline-4-carboxaldehyde.
[b] Inconsistent literature melting points.
[c] Product mixed with the corresponding carboxylic acid (18%).
[d] Product mixed with the corresponding carboxylic acid (52%).
[e] Product mixed with quinoline-2,3-dicarboxaldehyde.

Table 2. Quinoline-3-carboxaldehydes

Substituent(s)	Preparation	Yield(%)	M.p. (°C)	Deriv.*	M.p. (°C)	References
None	Q-3-CH=NNHCONH₂, HCl	65	70	SC	254	50, 51
				DNP	292	
	3-Me-Q, SeO₂	41	69.5			144
	Q-3-CONHNHTs, Na₂CO₃, (CH₂OH)₂, 160 °C	32.5	70			49
	Q-3-CN, SnCl₂, HCl; aq H₂SO₄	30	70	Ox	175–176	145
				DNP	292	
				Pic.	197–199	
2-Me-4-(2-FC₆H₄)	2-Me-4-(2-FC₆H₄)-Q-3-CH₂OH, MnO₂	83	111–112	Ox	212–214	146
2-Me-7, 8-diOMe	2-Me-7, 8-diMeO-Q-3-CONHNHSO₂Ph, Na₂CO₃		71	DNP	240	147
2-Me-4-NH₂	2-Me-4-NH₂-Q-3-CH₂OH, MnO₂	90	262	Ox	>300	137
7-Me	2-Cl-3-CH(OCH₂)₂-7-Me-Q, Zn, NaOH; HCl	97	119–120			30
7-Me-2-NH₂	2-I-7-Me-Q-3-CHO, NH₃	92	194–196			31
2-Et-6-Me	2-Et-3, 6-diMe-Q, CrO₃		56–57			148
2-Ph-7-Me	2-Cl-7-Me-Q-3-CHO, C₆H₆, hν	36	109.5–110.5			30
4-(2-FC₆H₄)-6-Cl	2-NH₂-5-Cl-C₆H₃-CO-C₆H₄-2-F, CH₂[CH(OMe)₂]₂, ZnCl₂	38	124–126	Ox	248–250	146
4-(1-napthyl)	See text, Section II.7					57
4-Me-4-iPr-imidazolon-2-yl	2-Ar-Q, BuLi; DMF	78	226–227			149, 150
2-Cl	PhNHAc, POCl₃, DMF	78	148–149	Ox	195–197	27–29, 151–153
2-Cl-6-Me	4-MeC₆H₄NHAc, POCl₃, DMF	74	124–125			27, 29
2-Cl-6, 7-diMe						152

Table 2. (*Contd.*)

Substituent(s)	Preparation	Yield(%)	M.p. (°C)	Deriv.*	M.p. (°C)	References
2-Cl-6, 8-diMe	2,4-diMeC$_6$H$_3$NHAc, POCl$_3$, DMF	32	110–111			27
2-Cl-7-Me	3-MeC$_6$H$_4$NHAc, POCl$_3$, DMF	66	144.5–145.5			26–29
2-Cl-8-Me	2-MeC$_6$H$_4$NHAc, POCl$_3$, DMF	67	137–138			27, 29
2,6-di-Cl	4-ClC$_6$H$_4$NHAc, POCl$_3$, DMF	13	190.5–191.5			29, 154
2,7-di-Cl	3-ClC$_6$H$_4$NHAc, POCl$_3$, DMF	35	159–160			27, 29
2-Cl-6-Br	4-BrC$_6$H$_4$NHAc, POCl$_3$, DMF	30	188–189			27
2-Cl-5,6,7-triOMe	3,4,5-triMeOC$_6$H$_4$NHAc, POCl$_3$, DMF	92	149–149.5			26, 27, 29
2-Cl-6-OMe	4-MeOC$_6$H$_4$NHAc, POCl$_3$, DMF	56	146–147	Ox	190–192	27, 29, 151
2-Cl-6, 7-diOMe	3,4-diMeOC$_6$H$_4$NHAc, POCl$_3$, DMF	72	215	Z-Ox	241	26–29
				E-Ox	249–250	151, 152
2-Cl-6, 7-diOMe-8-NO$_2$	2-Cl-6, 7-diMeO-Q-3-CHO, HNO$_3$	95	180–182			155
2-Cl-7-OMe	3-MeOC$_6$H$_4$NHAc, POCl$_3$, DMF	65	197–198			26–29
2-Cl-8-OMe	2-MeOC$_6$H$_4$NHAc, POCl$_3$, DMF	14	191.5–192.5			27, 29
2-Cl-6, 7-diOEt	3,4-diEtOC$_6$H$_3$NHAc, POCl$_3$, DMF	70	162–163	Ox	210–212	151, 156
2-Cl-6, 7-diOEt-8-NO$_2$	2-Cl-6, 7-diEtO-Q-3-CHO, HNO$_3$	94	181			155
2-Cl-6-SMe	4-MeSC$_6$H$_4$NHAc, POCl$_3$, DMF	92	195–196			152
2-Cl-7-SMe	3-MeSC$_6$H$_4$NHAc, POCl$_3$, DMF	75	300(d)			27, 29, 154
4-Cl-6-OMe	4-Cl-6-MeO-Q-3-CN, NH$_2$CONHNH$_2$, H$_2$, Ni; HCl	26	114			51
4-Cl-7-OMe	4-Cl-7-MeO-Q-3-CH=NNHCONH$_2$, H$_2$O					52
6-Cl	6-Cl-Q-3-CHN=NHCONH$_2$, HCl	20	108			51
2-I	2-Cl-Q-3-CHO, HI	85	150–152			30
2-I-7-Me	2-Cl-7-Me-Q-3-CHO, HI	93	180–181			30
2,4-diOMe	2,4-diMeO-Q-3-Li, DMF	94	82			157–159
	2,4-diCl-Q-3-CHCl$_2$, KOH		76			160
2,4,6-triOMe	2,4,6-triMeO-Q-3-Li, PhN(Me)CHO	68	110–111			157, 158

Substituent	Method / Reagents	Yield	mp			Ref
2,4,6,7-tetraOMe	2,4,6,7-tetraMeO-Q-3-Li; PhN(Me)CHO	69.2	166			161
2,4,7-triOMe	2,4,7-triMeO-Q-3-Li, PhN(Me)CHO	53	126			157
2,4,8-triOMe	2,4,8-triMeO-Q, BuLi; PhNHCHO	40.3	161			158
2,4-diMeO-6-OCH$_2$O-7	2,4-diMeO-6-OCH$_2$O-7-Q-3-Li, DMF	40	256			162
2,5,6,7-tetraOMe	2-Cl-5,6,7-triMeO-Q-3-CHO, MeO$^-$	70	158–159			163
2,6,7-triOMe	2-Cl-6,7-diMeO-Q-3-CHO, MeO$^-$	60	178–180			163
2,7-diOMe	2-Cl-7-MeO-Q-3-CHO, MeO$^-$	65	151–152			163
5,6,7-triOMe	2-Cl-5,6,7-triMeO-Q-3-CHO, reduce	48	132–134			163
6-OMe	6-MeO-Q-3-CN, NH$_2$CONHNH$_2$, H$_2$, Ni; HCl	70	102	SC	240–241	51
6,7-diOMe	2-Cl-6,7-diMeO-Q-3-CHO, reduce	45	159–161			163
7-OMe	2-Cl-7-MeO-Q-3-CHO, reduce	30	120–121			163
2-OCH$_2$COOH-4-Cl	2-OCH$_2$COOH-4-Cl-Q-3-CH(OMe)$_2$; HCl		162			164
2-SH	2-SCNC$_6$H$_4$CH=CHCHO, CH$_2$(COOEt)$_2$, EtONa	92	288	Ox	205	25, 165
2-S—S-2	3-CHO-2-quinolinethione, I$_2$		213.5			25, 166
2-SMe	2-SH-Q-3-CHO, MeI	58	110			167
2-SMe-4-OMe-7-Cl	2-NaS-4-MeO-7-Cl-Q-3-CHO, MeI		178–179			56
2-SCH$_2$Ph	2-SH-Q-3-CHO, NaOMe, PhCH$_2$Cl	88	108–110			168
2-SCH$_2$Ph-4-OMe-7-Cl	2-NaS-4-MeO-7-Cl-Q-3-CHO, PhCH$_2$Cl		144–146			56
2-SCH$_2$C$_6$H$_4$-3-F		92	127–129			168
2-SCH$_2$C$_6$H$_4$-4-F		90	128–129			168
2-SCH$_2$C$_6$H$_4$-4-Cl		91	143–145			168
2-SCH$_2$C$_6$H$_4$-4-OMe		92	108–110			168

Table 2. (*Contd.*)

Substituent(s)	Preparation	Yield(%)	M.p. (°C)	Deriv.*	M.p. (°C)	References
2-SEt-4-OMe-7-Cl	2-SNa-4-MeO-7-Cl-Q-3-CHO, EtI	80.7	115–116	Ox	153–154	56
2-SEt-4-NH_2-7-Cl	2-EtS-4-MeO-7-Cl-Q-3-CHO, NH_4OH		179–180			56
2-SEt-4-$NHCH_2CH_2NMe_2$-7-Cl	2-EtS-4-MeO-7-Cl-Q-3-CHO, $Me_2NCH_2CH_2NH_2$	20	121–123			56
2-SCH_2COOMe	3-CHO-2-quinolinethione, $ClCH_2COOMe$		109–110	Ox / DNP	150–151 / 274	25
2-SCH_2COOMe-7-Me	2-Cl-7-Me-Q-3-CHO, $HSCH_2COOMe$, Na_2CO_3	67	110–112			30
2-StBu-7-Me	2-Cl-7-Me-Q-3-CHO, tBuSH, K_2CO_3	47	120–121	Ox	206–207	30
2-SOMe	2-MeS-Q-3-CHO, $NaIO_4$	82	178–180			167
2-SO_2Me	2-MeS-Q-3-CHO, $KMnO_4$	17	151–153			167
2-CHO	See Table 1					
2-CHO-7-Me	See Table 1					
2-COOH	2-Li-7-Me-Q-3-CH(OCH$_2$)$_2$, CO_2; HCl	87	160(d)			169
2-COOH-7-Me						30
4-COOMe	Q-3, 4-diCOOMe, LAH	15	109			38, 39
2-COSEt	3-CHO-Q-2-COOH, $SOCl_2$; EtSH	64	106			169
2-SO_2Et-4-OMe-7-Cl	2-EtS-4-MeO-7-Cl-Q-3-CHO, $KMnO_4$	47	94–95			56
2-SO_2Et-4-NH_2-7-Cl	2-$EtSO_2$-4-MeO-7-Cl-Q-3-CHO, NH_4OH	63	222–223			56
6-NO_2	6-NO_2-Q-3-CH=CHCOOH, $KMnO_4$	40.6	188–189	Ox / DNP	235–237 / 295–296	170, 171
2-NH_2	2-NH_2-Q-3-CH_2OH, MnO_2	70	197	Ox	245	137
4-NH_2	4-NH_2-Q-3-CH_2OH, MnO_2	70	258	Ox	>300	137
6-NH_2	6-NO_2-Q-3-CH=NOH, H_2/Pd			Ox	196–197	170, 171

*Ox = oxime, DNP = 2, 4-dinitrophenylhydrazone, SC = semicarbazone, Pic. = picrate.

Table 3. Quinoline-4-carboxaldehydes

Substituent(s)	Preparation	Yield(%)	M.p. (°C)	Deriv.*	M.p. (°C)	References
None	Q-4-CHOHCHOH-4-Q, Pb(OAc)$_4$	100		PH	175–176	172
	Q-4-CH$_2$Br, 4,6-diPh-2-pyridone N-oxide	82	50–51.5			58
	4-Me-Q, (PhSeO)$_2$O, PhCl, 130 °C	80	48–50			12
	4-Me-Q, gas phase oxidation	68.5	51–52			16, 17, 82
	4-Me-Q, SeO$_2$	64	52	SC	244–245	5, 74, 173
	4-Me-Q, SeO$_2$, xylene	61	51–53	Ox	181–182	86
				NPH	262–263(d)	174
				DNP	256(d)	144
	4-Me-Q, SeO$_2$, HOAc, Ac$_2$O	50–60	48–49	Ox	180–181	175, 176
	4-Me-Q, SeO$_2$, dioxane		48–50			177
	Q-4-CH=CHCOOH, KMnO$_4$	58	52	PH	176	66, 178, 179
			b.p. 122–123°/ 4 mm	Pic.	179	
				HCl	206	
	Q, (HCHO)$_3$, FeSO$_4$, tBuOOH, TFA	34^a				81
	4-Me-Q N-oxide, Tl(OAc)$_3$	low				180
	3-Indolylacetaldehyde, hv, O$_2$	16.3				33
2-Me	4-Me-Q, AmONO, NaNH$_2$	66.8		TSC	237–238(d)^b	181
	2-NH$_2$C$_6$H$_4$COCH(OMe)$_2$, MeCOMe, EtONa	95	141	TSC	183^b	182, 183
	2,4-diMe-Q, EtONO, KNH$_2$	59	84–85	Ox	179–181	46
	2-Me-Q, (HCHO)$_3$, FeSO$_4$, tBuOOH, TFA	65		Ox	216–216.5	32, 184
						185, 186
						81

Table 3. (*Contd.*)

Substituent(s)	Preparation	Yield(%)	M.p. (°C)	Deriv.*	M.p. (°C)	References
2,6,8-triMe	2-Me-4-Br-Q, BuLi; DMF	34	72–75	Acetal		187
	2-NH$_2$-3,5-diMeC$_6$H$_2$CO-CH(OMe)$_2$, MeCOMe, EtONa					19
2-Me-3-NH$_2$	2-Me-3-NH$_2$-Q-4-CH$_2$OH, MnO$_2$	85	174	Ox	247–248	137, 141
2-CF$_3$	2-CF$_3$-4-Br-Q, BuLi; DMF	80				188
2,8-diCF$_3$	2,8-diCF$_3$-4-Br-Q, BuLi; DMF	63	91–92			189
2-CF$_3$-7-OMe	2-CF$_3$-4-Br-7-MeO-Q, BuLi; DMF	56	135–137			187
6-Me-2-S——S-2-Q-6-Me	(4,6-diMe-Q-2-S)$_2$, SeO$_2$, dioxane	65	179			190
6,7-diMe				PH	121	191
7-CF$_3$	4-Br-7-CF$_3$-Q, BuLi; DMF	23	63–64.5			187
2-CH=CHC$_6$H$_4$-Cl	2,6,8-triMe-Q-4-CH(OMe)$_2$, 4-ClC$_6$H$_4$CHO; HCl	50	167–168			19
2-Pr		60		DNP	243–244	192
2-Bu		65	b.p. 160–162°/3 mm	DNP	197–198	192
2-(CH$_2$)$_3$-3	2-NH$_2$C$_6$H$_4$COCH(OMe)$_2$, C$_5$H$_8$O, EtONa	95				184
2-(CH$_2$)$_3$-3-6, 8-diMe	2-NH$_2$-3, 5-diMeC$_6$H$_2$-COCH(OMe)$_2$, C$_5$H$_8$O, EtONa			Acetal		19
2-(CH$_2$)$_4$-3	2-NH$_2$C$_6$H$_4$COCH(OMe)$_2$, C$_6$H$_{10}$O, EtONa	97	81			184
2-(CH$_2$)$_5$-3	2-NH$_2$C$_6$H$_4$COCH(OMe)$_2$, C$_7$H$_{12}$O, EtONa	95	104			184
2-(CH$_2$)$_6$-3	2-NH$_2$C$_6$H$_4$COCH(OMe)$_2$, C$_8$H$_{14}$O, EtONa	85	98.5			184
2-Ph	2-Ph-4-Me-Q, SeO$_2$, dioxane	93	75			193, 194
	2-NH$_2$C$_6$H$_4$COCH(OMe)$_2$, MeCOPh, EtONa	91		Ph	130(d)	184
	2-Ph-Q-4-CH=CHCOOH, KMnO$_4$	74				179
	Rosenmund reduction of 2-Ph-Q-4-COCl			Ox	126–127(d)	195

Substituents	Preparation	Yield (%)	mp (°C)	Ox	170	SC	Ref.
2-Ph-3-NH_2	See text, Section II.6						42
2-(3,4-diClC_6H_3)-6, 8-diCl	2-(3,4-diClC_6H_3)-4-CHOHCH_2NBu_2-6, 8-diCl-Q, *hv*	91 / 80	135 / 195.5–197				197
2-(3,4-diClC_6H_3)-6-OMe	2-(3,4-diClC_6H_3)-4-CH_2OH-6-MeO-Q, CrO_3, Py, 250 °C	29.7[c]	163–164				198
	2-(3,4-diClC_6H_3)-4-CHOHCH_2NEt_2-6-MeO-Q, *hv*	7.2[c]					198
	2-(3,4-diClC_6H_3)-4-CHOHCH_2N-$(C_6H_{13})_2$-6-MeO-Q, *hv*	10.6[c]					198
	2-(3,4-diClC_6H_3)-4-CHOHCH_2N-$(C_8H_{17})_2$-6-MeO-Q, *hv*						198
2-(4-ClC_6H_4)	2-$NH_2C_6H_4$COCH$(OMe)_2$, MeCOC_6H_4-4-Cl, EtONa	95					184
2-(4-ClC_6H_4)-3-OMe-6, 8-diMe	2-NH_2-3, 5-diMeC_6H_2CO-CH$(OMe)_2$, MeONa, MeOCH_2COC_6H_4-4-Cl	88	136–137				20
2-(4-MeOC_6H_4)-6-OMe-7-Cl	2-(4-MeOC_6H_4)-4-CH_2OH-6-OMe-7-Cl-Q, CrO_3, Py, 250 °C	8.8[c]	202–203				198
	2-(4-MeOC_6H_4)-4-CHOHCH_2-NEt_2-6-Me-O-Q, *hv*	6.5[c]					198
	2-(4-MeOC_6H_4)-4-CHOHCH_2NBu_2-6-MeO-Q, *hv*	39.4[c]					198
	2-(4-MeOC_6H_4)-4-CHOHCH_2N-$(C_6H_{13})_2$-6-MeO-Q, *hv*						198
2-(3-$NO_2C_6H_4$)	2-$NH_2C_6H_4$COCH$(OMe)_2$, MeCOC_6H_4-3-NO_2, EtONa	89					184
2-(4-$NO_2C_6H_4$)	2-$NH_2C_6H_4$COCH$(OMe)_2$, MeCOC_6H_4-4-NO_2, EtONa	90					184
2-(3-$NH_2C_6H_4$)	2-(3-$NO_2C_6H_4$)-Q-4-CH$(OMe)_2$, $Na_2S_2O_4$	100					184
2-(3-NH_2-4-ClC_6H_3)	2-$NH_2C_6H_4$COCH$(OMe)_2$, MeCOC_6H_4-3-NH_2-4-Cl, EtONa	86					184
2-(4-$NH_2C_6H_4$)	2-(4-$NO_2C_6H_4$)-Q-4-CH$(OMe)_2$, $Na_2S_2O_4$	100					184
2-(4-MeOC_6H_4)-6-OMe-7-Cl	[2-(4-MeOC_6H_4)-6-MeO-7-Cl-Q-4-CHOH]$_2$, heat						199
2-thienyl	2-thienyl-4-Me-Q, SeO_2, dioxane, heat	89	111.1–112			252–253	7

Table 3. (*Contd.*)

Substituent(s)	Preparation	Yield(%)	M.p. (°C)	Deriv.*	M.p. (°C)	References
2-Cl	2-Cl-4-CHBr$_2$-Q, AgNO$_3$			NPH	241–243(d)	200
6-Cl	4-Me-6-Cl-Q, SeO$_2$, dioxane	54	152.5–153	SC	234	1, 201
6,8-diCl	4-Me-6, 8-diCl-Q, SeO$_2$, PhBr, reflux	62	133–139			202
7-Cl	4-Me-7-Cl-Q, SeO$_2$, dioxane	40	112–113.5	SC	263	1, 201, 202
7-Cl-2-Q-S—S-2-Q-7-Cl	(4-Me-7-Cl-Q-2-S)$_2$, SeO$_2$, dioxane	65	204	PH	112	190
8-Cl	4-Me-8-Cl-Q, SeO$_2$, PhBr, reflux	58	170–171	DNP	202	202
	4-Me-8-Cl-Q, SeO$_2$, toluene	42	150	TSC	210	1, 182, 183
6-Br-8-Ph	4-Me-6-Br-8-Ph-Q, SeO$_2$, dioxane	87.5	113–114			203
6,8-diBr	See text, Section II.7	32.5	158–159			55
8-OH	Q-8-OCH$_2$Ph, HCl	82	144–145	TSC	256–256.5(d)	124, 125
	Natural product		155–156			204
8-OAc	8-HO-Q-4-CHO, Ac$_2$O		100	Ox	187–188	204
5,8-diMeO	4-Me-5,8-diMeO-Q, SeO$_2$, dioxane	65	125–126	DNP	275–276	127
5,8-diMeO-6-NO$_2$	4-Me-5,8-diMeO-6-NO$_2$-Q, SeO$_2$, dioxane	36	184–186			127
5,8-diMeO-6-NHAc	4-Me-5,8-diMeO-6-AcNH-Q, SeO$_2$, dioxane	79	181–184			127
6-OMe	4-Me-6-MeO-Q, SeO$_2$, xylene	56	97–97.5	Ox	214–216	86, 205
	4-Me-6-MeO-Q, SeO$_2$	54.6	98.5			173
6-MeO-2-S—S-2-Q-6-OMe	4-Me-6-MeO-Q, SeO$_2$, 130–140 °C		172	TSC	238–240	1
	(4-Me-6-MeO-Q-2-S)$_2$, SeO$_2$, dioxane	65		PH	128	190
6,7-diMeO	4-Me-6,7-diMeO-Q, SeO$_2$, dioxane	71	171–171.5			205
6-OMe-8-NO$_2$	4-Me-6-MeO-8-NO$_2$, SeO$_2$, AcOH	80	190–195	TSC	246–247(d)	206
8-OMe	4-Me-8-MeO-Q, SeO$_2$, dioxane	62.5	109–110	TSC	220	125
8-OCH$_2$Ph	4-Me-8-PhCH$_2$O-Q, SeO$_2$, dioxane, AcOH, Ac$_2$O	95	148.5–149			125
2-OPr		51	b.p. 155–156°/1 mm	DNP	209–210	192

Substituent	Reagents / Conditions	Yield (%)	M.p. (b.p.)	Derivative	Derivative m.p.	Refs.
2-OBu		68	35–37; b.p. 170–173°/1.5 mm	DNP	210–211	192
2-OcycloC$_6$H$_{11}$		67	87–89; b.p. 180–183°/0.5 mm	DNP	253–254	192
5-O-(3-CF$_3$C$_6$H$_4$)-6-OMe-8-NO$_2$	4-Me-5-(3-CF$_3$C$_6$H$_4$)O-6-MeO-8-NO$_2$-Q, SeO$_2$, dioxane	81	204–205			207
2-CHO	See Table 1					
6-CHO	6-CH$_2$OH-Q-4-CH(OEt)$_2$, Ag$_2$CO$_3$; HCl		171			14, 37
7-CHO	7-CH$_2$OH-Q-4-CH(OEt)$_2$, Ag$_2$CO$_3$; HCl	60	154–155			14, 37
2-COOH	2-NH$_2$C$_6$H$_4$COCH(OMe)$_2$, MeCOCOONa, EtONa; HCl	97	190			208
2-COOH-6-OMe	2-NH$_2$-5-MeOC$_6$H$_3$COCH(OMe)$_2$, MeCOCOONa, EtONa; HCl	71	230–232			208
8-COOH	4-Me-Q-8-COOH, SeO$_2$, H$_2$O$_2$, dioxane	Trace	205–207	Ox	234–235(d)	209
3-COOMe	4-Me-Q-3-COOMe, SeO$_2$		182			38
6-COOMe	4-Me-Q-6-COOMe, SeO$_2$		153			37
7-COOMe	4-Me-Q-7-COOMe, SeO$_2$		132–133			37
3-NO$_2$	3-NO$_2$-4-CH$_2$Br-Q, Me$_3$N oxide	35	134–136			210
7-NO$_2$						211
8-NO$_2$	4-Me-8-NO$_2$-Q, SeO$_2$, EtOH	53	175–177			212, 213
	8-NO$_2$-Q-4-CHBr$_2$, AcOAg		175	TSC	225	182, 183, 214
	4-Me-8-NO$_2$-Q, SeO$_2$, dioxane			TSC	240–245	1
3-NH$_2$	3-NH$_2$-Q-4-CH$_2$OH, MnO$_2$	85	172	Ox	257	137, 141
3-N$_3$	See text, Section II.6		135(d)			41

*Ox = oxime, PH = phenylhydrazone, NPH = 4-nitrophenylhydrazone, DNP = 2,4-dinitrophenylhydrazone, SC = semicarbazone, TSC = thiosemicarbazone, Pic. = picrate.
a Product mixed with quinoline-2-carboxaldehyde (47%).
b Inconsistent literature melting points.
c The aldehyde was part of a multicomponent mixture.

Table 4. Quinoline-5-carboxaldehydes

Substituent(s)	Preparation	Yield (%)	M.p. (°C)	Deriv.*	M.p. (°C)	References
None	5-Br-Q, BuLi; DMF	70	94.5–96			215
	Q-5-CH$_2$OH, MnO$_2$	63	95.5–96.6			216
	Q-5-CONHNHSO$_2$C$_6$H$_4$-4-Me, Na$_2$CO$_3$	5	96			49
	5-Me-Q, SeO$_2$		95.5–96.5	Ox	188–190	4, 217
				Hy	110–111.5	
				PH	188–189	
				SC	212–215	
				HCl	218–219	
2-Me[a]	QCH=CHCOOH, KMnO$_4$		73	Pic.	182(d)	218, 219
2-Me-7-Cl-8-OH	2-Me-8-HO-Q-5-CHO, tBuOCl	68	253–254			10
2-Me-7-Br-8-OH	2-Me-8-HO-Q-5-CHO, Br$_2$/H$_2$O	67	255(d)			10
2-Me-8-OH	2-Me-8-TsO-Q-5-CHO, HCl	61–68	144–144.5	Ox	134.5–135	10, 220, 227
				TSC	233–234(d)	
2-Me-8-OTs	2-Me-8-TsO-Q-5-CH$_2$OH, Na$_2$Cr$_2$O$_7$	79–83	134			10, 220
7-Me-8-OH	7-Me-8-TsO-Q-5-CHO, HCl	83–88	183–183.5	Ox	206–207	10, 220
				TSC	> 249	
7-Me-8-OTs	7-Me-8-TsO-Q-5-CH$_2$OH, Na$_2$Cr$_2$O$_7$	85	178–179			10, 220
3-Ph-8-Me	3-Ph-5-Br-8-Me-Q, BuLi; DMF	90	oil	PH	154–156.5	221
7-Br-8-OH	8-HO-Q-5-CHO, Br$_2$, AcOH		232			59
7-I-8-OH	8-HO-Q-5-CHO, I$_2$, AcOH		240			59
6-OH	Reimer–Tiemann reaction	42	136–137	Ox	235	222, 223
				PH	232–234(d)	
8-OH	8-HO-Q-5-CH$_2$OH, MnO$_2$	59	179.5–180	Pic.	234.5(d)	9
	8-HO-Q-5-CHOHCCl$_3$, NaOMe	50	178			59, 60

Substituents	Preparation	Yield (%)	m.p.	Derivative	Derivative m.p.	References
	8-HO-Q-5-CHOHCCl₃, NaCH₂COCH₃	67		HCl	274(d)	61
	Reimer–Tiemann reaction	10.5	180.5–181	Ox	196	124, 125, 220, 224, 225, 226, 228
				PH	132	
				TSC	260.5–261.5	
	5-CH₂N(CH₂)₅-8-HO-Q,(CH₂)₆N₄, AcOH			DNP	296–297(d)	229
8-OAc	8-HO-Q-5-CHO, Ac₂O		118			224
8-OSO₂Ph	8-PhSO₂O-Q-CH₂OH, Na₂CrO₄	86	101–102	TSC	220–221(d)	230
8-OTs						231
6-OMe	6-HO-Q-5-CHO, Me₂SO₄, K₂CO₃					222
8-OMe						232
7-CHO-8-OH	8-TsO-Q-5,7-diCHO, 15% HCl	82	299–301(d)	diOx	220–223(d)	230
7-CHO-8-OTs	8-TsO-Q-5,7-diCH₂OH, Na₂CrO₄, AcOH	66	165–167			230
6-NO₂	6-NO₂-Q, MeNO₂; KMnO₄	39	152–154			233
7-NO₂-8-OH	8-HO-Q, HNO₃		301–311(d)			59

*Ox = oxime, Hy = hydrazone, PH = phenylhydrazone, DNP = 2,4-dinitrophenylhydrazone, SC = semicarbazone, TSC = thiosemicarbazone, Pic. = picrate.

aThe identity of this compound is in doubt. It may be 2-methylquinoline-7-carboxaldehyde.

Table 5. Quinoline-6-carboxaldehydes

Substituent(s)	Preparation	Yield (%)	M.p. (°C)	Deriv.*	M.p. (°C)	References
None	1,2,3,4-tetraH-Q-6-CHO, chloranil, xylene	48.3	74–75			62
	6-Me-Q, SeO_2	45	52–54	SC	234–237	4,234
	6-Me-Q, SeO_2, 265–270 °C		75	TSC	247–248	181,235,236
	Q-6-$CHBr_2$, H_2O		75–76	Ox	191	237
				PH	185	
				SC	239	
2-Me	Q-6-CONHNHTs, Na_2CO_3	45	72			49
2-Me-4,8-diOMe	2-Me-Q-6-CH=CHCOOH, $KMnO_4$		106	PH	160	238
2-Me-4-NH_2	2-Me-4,8-diMeO-6-Br-Q, BuLi; DMF	60				187
	(2-Me-4-NH_2-Q-6-CH)$_2$, O_3		215–216(d)			239
2-Me-4-NMe_2-8-OMe	2-Me-4-Me_2N-6-Br-8-MeO-Q, BuLi; DMF		127–129			187
7-Me	6,7-diMe-Q, SeO_2	a				14
8-Me	6-Br-8-Me-Q, BuLi; DMF	76	98–98.5			215
8-Ph	6-Br-8-Ph-Q, BuLi; DMF	81.5	144			215
2-Ph-8-Me	2-Ph-6-Br-8-Me-Q, BuLi; DMF	71.5	108–109.5			215
2-Ph-8-Cl	2-Ph-6-Br-8-Cl-Q, BuLi; DMF	71	133.5–134			215
2-(4-MeC_6H_4)-8-Me	Sommelet reaction	46	133.5–134[b]	TSC	240–241	203,215
	2-(4-MeC_6H_4)-6-Br-8-Me-Q, BuLi; DMF	74	158.5–160.5			203
2-(4-ClC_6H_4)	2-(4ClC_6H_4)-6-Br-8-Me-Q, BuLi; DMF	78	154–155			215

Substituent	Reagents and conditions	Yield (%)	M.p. (°C)	Ox		Ref.
3-Br	3-Br-6-CHBr$_2$-Q, H$_2$O, K$_2$CO$_3$		139		217	237
4-CHO	See Table 3					
7-CHO	6,7-diMe-Q, SeO$_2$	15	117			14
5-NO$_2$	5-NO$_2$-Q, MeNO$_2$; KMnO$_4$	28	156–158			233
2-NMe$_2$-4-Me	2-Me$_2$N-4-Me-6-Br-Q, BuLi; DMF	17	116–117.8			187
2-N(CH$_2$CH$_2$)$_2$O-4,8-diMe	2-N(CH$_2$CH$_2$)$_2$O-4,8-diMe-6-Br-Q, BuLi; DMF	59.3	161.5–162.5			187

* Ox = oxime, PH = phenylhydrazone, SC = semicarbazone, TSC = thiosemicarbazone.
a Obtained as a mixture with 6-methylquinoline-7-carboxaldehyde.
b A misprint in the original reference[203] gave this melting point as 153–154 °C. The figure in the Table has been confirmed.[240]

Table 6. Quinoline-7-carboxaldehydes

Substituent(s)	Preparation	Yield (%)	M.p. (°C)	Deriv.*	M.p. (°C)	References
None	7-Me-Q, SeO$_2$ 7-Me-Q, SeO$_2$, 265–270°C	91	85–86	Ox Hy PH SC HCl	185–187 123–124 170–171 234 187–189	4, 217
2-Me	See footnote to Table 4		86	TSC	240–241	181, 235
2-Me-5-Br-8-OH	2-Me-5-Br-7-CH=CHMe-8-HO-Q, O$_3$	70	172			241
2-Me-5-I-8-OH	2-Me-5-I-7-CH=CHMe-8-HO-Q, O$_3$	61	189			241
2-Me-8-OH	2-Me-7-CH=CHMe-8-HO-Q, O$_3$	34	122			241
4-Me-8-OH	4-Me-7-CH=CHMe-8-HO-Q, O$_3$	30	161			241
5-Me-8-OH	5-Me-8-TsO-Q-7-CHO, HCl	81–87	182–182.5	Ox NPH TSC Pic.	240(d) 302–303(d) 235–235.5 193.5–194.5(d)	10, 220
5-Me-8-OTs	Reimer–Tiemann reaction 5-Me-7-CH$_2$OH-8-TsO-Q, Na$_2$Cr$_2$O$_7$	88	168	TSC	238–238.5	228
6-Me	6,7-diMe-Q, SeO$_2$	a	153.5–154			10 14
2-(4-ClC$_6$H$_4$)	Sommelet reaction from the 7-CH$_2$Br-Q	26	163–164			242
2-(4-ClC$_6$H$_4$)-6,8-diCl	Sommelet reaction from the 7-CH$_2$Br-Q	60	200–201			242
5-Cl-8-OH	Sommelet reaction from the 7-CH$_2$Br-Q	27	208–209			230
5-Br-8-OH	8-HO-Q-7-CHO, Br$_2$, AcOH	74	213–214			230
5-OH	5-HO-Q-7-CH$_2$N(CH$_2$)$_5$, (CH$_2$)$_6$N$_4$, AcOH			DNP	288–290	229

Compound	Method	%	mp	Derivative	Derivative mp	Refs
8-OH	8-HO-Q-7-CH=CHMe, O$_3$	62	178–178.5	Ox Ph DNP TSC Pic.	233–234(d) 200–202(d) 303.5–304(d) 233.5–234.5(d) 167.5–168	220, 230, 243
	7-CHO-8-HO-Q-5-SO$_3$H, 50% H$_2$SO$_4$		> 250[b]			230 226
8-OH[c]	Reimer–Tiemann reaction on 8-HO-Q		177–178	PH	143.5–144	244
2-OMe-4-Me	2-MeO-4-Me-7-Br-Q, BuLi; DMF	76	111–112			187
8-OEt	7-CH=CHMe-8-EtO-Q, KMnO$_4$	31	74–75			245
4-CHO	See Table 3					
6-CHO	See Table 5					
5-SO$_3$H-8-OH	Reimer–Tiemann reaction on the 8-HO-Q	42	287–288(d)	TSC	297–298	246
	Sommelet reaction on the 7-CH$_2$N(CH$_2$CH$_2$)$_2$O-8-HO-Q					230, 243
(5-SO$_3$H-8-O)$_2$-SbOH	5-SO$_3$H-7-CHO-8-HO-Q, SbCl$_3$, conc. HCl		> 360(d)			246
5-NO$_2$-8-OH	Sommelet reaction on the 7-CH$_2$N(CH$_2$CH$_2$)$_2$O-Q	50	255(d)	DNP	297(d)	241
8-NO$_2$	8-NO$_2$-Q, MeNO$_2$; KMnO$_4$	47	172–174			233

*Ox = oxime, Hy = hydrazone, PH = phenylhydrazone, NPH = 4-nitrophenylhydrazone, DNP = 2,4-dinitrophenylhydrazone, SC = semicarbazone, TSC = thiosemicarbazone, Pic. = picrate.

[a] Obtained as a mixture with 7-methylquinoline-6-carboxaldehyde.

[b] The structure assigned to this compound is obviously wrong; the isolated product is probably a polymer.

[c] This is probably 8-hydroxyquinoline-5-carboxaldehyde; see Table 4. The melting point of the phenylhydrazone is similar to that of the derivative from the 5-aldehyde.

Table 7. Quinoline-8-carboxaldehydes

Substituent(s)	Preparation	Yield (%)	M.p. (°C)	Deriv.*	M.p. (°C)	References
None	Q-8-CH$_2$I, HNO$_3$	90	94–95			247
	8-Me-Q, SeO$_2$	70.3	94–95			4, 234
	8-Me-Q, SeO$_2$, 180–190 °C	49	92–93			96
	8-Me-Q, SeO$_2$, 265–270 °C		95	Ox	121	76, 79, 235, 248
				SC	238	
				TSC	238–240	
	Q-8-Li, DMF or (CD$_3$)$_2$NCDO[a]	42				249
	Q-8-CONHNHTs, Na$_2$CO$_3$	25				49
	Q-8-CH$_2$OH, HNO$_3$		94–95	PH	176	250
				HCl	213	
4-Me-5-Cl	4-Me-5-Cl-8-CH$_2$OH, SO$_3$	67	140–142			251
4-Me-6-Cl	4-Me-6-Cl-8-CH$_2$OH, SO$_3$	70.7	153–153.5			251
5,6,7-triMe	5,6,7-triMe-8-Br-Q, BuLi; C$_5$H$_{10}$NCHO		162			252
2-Ph-6-Me	2-Ph-6-Me-8-Br-Q, BuLi; DMF	72.5	135.5–137			212
2-(4-ClC$_6$H$_4$)	2-(4-ClC$_6$H$_4$)-8-Me-Q, SeO$_2$, 200 °C	63.9	142–146			251
2-(4-ClC$_6$H$_4$)-4-Me-5-Cl	2-(4-ClC$_6$H$_4$)-4-Me-5-Cl-8-CH$_2$OH, SO$_3$	82.4	192–193			251
2-(4-ClC$_6$H$_4$)-4-Me-6-Cl	2-(4-ClC$_6$H$_4$)-4-Me-6-Cl-8-CH$_2$OH, SO$_3$	91.2	242–245			251
2-(4-ClC$_6$H$_4$)-6-Cl	2-(4-ClC$_6$H$_4$)-6-Cl-8-Me-Q, SeO$_2$, 200 °C	88	195–196			251
6-Ph	6-Ph-8-Br-Q, BuLi; DMF	68.5	133–134			215
6-F	6-F-8-Br-Q, BuLi; DMF	80	116–117			215
3,5-diCl	3,5-diCl-Q-8-CHCl$_2$, H$_2$SO$_4$		170			253

Substituent	Synthesis	Yield (%)	mp	Derivative	Derivative mp	Ref.
3,7-diCl	3,7-diCl-Q-8-CHCl$_2$, H$_2$SO$_4$	87	208	Ox SC DNP	202 > 280 > 280	253, 254
6-Cl	6-Cl-8-Me-Q, SeO$_2$, 200 °C	31.9	151–155			251
	6-Cl-Q-8-Li, DMF	83	158–159			215
7-Cl	7-Cl-Q-8-CH$_2$OH, CrO$_3$, Py		168	Ox	118	222 250
3-Br	3-Br-Q-8-CH$_2$OH, HNO$_3$			SC	250	231
7-OH	Reimer–Tiemann reaction on 7-HO-Q	55	135–135.5	Ox	201(d)	222, 255
7-OMe	7-HO-Q-8-CHO, Me$_2$SO$_4$, K$_2$CO$_3$					222
5-NO$_2$	5-NO$_2$-Q-8-CH$_2$I, HNO$_3$	100	146–147			256

* Ox = oxime, PH = phenylhydrazone, DNP = 2,4-dinitrophenylhydrazone, SC = semicarbazone, TSC = thiosemicarbazone.
ᵃ Gave Q-8-CDO.

to be the lactic acid **45**, R = H, was in fact 3-acetyl-1, 2-dihydroquinoline-2-carboxylic acid. The oxidation therefore produced 3-acetylquinoline, which was wrongly identified as 2-quinolylacetaldehyde.[263,264]

The acetal **48**, R = Me was formed by treating 2-ethynylquinoline with sodium methoxide; 4-ethynylquinoline similarly gave 4-quinolylacetaldehyde dimethylacetal (85%, b.p. 134–136 °C/1 mm), but 3-ethynylquinoline needed a prolonged reaction time and the product was the enol ether **49**.[265] Treatment of quinoline N-oxide with methyl vinyl ether and benzoyl chloride was claimed to give compound **48**, R = COPh which was converted into the acetal **48**, R = Me by p-toluenesulphonic acid in methanol. Neither intermediate was isolated; both were pyrolysed to 2-(2-methoxyvinyl)quinoline.[266]

48, R=Me,89%, b.p. 145 − 148 °C/1 mm   **49**, 83%, b.p. 135 −139 °C/1 mm

The Vilsmeier reagent reacted with 2-methylquinoline to give the salt **50**, which was hydrolysed in water to the dialdehyde **51**.[267] More recently, the partial hydrolysis product **52** was described as a starting material in a patent, but details of its preparation and characterization were not given.[268]

50   Cl⁻   **51**, 70%, m.p. 201−202 °C

52

Quinoline N-oxide and propiolic esters reacted to give the aldehydo-esters **54a**, or **54b** presumably via the intermediates **53**.[269] Ethyl formate and 2-quinolylacetonitrile gave the aldehyde **54c** on boiling or in the cold under basic catalysis.[258]

53 a R=COOMe   **54 a** R=COOMe, 33%, m.p. 170−171 °C
   b R=COOEt   **b** R=COOEt, 14.5%, m.p. 120−121 °C
                            c R=CN, m.p. 231 °C

The quinoline *N*-oxides **55** and the enamines **56** reacted under the influence of benzoyl chloride to give the substituted aldehydes **57**.[270] Similar reactions can be used to prepare ketones; see Chapter 2, Section V.3.

55   **56**   **57**

R=H, 89%, b.p. 138−139 °C/5 mm
R=Me, 67%, b.p. 154−156 °C/2 mm
R=Cl, 92%, b.p. 143−147 °C/2 mm

The quinolylpyruvic acid **58** reacted readily with simple aldehydes to give the hydroxyaldehydes **59**, but the ethyl ester of **58** gave the unsaturated ketones **60a** and **b**; 4-quinolylpyruvic acid reacted like compound **58** to give hydroxyaldehydes **61**.[271]

58   **59**

R=Ph, 74%, m.p. 245−246 °C
R=3-NO$_2$C$_6$H$_4$, 51%, m.p. 256−257 °C
R=4-NO$_2$C$_6$H$_4$, 71%, m.p. 265−266 °C
R=4-Quinolyl, 50%, m.p. 236−237 °C

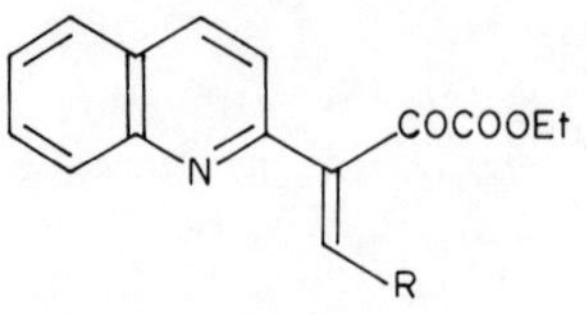

60 a R = 3-NO$_2$C$_6$H$_4$, 50%, m.p. 218 – 219 °C
   b R = 4-NO$_2$C$_6$H$_4$, 75%, m.p. 198 – 199 °C

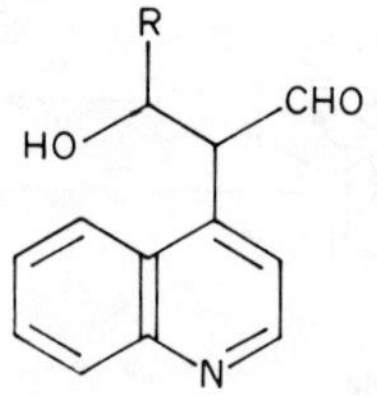

61 R = 3-NO$_2$C$_6$H$_4$, 50%, m.p. 260 – 261 °C
   R = 4-NO$_2$C$_6$H$_4$, 56%, m.p. 249 – 250 °C

Pentyl nitrite and 4-acetylquinoline gave with sodium ethoxide the aldehyde oxime **62**;[272] 3-acetylquinoline reacted similarly.[273]

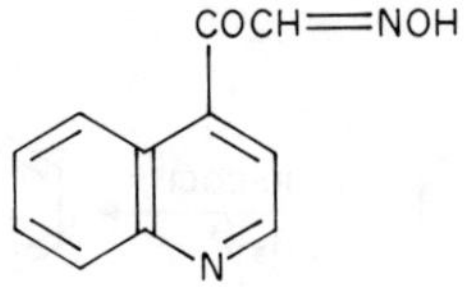

62, m.p. 237 – 242 °C (d)

The vinylogous amidine **63** was hydrolysed by alkali to the malondialdehyde **64**.[267,274] On mild hydrolysis with aqueous potassium carbonate the amino-aldehyde **65** was isolated.[267]

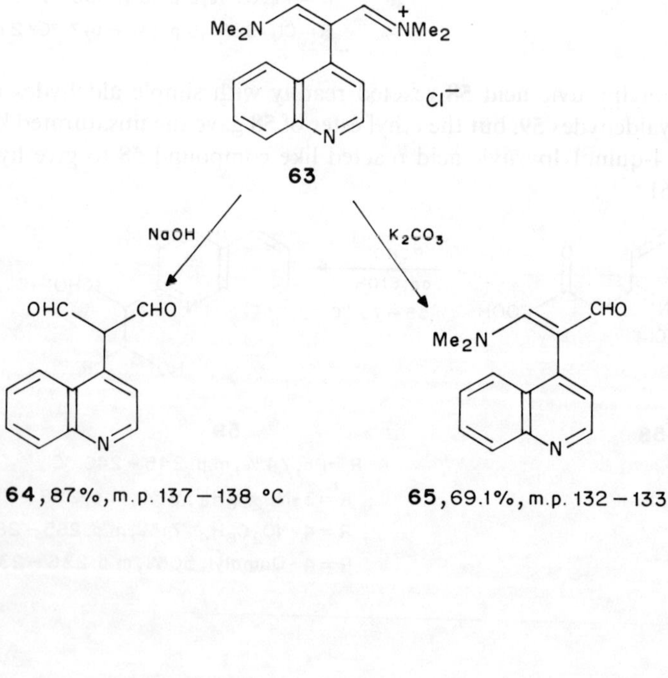

63

NaOH   K$_2$CO$_3$

64, 87%, m.p. 137 – 138 °C   **65**, 69.1%, m.p. 132 – 133 °C

Selenium dioxide oxidation of 3-acetylquinoline in dioxane gave the ketoaldehyde **66**[275], and of 6-acetylquinoline in acetic acid at 90 °C the keto-aldehyde **67**,[276-278] but 4-acetylquinoline with selenium dioxide in toluene at 100 °C gave cinchoninic acid (quinoline-4-carboxylic acid).[257]

66, m.p. 134 – 136 °C   **67**, m.p. 145 °C

The anion from 2-methylquinoline reacted with bromoacetaldehyde dimethylacetal to give the acetal **68**. The free aldehyde was not isolated.[279]

68, 30%, b.p. 145 – 150 °C/0.04 mm
Pic. m.p. 96 °C

The tosylates **69** dissolved in warm pyridine to give, presumably via intermediates **70** and **71**, the quinoline aldehydes **72**, which were further reacted as shown in Scheme 9. Alternatively, the appropriate phenol could be treated with p-toluenesulphonyl chloride in pyridine to give the aldehydes **72** directly. In one case **70**, $R^1 = NO_2$, $R^2 = H$ the intermediate pyridinium salt was isolated. Under similar conditions, the ketones **73** both gave the same 4-phenylquinoline **74**.[280]

Quinoline-4-carboxaldehyde and (triphenylphosphoranylidene)acetaldehyde gave the unsaturated aldehyde **75a**;[281] 8-nitroquinoline-4-carboxaldehyde underwent aldol condensation with acetaldehyde in sodium hydroxide to give compound **75b**.[182,183]

An aldol condensation between quinoline-6-carboxaldehyde and acetaldehyde gave the olefinic aldehyde **76**.[282]

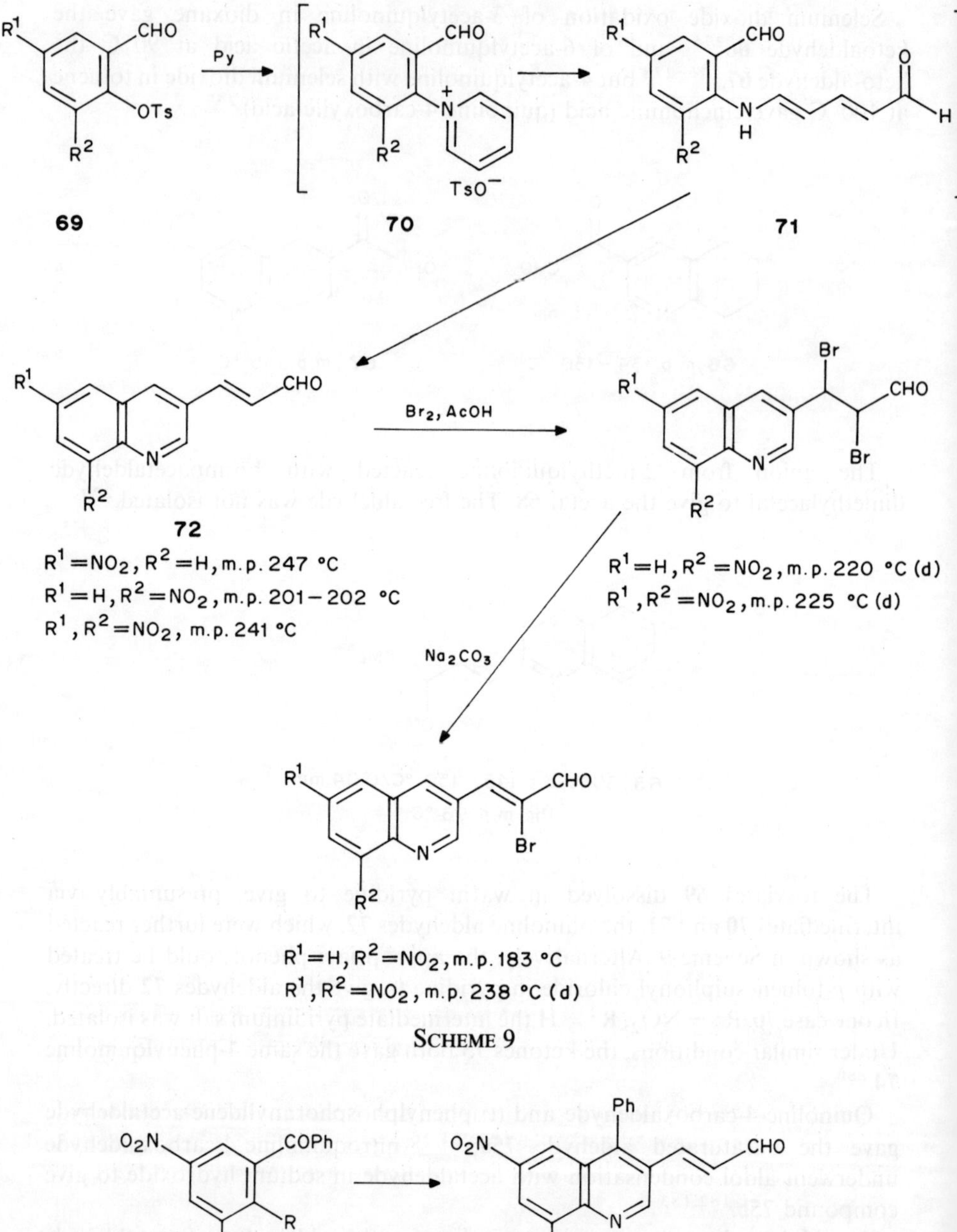
69
Py
70
71
Br2, AcOH
72
$R^1 = NO_2, R^2 = H$, m.p. 247 °C
$R^1 = H, R^2 = NO_2$, m.p. 201–202 °C
$R^1, R^2 = NO_2$, m.p. 241 °C
$R^1 = H, R^2 = NO_2$, m.p. 220 °C (d)
$R^1, R^2 = NO_2$, m.p. 225 °C (d)
Na_2CO_3
$R^1 = H, R^2 = NO_2$, m.p. 183 °C
$R^1, R^2 = NO_2$, m.p. 238 °C (d)

SCHEME 9

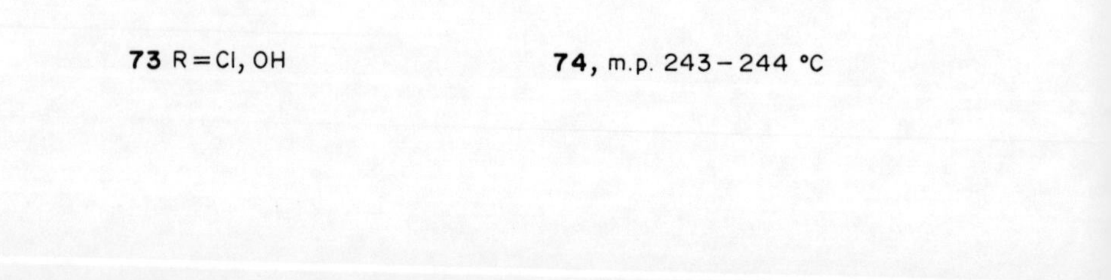

73 R = Cl, OH

74, m.p. 243–244 °C

75 a R=H, 55%, m.p. 92−92.5 °C   **76**, 33%, m.p. 162−163 °C
b R=NO$_2$, 45%, m.p. 286 °C

The 6-quinolylmethylmagnesium bromides **77** reacted with the oxazolium salt **78** to give the quinoline acetaldehydes **79**.[283,284]

R^1	R^2
4-FC$_6$H$_4$	Me
4-ClC$_6$H$_4$	H
4-ClC$_6$H$_4$	Me
4-MeOC$_6$H$_4$	Me
2-Thienyl	H

IV. Quinoline Aldehydes with Partially or Fully Saturated Rings

1. Quinoline Aldehydes with Partially Saturated Pyridine Rings

When the 4-dibromomethylquinolines **80** were hydrolysed to aldehydes **81** in aqueous alcoholic sodium hydroxide, the bromine at C-3 was reductively removed. It was suggested that the neighbouring aldehyde group may have assisted in this.[285,286] Bromine in concentrated sulphuric acid with 1,2-dihydro-2,2,4-trimethylquinoline gave the biquinolyl **82**, which was hydrolysed, again with loss of the C-3 bromine atoms, to the dialdehyde **83**.[286]

Quinoline-3-carboxaldehyde was reduced to its 1,4-dihydro derivative (74%; m.p. 171–172 °C) by triethylammonium formate.[287] Quinoline was reduced by sodium hydride in HMPT and the product treated with methyl chloroformate

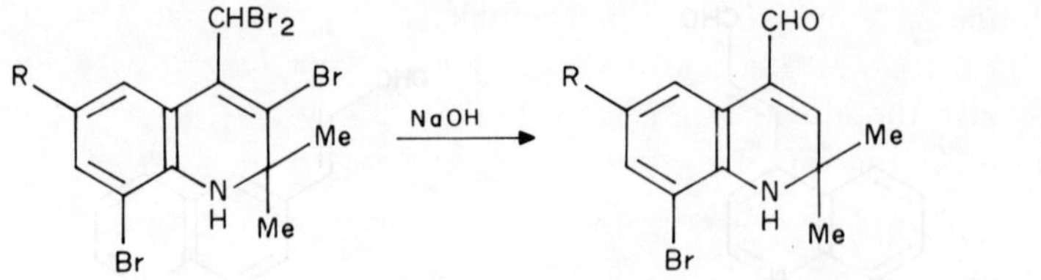

80

81, R = Br, m.p. 244 – 245 °C
R = NO$_2$, 99.8%, m.p. 258 °C(d)

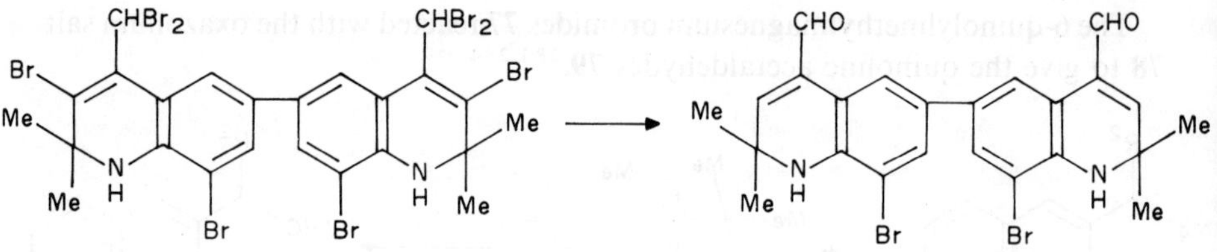

82

83, m.p. >300 °C

to give a mixture of the dihydro derivative **84** and the equivalent 1, 2-dihydroquinoline. With the Vilsmeier reagent the mixture gave the aldehyde **85** as a single product.[288] Careful hydrolysis of compound **86** gave a mixture of which the aldehyde **87** was one component. More vigorous hydrolysis gave the 2-quinolone; see Section VII.[289]

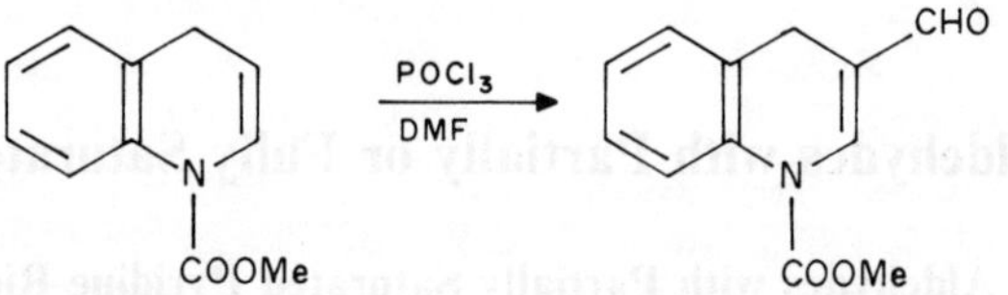

84

85, 22%, oil

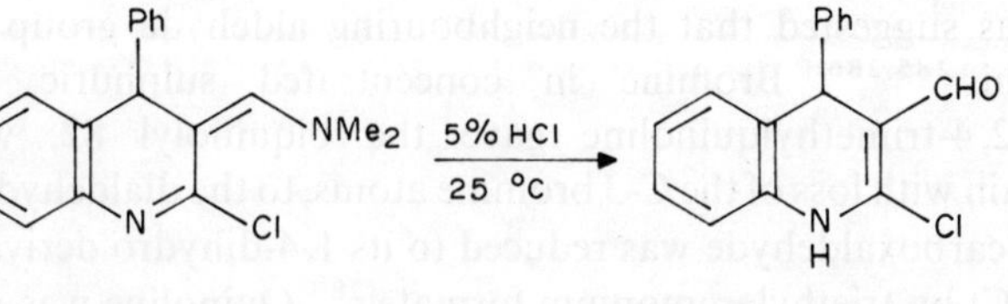

86

87, 39%, m.p. 192 – 193 °C

A solution of ethylquinolinium bromide, isopropyl isonitrile and sodium formate in methanol was allowed to stand at room temperature for several weeks to give the aldehyde **88**; see Chapter 2, Section X.4 for the suggested mechanism.[290]

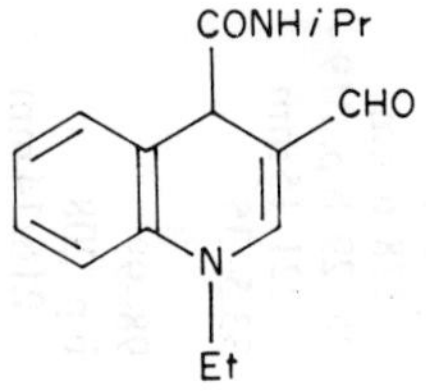

88, 11%, m.p. 226—227 °C

In the presence of zinc chloride at room temperature 5-chloro-2-methylaminobenzophenone underwent Michael addition to acrolein. Spontaneous ring closure gave the aldehyde **89**; see Chapter 2, Section X.2.[291]

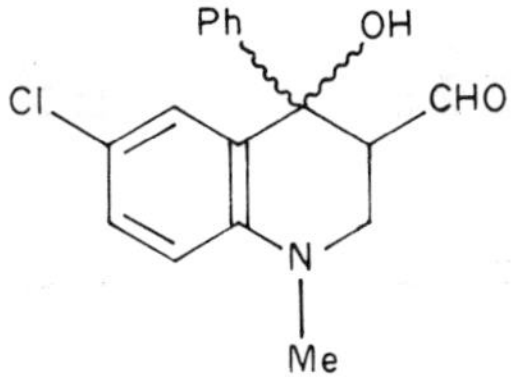

89, m.p. 95—97 °C

The ethylene glycol acetal from quinoline-4-carboxaldehyde was hydrogenated over platinum in acetic anhydride to give, after acid hydrolysis, 1-acetyl-1,2,3,4-tetrahydroquinoline-4-carboxaldehyde. The compound was not characterized, but was converted into its phenylhydrazone (67%; m.p. 183–185 °C).[292] The 1,2,3,4-tetrahydroquinoline-6-carboxaldehydes, e.g. **90a**, have been prepared by Vilsmeier reactions unless otherwise stated. They are listed in Table 8. After Vilsmeier preparation of **90a**, a small yield of a dialdehyde, **90b**, was isolated as a by-product.[293]

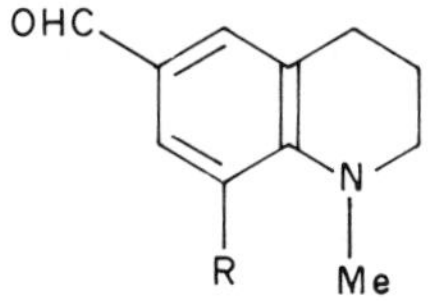

90a, R = H

b, R = CHO

Table 8. 1,2,3,4-Tetrahydroquinoline-6-carboxaldehydes
Prepared by Vilsmeier reactions except where noted

R^1	R^2	R^3	R^4	R^5	R^6	Yield (%)	M.p. (°C)	References
H	H	H	H	H	H	39[a]	80–82	294
H	CH_2NEt_2	H	H	NO_2	H	a	94–95	101
H	CH_2NHiPr	H	H	NO_2	H			101
Me	H	H	H	H	H	80	b.p. 175–178/6 mm	295
Me	H	H	H	H	H	80	28–29; b.p. 219–221/15 mm	296, 297
Me	H	H	H	H	H	69	33.5–35	293
Me	H	H	H	H	CHO	67		298
Me	H	H	H	Me	H		98–99	293
Me	H	H	H	H	H	3	b.p. 208–216/14 mm	299
Me	Me	Me	Me	H	H			300
Me	Me	Me	Me	OH	H	80[b]	68–70	301
Me	Me	H	Me	OMe	H	94	105–107	301
CH_2Ph	H	H	H	H	H			302
CH_2Ph	Me	Me	Me	OMe	H			302
CH_2Ph	Me	Me	Me	Me	H			301
$CH_2C_6H_4$-4-COOMe	Me	Me	Me	Me	H			303
CH_2S-(benzothiazol-2-yl)	Me	Me	Me	Me	H			304, 305

Substituent						m.p./b.p. (°C)	Note	Ref.
Et	Me	Me	Me	H	H			306
Et	Me	Me	Me	H	H			307
CH$_2$CH$_2$Cl	Me	H	H	Me	H			308
CH$_2$CH$_2$Cl	Me	Me	Me	Me	H			304
CH$_2$CH$_2$Cl	Me	Me	Me	H	H			302
CH$_2$CH$_2$OH	Me	Me	Me	OH	H			309
CH$_2$CH$_2$OH	Me	Me	Me	H	H			302
CH$_2$CH$_2$OAc	Me	Me	Me	Me	H	70–75	c	301
CH$_2$CH$_2$OC$_6$H$_4$-4-cyclo C$_6$H$_{11}$	Me	H	Me	H	H	95	d	302
CH$_2$CH$_2$OCOPh	Me	H	Me	Me	H			304, 305
CH$_2$CH$_2$OCOPh	Me	H	Me	H	H	Oil		310
CH$_2$CH$_2$OCOC$_6$H$_4$-3-COOCH$_2$CH$_2$R[e]	Me	Me	Me	Me	H	Oil		310
CH$_2$CH$_2$OCONHPh	Me	H	Me	H	H	122–124		311
CH$_2$CH$_2$OCONHPh	Me	H	Me	Me	H	Oil		310
CH$_2$CH$_2$OCONHPh	Me	Me	Me	Me	H	Oil		310
CH$_2$CH$_2$OCONHPh	Me	Me	Me	OH	H	96–97		310, 312
CH$_2$CH$_2$OCONHPh	iPr	H	H	OEt	H	96–97		301
CH$_2$CH$_2$OCONHPh	iPr	H	OEt	OEt	H		d	310
[benzoxazol-2-yl] CH$_2$CH$_2$S	Me	Me	Me	Me	H			310
[benzothiazol-2-yl] CH$_2$CH$_2$S	Me	Me	Me	Me	H			309, 313
[benzothiazol-2-yl-S,S-dioxide] CH$_2$CH$_2$S	Me	Me	Me	Me	H			309
CH$_2$CH$_2$NHCOMe	Me	H	Me	Me	H	Oil		313
CH$_2$CH$_2$NHCOMe	Me	H	H	H	H	Oil		310
CH$_2$CH$_2$NHCOOEt	Me	H	Me	Me	H	Oil		310
CH$_2$CH$_2$NHCOOEt	Me	H	H	H	H	Oil		310
CH$_2$CH$_2$NHCONHPh	Me	H	Me	Me	H	Wax		310
CH$_2$CH$_2$NHCONHPh	Me	H	Me	H	H	Wax		310

Table 8. (*Contd.*)

R¹	R²	R³	R⁴	R⁵	R⁶	Yield (%)	M.p. (°C)	References
CH₂CH₂-succinimido	Me	H	H	H	H			314
CH₂CH₂-succinimido	Me	Me	Me	Me	H			314
CH₂CH₂-glutarimido	Me	H	H	Me	H			314
CH₂CH₂-phthalimido	Me	H	H	Me	H		160–161	314
CH₂CH₂N(SO₂Bu)Me	Me	Me	H	Me	H			315
CH₂CH₂N(SO₂C₆H₄-4-Me)Me	Me	Me	Me	Me	H		138–139	315
CH₂CH₂N(SO₂Ph)CH₂CH₂-pyrrolidonamido	Me	Me	Me	Me	H		161–162	315
CH₂CH₂N(SO₂C₆H₄-4-Me)CH₂CH₂-pyrrolidonamido	Me	Me	Me	Me	H			315
CH₂CH₂N(SO₂C₆H₄-4-Me)CH₂CH₂-(CH₂)₃OMe	Me	Me	Me	Me	H		125–126	315
CH₂CH₂N(SO₂Me)Ph	Me	Me	Me	Me	H			315
CH₂CH₂N(SO₂Bu)Ph	Me	Me	Me	Me	H			315
CH₂CH₂N(SO₂Me)C₆H₄-4-NO₂	Me	Me	Me	Me	H		170–171	315
CH₂CH₂N– (ring structure)	Me	Me	Me	Me	H		177–178	309, 313
CH₂CH₂CN	H	H	H	H	H		76.5–77.5	316
CH₂CH₂CN	Me	H	H	Me	H	64	119–121	316
CH₂CH₂CN	Me	Me	Me	Me	H			316
(CH₂)₃-succinimido	Me	H	H	Me	H			314
(CH₂)₃-succinimido	Me	Me	Me	Me	H			314
CH₂CH=CH₂	H	H	H	H	H			302
Bu	Me	Me	Me	OMe	H			301
C₆H₁₃	Me	Me	Me	OMe	H			301
COOMe	H	H	H	H	H		186–188	317

a Prepared from the appropriate 6-hydroxymethyl compound and manganese dioxide.
b Prepared from the 7-methoxy compound with aluminium chloride in carbon disulphide.
c Method of preparation not given.
d The preparations are not given, but the patent implies that the method of footnote *a* was used.

Treatment of 1,2,3,4-tetrahydroquinoline with boron trichloride, an alkyl isonitrile and triethylamine followed by aqueous acid gave the 8-carboxaldehyde **91**. The suggested rationalization is shown in Scheme 10.[318]

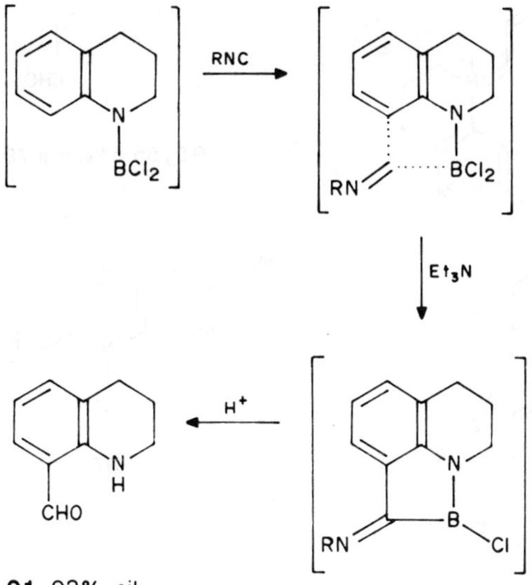

91, 92%, oil

Oxime m.p. 69–70 °C

SCHEME 10

A peroxidase enzyme and julolidine **92** gave a small yield of the dialdehyde **93**.[319]

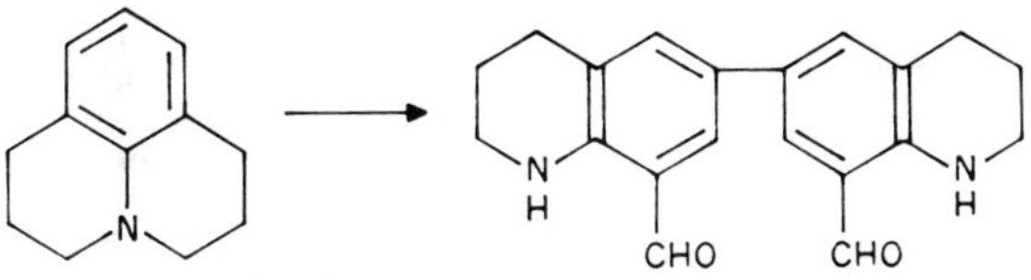

92   **93**, 1.7%, m.p. 230 °C

Propargyl aldehyde acetal, 1,2,3,4-tetrahydroquinoline, and hydrochloric acid gave **94**, which was hydrolysed to the aldehyde **95**.[320] Compounds **96a**, **96b**[321] and **97**[322] were prepared via similar hydrolyses. Diphenylformamidine was fused with 1,2,6-trimethylquinolinium iodide to give compound **98**, R = Me, Hal = I. Hydrolysis of such compounds with alkali gave the enaminals **99**. Compound **100** was prepared similarly, but was unstable and had to be used immediately.[323–325] The aldehydes **99**, R^1 = Me, R^2 = H[326] and **99**, R^1 = Et, R^2 = H[327] have been used as starting materials for further reactions.

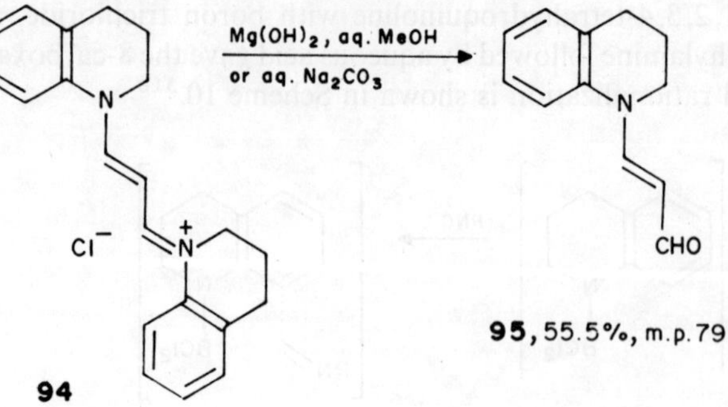

Mg(OH)₂, aq. MeOH
or aq. Na₂CO₃

94

95, 55.5%, m.p. 79 – 80 °C

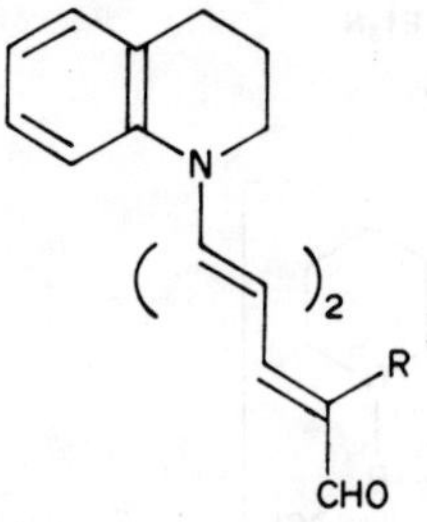

96 a, R = H, m.p. 152 °C
b, R = OAc

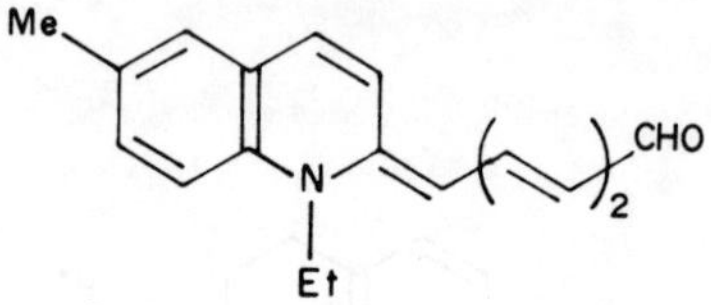

97

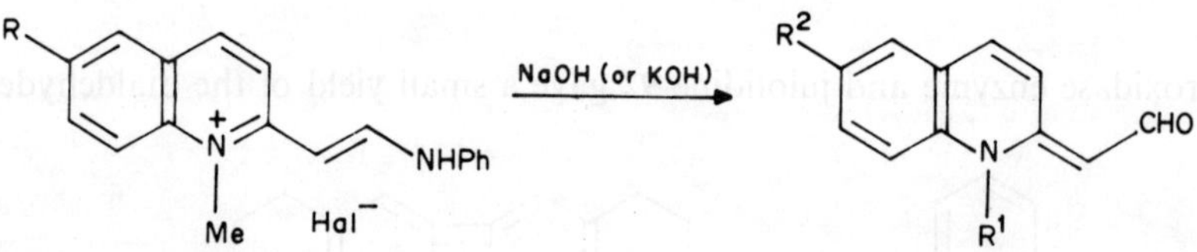

NaOH (or KOH)

98

99

R¹ = Me, R² = H, 27%, m.p. 90 °C
R¹, R² = Me, m.p. 118 – 120 °C

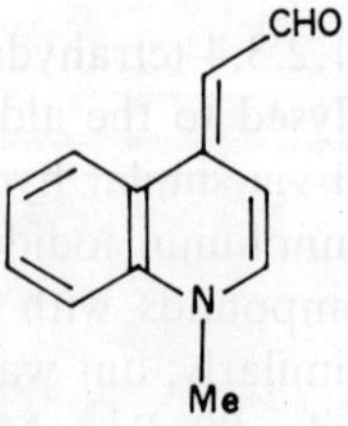

100

The Vilsmeier reagent reacted with 1-ethyl-2-methylquinolinium iodide to give, via an intermediate vinylogous amidine (see compound **50**, Section III), the dialdehyde **101**;[267] 1-methyl-2-quinolinylideneacetonitrile and the Vilsmeier reagent gave the unsaturated aldehyde **102**.[328]

101, 40.7%, m.p. 186 − 187 °C   **102**, m.p. 177 − 178 °C

The vinylogous amidine **103** was hydrolysed to the dialdehyde **104** (see compounds **63** and **64** Section III).[267,329]

103   **104**

R = Me, no data
R = Et, 40%, m.p. 219 °C

The salt **105**, R^1 = Me reacted with the trialdehyde **106**, R^2 = CHO in pyridine at room temperature to give the unsaturated aldehyde **107a**. If the reaction was carried out in boiling ethanol containing triethylamine, the aldehyde **108** resulted. Similarly, from 1,4-dimethylquinolinium tetrafluoroborate were obtained aldehydes **109a** or **110**.[330,331] The sodium salt of the nitro-dialdehyde

105   **107**

a, R^1 = Me, R^2 = CHO, 93%, m.p. 235 − 236 °C
b, R^1 = Me, R^2 = NO_2, 74%, m.p. 246 − 248 °C
c, R^1 = Et, R^2 = NO_2, 85%, m.p. 223 − 224 °C
d, R^1 = $C_{18}H_{37}$, R^2 = NO_2, no data

106, $R^2 = NO_2$ gave aldehyde **107b** from compound **105**, $R^1 = Me$ and aldehydes
107c and **109b** from compound **105**, $R^1 = Et$ and from the 1,4-dimethyl-
quinolinium salt, respectively, in ethanol at room temperature.[332,333]
Compound **107d** was prepared similarly, but no data were reported.[334]

108, 56%, m.p. 279 – 279.5 °C

109

a , $R^1 =$ Me, $R^2 =$ CHO, 70%, m.p. 226 – 227 °C
b , $R^1 =$ Et, $R^2 =$ NO$_2$, 95%, m.p. 234 – 235 °C

110, 26%, m.p. 263 – 263.5 °C

The malondialdehydes **111** reacted with 1,2-dimethylquinolinium iodide in
the presence of piperidine to give the conjugated aldehydes **112**, X = O, S; see
Scheme 11.[335]

The appropriate quinoline (see Section III) was reduced (H$_2$/Pt) to the acetal

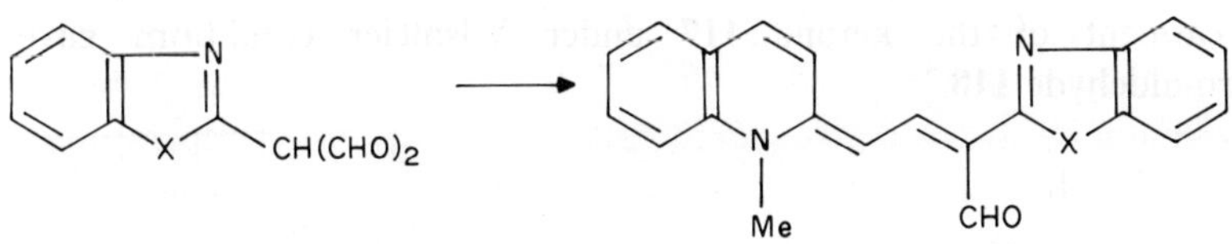

111

X = O, S

112

SCHEME 11

113a, the nitrosamine of which was reduced (Zn, AcOH) to compound **113b**. No free aldehydes were isolated, but when compound **113b** was added to oxalic acid in ethanol it gave the expected pyridazinoquinoline directly.[279]

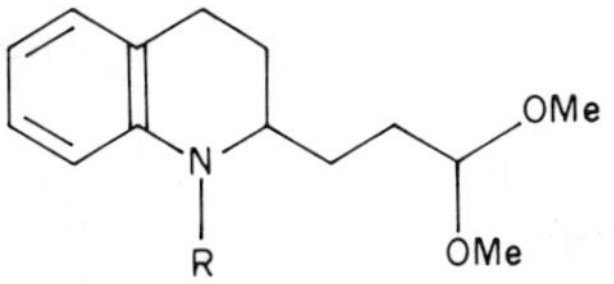

113

a , R=H, b.p. 145−150 °C/0.02 mm

b , R=NH$_2$, b.p. 165−168 °C/0.06 mm

2. Quinoline Aldehydes with Partially Saturated Benzene Rings

The ester **114** was converted in three steps to the aldehyde **115**.[336]

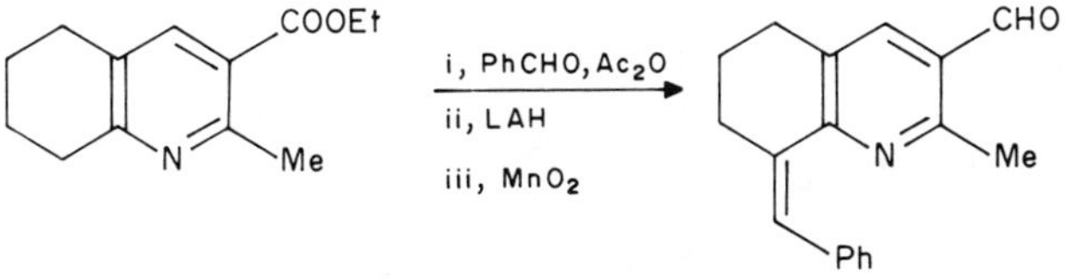

i, PhCHO, Ac$_2$O

ii, LAH

iii, MnO$_2$

114

115 , m.p. 217−219 °C

The ester **116a** was reduced (LAH) and re-oxidized (CrO$_3$, Py) to the aldehyde **116b**.[337]

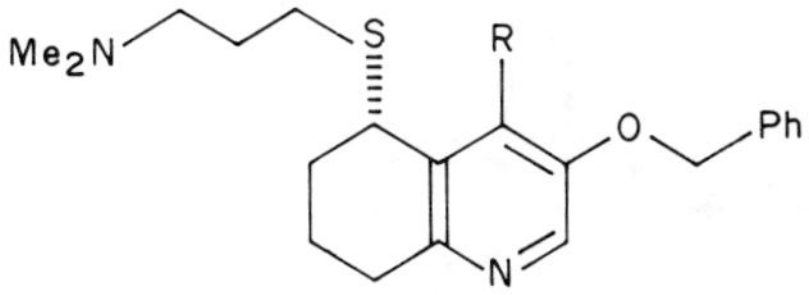

116

a , R =COOEt

b , R = CHO

Treatment of the ketone **117** under Vilsmeier conditions gave the chloro-aldehyde **118**.[52]

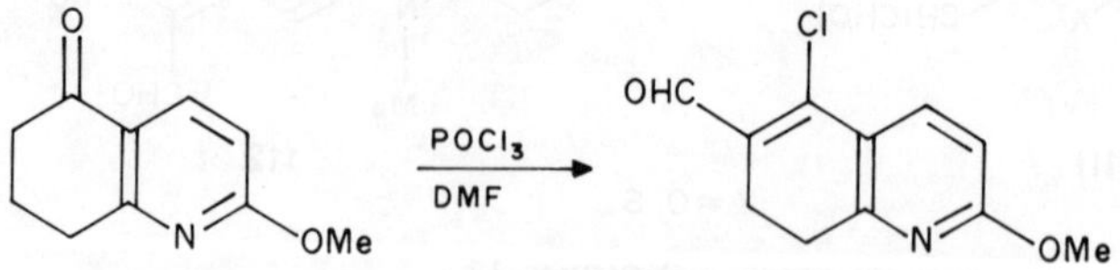

117   **118**, 48%, m.p. 77 °C

The nitrile **119a** was reduced with DIBAH and hydrolysed with aqueous acid to give the aldehyde **119b**.[338] The side-chain aldehyde **120** was made in an exactly similar way.[339]

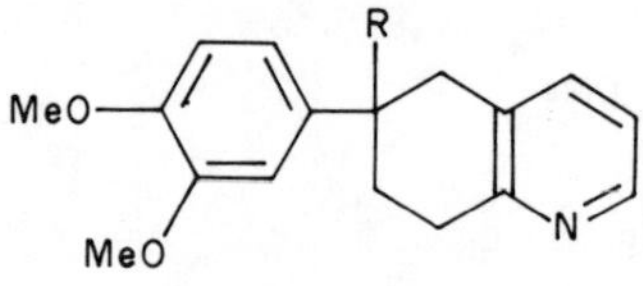

119 a, R=CN
     b, R=CHO, 88%, oil

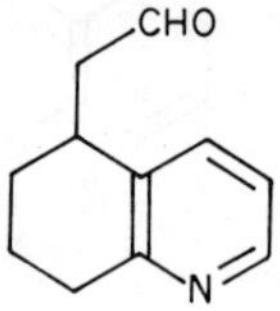

120, 45.5%, b.p. 103 – 105 °C/0.2 mm

3. Quinoline Aldehydes with Both Rings Saturated

The octahydro- and decahydro-quinoline-5-carboxaldehydes shown in Scheme 12a have been prepared as synthetic intermediates. Compound **122** was

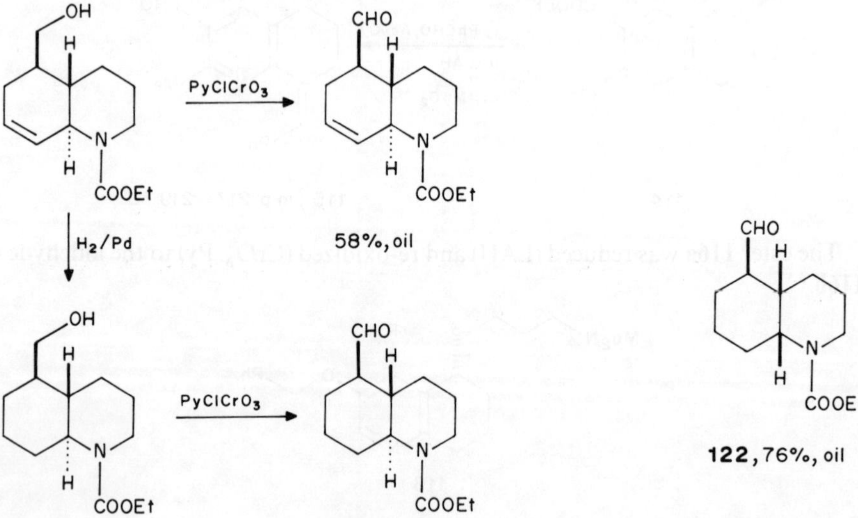

121, 86%, oil

122, 76%, oil

SCHEME 12a

prepared similarly to compound **121**[340] Scheme 12b shows the preparation of a related intermediate.[341]

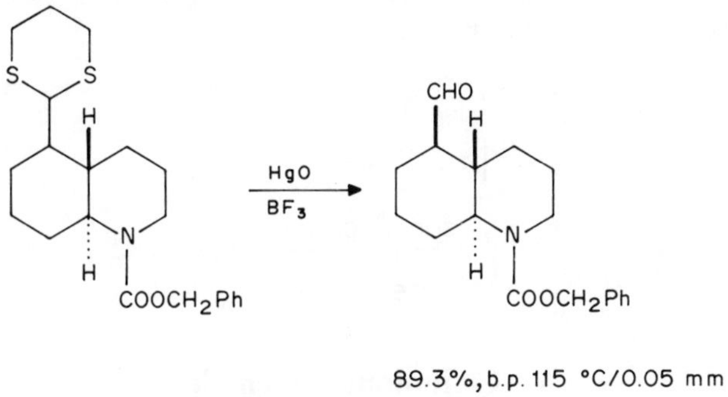
HgO
BF_3
CHO
H
N
H
$COOCH_2Ph$
89.3%, b.p. 115 °C/0.05 mm

SCHEME 12b

The protected aldehyde **123** was made by ring closure and converted into compound **124** and **125**, intermediates in the preparation of gephyrotoxin.[342]

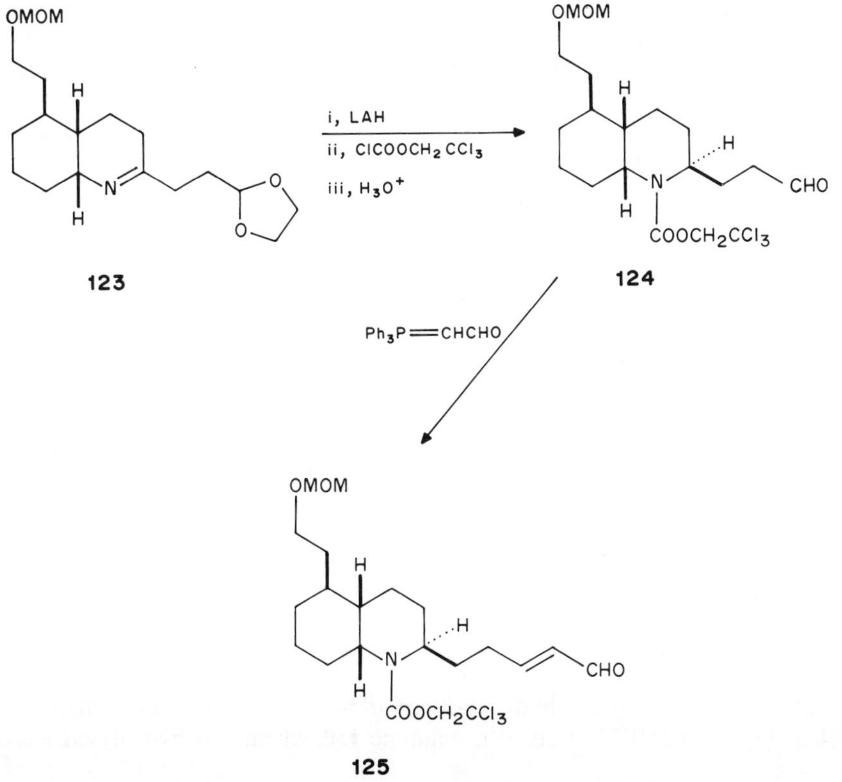
OMOM
i, LAH
ii, $ClCOOCH_2CCl_3$
iii, H_3O^+
CHO
$COOCH_2CCl_3$
123
124
$Ph_3P{=}CHCHO$
125

The aldehyde **126**, C2–C3 single bond, as its acetal was converted to **126**, C2–C3 double bond, as part of a synthetic scheme.[343]

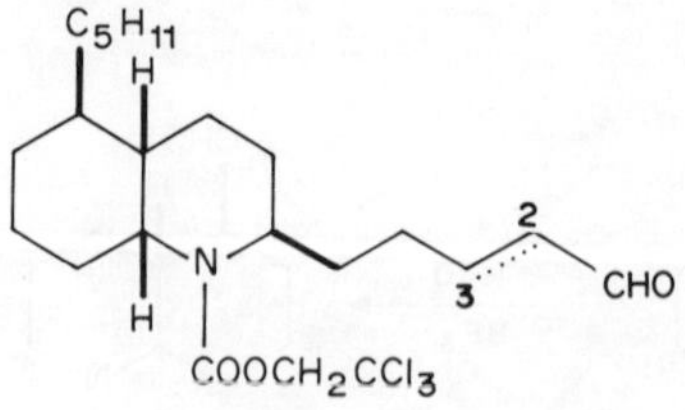

126

V. Quinolone Aldehydes

1. 3-Formyl-2-quinolones

Many of these aldehydes have been prepared by standard Reimer–Tiemann or Vilsmeier reactions; see Table 9 (p. 58).

When halogeno-acetanilides were reacted under Vilsmeier conditions 3-formyl-2-quinolones **128**, R = 6-, 7-, or 8-Cl or 8-Br, could be obtained.[29] This reaction usually produces 2-chloroquinoline-3-carboxaldehydes by further reaction with the phosphorus oxychloride; see Section II.3. Two of these, compounds **127**, R = 6-Cl or 7-Me, have been hydrolysed to the quinolones **128**, R = 6-Cl or 7-Me and another, **127**, R = H gave the quinolthione **129**, with sodium hydrosulphide.[30]

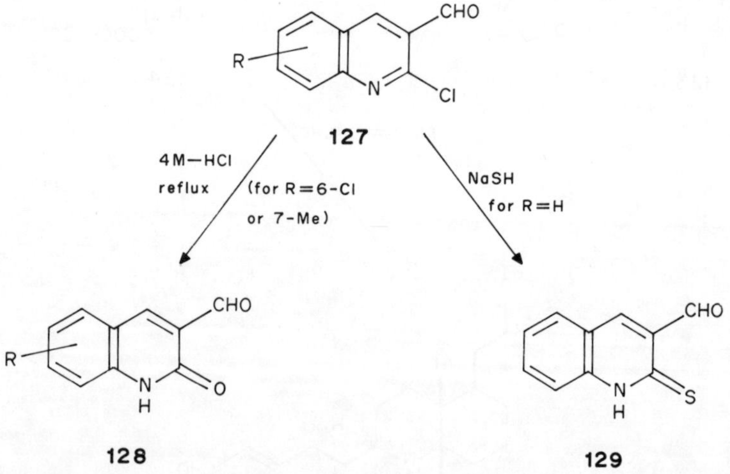

127

4M–HCl
reflux
(for R = 6-Cl
or 7-Me)

NaSH
for R = H

128   **129**

Triethyl orthoformate, 4-hydroxy-2-quinolone, and aniline condensed in ethylene glycol at 180 °C to give the enamine **130**, which was hydrolysed with dilute sodium hydroxide to the aldehyde **131a**.[344] The alkaloid *N*-methyl-flindersine **132** was ozonolysed to the aldehyde **131b**. The structure of compound **131b** was confirmed by a Reimer–Tiemann reaction on 4-hydroxy-1-methyl-2-quinolone. The same aldehyde was obtained by lead tetraacetate oxidation of the *cis* glycol from compound **132**.[345]

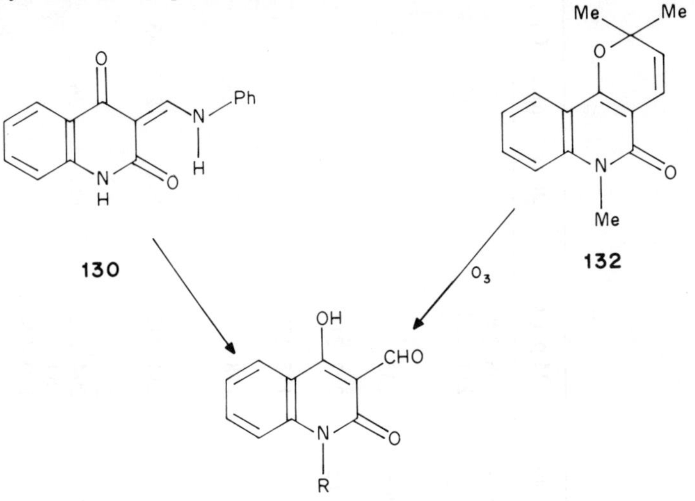

130

132

131

a , R=H   **b** , R=Me

The aldehydes **134** were prepared by oxidation of furanoquinolines **133**, for the compound with R = H by ozonolysis and for R = Me by potassium permanganate.[346] The latter method was used [347] to make aldehydes **135** and **136**.

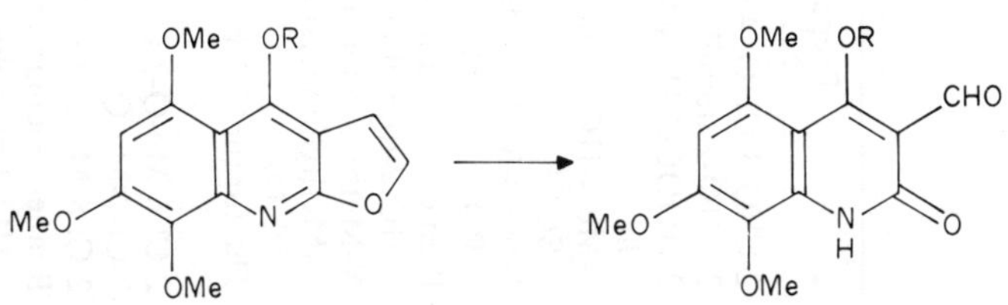

133

134 , R=H, Me

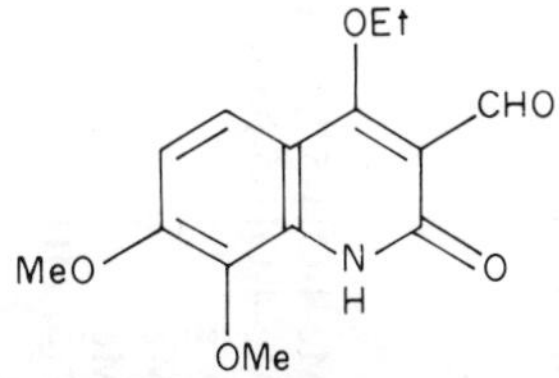

135

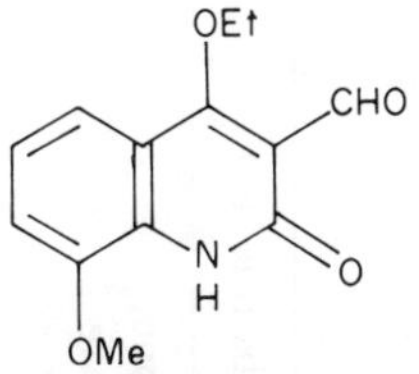

136

Table 9. 3-Formyl-2-quinolones

Substituent(s)	Preparation	Yield(%)	M.p. (°C)	Deriv.*	M.p. (°C)	References
None	2-Cl-Q-3-CHO, 4M-HCl	100	308–309			153, 349
1-Me	3-CHO-2-quinolone, NaH, MeI	24.6	211–214			153
	2-HOOCC$_6$H$_4$N(Me)CO-(3-furyl), hv	7.5				350
1-Me-4-Cl	1-Me-4-HO-2-quinolone, DMF, SOCl$_2$	20^a	135			351
1-Me-4-Cl-7-NMe$_2$	1-Me-4-MeO-7-Me$_2$N-2-quinolone, POCl$_3$, DMF	79.5	217–218			352
1-Me-4-OH	Reimer–Tiemann reaction on 1-Me-4-HO-2-quinolone		175–177			345, 352, 353
	N-Methylflindersine, O$_3$		178	PH	235	345
	N-Methylflindersine cis-glycol, Pb(OAc)$_4$					345
1-Me-4-OH-5,7,8-triOMe	The furano[2,3-b]quinolone, O$_3$		214–215	PH	242–243	346
1-Me-4-OH-6-NO$_2$						354
6-Me	2-Cl-6-Me-Q-3-CHO, 4M-HCl	96	275			349
7-Me	2-Cl-7-Me-Q-3-CHO, 4M-HCl	96	294–295(d)	Ox	268(d)	30
8-Me	2-Cl-8-Me-Q-3-CHO, 4M-HCl	92	284			349
1-Et-4-OH	Reimer–Tiemann reaction on 1-Et-4-HO-2-quinolone		130–134			353
1-Ph-4-OH						354

Substituent	Method	Yield (%)	m.p. (°C)	Derivative	Deriv. m.p. (°C)	References
4-Cl	3-CH(OMe)$_2$-4-Cl-2-quinolone, HCl	94	268	PH	233	164
6-Cl	4-ClC$_6$H$_4$NHCOCH(CHO)$_2$, PPA, 150 °C		357–358(d)			29
7-Cl	2,6-diCl-Q-3-CHO, 4M-HCl, reflux	87				30
	3-ClC$_6$H$_4$NHCOCH(CHO)$_2$, PPA, 150 °C	98	338–340(d)			29, 349
8-Cl	2-ClC$_6$H$_4$NHCOCH(CHO)$_2$, PPA, 150 °C	91	259(d)			29
6-Br	4-BrC$_6$H$_4$NHCOCH(CHO)$_2$, PPA, 150 °C	96	342–344(d)			29
4-OH	4-OH-2-quinolone, (EtO)$_3$CH, PhNH$_2$; (CH$_2$OH)$_2$, 180 °C; NaOH	78	300			344
	Reimer–Tiemann reaction on 4-HO-2-quinolone	40	>350	PH	235	355, 356
	3-CHO-4-OMe-2-quinolone, HBr, AcOH					355, 356
	3-CHO-4-OC(Me)$_2$CHO-2-quinolone, H$_2$SO$_4$					345
4-OH-5,7,8-triOMe	See text, Section V.1		251–252(d)	PH	241–242	346
4-OMe (dictamnal)	Dictamnine, O$_3$ or KMnO$_4$	65	260	PH	228	356, 357
	3-CHO-4-Cl-2-quinolone, MeONa		260	PH	228	164
4,7,8-triOMe	Skimmianine, KMnO$_4$	15	238	PH	210	358
4,5,7,8-tetraOMe	See text, Section V.1	10	219.5–220.5	DNP	302–304	346
6,7-diOMe	2-Cl-6,7-diMeO-Q-3-CHO, 4M-HCl	83	285			349
7-OMe	2-Cl-7-MeO-Q-3-CHO, 4M-HCl	86	263			349
4-OEt-7,8-diOMe	See text, Section V.1	16	212	PH	178.5	347
4-OEt-8-OMe	γ-Fagarine, KMnO$_4$		192–193	PH	185–186	347
4-OC(Me)$_2$CHO	Flindersine cis-glycol, HIO$_4$		126			345
4-NMe$_2$	PhN=C=O, Me$_2$NC≡CCHO	75	210–211			359

* Ox = oxime, PH = phenylhydrazone, DNP = 2,4-dinitrophenylhydrazone.
a Obtained as a mixture with 4-chloro-1-methyl-2-quinolone.

When 2-aminobenzaldehyde was heated with the pyrazolone **137** it gave the quinolone aldehyde derivative **139**, a reaction presumed to go via the hydrazone **138**; see Scheme 13.[348]

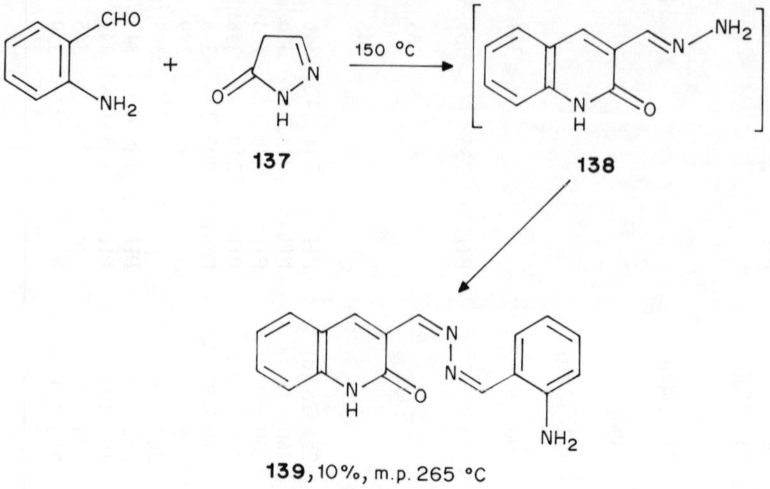

SCHEME 13

Sodium diethyl malonate reacted with 2-isothiocyanato-*trans*-cinnamaldehyde, **140**, to give the thione aldehyde **141** which was oxidized to the disulphide. The suggested mechanism is included in Scheme 14.[25]

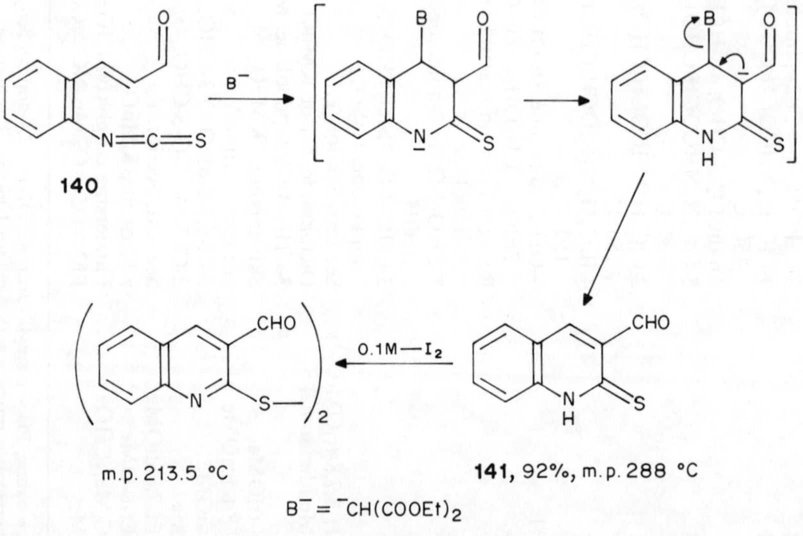

$$B^- = {}^-CH(COOEt)_2$$

SCHEME 14

When 7-chloro-4-methoxyquinoline was *N*-acylated with thiophosgene and subsequently treated with barium carbonate in a water–dichloromethane mixture, compound **143** was obtained, presumably via intermediate **142**. The intermediate **143** was isolated crude, but on standing changed to the aldehydo-thione **144**. This in turn rearranged to the 3-formyl-2-mercapto-4-quinolone **145**, R = Me, when its sodium salt was refluxed in DMF. Compound **144** was an intermediate for several other quinoline-3-carboxaldehydes (Scheme 15 and Table 2), and for a number of fused quinoline heterocycles.[56]

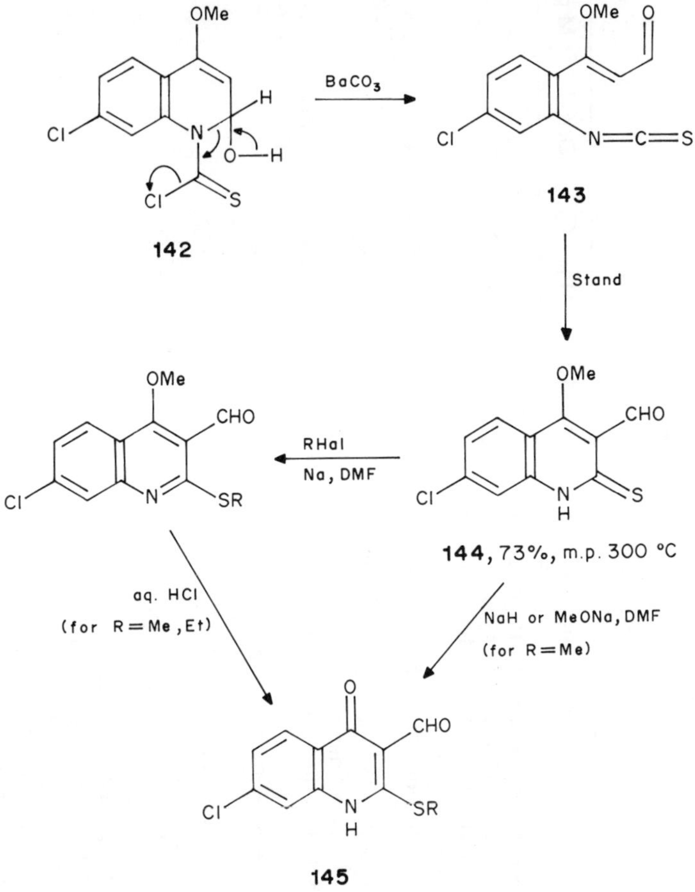

SCHEME 15

2. 4-Formyl-2-quinolones

For the oxidation of the 4-methyl groups of quinolones **146**, selenium dioxide was used either without solvent at 175 °C or in diphenyl ether at about 185 °C;

Quinoline Aldehydes

Table 10. 4-Formyl-2-quinolones

Substituent(s)	Preparation	Yield (%)	M.p. (°C)	Deriv*	M.p. (°C)	References
None	4-Me-2-quinolone, SeO_2, Ph_2O, 185°C	72.7	267			360
	4-Me-2-quinolone, SeO_2, xylene	32	261–262			361, 362
	2-AcNHC$_6$H$_4$COCH(OMe)$_2$, EtONa; H$_2$O		257			184
1-Me	1,4-diMe-2-quinolone, SeO_2, 175°C	70	180–182.5	Ox DNP	246–247(d) 315(d)	177, 363–365
1,6-diMe	1,4,6-triMe-2-quinolone, SeO_2, 175°C	38	181–183			366
1,7-diMe	1,4,7-triMe-2-quinolone, SeO_2, 175°C	69	185–187			366
8-Me	4,8-diMe-2-quinolone, SeO_2	36.5	249–250	NPH DNP	277–278 317–318	130
1-Et	1-Et-4-Me-2-quinolone, SeO_2, 175°C	56	117.5–119			364, 365
1-Pr	1-Pr-4-Me-2-quinolone, SeO_2, 175°C	39	104–105.5			364
1-Bu	1-Bu-4-Me-2-quinolone, SeO_2, 175°C	32	112–113.5			364
1-C$_5$H$_{11}$	1-C$_5$H$_{11}$-4-Me-2-quinolone, SeO_2, 175°C	19	74–75.5			364
1-iPrCH$_2$CH$_2$	1-iPrCH$_2$CH$_2$-4-Me-2-quinolone, SeO_2, 175°C	27	71–72.5			364
1-PhCH$_2$	1-PhCH$_2$-4-Me-2-quinolone, SeO_2, 175°C	56	161.5–162.5			364
6-Cl	4-CH(SMe)OAc-6-Cl-2-quinolone, MeOH, I$_2$; HCl	65	304–305			367
3-Ac	Friedländer synthesis					184
3-COOEt	Friedländer synthesis					184
3-CN	Friedländer synthesis					184

*Ox = oxime, NPH = 4-nitrophenylhydrazone, DNP = 2,4-dinitrophenylhydrazone.

see Table 10. Alternatively, compound **147**, R^1, R^2 = H was made by ring closure of the anilide **148** with sodium ethoxide.[184]

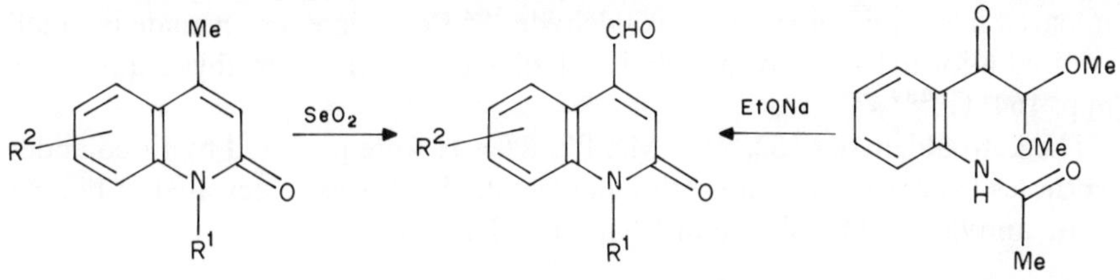
Me
SeO_2
R^2
N
O
R^1
CHO
EtONa
OMe
OMe
H
Me

146 **147** **148**

3. 5-Formyl-2-quinolones

The amino-alcohol **149** was oxidized directly to the aldehyde **151**. Alternatively, the 8-hydroxy group was protected and oxidation achieved in 80% yield to give aldehyde **150**. Acetal formation followed by hydrogenation and hydrolysis converted the ether **150** to the phenol **151** in 92% yield.[368] Compounds **150**, **151** and the 8-*O*-methyl derivative of the phenol **151** have been prepared via Vilsmeier procedures.[369,370]

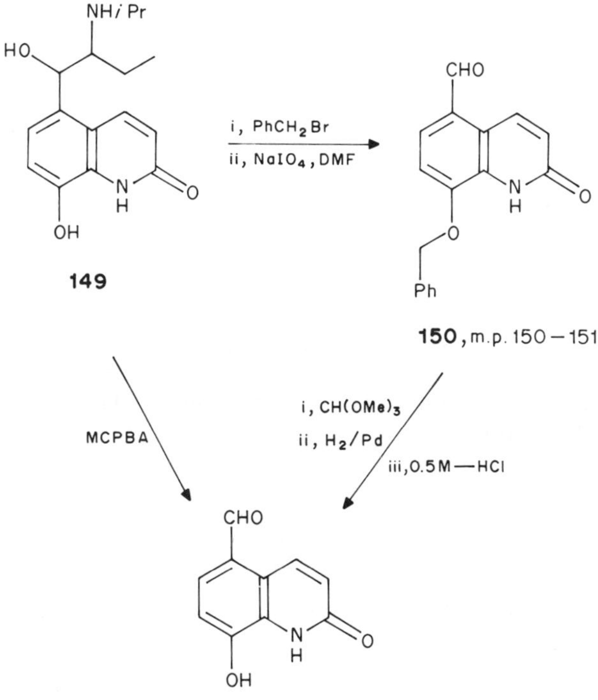
NH*i* Pr
HO
OH
i, $PhCH_2Br$
ii, $NaIO_4$, DMF
CHO
N
H
O
Ph
MCPBA
i, $CH(OMe)_3$
ii, H_2/Pd
iii, 0.5M—HCl

149

150, m.p. 150 – 151 °C

151, m.p. 315 – 317 °C (d)

4. 6-Formyl-2-quinolones

By oxidation with quinoline dehydrogenase, 6-formyl-2-quinolone has been prepared from quinoline-6-carboxaldehyde.[362] Potassium ferricyanide in alkali oxidized 6-formyl-1-methylquinolinium iodide to 6-formyl-1-methyl-2-quinolone (m.p. 164 °C).[237]

The keto-aldehydes **152**, R^1 = Me, Ph; R^2 = H were prepared by Friedländer syntheses from 4, 6-diaminoisophthalaldehyde. Each was N-acylated in boiling acetic anhydride. Details are in Chapter 2, Table 47.[371,372]

152, R^1 = Me, Ph; R^2 = H, Ac

The acid chloride **153** was reduced by bis(triphenylphosphine)copper boranate to the aldehyde **154**.[373]

153   **154**

5. 2-Formyl-4-quinolone

The alcohol **155**, R = CH_2OH was oxidized (MnO_2, $CHCl_3$, reflux) to the aldehyde **155**, R = CHO.[374]

155, R = CHO, 68%, m.p. 162 − 163 °C

6. 3-Formyl-4-quinolones

Most examples in this category were obtained by Reimer–Tiemann reactions
or by hydrolysis of 4-methoxy-quinolines etc. as shown in Table 11 (page 66);
see also compound **33**, Section II.7.

Compound **156**, R = CHO was prepared from 2, 3-dihydro-4-quinolone and
ethyl formate in the presence of sodium ethoxide.[375] The hydroxymethylene
compounds **156**, R = Me, Ph were oxidized to aldehydes **157**, R = Me, Ph when
ether solutions were left exposed to the atmosphere at 0 °C for 2 months.[376] A
di-decarboxylation of the acid **158** over copper bronze also provided the
aldehyde **157**, R = Me.[377] See Chapter 2, Section XIV.9, Scheme 69.

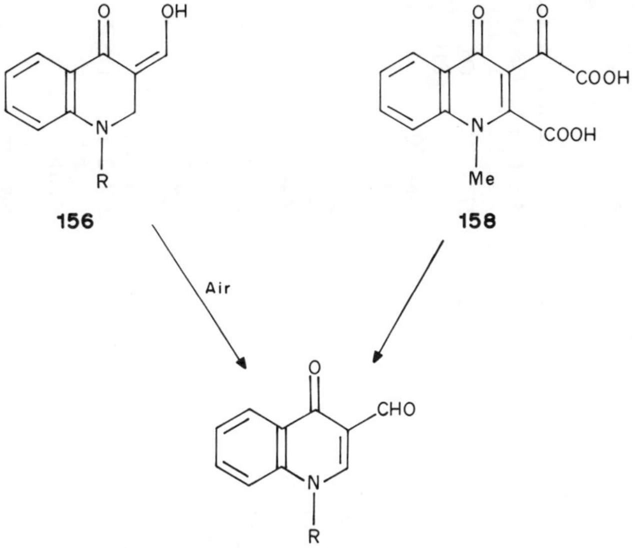

156

158

Air

157

R = Me, 70%, m.p. 210 °C
R = Ph, 31%, m.p. 187 °C

Cyclohexylamine and ethoxymethylene diethyl malonate in methanol gave
an intermediate which on flash vacuum pyrolysis (410 °C, 20 mm) cyclized to
the aldehyde **159**.[379]

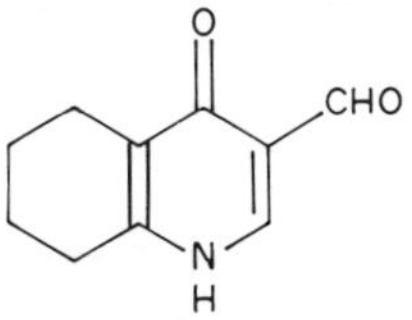

159, 90%, m.p. 212 °C

Table 11. 3-Formyl-4-quinolones

Substituent(s)	Preparation	Yield (%)	M.p. (°C)	Deriv*	M.p. (°C)	References
None	Reimer–Tiemann reaction on 4-quinolone	12	273	Ox PH	222–223 244–245(d)	380, 381
1-Me	See text, Section V.6, Scheme 16		268–270			378
	1-Me-2-H-3=CHOH-4-quinolone, air or MnO_2	70	210			376, 378
1-Me-6-CF_3	See text, Section V.6, Scheme 16		205			378
1-Me-6-iPr	See text, Section V.6, Scheme 16		173–175			378
1-Me-6-Bu	See text, Section V.6, Scheme 16		164–166			378
1-Me-6-Cl	See text, Section V.6, Scheme 16		250			378
1-Me-7-Cl	See text, Section V.6, Scheme 16		266			378
1-Me-8-Cl	See text, Section V.6, Scheme 16		224–226			378
1-Me-6-OMe	See text, Section V.6, Scheme 16		228–232			378
1-Me-6-OBu	See text, Section V.6, Scheme 16		159			378
1-Me-6-SMe	See text, Section V.6, Scheme 16		220–222			378
	1-Me-2-COOH-3-COCOOH-4-quinolone, Cu bronze		217–218	DNP	318–319	377
1-Me-6-NO_2	1-Me-3-CHOH-6-NO_2-4-quinolone, $K_2Cr_2O_7$, AcOH		276(d)			382
2-Me	See Text, Section V.6, Scheme 16		268–272			378
	Reimer–Tiemann reaction on 2-Me-4-quinolone	58 38	278–280(d) 273–275	Ox PH	300 258–260	383–386

Compound	Method	Yield (%)	mp	Derivative	mp (deriv.)	Ref.
2,5,6,8-tetraMe	Reimer–Tiemann reaction on 2,5,6,8-tetraMe-4-quinolone					385
2,6-diMe	Reimer–Tiemann reaction on 2,6-diMe-4-quinolone	35	300(d)			383
2,6-diMe-8-OMe						387
2,8-diMe						387
2-Me-6-OMe						387, 388
1-Bu-6,7-diMe	See text, Section V.6, Scheme 16		168–170			378
1-Bu-7-CF$_3$	See text, Section V.6, Scheme 16		148–150			378
6-Bu	See text, Section V.6, Scheme 16		215–217			378
6-OMe	See text, Section V.6, Scheme 16		290			378
6,8-diOMe	See text, Section V.6, Scheme 16		>260			378
1-Ph	1-Ph-2-H-3=CHOH-4-quinolone, air	31	187			376
2-Ph	Reimer–Tiemann reaction on 2-Ph-4-quinolone	37	250–252	DNP	275–277.5(d)	389
2-OMe	2,4-diMeO-Q-3-CHO, 5% HCl or aq. KOH	60	104–105[a]			53, 54, 159
6-OMe	2,4-diCl-3-CHCl$_2$-Q, MeONa; HCl	72	82–83[a]			164
	3-CH=NNHCONH$_2$-6-MeO-4-quinolone, HCl		300			51
2-SMe-7-Cl	3-CHO-4-MeO-7-Cl-2-quinolinethione, MeONa, MeI	88	266–267			56
	2-MeS-4-MeO-7-Cl-Q-3-CHO, acid					56
2-SEt-7-Cl	2-EtS-4-MeO-7-Cl-Q-3-CHO, acid	74	192–193			56

* Ox = oxime, PH = phenylhydrazone, DNP = 2,4-dinitrophenylhydrazone.
[a] Inconsistent literature melting points.

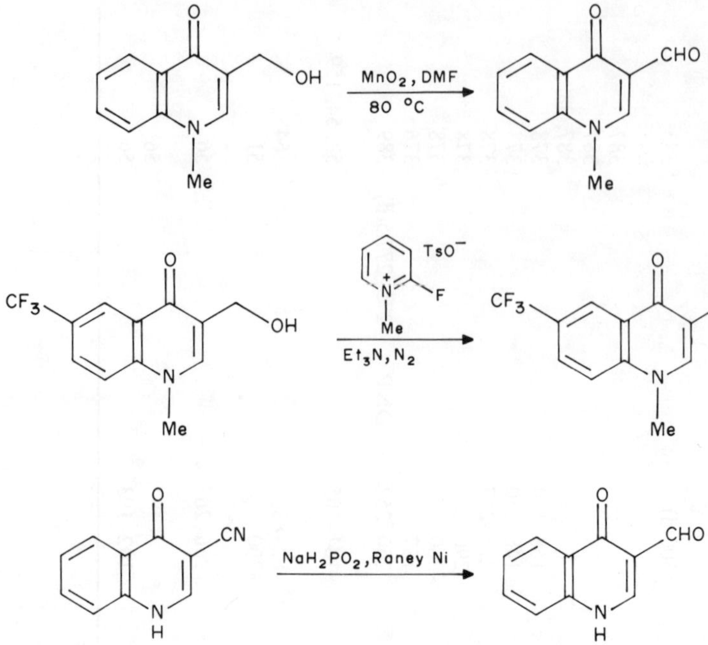

SCHEME 16

Other methods for the production of such aldehydes are summarized in Scheme 16. The reference does not specify which method was used for each of the examples given in Table 11.[378]

VI. Quinolone Side-chain Aldehydes

The aldehydes **160** were prepared from reactions of formamides with the lithium derivatives of 2, 4-dimethoxy-quinolines (Table 2). They were converted to the quinolyl acetaldehydes **161** as shown in Scheme 17.[157,158,161,162] However, hydrolysis of compound **160**, $R_1, R_2 = H$ either with aqueous acid or with potassium hydroxide in dimethyl sulphoxide gave 3-formyl-2-methoxy-4-quinolone; see Section II.7.[54,159]

Ozonolysis of the butenylquinolines **162** gave the aldehydes **163**. It was noted that compound **163a** was not purified further. Compound **163b** has also been prepared by the method shown in Scheme 17 as an intermediate in the preparation of γ-fagarine.[158] N-Methylpreskimmianine was similarly oxidized with osmium tetroxide to aldehyde **164**.[390] The alkaloid (±)edulinine **165**, $R^1, R^2 = H$ was oxidized to the aldehyde **166a** (the N-methyl derivative of compound **163a**).[391] Compounds **166b**[392] and **166c**[393,394] have been made in the same way, but were derivatized without isolation.

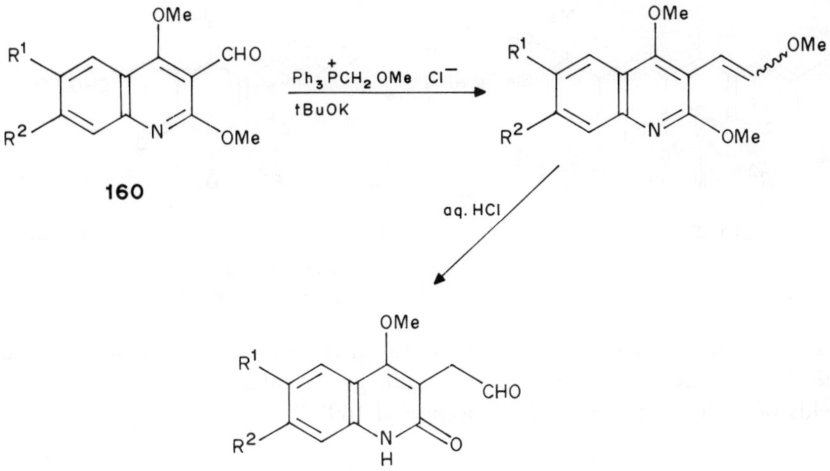

160

$$\xrightarrow[\text{tBuOK}]{\overset{+}{\text{Ph}_3\text{PCH}_2\text{OMe}} \ \text{Cl}^-}$$

aq. HCl

161

R^1, R^2 = H, 60%, m.p. 209 – 210 °C
R^1 = H, R^2 = OMe, 81%, m.p. 206 – 207 °C
R^1 = OMe, R^1 = H, 83%, m.p. 198 – 199 °C
R^1, R^2 = OMe, 81%, m.p. 139 °C
R^1, R^2 = OCH$_2$O, 78%, m.p. 159 – 161 °C

SCHEME 17

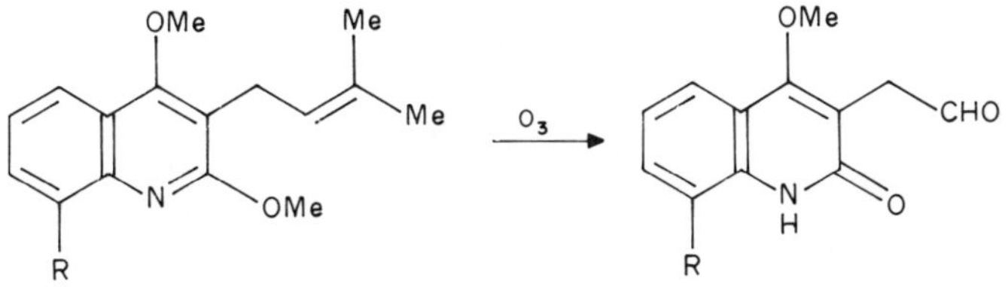

162

$\xrightarrow{\text{O}_3}$

163 a , R = H, 92%, m.p. 154 – 157 °C
b , R = OMe, 82%, m.p. 148 – 151 °C

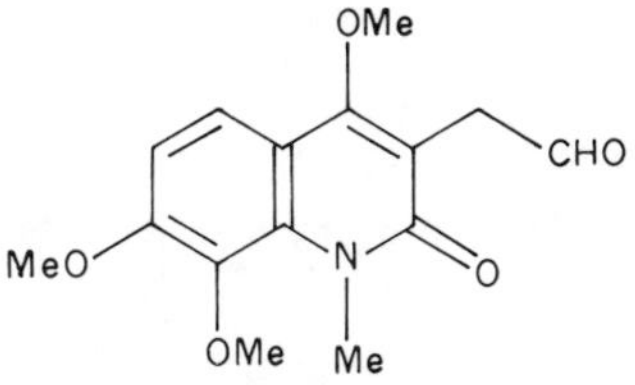

164 , m.p. 138 – 139 °C

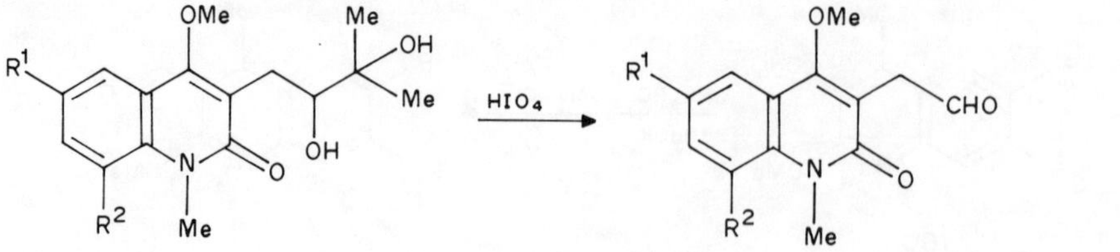

(±) 165

166 a , R^1, R^2 =H, 94%, m.p. 113−115 °C
   b , R^1 =OMe, R^2 =H
   c , R^1 =H, R^2 =OMe

Irradiation of the 3-furoic acid amide **167** gave a mixture of the two aldehydes **168**. The corresponding 2-furoic acid amide gave a mixture which included 10% yields of each of the epimers of compound **169**.[350]

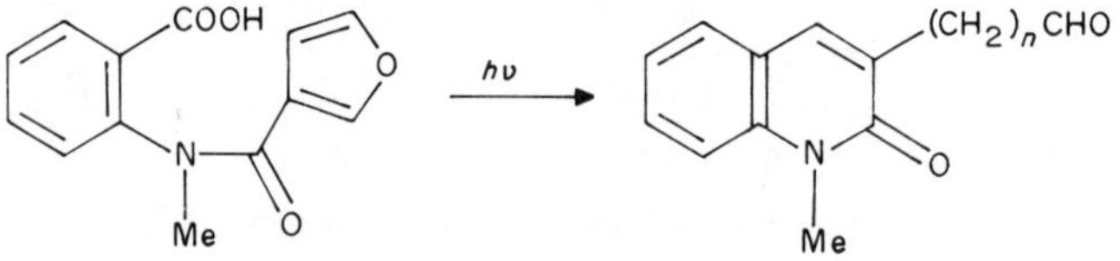

167

168

n = 0, 7.5%
n = 1, 25%

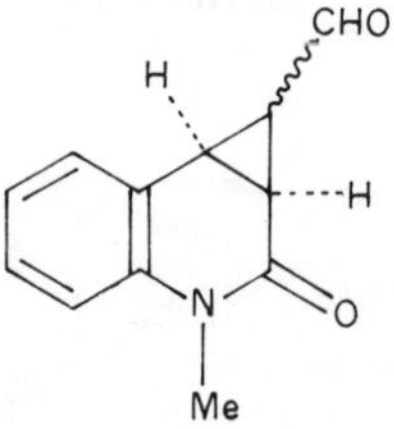

169

Photochemical addition of vinyl acetate to 4-hydroxy-3-methyl-2-quinolone gave a mixture of compound **170** and its epimer. Either compound could be oxidized to the aldehyde **171**.[395]

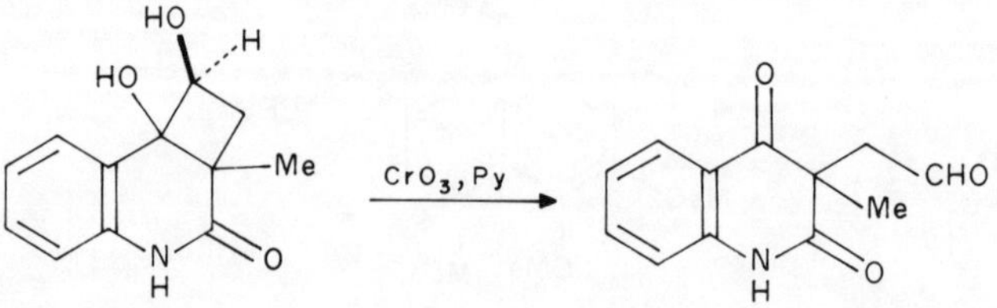

170

171, m.p. 140−141 °C

Selenium dioxide oxidized 6-acetyl-1-alkyl-4-quinolones and 7-acetyl-1-methyl-4-quinolone to the keto-aldehydes **172** and **173** respectively.[278]

172, R=H, Me, Et, Bu, CH₂CH=CH₂   **173**

The olefines **174** were ozonolyzed to the acetaldehydes **175**.[396]

174		**175**

$R^1, R^2 = H$	30%, m.p. 230 °C
$R^1 = OMe$, $R^2 = H$,	30%, m.p. 235 °C
$R^1 = H$, $R^2 = Me$,	40%, m.p. 233 °C

VII. Partially and Fully Saturated Quinolone Aldehydes

The appropriate quinolone and dimethyl oxalate (NaH, DMF) at 130–140 °C gave the enol ether **176** rather than the expected keto-ester, which was only formed at lower temperatures; see Chapter 2, Section XVI.1. Hydrolysis and decarboxylation then gave aldehyde **177**.[397]

176   **177**, 9%, oil

Reaction between 3,4-dihydro-4-phenyl-2-quinolone and the Vilsmeier reagent gave compound **86** which, with 50% sulphuric acid at room temperature or refluxing 10% hydrochloric acid, gave the aldehydo-quinolone **178**; see Section IV.1.[289]

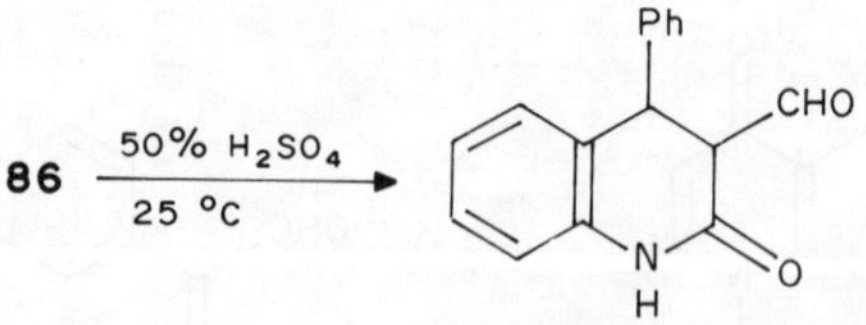

178, 99%, m.p. 210−212 °C

The pyrrolidinone **179** rearranged to the aldehydoquinolone **180** either with toluenesulphonic acid in benzene[398] or with a mixed silica/alumina catalyst in refluxing dioxane.[399]

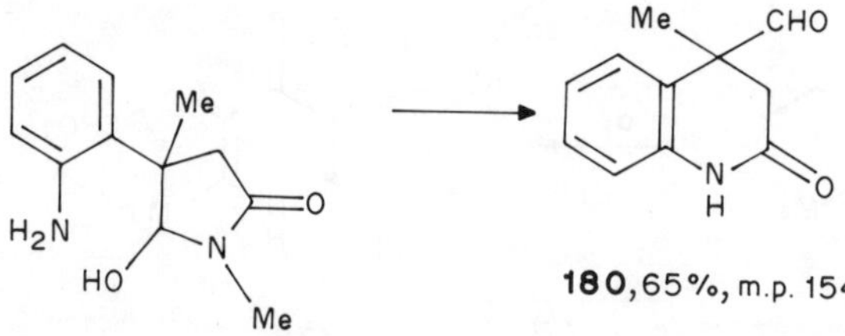

180, 65%, m.p. 154−156 °C

179

Aldehyde **181** was treated with ethylene glycol before hydrogenation over Raney nickel. The resulting acetal **182** rearranged in ethanolic hydrogen chloride or in ethanolic potassium hydroxide to indolylacetic acid.[361]

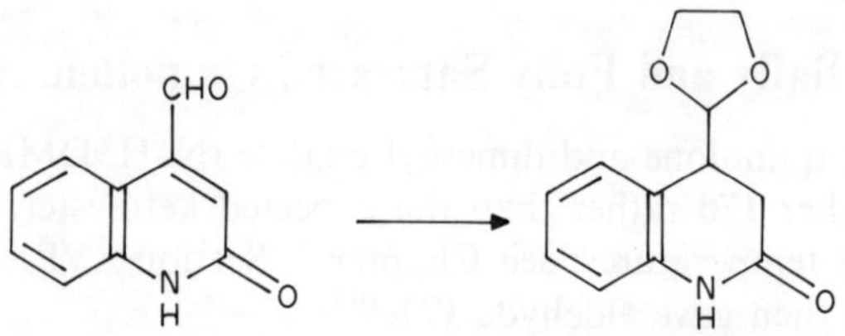

181   **182**

The 5-formyl dihydroquinolone **183** was apparently produced in a Vilsmeier reaction.[370]

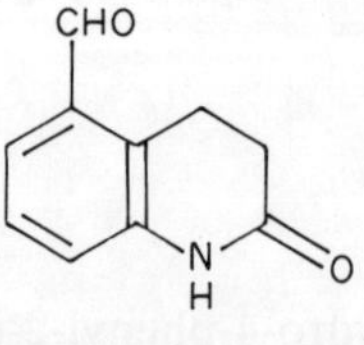

183

A series of patents describes the reactions shown in Scheme 18. For the most part, the intermediate aldehydes **184**, aldehydo-ketones **185** and ketones **186** were not characterized. About 450 examples of compounds of formulae **185** and **186** were claimed. Many were further elaborated to the benzo[c]quinolinones **187** which were characterized.[400-406]

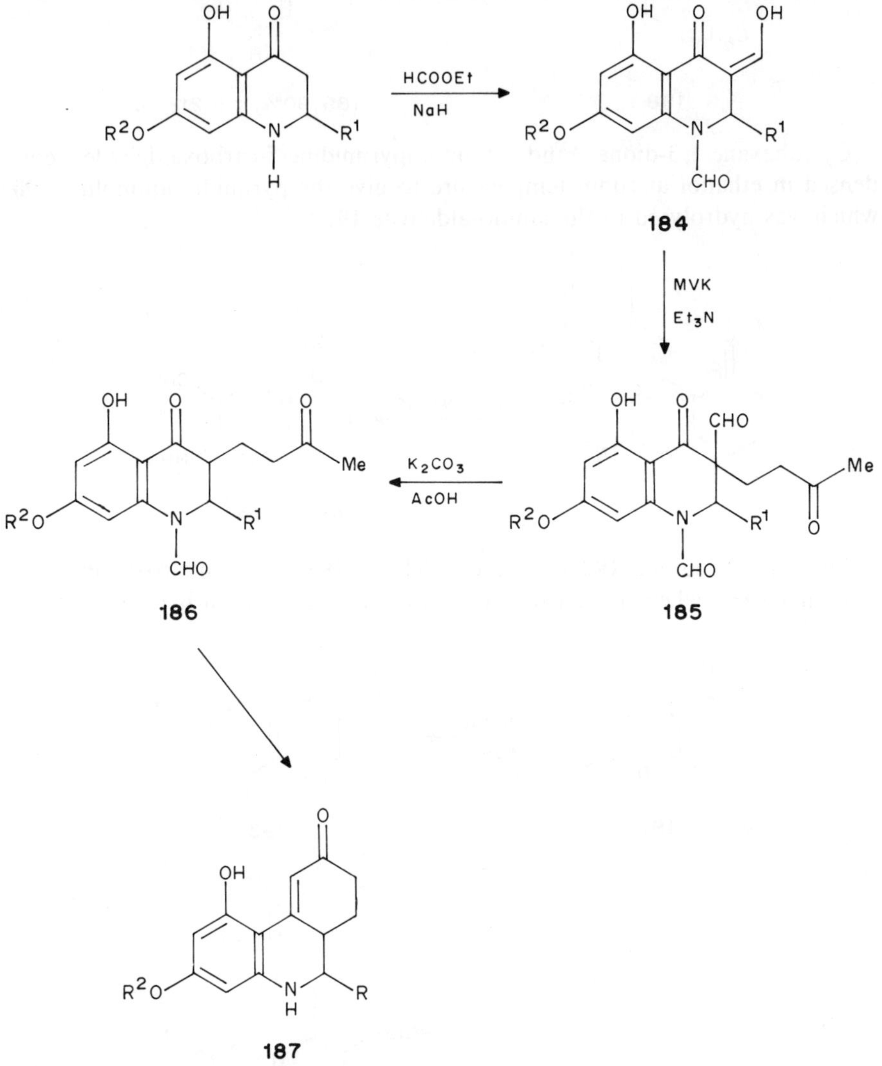

187

R^1 = H, Me, Pr etc.

R^2 = Me(CH$_2$)$_3$, Me(CH$_2$)$_4$, Me(CH$_2$)$_3$CHMe, Ph(CH$_2$)$_4$, Ph(CH$_2$)$_3$CHMe etc.

SCHEME 18

Flash vacuum pyrolysis converted the enamine **188** into the quinolone aldehyde **189**.[379]

188   **189**, 90%, m.p. 212 °C

Cyclohexane-1,3-dione and 4-aminopyrimidine-5-carboxaldehyde condensed in ethanol at room temperature to give the pyrimidinoquinoline **190**, which was hydrolysed to the amino-aldehyde **191**.[407]

190   **191**, m.p. 236–237 °C

The quinolinedione **192** gave the aldehyde **193** (as its hydroxymethylene tautomer **193a**) when treated with ethyl formate and sodium hydride.[408]

192   **193**

193a

The *trans*-6-oxodecahydroquinoline **194** was formylated to the keto-aldehyde **195**, which was normally treated crude with hydrazine to give an octahydropyrazolo[3, 4-*g*]quinoline, but could be isolated as the sodium salt.[409]

194   **195**

VIII. Appendix

Abbreviations used in Text and Tables

AA	Pentane-2,4-dione
DCCI	Dicyclohexylcarbodiimide
DCM	Dichloromethane
DDQ	Dichlorodicyano-*p*-benzoquinone
DEAD	Diethyl acetylenedicarboxylate
DIBAH	Diisobutyl aluminium hydride
DMAD	Dimethyl acetylenedicarboxylate
DME	Dimethoxyethane
DMF	Dimethylformamide
DMSO	Dimethyl sulphoxide
DNP	2,4-Dinitrophenylhydrazone
EAA	Ethyl acetoacetate
HMPT	Hexamethyl phosphoric triamide
Hy	Hydrazone
LAH	Lithium aluminium hydride
LDA	Lithium diisopropylamide
MAA	Methyl acetoacetate
MCPBA	*m*-Chloroperbenzoic acid
MVK	Methyl vinyl ketone
NPH	4-Nitrophenylhydrazone
Ox	Oxime
PH	Phenylhydrazone
Pic.	Picrate
PPA	Polyphosphoric acid
Py	Pyridine
$PyClCrO_3$	Pyridinium chlorochromate
Q	Quinoline

SC   Semicarbazone
SDA   Sodium diisopropylamide
TFA   Trifluoroacetic acid
TFAA   Trifluoroacetic anhydride
Ts $p\text{-}CH_3C_6H_4SO_2$
TSC   Thiosemicarbazone

IX. References

1. D. H. Jones, R. Slack, S. Squires, and K. R. H. Woolridge, *J. Med. Chem.*, **8**, 676 (1965).
2. T. Ishiguro and I. Utsumi, *Yakugaku Zasshi*, **72**, 865 (1952); *Chem. Abstr.*, **47**, 6416 (1953).
3. D. Jerchel, J. Heider, and H. Wagner, *Justus Liebigs Ann. Chem.*, **613**, 153 (1958).
4. V. M. Rodionov and M. A. Berkengeun, *J. Gen. Chem USSR (Engl. Transl.)*, **14**, 330 (1944); *Chem. Abstr.*, **39**, 4076 (1945).
5. H. Kaplan, *J. Amer. Chem. Soc.*, **63**, 2654 (1941).
6. V. G. Ramsey, *J. Amer. Pharm. Assoc.*, **40**, 564 (1951).
7. J. Schaefer, K. S. Kulkarni, R. Costin, J. Higgins, and L. Honig, *J. Heterocycl. Chem.*, **7**, 607 (1970).
8. E. Ziegler and G. Zigeuner, *Monatsh. Chem.*, **79**, 42, 89 (1948).
9. H. Zinner and H. Fiedler, *Arch. Pharm. Ber. Dtsch. Pharm. Ges.*, **291**, 493 (1958); *Chem. Abstr.*, **53**, 15080 (1959).
10. H. Fiedler, *Arch. Pharm. Ber. Dtsch. Pharm. Ges.*, **293**, 609 (1960); *Chem. Abstr.*, **54**, 24743 (1960).
11. F. Nerdel and J. Kleinwachter, *Naturwissenschaften*, **42**, 577 (1955); *Chem. Abstr.*, **51**, 11151 (1957).
12. D. H. R. Barton, R. A. H. F. Hui, and S. V. Ley, *J. Chem. Soc., Perkin Trans. I*, **1982**, 2179.
13. M. Weissenfels, B. Ulrici, and S. Kaufisch, *Z. Chem.*, **18**, 138 (1978).
14. P. Duballet, A. Godard, G. Quéguiner, and P. Pastour, *J. Heterocycl. Chem.*, **10**, 1079 (1973).
15. M. Seyhan, *Rev. Faculte Sci. Univ. Istanbul*, **16A**, 252 (1951); *Chem. Abstr.*, **47**, 3312 (1953).
16. W. Mathes and W. Sauermilch, *U.S. Pat.* 2 798 071 (1957); *Chem. Abstr.*, **52**, 451 (1958).
17. F. Raschig, *Brit. Pat.* 772520 (1957); *Chem. Abstr.*, **52**, 449 (1958).
18. F. Komatsu, *Muroran Kogyô Daigaku Kenkyu Hôkoku*, **1958**, 41; *Chem. Abstr.*, **53**, 13149 (1959).
19. R. E. Lutz and J. M. Sanders, *J. Med. Chem.*, **19**, 407 (1976).
20. H. R. Munson, Jr., R. E. Johnson, J. M. Sanders, C. J. Ohnmacht, and R. E. Lutz, *J. Med. Chem.*, **18**, 1232 (1975).
21. T. W. J. Taylor, D. H. G. Winckles, and M. S. Marks, *J. Chem. Soc.*, **1931**, 2778.
22. W. Pfitzinger, *J. Prakt. Chem.*, **66**, 263 (1902); *J. Chem. Soc. Abstr.*, **84**, 53 (1903).
23. W. Borsche and W. Ried, *Justus Liebigs Ann. Chem.*, **554**, 269 (1943).
24. A. Kaufmann and L. G. Vallette, *Ber.*, **46**, 49 (1913).
25. R. Hull, *J. Chem. Soc., Perkin Trans. I*, **1973**, 2911.
26. O. Meth-Cohn and B. Narine, *Tetrahedron Lett.*, **1978**, 2045.
27. O. Meth-Cohn, B. Narine, and B. Tarnowski, *Tetrahedron Lett.*, **1979**, 3111.

28. O. Meth-Cohn and B. Narine, *PCT Int. Appl.*, 79 00,540 (1979); *Chem. Abstr.*, **92**, 111035 (1980); *U.S. Pat.* 4 375 544 (1983).
29. O. Meth-Cohn, B. Narine, and B. Tarnowski, *J. Chem. Soc., Perkin Trans. I*, **1981**, 1520.
30. O. Meth-Cohn, B. Narine, B. Tarnowski, R. Hayes, A. Keyzad, S. Rhonati, and A. Robinson, *J. Chem. Soc., Perkin Trans. I*, **1981**, 2509.
31. R. Hayes and O. Meth-Cohn, *Tetrahedron Lett.*, **23**, 1613 (1982).
32. G. P. Gardini, *Tetrahedron Lett.*, **1972**, 4113.
33. M. Ihara, K. Noguchi, K. Fukumoto, and T. Kametani, *Heterocycles*, **20**, 421 (1983).
34. A. Godard, G. Quéguiner, and P. Pastour, *Bull. Soc. Chim. Fr.*, **1971**, 906.
35. G. Quéguiner, G. Joly, and P. Pastour, *Compt. Rend.*, **263C**, 307 (1966).
36. G. Quéguiner, G. Joly, and P. Pastour, *Compt. Rend.*, **263C**, 307 (1966).
37. A. Godard, P. Duballet, G. Quéguiner, and P. Pastour, *Bull. Soc. Chim. Fr.*, **1976**, 789.
38. A. Godard and G. Quéguiner, *Tetrahedron Lett.*, **22**, 4813 (1981).
39. A. Godard and G. Quéguiner, *J. Heterocycl. Chem.*, **21**, 27 (1984).
40. W. Ried and G. Neidhardt, *Justus Liebigs Ann. Chem.*, **666**, 148 (1963).
41. D. W. Ockenden and K. Schofield, *J. Chem. Soc.*, **1953**, 1915.
42. C. M. Atkinson and A. R. Mattocks, *J. Chem. Soc.*, **1962**, 1671.
43. W. König, *Ber.*, **56**, 1543 (1923).
44. C. W. Muth, J. C. Patton, B. Bhattacharya, D. L. Giberson, and C. A. Ferguson, *J. Heterocycl. Chem.*, **9**, 1299 (1972).
45. M. Nishikawa, S. Saeki, M. Hamana and H. Noda, *Chem. Pharm. Bull.*, **28**, 2436 (1980).
46. T. Kato, Y. Goto, and M. Kondo, *Yakugaku Zasshi*, **84**, 290 (1964); *Chem. Abstr.*, **61**, 3070 (1964).
47. M. Wilk, H. Schwab, and J. Rochlitz, *Justus Liebigs Ann. Chem.*, **698**, 149 (1966).
48. S. Blechert, R. Gericke, and E. Winterfeldt, *Chem. Ber.*, **106**, 355 (1973).
49. A. H. Cook, I. M. Heilbron, and L. Steger, *J. Chem. Soc.*, **1943**, 413.
50. K. K. Hsu, S. W. Sun, and Y. H. Chen., *J. Chin. Chem. Soc. (Taipei)*, **29**, 29 (1982); *Chem. Abstr.*, **96**, 181118 (1982).
51. F. Zymolkowski and P. Tinapp, *Justus Liebigs Ann. Chem.*, **699**, 98 (1966).
52. E. Schroder, M. Lehmann, and I. Bottcher, *Eur. J. Med. Chem. – Chim. Ther.*, **14**, 499 (1979).
53. T. Ohta and Y. Mori, *Ann. Rep. Tokyo Coll. Pharm.*, **4**, 255 (1954); *Chem. Abstr.*, **50**, 998 (1956).
54. N. S. Narasuinham and S. D. Joag, *Indian J. Chem.*, **20B**, 543 (1981).
55. A. Cromarty, G. R. Proctor, and M. Shabfir, *J. Chem. Soc., Perkin Trans. I*, **1972**, 2012.
56. R. Hull, P. J. Van den Broek, and M. M. L. Swain, *J. Chem. Soc., Perkin Trans. I*, **1975**, 2271.
57. A. I. Meyers and D. G. Wettlaufer, *J. Amer. Chem. Soc.*, **106**, 1135 (1984).
58. A. R. Katritzky, A. V. Chapman, and H. M. Dowlatshahi, *Acta Chim. Acad. Sci. Hung.*, **107**, 315 (1981).
59. K. Matsumura and M. Ito, *J. Amer. Chem. Soc.*, **77**, 6671 (1955).
60. K. Matsumura, T. Kasai, and H. Tashiro, *Bull. Chem. Soc. Japan*, **42**, 1741 (1969).
61. Y. Matsumura, M. Ito, and T. Nagano, *Japan Pat.* 585637; *Chem. Abstr.*, **53**, 18065 (1959).
62. A. Terent'ev, I. G. Il'ina, L. G. Yudin, N. B. Kazennova, and E. I. Levkoeva, *Khim. Geterotsikl. Soedin.*, **1970**, 1663; *Chem. Heterocycl. Compd. (Engl. Transl.)* **6**, 1553 (1970); *Chem. Abstr.*, **74**, 53474 (1971).

63. E. Ochiai and M. Ikehara, *Yakugaku Zasshi*, **70**, 265 (1950); *Chem. Abstr.*, **45**, 2945 (1951).
64. D. L. Hammick, *J. Chem. Soc.*, **1926**, 1302.
65. L. K. Sharp, *J. Pharm. Pharmacol.*, **1**, 395 (1949).
66. F. M. Hamer, *J. Chem. Soc.*, **1952**, 3197.
67. K. E. Cooper and J. B. Cohen, *J. Chem. Soc.*, **1932**, 723.
68. W. von Miller and J. Spady, *Ber.*, **18**, 3405 (1885).
69. W. von Miller and J. Spady, *Ber.*, **19**, 130 (1886).
70. A. Einhorn, *Ber.*, **19**, 904 (1886); *J. Chem. Soc. Abstr.*, **50**, 721 (1886).
71. S. Sugasawa and K. Mizukaini, *Pharm. Bull. (Japan)*, **3**, 393 (1955).
72. E. V. Brown and M. G. Frazer, *J. Heterocycl. Chem.*, **6**, 567 (1969).
73. J. Miochowski, *Pr. Nauk. Inst. Chem. Org. Fiz. Politech. Wroclaw.*, **9**, 3 (1975); *Chem. Abstr.*, **86**, 43584 (1977).
74. H. Kaplan and H. G. Lindwall, *J. Amer. Chem. Soc.*, **65**, 927 (1943).
75. K. K. Hsu and S. F. Chang, *T'ai-wan K'o Hsueh*, **29**, 51 (1975); *Chem. Abstr.*, **84**, 164725 (1976).
76. J. Reihsig and H. W. Krause, *J. Prakt. Chem.*, **31**, 167 (1966).
77. M. Heuzle, *Ber.*, **67**, 750 (1934).
78. Höchster Farb., *D.R. Pat.* 36964; Frdl. **1**, 194; *Beilstein*, **21**, 322.
79. K. A. Jensen and P. H. Nielsen, *Acta Chem. Scand.*, **18**, 1 (1964).
80. D. H. R. Barton, R. A. H. F. Hui, D. J. Lester, and S. V. Ley, *Tetrahedron Lett.*, **1979**, 3331.
81. G. Giordano, *J. Org. Chem.*, **51**, 536 (1986).
82. W. Mathes and W. Sauermilch, *Chem. Ber.*, **87**, 1179 (1954).
83. H. Saito, H. Muro, S. Saeki, and M. Hamana, *Heterocycles*, **5**, 331 (1976).
84. D. Oda, *Mem. Defense Acad., Math Phys. Chem. Eng.*, **4**, 355 (1965); *Chem. Abstr.*, **63**, 13205 (1965).
85. W. M. Schubert, H. Burkett, and A. L. Schy, *J. Amer. Chem. Soc.*, **86**, 2520 (1964).
86. C. E. Kwarther and H. G. Lindwall, *J. Amer. Chem. Soc.*, **59**, 524 (1937).
87. S. Skidmore and E. Tidd, *J. Chem. Soc.*, **1961**, 1098.
88. J. B. Adams, J. Cymerman-Craig, C. Ralph, and D. Willis, *Aust. J. Chem.*, **8**, 392 (1955).
89. M. Seyhan, *Chem. Ber.*, **85**, 425 (1952).
90. M. Seyhan, *Chem. Ber.*, **90**, 1386 (1957).
91. B. R. Brown and D. L. Hammick, *J. Chem. Soc.*, **1950**, 628.
92. A. Burger and L. R. Modlin, Jr., *J. Amer. Chem. Soc.*, **62**, 1079 (1940).
93. T. Sakamoto, T. Sakasai, and H. Yamanaka, *Chem. Pharm. Bull.*, **29**, 2485 (1981).
94. I. A. Krasavin, V. M. Dziomko, and T. N. Egorova, *Metody Poluch. Khim. Reakt. Prep.*, **1965**, (13), 34; *Chem. Abstr.*, **65**, 5437 (1966).
95. I. A. Krasavin, Yu. P. Radin, Yu. S. Ryabokobylko, B. V. Parusuikov and V. M. Dziomko, *Khim. Geterotsikl. Soedin.*, **1978**, 235; *Chem. Heterocycl. Compd. (Engl. Transl.)*, **1978**, 190; *Chem. Abstr.*, **88**, 190566 (1978).
96. M. Seyhan and W. C. Fernelius, *J. Org. Chem.*, **22**, 217 (1957).
97. B. P. Lugovkin, *Zh. Obshch. Khim.*, **46**, 2498 (1976); *J. Gen. Chem. U.S.S.R. (Engl. Transl.)*, **46**, 2390 (1976); *Chem. Abstr.*, **86**, 106726 (1977).
98. C. A. Buehler and S. P. Edwards, *J. Amer. Chem. Soc.*, **74**, 977 (1952).
99. G. Panajotow, *Ber.*, **23**, 1471 (1890); *J. Chem. Soc. Abstr.*, **58**, 1158 (1890).
100. H. C. Richards, *Ger. Offen.* 1 901 262 (1968); *Chem. Abstr.*, **72**, 21713 (1970); *U.S. Pat.* 3 899 490 (1975).
101. H. C. Richards, *U.S. Pat.* 3 821 228 (1974); *Chem. Abstr.*, **81**, 120498 (1974); *U.S. Pat.* 3 929 784 (1975); *Chem. Abstr.*, **85**, 32 869 (1976).

102. C. A. R. Baxter and H. C. Richards, *J. Med. Chem.*, **14**, 1033 (1971).
103. M. Seyhan and W. C. Fernelius, *Chem. Ber.*, **89**, 2212 (1956).
104. W. Mathes and W. Sauermilch, *Chem. Ber.*, **90**, 758 (1957).
105. A. J. Deeming, I. P. Rothwell, M. B. Hursthouse, and K. M. A. Malik, *J. Chem. Soc., Dalton Trans.*, **1979**, 1899.
106. Y. Morisawa and T. Sakamoto, *Japan Kokai*, 75 89,378 (1975); *Chem. Abstr.*, **84**, 4825 (1976).
107. Y. Morisawa, T. Sakamoto, and H. Takagi, *Japan Kokai* 75 101 397 (1975); *Chem. Abstr.*, **84**, 59425 (1976).
108. R. A. Glenn and J. R. Bailey, *J. Amer. Chem. Soc.*, **63**, 639 (1941).
109. E. A. Fehnel, *J. Org. Chem.*, **31**, 2899 (1966).
110. A. L. Gershuns, A. A. Verezubova, and L. M. Ptyagina, *Izv. Vyssh. Uchebn. Zaved., Khim. Khim. Tekhnol.*, **18**, 869 (1975); *Chem. Abstr.*, **83**, 178995 (1975).
111. E. A. Fehnel, *J. Org. Chem.*, **23**, 432 (1958).
112. R. R. G. Haber and E. Schoenberger, *S. African Pat.* 67 03,320 (1968); *Chem. Abstr.*, **71**, 91337 (1969).
113. R. R. G. Haber and E. Schoenberger, *Israel Pat.* 27993 (1972); *Chem. Abstr.*, **78**, 147821 (1973).
114. A. L. Gershuns, A. N. Brizitskaya, *Khim. Geterotsikl. Soedin.*, **1970**, 835; *Chem. Heterocycl. Compd.* (*Engl. Transl.*), **6**, 775 (1970); *Chem. Abstr.*, **73**, 109738 (1970).
115. J. Mlochowski, *Rocz. Chem.*, **44**, 1331 (1970); *Chem. Abstr.*, **73**, 87761 (1970).
116. M. V. Rubtsov, V. M. Berenfel'd, and L. N. Yakhontov, *Zh. Obshch. Khim.*, **34**, 1121 (1964); *J. Gen. Chem. U.S.S.R.* (*Engl. Transl.*), **34**, 1112 (1964); *Chem. Abstr.*, **61**, 1826 (1964).
117. W. M. Tadros, H. A. Shoeb, M. A. Kira, F. Yousif, E. M. Ekladios, and S. A. Ibrahim, *Indian J. Chem.*, **13**, 1366 (1975).
118. M. Carissimi, P. G. De Meglio, F. Ravenna, and G. Riva, *Farmaco, Ed. Sci.*, **24**, 478 (1969); *Chem. Abstr.*, **71**, 124175 (1969).
119. M. Carissimi and F. Ravenna, *U.S. Pat.* 3 682 927 (1972); *Chem. Abstr.*, **77**, 164525 (1972).
120. M. G. Vaidya and J. G. Cannon, *J. Med. Pharm. Chem.*, **5**, 389 (1962).
121. A. Markovac, C. L. Stevens, and A. B. Ash, *J. Med. Chem.*, **15**, 490 (1972).
122. K.-K. Hsu, S.-F. Chang, and S.-W. Sun, *T'ai-wan K'o Hsueh*, **34**, 104 (1980); *Chem. Abstr.*, **94**, 208655 (1981).
123. T. Hata and T. Uno, *Bull. Chem. Soc. Japan*, **45**, 477 (1972).
124. F. Fukujiro, H. Kunio, T. Takeshi, T. Ryusuke, K. Sai, N. Masukazu, and T. Shuynji, *Yakugaku Zasshi*, **87**, 844 (1967); *Chem. Abstr.*, **68**, 29557 (1968).
125. J. Büchi, A. Aebi, A. Deflorin, and H. Hurni, *Helv. Chim. Acta*, **39**, 1676 (1956).
126. C. Antonello, *Gazz. Chim. Ital.*, **91**, 926 (1961).
127. P. Nickel, R. Zimmerman and E. Fink, *Arch. Pharm.* (*Weinheim*), **310**, 529 (1977).
128. K. K. Hsu and T. S. Wu, *J. Chin. Chem. Soc.* (*Taipei*), **26**, 17 (1979); *Chem. Abstr.*, **91**, 91475 (1979).
129. H. Irving and A. R. Pennington, *J. Chem. Soc.*, **1954**, 3782.
130. M. Seyhan, *Chem. Ber.*, **92**, 1480 (1959).
131. M. Seyhan and J. Sargin, *Chem. Ber.*, **99**, 2072 (1966).
132. M. J. Haddadin, G. E. Zahr, T. N. Rawdah, N. E. Chelhot, and C. H. Issidorides, *Tetrahedron*, **30**, 659 (1974).
133. A. Godard, G. Quéguiner, and P. Pastour, *Compt. Rend.*, **281C**, 941 (1975).
134. G. Quéguiner and A. Godard, *Compt. Rend.*, **269C**, 1646 (1969).
135. N. M. Sukhova, I. Sprunka, M. Lidaka and A. Zidermane, *Khim.-Farm. Zh.*, **16**, 169 (1982); *Chem. Abstr.*, **96**, 199484 (1982).

136. E. Besthorn and B. Geisselbrecht, *Ber.*, **53**, 1017 (1920).
137. A. Godard and G. Quéguiner, *J. Heterocycl. Chem.*, **17**, 465 (1980).
138. H. A. Shoeb, M. I. Korkar, and G. H. Tammam, *Pharmazie*, **33**, 581 (1978).
139. S. Britwell and W. Hepworth, *Brit. Pat.* 874 980 (1958); *Chem. Abstr.*, **56**, 4739 (1962).
140. V. M. Dziomko, Z. S. Sidenki, and G. S. Chizhova, *Metody Poluch. Khim. Reakt. Prep.*, **1967** (15), 114; *Chem. Abstr.*, **69**, 18989 (1968).
141. A. Godard, D. Brunet, G. Quéguiner, and P. Pastour, *Compt. Rend.*, **284C**, 459 (1977).
142. T. Kato, N. Katagini, and A. Wagai, *Chem. Pharm. Bull*, **29**, 1069 (1981).
143. G. R. Clemo and G. A. Swan, *J. Chem. Soc.*, **1945**, 867.
144. B. R. Brown, D. L. Hammick, and B. H. Thewlis, *J. Chem. Soc.*, **1951**, 1145.
145. A. Furst, H. A. Harper, R. J. Seiwald, M. D. Morris, and R. A. Neve, *Arch. Biochem. Biophys.*, **31**, 190 (1951).
146. A. Walser, T. Flynn, and R. I. Fryer, *J. Heterocycl. Chem.*, **12**, 737 (1975).
147. W. Ried, A. Berg, and G. Schmidt, *Chem. Ber.*, **85**, 204 (1952).
148. K. Harz, *Ber.*, **18**, 3384 (1885); *J. Chem. Soc. Abstr.*, **50**, 261 (1886).
149. P. L. Orwick and R. A. Templeton, *Eur. Pat. Appl.* 41624 (1981); *Chem. Abstr.*, **97**, 105607 (1982).
150. M. Los, *Eur. Pat. Appl.* 41623 (1981); *Chem. Abstr.*, **96**, 199687 (1982).
151. B. B. Neelima and A. P. Bhaduri, *Indian J. Chem.*, **23B**, 431 (1984).
152. T. L. Wright, *Eur. Pat. Appl.* 120484 (1984); *Chem. Abstr.*, **102**, 113500 (1985).
153. M. Uchida, M. Komatsu, and K. Nakagawa, *Ger. Pat.* 3 324 034 (1984); *Chem. Abstr.*, **101**, 54936 (1984).
154. T. L. Wright, *Eur. Pat. Appl.* 120483 (1984); *Chem. Abstr.*, **102**, 95649 (1985).
155. B. Bhat and A. P. Bhaduri, *Indian J. Chem.*, **23B**, 33 (1984).
156. B. Bhat and A. P. Bhaduri, *Indian J. Chem.*, **21B**, 729 (1982).
157. N. S. Narasimhan and R. S. Mali, *Tetrahedron Lett.*, **1973**, 843.
158. N. S. Narasımhan and R. S. Mali, *Tetrahedron*, **30**, 4153 (1974).
159. N. S. Narasimhan and S. P. Bhagwat, *Synthesis*, **1979**, 903.
160. T. Ohta and Y. Mori, *Ann. Rep. Tokyo Coll. Pharm.*, **4**, 255 (1954); *Chem. Abstr.*, **50**, 998 (1956).
161. N. S. Narasimhan, R. S. Mali, and A. M. Gokhale, *Indian J. Chem.*, **18B**, 115 (1979).
162. A. C. Ranade, R. S. Mali, and V. M. Kurnawal, *Indian J. Chem.*, **21B**, 528 (1982).
163. R. Gatti, V. Cavrini, R. Poveri, D. Matteuzzi, and P. Brigidi, *Eur. J. Med. Chem. – Chim. Ther.*, **19**, 468 (1984).
164. T. Sato and M. Ohta, *Bull. Chem. Soc. Japan*, **29**, 817 (1956).
165. I. Iijima and K. C. Rice, *J. Heterocycl. Chem.*, **15**, 1527 (1978).
166. K. J. Brown and O. Meth-Cohn, *Tetrahedron Lett.*, **1974**, 4069.
167. D. Griffiths and R. Hull, *J. Heterocycl. Chem.*, **14**, 1097 (1977).
168. E. A. Harrison, Jr., K. C. Rice, and M. E. Rogers, *J. Heterocycl. Chem.*, **14**, 909 (1977).
169. L. Garuti, G. Giovanninetti, M. Baserga, and A. M. Palenzona, *Farmaco, Ed. Sci.*, **36**, 779 (1981).
170. J. Kotler-Brajtburg, *Acta Pol. Pharm.*, **25**, 239 (1968); *Chem. Abstr.*, **70**, 47251 (1969).
171. J. Brajtburg, *Pol. Pat.* 56421 (1968); *Chem. Abstr.*, **70**, 115023 (1969).
172. A. P. Phillips, *J. Amer. Chem. Soc.*, **68**, 2568 (1946).
173. O. E. Schultz and U. Anschler, *Justus Liebigs Ann. Chem.*, **740**, 192 (1970).
174. M. Seyhan, *Chem. Ber.*, **86**, 572 (1953)
175. S. F. MacDonald, *J. Amer. Chem. Soc.*, **69**, 1219 (1947).
176. S. Ginsburg and I. B. Wilson, *J. Amer. Chem. Soc.*, **79**, 481 (1957).
177. R. M. Forbis and K. L. Rinehart, Jr., *J. Amer. Chem. Soc.*, **95**, 5003 (1973).

178. G. R. Clemo and E. Hoggarth, *J. Chem. Soc.*, **1939**, 1241.
179. E. Schering, *Brit. Pat.* 240051 (1925); *Chem. Abstr.*, **20**, 2167 (1926); *D.R. Pat.* 421 088; *Chem. Zent.*, 1926, **1**, 2054.
180. H. Saito, H. Muro, H. Noda, S. Saeki, and M. Hamana, *Heterocycles*, **9**, 110 (1978).
181. J. Bernstein, H. Yale, K. Losee, M. Holsing, J. Martins, and W. A. Lott, *J. Amer. Chem. Soc.*, **73**, 906 (1951).
182. H. A. Shoeb and M. I. Korkor, *Res. Commun. Chem. Pathol. Pharmacol.*, **23**, 359 (1979); *Chem. Abstr.*, **90**, 168423 (1979).
183. H. A. Shoeb and M. I. Korkor, *Egypt. J. Chem.*, **22**, 329 [1979 (Pub. 1980)], *Chem. Abstr.*, **95**, 42856 (1981).
184. E. Ziegler, T. Kappe, and H. G. Foraita, *Monatsh. Chem.*, **97**, 409 (1966).
185. T. Sakamoto, S. Konno, and H. Yamanaka, *Heterocycles*, **6**, 1616 (1977).
186. H. Yamanaka, H. Abe, T. Sakamoto, H. Hiranuma, and A. Kamata, *Chem. Pharm. Bull.*, **25**, 1821 (1977).
187. S. M. Daluge, P. M. Skonezuy, B. Roth, and B. S. Rauckman, *Eur. Pat. Appl.* 96214 (1983); *Chem. Abstr.*, **100**, 139135 (1984).
188. D. Monti, P. Gramatica, and P. Manitto, *Farmaco, Ed. Sci.*, **36**, 412 (1981).
189. P. Blumbergs, M.-S. Ao, M. P. LaMontagne, A. Markovac, J. Novotny, C. H. Collins, and F. W. Starks, *J. Med. Chem.*, **18**, 1122 (1975).
190. A. Zayed, H. H. Zoorob, and M. T. El-Wassimi, *Pharmazie*, **33**, 572 (1978).
191. F. Meitzsch, *Ger. Pat.* 958 832 (1957); *Chem. Abstr.*, **54**, 2367 (1960).
192. S. Rossi, A. Salvatori, and G. Peruzzi, *Farmaco, Ed. Sci.*, **34**, 486 (1979).
193. F. Zymalkowski, *Arch. Pharm. Ber. Dtsch. Pharm. Ges.*, **288**, 162 (1955).
194. D. S. Tarbell, J. F. Bunnett, R. B. Carlin, and V. P. Wystrach, *J. Amer. Chem. Soc.*, **67**, 1582 (1945).
195. C. A. Rojahn and J. Schulten, *Arch. Pharm. Ber. Dtsch. Pharm. Ges.*, **264**, 348 (1926).
196. H. Gilman, F. J. Marshall, and R. A. Benkeser, *J. Amer. Chem. Soc.*, **68**, 1849 (1946).
197. H. Okada, V. Stella, J. Haslam, and N. Yata, *J. Pharm. Sci.*, **64**, 1665 (1975).
198. G. A. Epling, U. C. Yoon, and N. K. N. Ayengar, *Photochem. Photobiol.*, **39**, 469 (1984).
199. G. A. Epling and N. K. Ayengar, *Tetrahedron Lett.*, **1976**, 3009.
200. M. Hasegawa, *Pharm. Bull. (Japan)*, **1**, 47 (1953).
201. K. N. Campbell, A. H. Sommers, J. F. Kerwin, and B. K. Campbell, *J. Amer. Chem. Soc.*, **68**, 1851 (1946).
202. D. R. Bender and D. L. Coffen, *J. Heterocycl. Chem.*, **8**, 937 (1971).
203. J. B. Wommack, Jr., T. G. Barbee, Jr., K. N. Subbaswami, and D. E. Pearson, *J. Med. Chem.*, **14**, 1218 (1971).
204. A. A. L. Gunatilaka, J. S. H. Q. Perera, M. U. S. Sultanbawa, P. M. Brown, and R. H. Thompson, *J. Chem. Res., Synop.*, **1979**, 61; *Miniprint*, p. 779.
205. M. Levitz and M. T. Bogert, *J. Org. Chem.*, **10**, 341 (1945).
206. F. I. Carroll, B. D. Berang, and C. P. Linn, *J. Med. Chem.*, **22**, 1363 (1979).
207. F. I. Carroll, B. D. Berang, and C. P. Linn, *J. Med. Chem.*, **28**, 1564 (1985).
208. O. Schmut and T. Kappe, *Z. Naturforsch.*, **30B**, 140 (1975).
209. H. Arai, K. Saito and N. Murata, *Kogyo Kagaku Zasshi*, **63**, 319 (1960); *Chem. Abstr.*, **59**, 6362 (1963).
210. H. E. Baumgarten, R. P. Barkley, S-H. L. Chiu, and R. D. Thompson, *J. Heterocycl. Chem.*, **18**, 925 (1981).
211. H. Meyer, F. Bossert, W. Vater, and K. Stoepel, *Ger. Offen.* 2 210 667 (1973); *Chem. Abstr.*, **79**, 146412 (1973).
212. M. Ishikawa and I. Kikkawa, *Yakugaku Zasshi*, **75**, 33 (1955); *Chem. Abstr.*, **50**, 1007 (1956).
213. O. H. Johnson and C. S. Hamilton, *J. Amer. Chem. Soc.*, **63**, 2864 (1941).

214. W. Koenigs, *Ber.*, **31**, 2364 (1898).
215. J. B. Wommack, T. G. Barbee, Jr., D. J. Thoennes, M. A. McDonald, and D. E. Pearson, *J. Heterocycl. Chem.*, **6**, 243 (1969).
216. C. Kaiser, P. A. Dandridge, E. Garvey, K. E. Flaim, R. L. Zeid, and J. P. Hieble, *J. Med. Chem.*, **28**, 1803 (1985).
217. V. M. Rodionov and M. A. Berkenheim, *J. Gen. Chem. U.S.S.R.* (*Engl. Transl.*), **16**, 483 (1946); *Chem. Abstr.*, **41**, 966 (1947).
218. F. Eckhardt, *Ber.*, **22**, 277 (1889).
219. H. Decker and P. Remfry, *Ber.*, **38**, 2773 (1905).
220. H. Fiedler and U. Kaben, *Pharmazie*, **21**, 233 (1966).
221. A. Makriyannis, J. S. Frazee, and J. W. Wilson, *J. Med. Chem.*, **16**, 118 (1973).
222. R. C. Schuur, *French Demande* 2 487 350 (1982); *Chem. Abstr.*, **97**, 23775 (1982); *U.S. Pat.* 4 332 952 (1982).
223. B. Bobrański, *J. Prakt. Chem.*, **134**, 141 (1932).
224. G. R. Clemo and R. Howe, *J. Chem. Soc.*, **1955**, 3552.
225. G. M. Badger and A. G. Moritz, *J. Chem. Soc.*, **1958**, 3437.
226. R. N. Sen and S. K. Roy, *J. Indian Chem. Soc.*, **9**, 173 (1932).
227. J. P. Phillips, E. M. Barrall, and R. Breese, *Trans. Kentucky Acad. Sci.*, **17**, 135 (1956); *Chem. Abstr.*, **51**, 11349 (1957).
228. E. Senaga, *Nippon Kagaku Zasshi*, **82**, 1059 (1961); *Chem. Abstr.*, **58**, 11326 (1963).
229. B. Venkataramani, *Current Sci.* (*India*), **32**, 302 (1963); *Chem. Abstr.*, **59**, 11419 (1963).
230. H. Fiedler, *Arch. Pharm. Ber. Dtsch Pharm. Ges.*, **297**, 108 (1964).
231. H. Fiedler and U. Kaben, *Pharmazie*, **21**, 233 (1966).
232. S. M. Daluge and P. M. Skonezny, *Eur. Pat. Appl.* 51879 (1982); *Chem. Abstr.*, **97**, 127652 (1982).
233. W. Danikiewicz and M. Makosza, *Tetrahedron Lett.*, **26**, 3599 (1985).
234. R. C. Elderfield and M. Siegel, *J. Amer. Chem. Soc.*, **73**, 5622 (1951).
235. F. Gialdi and R. Ponci, *Farmaco* (*Pavia*), **6**, 332 (1951); *Chem. Abstr.*, **45**, 9117 (1951).
236. R. Behnisch, F. Mietzsch, and H. Schmidt, *U.S. Pat.* 2 775 593 (1956); *Chem. Abstr.*, **51**, 8804 (1957).
237. J. Howitz and J. Philipp, *Justus Liebigs Ann. Chem.*, **39b**, 23 (1913).
238. W. von Miller and F. Kinkelin, *Ber.*, **18**, 3234 (1885); *J. Chem. Soc. Abstr.*, **50**, 265 (1886).
239. H. Jensch, *Ger. Pat.* 831100 (1952); *Chem. Abstr.*, **52**, 11961 (1958).
240. D. E. Pearson, personal communication.
241. F. Przystal and J. P. Phillips, *J. Heterocycl. Chem.*, **4**, 131 (1967).
242. L. C. Washburn, T. G. Barbee, Jr. and D. E. Pearson, *J. Med. Chem.*, **13**, 1004 (1970).
243. H. Fiedler, *Z. Chem.*, **3**, 27 (1963), *Chem. Abstr.*, **59**, 5134 (1963).
244. C. Hamada, Y. Hirano, and T. Iida, *Nippon Kagaku Zasshi*, **77**, 1107 (1956); *Chem. Abstr.*, **53**, 5267 (1959).
245. S. A. Osadchii, *Izv. Sib. Otd. Akad. Nauk SSSR, Ser. Khim. Nauk*, **1978**, 120; *Chem. Abstr.*, **89**, 146742 (1978).
246. W. M. Tadros, H. A. Shoeb, M. A. Kira, and E. M. Ekladios, *Indian J. Chem.*, **14B**, 466 (1976).
247. J. Howitz, *Ber.*, **35**, 1274 (1902).
248. F. Gialdi and R. Ponci, *Farmaco.* (*Pavia*), **6**, 694 (1951); *Chem. Abstr.*, **46**, 11196 (1952).
249. W. J. Suggs and G. D. N. Pearson, *J. Org. Chem.*, **45**, 1514 (1980).
250. J. Howitz and W. Schwenk, *Ber.*, **38**, 1280 (1905).

251. J. S. Gillespie, Jr., S. P. Acharya, R. E. Davis, and B. K. Barman, *J. Med. Chem.*, **13**, 860 (1970).
252. G. J. Bird, G. J. Farquharson, and K. G. Watson, *Eur. Pat.* 124992 (1984); *Chem. Abstr.*, **103**, 5933 (1985).
253. H. Hagen, J. Markert, and B. Würzer, *Ger. Offen.* 3 108 873 (1982); *Chem. Abstr.*, **98**, 34511 (1983).
254. J. Markert, H. Hagen, R.-D. Kohler, and B. Würzer, *Ger. Offen.* 3 229, 175 (1984); *Chem. Abstr.*, **101**, 23360 (1984).
255. L. Kochańska and B. Bobrański, *Ber.*, **69**, 1807 (1936).
256. J. Howitz and P. Nöther, *Ber.*, **39**, 2705 (1906).
257. J. Kenner and B. K. Nandi, *Ber.*, **69**, 635 (1936).
258. W. Borsche and R. Manteuffel, *Justus Liebigs Ann. Chem.*, **526**, 22 (1936); *Chem. Abstr.*, **31**, 405 (1937).
259. E. Carlier and A. Einhorn, *Ber.*, **23**, 2894 (1890).
260. A. Einhorn, *Ber.*, **18**, 3465 (1885); *J. Chem. Soc. Abstr.*, **50**, 264 (1886).
261. A. Einhorn and P. Sherman, *Justus Liebigs Ann. Chem.*, **287**, 26 (1895); *J. Chem. Soc. Abstr.*, **70**, 61 (1896).
262. A. Hupe and A. Schramme, *Z. Physiol. Chem.*, **177**, 315 (1928).
263. K. Eiter and E. Mrazek, *Monatsh. Chem.*, **83**, 1491 (1952).
264. R. B. Woodward and E. C. Kornfeld, *J. Amer. Chem. Soc.*, **70**, 2508 (1948).
265. T. Sakamoto, Y. Kondo, M. Shiraiwa, and H. Yamanaka, *Synthesis*, **1984**, 245.
266. M. Hamana and H. Noda, *Chem. Pharm Bull.*, **18**, 26 (1970).
267. J. Čiernik, *Collect. Czech. Chem. Commun.*, **37**, 2273 (1972).
268. A. Brack, *Ger. Offen.* 2 721 190 (1978); *Chem. Abstr.*, **90**, 88746 (1979); *Brit. Pat.* 1 569 741 (1980).
269. M. Hamana, K. Funakoshi, H. Shigyo, and Y. Kuchino, *Chem. Pharm. Bull.*, **23**, 346 (1975).
270. M. Hamana and H. Noda, *Yakugaku Zasshi*, **89**, 641 (1969); *Chem. Abstr.*, **71**, 61165 (1969).
271. F. Al-Tai, G. Y. Sarkis, and F. A. Al-Najjar, *Bull. Coll. Sci., Univ. Baghdad*, **10**, 93 (1967); *Chem. Abstr.*, **72**, 43386 (1970).
272. A. Kaufmann, M. Kunkler, and H. Peyer, *Ber.*, **46**, 57 (1913).
273. G. Koller and H. Ruppersberg, *Monatsh. Chem.*, **58**, 238 (1931).
274. J. Čiernik and V. Vystavel, *Czech. Pat.* 150785 (1973); *Chem. Abstr.*, **80**, 132285 (1974).
275. A. Stener, *Gazz. Chim. Ital.*, **90**, 1365 (1960).
276. M. S. Chodnekar, A. F. Crowther, W. Hepworth, R. Howe, B. J. McLoughlin, A. Mitchell, B. S. Rao, R. P. Slortcher, L. H. Smith, and M. A. Stevens, *J. Med. Chem.*, **15**, 49 (1972).
277. W. Hepworth, A. Mitchell, M. S. Chodnekar and R. Howe, *French Pat.* M3697 (1965); *Chem. Abstr.*, **67**, 82119 (1967).
278. Imperial Chemical Industries Ltd., *Belg. Pat.* 633973 (1963); *Chem. Abstr.*, **61**, 672 (1964); *Brit. Pat.* 1 013 224 (1965).
279. M. Nagata, *Yakugaku Zasshi*, **86**, 608 (1966); *Chem. Abstr.*, **65**, 15356 (1966).
280. D. Allan and J. D. Loudon, *J. Chem. Soc.*, **1949**, 821.
281. P. Čarsky, S. Hunig, I. Stemmler, and D. Schentzow, *Liebigs Ann. Chem.*, **1980**, 291.
282. S. G. Waley, *J. Chem. Soc.*, **1948**, 2008.
283. R. D. Göschke, *Ger. Pat.* 2 310 773 (1973); *Chem. Abstr.*, **80**, 27119 (1974); *Brit. Pat.* 1 404 392; *U.S. Pat.* 3 897 436.
284. Ciba-Geigy A.-G., *Swiss Pat.* 576963 (1976); *Chem. Abstr.*, **85**, 177267 (1976).
285. J. P. Brown and L. M. Jackman, *J. Chem. Soc.*, **1964**, 3132.
286. J. P. Brown and O. Meth-Cohn, *J. Chem. Soc. (C)*, **1971**, 3631.

287. M. Ferles and O. Kocian, *Collect. Czech. Chem. Commun.*, **46**, 1518 (1981).
288. M. Natsume, S. Kumadaki, Y. Kanda, and K. Kiuchi, *Tetrahedron Lett.*, **1973**, 2335.
289. M. R. Chandramochan and S. Seshadri, *Indian J. Chem.*, **11**, 1108 (1973).
290. I. Ugi and E. Böttner, *Justus Liebigs Ann. Chem.*, **670**, 74 (1963).
291. A. Walser, G. Zenchoff, and R. I. Fryer, *J. Heterocycl. Chem.*, **13**, 131 (1976).
292. T. S. T. Wang, *Tetrahedron Lett.*, **1975**, 1637.
293. A. Šilhánková, M. Ferles and J. Malý, *Collect. Czech. Chem. Commun.*, **43**, 1484 (1978).
294. N. Roh, *Ger. Pat.* 660693 (1938); *Chem. Abstr.*, **32**, 7056 (1938).
295. G. L. Florvall, S. B. Ross and S.-O. Ögren, *U.S. Pat.* 4 000 280 (1976); *Chem. Abstr.*, **86**, 115371 (1977).
296. D. M. Brown and G. A. R. Kon, *J. Chem. Soc.*, **1948**, 2147.
297. G. L. Florvall, S. B. Ross and S.-Ö. Ögren, *Can. Pat.* 1 051 436 (1979); *Chem. Abstr.*, **92**, 76083 (1980); *Brit. Pat.* 1 461 667 (1977).
298. A. N. Prilepskaya, A. K. Sheinkman and S. N. Baranov, *Metody Poluch. Khim. Reakt. Prep.*, **1971** (23), 84; *Chem. Abstr.*, **78**, 4100 (1973).
299. F. H. C. Stewart, *J. Org. Chem.*, **26**, 3604 (1961).
300. M. R. Detty and B. J. Murray, *J. Org. Chem.*, **47**, 5235 (1982).
301. H. Scheuermann and D. Augart, *Ger. Offen.* 2 126 811 (1972); *Chem. Abstr.*, **78**, 85929 (1973).
302. H. Beecken, *Eur. Pat. Appl.* 37491; *Ger. Offen.* 3 012 599 (1981); *Chem. Abstr.*, **96**, 36877 (1982).
303. M. A. Weaver, *U.S. Pat.* 3 869 495 (1975); *Chem. Abstr.*, **82**, 172608 (1975).
304. H. Beecken, *Ger. Offen.* 2 446 759 (1976); *Chem. Abstr.*, **85**, 22771 (1976).
305. H. Beecken, *Ger. Offen.* 2 714 653 (1978); *Chem. Abstr.*, **90**, 40206 (1979).
306. A. G. Anderson, *Eur. Pat. Appl.* 39025 (1981); *Chem. Abstr.*, **96**, 77549 (1982); *U.S. Pat.* 4 268 667 (1981); *Chem. Abstr.*, **95**, 33394 (1981).
307. R. Zink and I. J. Fletcher, *Ger. Offen.* 3 423 369 (1985); *Chem. Abstr.*, **102**, 133553 (1985).
308. D. Mullen and A. W. McCann, *Brit. U.K. Pat. Appl.* 2 039 541 (1980); *Chem. Abstr.*, **94**, 123038 (1981).
309. C. A. Coates, Jr. and M. A. Weaver, *U.S. Pat.* 4 161 601 (1979); *Chem. Abstr.*, **91**, 142088 (1979).
310. M. A. Weaver, D. J. Wallace, and J. M. Straley, *U.S. Pat.* 3 247 211 (1966); *Chem. Abstr.*, **68**, 96804 (1968).
311. M. A. Weaver, *U.S. Pat.* 3 879 494 (1975); *Chem. Abstr.*, **83**, 195235 (1975).
312. Sumitomo Chemical Co. Ltd., *Japan Kokai Tokkyo Koho*, 81 133,220 (1981); *Chem. Abstr.*, **96**, 68596 (1982).
313. C. A. Coates, Jr. and M. A. Weaver, *U.S. Pat.* 3 595 863 (1971); *Chem. Abstr.*, **76**, 114848 (1972).
314. J. M. Straley, D. J. Wallace, and M. A. Weaver, *U.S. Pat.* 3 240 783 (1966); *Chem. Abstr.*, **64**, 19849 (1966).
315. M. A. Weaver and D. J. Wallace, *U.S. Pat.* 3 453 280 (1969); *Chem. Abstr.*, **71**, 92662 (1969).
316. D. J. Wallace and M. A. Weaver, *U.S. Pat.* 3 398 152 (1968); *Chem. Abstr.*, **70**, 12672 (1969).
317. L. Burgardt and O. Wahl, *U.S. Pat.* 2 968 557 (1961); *Chem. Abstr.*, **55**, 20738 (1961).
318. T. Sugasawa, H. Hamana, T. Toyoda, and M. Adachi, *Synthesis*, **1979**, 99.
319. V. R. Holland and B. C. Saunders, *Tetrahedron*, **27**, 2851 (1971).
320. I. L. Knunyants and A. K. Shillegodskii, *Zh. Obshch. Khim.*, **18**, 184 (1948); *Chem. Abstr.*, **43**, 2616 (1949).

321. I.G. Farbenindustrie A.-G., *Fr. Pat.* 828384 (1938); *Chem. Abstr.*, **33**, 400 (1939).
322. I.G. Farbenindustrie A.-G., *Fr. Pat.* 824 565 (1938); *Chem. Abstr.*, **32**, 6072 (1938).
323. Eastman Kodak Co., *Brit. Pat.* 466268 (1937); *Chem. Abstr.*, **31**, 8215 (1937).
324. L. G. S. Brooker, *U.S. Pat.* 2 165 218 (1939); *Chem. Abstr.*, **33**, 8421 (1939).
325. L. G. S. Brooker, *U.S. Pat.* 2 165 692 (1939); *Chem. Abstr.*, **33**, 8213 (1939).
326. F. S. Babichev and N. N. Romanov, *Ukr. Khim. Zh.*, **41**, 719 (1975); *Chem. Abstr.*, **83**, 165792 (1975).
327. V. P. Khibya and G. I. Galatsan, *Khim. Geterotsikl. Soedin.*, **1973**, 1282; *Chem. Abstr.*, **80**, 38347 (1974).
328. O. Weissel, *Ger. Pat.* 1 139 121 (1962); *Chem. Abstr.*, **58**, 9028 (1963).
329. J. Čiernik and V. Vystavel, *Czech. Pat.* 150787 (1973); *Chem. Abstr.*, **80**, 133286 (1974).
330. C. Reichardt and K. Halbritter, *Chem. Ber.*, **104**, 822 (1971).
331. C. Reichardt and W. Mormann, *Chem. Ber.*, **105**, 1815 (1972).
332. S. M. Kirtko and V. V. Perekalin, *Zh. Obshch. Khim.* **32**, 3298 (1962); *Chem. Abstr.*, **58**, 12531 (1963).
333. G. Reichardt, *Justus Liebigs Ann. Chem.*, **715**, 74 (1968); *Tetrahedron Lett.*, **1965**, 429.
334. H. Bücher, J. Wiegard, B. B. Snavely, K. H. Beck, and H. Kuhn, *Chem. Phys. Lett.*, **3**, 508 (1969).
335. N. Mishra, B. Bhuyan, L. N. Patnaik, and M. K. Rout, *Indian J. Chem.*, **14B**, 806 (1976).
336. H. W. Smith, *Eur. Pat. Appl.* 161867 (1985); *Chem. Abstr.*, **104**, 224839 (1986).
337. S. C. Zimmerman and R. Breslow, *J. Amer. Chem. Soc.*, **106**, 1490 (1986).
338. J. Koyama, T. Sugita, Y. Suzuta, and H. Irie, *Heterocycles*, **16**, 969 (1981).
339. E. Reimann and W. Dammertz, *Arch. Pharm. (Weinheim)*, **316**, 297 (1983).
340. D. T. Witiak, K. Tomita, and R. J. Patch, *J. Med. Chem.*, **24**, 788 (1981).
341. D. T. Witiak, R. J. Patch, S. J. Enna, and Y. K. Fung, *J. Med. Chem.*, **29**, 1 (1986).
342. L. E. Overman, D. Tesuisse and M. Hashimoto, *J. Amer. Chem. Soc.*, **105**, 5373 (1983).
343. L. E. Overman and R. L. Freeks, *J. Org. Chem.*, **46**, 2833 (1981).
344. F. A. L'Eplattenier, L. Vuitel, H. Junek and O. S. Wolfbeis, *Synthesis*, **1976**, 543.
345. R. F. C. Brown, J. J. Hobbs, G. K. Hughes, and E. Ritchie, *Aust. J. Chem.*, **7**, 348 (1954).
346. F. N. Lahey, J. A. Lamberton and J. R. Price, *Aust. J. Sci. Res.*, **3A**, 155 (1950).
347. B. Berinzaghi, V. Denlofeu, R. Labriola and A. Muruzabal, *J. Amer. Chem. Soc.*, **65**, 1357 (1943).
348. D. Tomasik, P. Tomasik, and R. A. Åbramovitch, *J. Heterocycl. Chem.*, **20**, 1539 (1983).
349. T. Tilakraj and S. Y. Ambekar, *J. Indian Chem. Soc.*, **62**, 251 (1985).
350. I. Ninomiya, T. Kiguchi, and T. Naito, *Heterocycles*, **9**, 1023 (1978).
351. K. A. Khan and A. Shoeb, *Indian J. Chem.*, **24B**, 62 (1985).
352. H. Harnisch and A. Brack, *Justus Liebigs Ann. Chem.*, **740**, 164 (1970).
353. K. Tomita, *Yakugaku Zasshi*, **71**, 1100 (1951); *Chem. Abstr.*, **46**, 5044 (1952).
354. T. R. Chamberlain, C. D. Campbell and J. M. McCrae, *U.S. Pat.* 4 153 601 (1979); *Brit. Pat.* 1 534 787 (1978); *Chem. Abstr.*, **91**, 58683 (1979); *Ger. Offen.* 2 728 863 (1976); *Chem. Abstr.*, **88**, 122660 (1978); *Ger. Offen.* 2 728 864 (1976); *Chem. Abstr.*, **88**, 154316 (1978).
355. Y. Asahina and M. Inubuse, *Yakugaku Zasshi*, **52**, 120 (1932).
356. Y. Asahina and M. Inubuse, *Ber.*, **65**, 61 (1932).
357. Y. Asahina, T. Ohta, and M. Inubuse, *Ber.*, **63**, 2045 (1930).
358. Y. Asahina and M. Inubuse, *Ber.*, **63**, 2052 (1930).
359. H.-J. Gais, K. Hafner, and M. Neunschwander, *Helv. Chim. Acta*, **52**, 2641 (1969).

360. D. J. Cook and R. N. Pierce, *Proc. Indiana Acad. Sci.*, **70**, 115 (1960); *Chem. Abstr.*, **56**, 10096 (1962).
361. E. Ochiai and T. Dodo, *Itsuu Kenkyusho Nempo*, **1965** (14), 33; *Chem. Abstr.*, **67**, 64159 (1967).
362. M. Hijikota, *Nichidai Igaku Zasshi*, **18**, 1105 (1959); *Chem. Abstr.*, **61**, 10937 (1964).
363. D. J. Cook and M. Stamper, *J. Amer. Chem. Soc.*, **69**, 1467 (1947).
364. D. J. Cook and C. K. Bjork, *J. Amer. Chem. Soc.*, **74**, 543 (1952).
365. D. J. Cook, R. W. Sears, and D. Dock, *Proc. Indiana Acad. Sci.*, **58**, 145 (1948); *Chem. Abstr.*, **44**, 4473 (1950).
366. D. J. Cook, R. S. Yunghans, T. R. Moore, and B. E. Hoogenboom, *J. Org. Chem.*, **22**, 211 (1957).
367. M. Von Strandtmann, D. Connor, and J. Shavel, Jr., *J. Heterocycl. Chem.*, **9**, 175 (1972).
368. S. Yoshizaki, S. Tamada, and E. Yo, *Chem. Pharm. Bull.*, **26**, 2267 (1978).
369. S. Yoshizaki, S. Tamada, K. Yo, and K. Nakagawa, *Japan Kokai*, 78 12,872 (1978); *Chem. Abstr.*, **88**, 190612 (1978).
370. S. Yoshizaki, S. Tamada, N. Yo, and R. Nakagawa, *Japan Kokai Tokkyo Koho*, 79 98,779 (1979); *Chem. Abstr.*, **92**, 41783 (1980).
371. P. Ruggli, P. Hindermann, and H. Frey, *Helv. Chim. Acta*, **21**, 1066 (1938).
372. P. Ruggli and H. Frey, *Helv. Chim. Acta*, **22**, 1413 (1939).
373. R. Goschke, P. G. Ferrini, and A. Sallmann, *Eur. Pat. Appl.* 62001 (1982); *Chem. Abstr.*, **98**, 160599 (1983).
374. J. R. Bantick, *Eur. Pat. Appl.* 79637 (1983); *Chem. Abstr.*, **99**, 175604 (1983).
375. W. E. Edmiston and K. Wiesner, *Can. J. Chem.*, **29**, 105 (1951).
376. L. Mosti, P. Schenone, and G. Menozzi, *J. Heterocycl. Chem.*, **16**, 177 (1979).
377. W. D. Crow and J. R. Price, *Aust. J. Sci. Res.*, **2A**, 282 (1949).
378. J. Goldsworthy, W. J. Ross, and J. P. Verge, *Eur. Pat. Appl.* 55068 (1982); *Chem. Abstr.*, **97**, 162848 (1982).
379. F. Arya, J. Bouquant, and J. Chuche, *Synthesis*, **1983**, 946.
380. B. Bobrański, *Ber.*, **69**, 1113 (1936).
381. D. J. Evans and F. Eastwood, *Aust. J. Chem.*, **27**, 537 (1974).
382. N. Barton, A. F. Crowther, W. Hepworth, D. N. Richardson, and G. W. Driver, *Brit. Pat.* 830832 (1960); *Chem. Abstr.*, **55**, 7442 (1961).
383. T. A. Geissman, M. J. Schlatter, I. D. Webb, and J. D. Roberts, *J. Org. Chem.*, **11**, 741 (1946).
384. F. Eiden, R. Wendt, and H. Fenner, *Arch Pharm. (Weinheim)*, **311**, 561 (1978).
385. M. Conrad and L. Limpach, *Ber.*, **21**, 1965 (1888).
386. R. J. Chudgar and K. N. Trivedi, *J. Indian Chem. Soc.*, **49**, 41 (1972).
387. R. J. Chudgar and K. N. Trivedi, *J. Indian Chem. Soc.*, **48**, 739 (1971).
388. P. V. Thakore and K. N. Trivedi, *J. Indian Chem. Soc.*, **57**, 536 (1980).
389. L. R. Morgan, Jr., R. J. Schunior, and J. H. Boyer, *J. Org. Chem.*, **28**, 260 (1963).
390. J. F. Ayafor, B. L. Sondengam, and B. T. Ngadjui, *Phytochemistry*, **21**, 2733 (1982).
391. J. F. Collins, G. A. Gray, M. F. Grundon, D. M. Harrison, and C. G. Spyropoulos, *J. Chem. Soc., Perkin Trans. 1*, **1973**, 94.
392. R. A. Corral and O. O. Orazi, *Tetrahedron*, **22**, 1153 (1966).
393. S. Goodwin, J. N. Shoolery, and E. C. Horning, *J. Amer. Chem. Soc.*, **81**, 3736 (1959).
394. H. Rapoport and K. G. Holden, *J. Amer. Chem. Soc.*, **81**, 3738 (1959).
395. T. Naito and C. Kaneko, *Chem. Pharm. Bull.*, **28**, 3150 (1980).
396. R. J. Chudgar and K. N. Trivedi, *J. Indian Chem. Soc.*, **49**, 513 (1972).
397. E. J. Reist, H. P. Hamlow, I. G. Junga, R. M. Silverstein, and B. R. Baker, *J. Org. Chem.*, **25**, 1368 (1960).

398. J. B. P. A. Wijnberg and W. N. Speckamp, *Tetrahedron Lett.*, **1975**, 4035.
399. J. B. P. A. Wijnberg and W. N. Speckamp, *Tetrahedron*, **34**, 2399 (1978).
400. Pfizer Inc., *Brit. Pat.* 1 579 228 (1980).
401. Pfizer Inc., *Brit. Pat.* 1 579 230 (1980).
402. M. R. Johnson, *Ger. Offen.* 2 722 383 (1977); *Chem. Abstr.*, **88**, 89533 (1978).
403. M. R. Johnson and G. M. Milne, *U.S. Pat.* 4 228 169 (1980); *Chem. Abstr.*, **94**, 192166 (1981).
404. M. R. Johnson, *Can. Pat.* 1 101 860 (1981); *Chem. Abstr.*, **95**, 150473 (1981).
405. M. R. Johnson, *U.S. Pat.* 4 260 764 (1981); *Chem. Abstr.*, **95**, 187099 (1981).
406. M. R. Johnson, *U.S. Pat.* 4 309 545 (1982); *Chem. Abstr.*, **96**, 162554 (1982).
407. T. G. Majewicz and P. Caluwe, *J. Org. Chem.*, **41**, 1058 (1976).
408. H. Juraszyk, H. J. Enenkel, K. O. Minck, H. J. Schliep, and J. Piulats, *Eur. Pat. Appl.* 154190 (1985); *Chem. Abstr.*, **104**, 224890 (1986).
409. J. M. Schaus, *Brit. Pat.* 2 130 576 (1984); *Chem. Abstr.*, **101**, 191892 (1984).

CHAPTER 2

Quinoline Ketones

JOHN V. GREENHILL*

Pharmaceutical Chemistry, School of Pharmacy,
University of Bradford
Bradford BD7 1DP, England

---

*Present address: Department of Chemistry, University of Florida, Gainesville, Florida 32611, USA.

List of Tables

I. Introduction

This chapter sets out to review the literature on the preparation of all quinoline ketones known to the end of *Chemical Abstracts* Volume 104 (1986). It seemed easy at the outset—five long years ago. Up to and including Collective Index 8, the relevant compounds appear under 'ketones'. One had to read through 62 pages of Index 8, but the quinoline derivatives were easily identified. Since then indexing policy has changed, so that now ketones appear under separate headings—methanone, ethanone etc. It is no longer possible, therefore, to be comprehensive via a manual search.

CAS online came to the rescue, although at some cost to a university impoverished by recent government policy. However, the online service is designed to pull out named individual compounds rather than give blanket coverage of a class. I am grateful to Dr John Thomas of The Royal Society of Chemistry and Mr Ken Tideswell of Bradford University library for producing the necessary programs and carrying out the searches. The comprehensive nature of the review does, of course, depend on these search programs, and I have found nothing to show that any examples were missed. In one area of marginal interest, I understand that it was not possible to make the programs fully reliable. I had decided to include fused carbocyclic ketones such as **1** on the grounds that their chemistry would be similar to that of simple ketones such

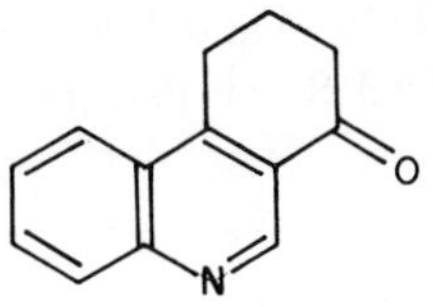

1 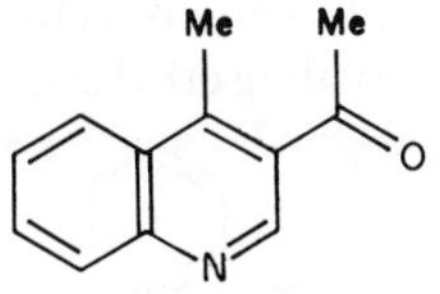   **2**

as **2**. The fused structures that have been found are mentioned, but there may well be examples in the literature that have been missed.

II. The Preparation of Quinolyl Ketones

1. General

Ketones with the carbonyl group attached to a fully aromatic quinoline ring system are listed in Tables 1–18. Those recorded in the primary literature with hydroxy groups in the 2- or 4-positions are treated as quinolones and listed in Section XIV. The ketonic synthetic intermediates to the cinchona alkaloids are prepared by distinctive methods, so are dealt with separately in Section IV. Quinaldoins, Section III, are also treated separately. Many of the tabulated ketones were prepared by standard methods, as indicated in the tables, and are not discussed further. Preparative methods of particular interest to quinoline chemistry are described below.

A valuable new organometallic procedure uses 2-, 3- and 4-trimethyltin quinolines in reaction with acid chlorides, e.g. Scheme 1. The 2-substituted quinolines reacted at room temperature. The 3-substituted compounds required eight hours' reflux in the presence of a catalyst [$PdCl_2$ or $PdCl_2(PPh_3)_2$]. These gave good yields, but 4-substituted quinolines had to be refluxed for at least four days with the catalyst and yields were generally lower.[1,2]

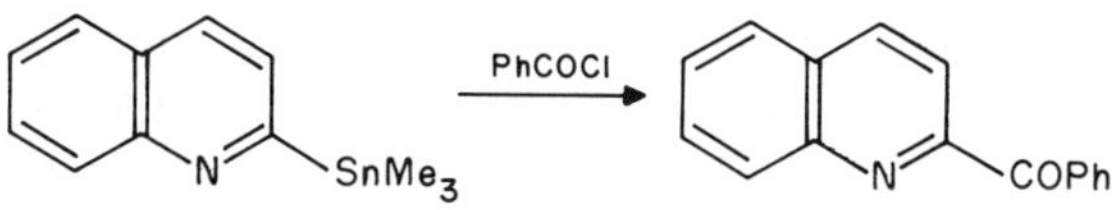

Scheme 1

2. 2-Keto and 4-Keto-quinolines

A. Organometallic Reactions

Some reagents, particularly 2-pyridyllithiums, react satisfactorily with quinolinecarboxylic acids to give ketones; see Table 8. When 7-trifluoromethyl-quinoline-4-carboxylic acid was treated with 2-pyridyllithium in THF/ether at

$- 70 \,^{\circ}$C it gave a mixture of ketones 3, R = H (16%) and 3, R = 2-pyridyl (12%). With the corresponding ethyl ester, only ketone 3, R = H (67%) was obtained.[3]

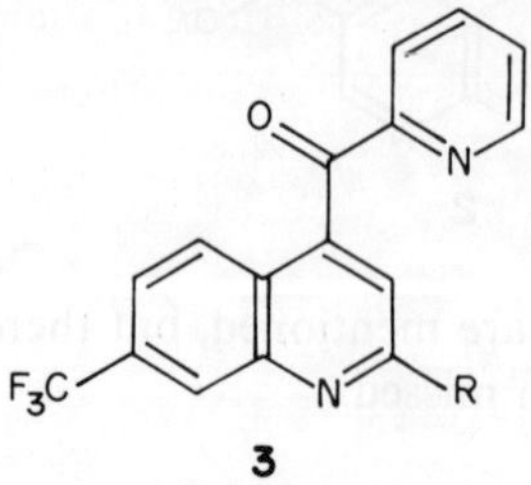

The alkyl groups of the three methylpyridines, 2- and 4-methylquinolines and 1-methylisoquinoline were deprotonated by phenyllithium and treated with methyl quinoline-2-carboxylate to give good yields of ketones, e.g. 4. These products could be hydrolysed to quinoline-2-carboxylic acid and the starting methyl heterocycle by hot alkali or mineral acid.[4]

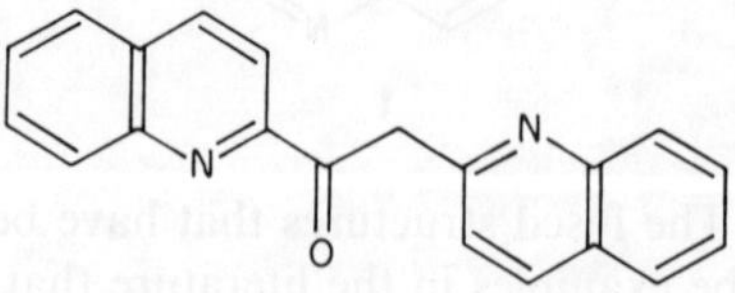

B. Carbanion Reactions

Ziegler-type condensation between 2- and 4-quinolinecarbonitriles and phenylacetonitrile gave imines, e.g. 5, which were hydrolysed to ketones; see Scheme 2.[5,6] In some cases milder hydrolysis allowed isolation of the intermediate keto-amide.[6]

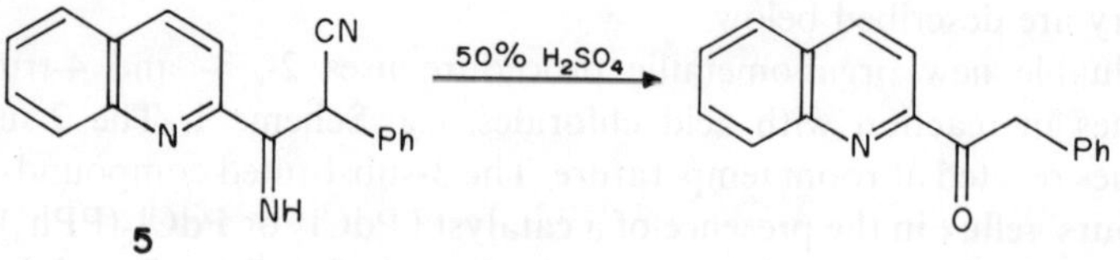

SCHEME 2

Di- and tri-carboxylic acids were protected with dihydropyran and the derivatives 6 were deprotonated in situ and heated with quinoline-2-carbonylchloride to give the ketones 7; see Scheme 3 and Table 1.[7]

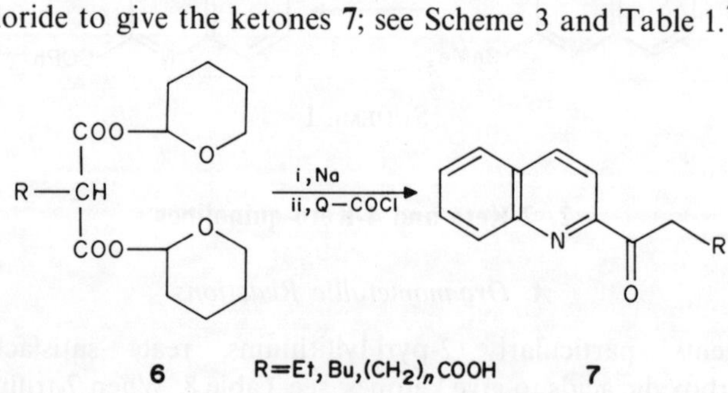

6   R=Et, Bu, (CH$_2$)$_n$COOH   7

SCHEME 3

Decarboxylation of quinoline-2-carboxylic acid by boiling in ethyl or methyl benzoate gave the benzoylquinoline **9** in low yield, presumably via the anion **8**.[8] In the Hammick reaction, ketone **9** and derivatives substituted in the phenyl ring were similarly prepared from quinoline-2-carboxylic acid and aromatic aldehydes.[9] With benzaldehyde the intermediate alcohol could be isolated but was readily oxidized by air.[10] Decarboxylation of benzoic acid in quinoline also gave a small yield of 2-benzoylquinoline **9** along with several other products.[11]

A mixture of quinoline-2-carboxylic acid and phthalic anhydride in nitrobenzene at 160–165° gave the keto-acid **10**, also via anion **8**. Keto-acids **12** (in the hydroxylactone form) and **13** were made similarly. The solid **10**, in nujol, was essentially in the keto form, with $v = 1690$ and $1671 \, cm^{-1}$, but in dioxane solution an equilibrium with the lactone form **11** was demonstrated, with $v = 1778$, 1720 and $1684 \, cm^{-1}$. Compound **12** existed as a six-membered lactone ring in both nujol, with $v = 1715 \, cm^{-1}$ and dioxane, with $v = 1732 \, cm^{-1}$. No other carbonyl band was seen, but $v_{OH} = 3270 \, cm^{-1}$ was reported. For compound **13** to form a lactone a seven-membered ring would be needed. This was not observed in either solvent.[12]

10,43%,m.p. 225−226 °C
HCl, 182− 183 °C

11

12,43%, m.p. 179−180 °C
HCl,196−198 °C

13,24%,m.p. 185−186 °C
HCl, 156−157 °C

Condensation between 2-acetylquinoline **14** and cinnamaldehydes **15** in the presence of dilute sodium hydroxide solution gave the dienones **16** in a reaction reminiscent of chalcone formation; see Section VI.5.[13]

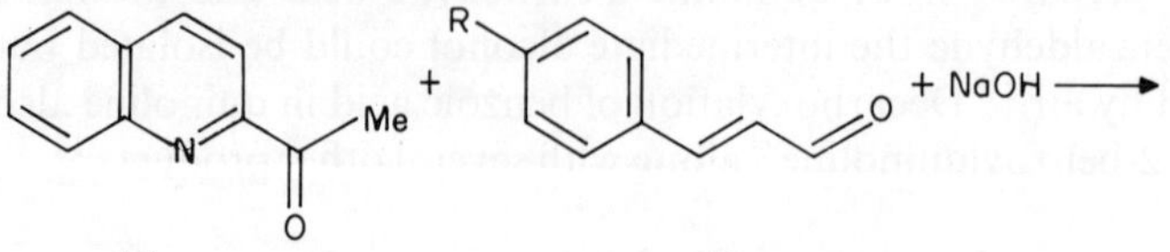

14   **15**

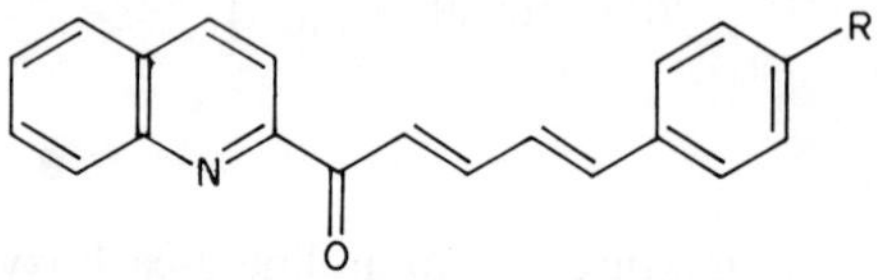

16, R=H, 96%, m.p. 120—121 °C
R=NMe$_2$, 97%, m.p. 132.5 °C

C. Radical Methods

Either butane-2,3-dione or 3-hydroxybutanone in the presence of silver persulphate gave acetyl free radicals, which reacted with quinoline to give a mixture of 2-acetyl-quinoline (30%), 4-acetylquinoline (32%) and 2,4-diacetylquinoline (38%); 4-methylquinoline gave only 2-acetyl-4-methylquinoline (100%) and with the benzoyl free radical, similarly produced from benzoin, 2-benzoyl-4-methylquinoline was obtained (21%).[14]

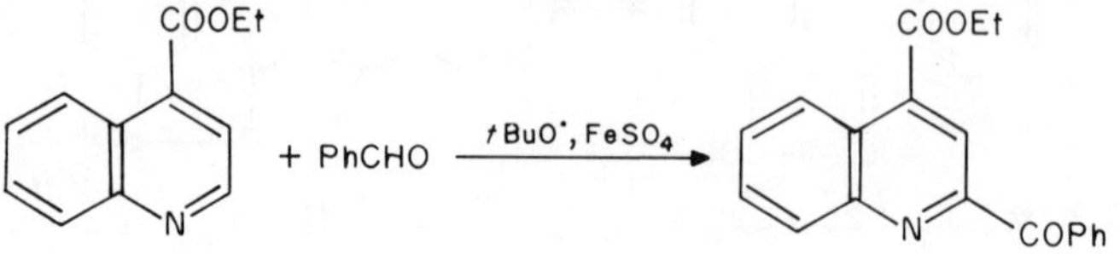

SCHEME 4

Aldehydes and *t*-butyl peroxide gave radicals which were used to prepare 2- and 4-ketoquinolines, e.g. Scheme 4. An alternative source of the required radicals was silver catalysed decarboxylation of 2-keto acids.[15] These procedures also allowed the preparation of several 2,4-diketoquinolines;[16] see Table 1.

D. Ring Closure Methods

Doebner reaction using aryl keto-aldehydes gave aryl 2-quinolyl ketones **17**. Two of the products were synthesized unequivocally by the Pfitzinger synthesis from the isatin **18**; see Scheme 5.[17]

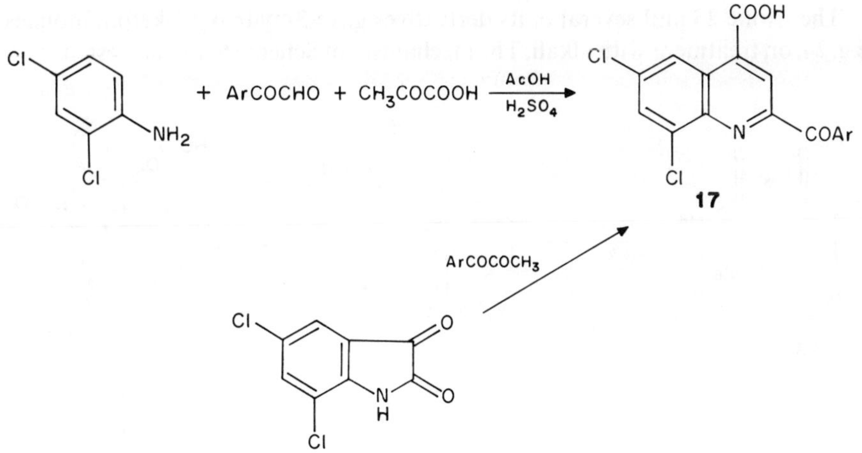

SCHEME 5

Friedländer condensation of 2-aminoacetophenone with cyclohexane-1,2-dione gave the 2-ketoquinoline **19**.[18]

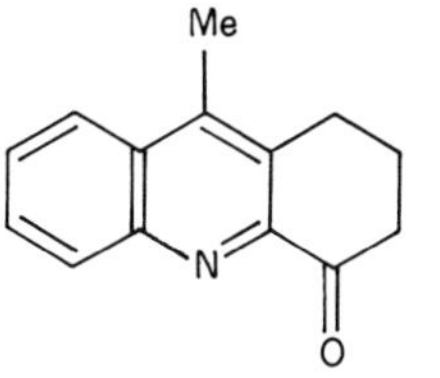

19,61%, m.p. 183 −184 °C

The amino-aldehydes **20a**, amino-ketones **20b** and **c** and amino-nitrile **20d** all reacted with dibenzoylacetylene to give the 2,3-dibenzoylquinolines **22a–d** via the enamines **21**[19−21] A product identical with compound **22a** was obtained by oxidation of 3-benzoyl-2-benzylquinoline with chromic acid.[19]

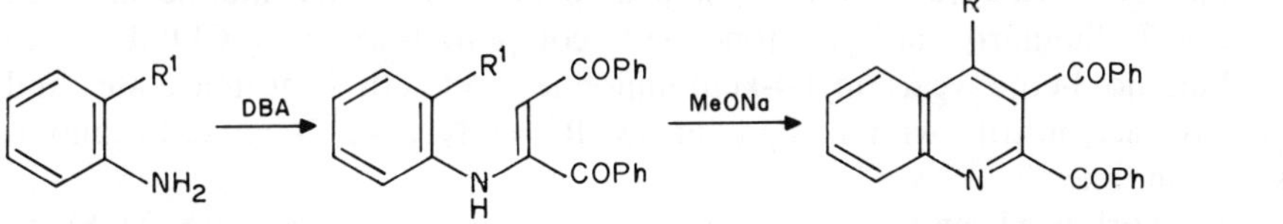

20 a, R^1=CHO
   b, R^1=COMe
   c, R^1=COPh
   d, R^1=CN

21

22 a, R^2=H, 58%, m.p. 118 −120 °C
   b, R^2=Me
   c, R^2=Ph
   d, R^2=NH$_2$

The trione **23** and several of its derivatives gave 3-hydroxy-2-ketoquinolines, e.g. **24**, on treatment with alkali. The mechanism of Scheme 6 was suggested.[22]

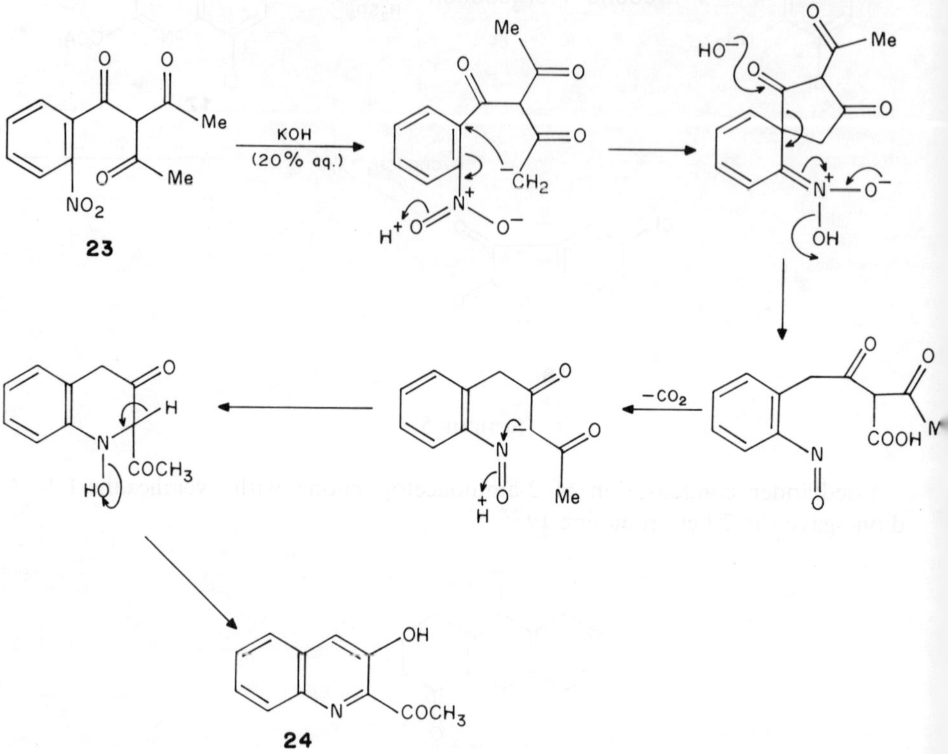

23

24

SCHEME 6

E. Rearrangements

The pyrazolones **26** were reported to react with the isatin derivatives **25** to give ketones **28**. Structure **27** was presumed to be that of the intermediate; see Scheme 7. Similarly, acetylacetone with compound **25**, $R^1 = COOEt$ gave a 2-amino-3-ethoxycarbonyl-4-ketoquinoline and ethyl acetoacetate and benzoyl acetonitrile with compound **25**, $R^1 = CN$ gave 2-amino-3-cyano-4-ketoquinoline.[23]

This work has been re-examined (E. A. A. Hafez, F. M. A. Gahil, S. M. Sherif and M. H. Elnagdi, *J. Heterocycl. Chem.*, 1986, **23**, 1375) and compounds formerly assigned structures **28** were shown to be spiro compounds **28a**.

The tryptophan derivative **29** rearranged in the presence of peracetic acid to the quinolyl ketone **30**, Scheme 8.[24]

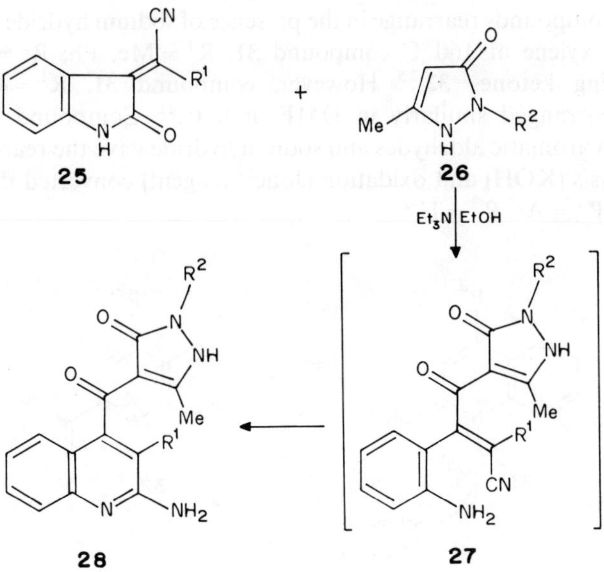

25 + 26

Et₃N|EtOH

28 ← 27

R¹=CN,COOEt; R²=H,Ph

SCHEME 7

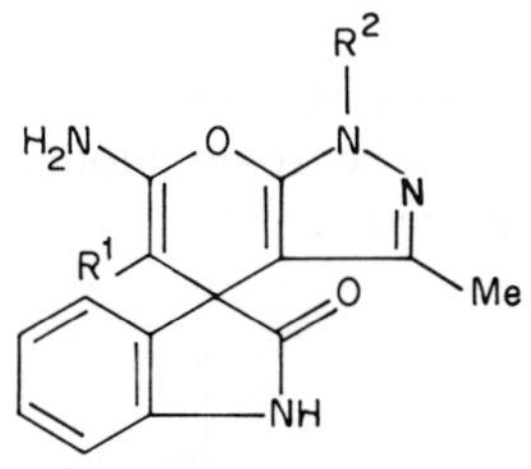

28a

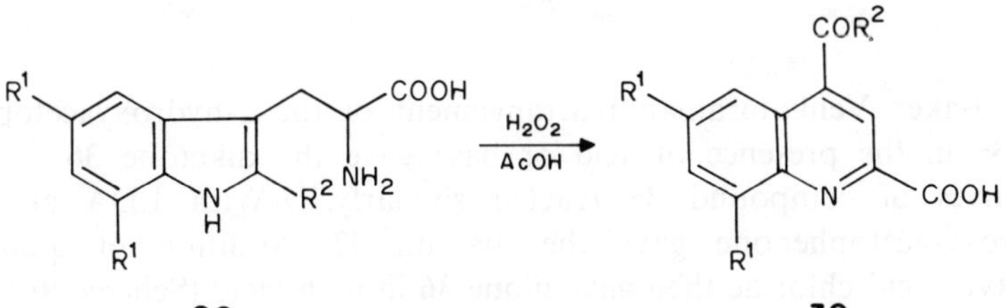

29 → 30

R¹=CH₂CH₂CMe₂, R²=C(Me)₂Et

SCHEME 8

Reissert compounds rearrange in the presence of sodium hydride to 2-quinolyl ketones. In xylene at $160\,^{\circ}$C compound **31**, $R^1 = $ Me, Ph; $R^2 = $ H gave the corresponding ketones **32**.[25] However, compound **31**, $R^1 = 4\text{-MeOC}_6\text{H}_4$, $R^2 = $ Me rearranged similarly in DMF at $0\,^{\circ}$C.[26] Compound **31**, $R^1 = $ Ph, $R_2 = $ H with aromatic aldehydes and sodium hydride gave the rearranged esters **33**. Hydrolysis (KOH) and oxidation (Jones' reagent) converted the esters into ketones **32** $R^1 = $ Ar; $R^2 = $ H.[27]

Either the cyclopropane alcohol **34** or the corresponding ketone reacted with ethyl azodicarboxylate to give 2-benzoylquinoline (**9**). A plausible mechanism for the rearrangement was suggested.[28]

The Baker–Venkataraman rearrangement of the 2-hydroxyacetophenone ester **35** in the presence of acid or base gave the diketone **36**.[29,30] Some derivatives of compound **35** reacted similarly.[31] With LDA at $-25\,^{\circ}$C 2-hydroxy-acetophenone gave the dianion **37**. Addition of quinoline-2-carboxylic acid chloride then gave dione **36** in high yield (Scheme 9).[32]

Irradiation of the amides **38**, R = H, Me, Cl caused rearrangement to the secondary amino-ketones **39** and other non-ketonic products; see Scheme 10. Compounds **39**, R = H, Me were isolated in useful yields; see Table 1.[33]

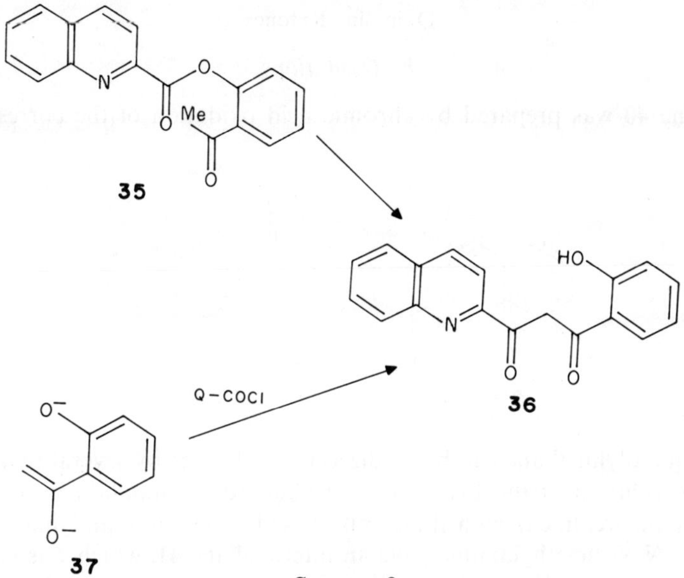

35

SCHEME 9

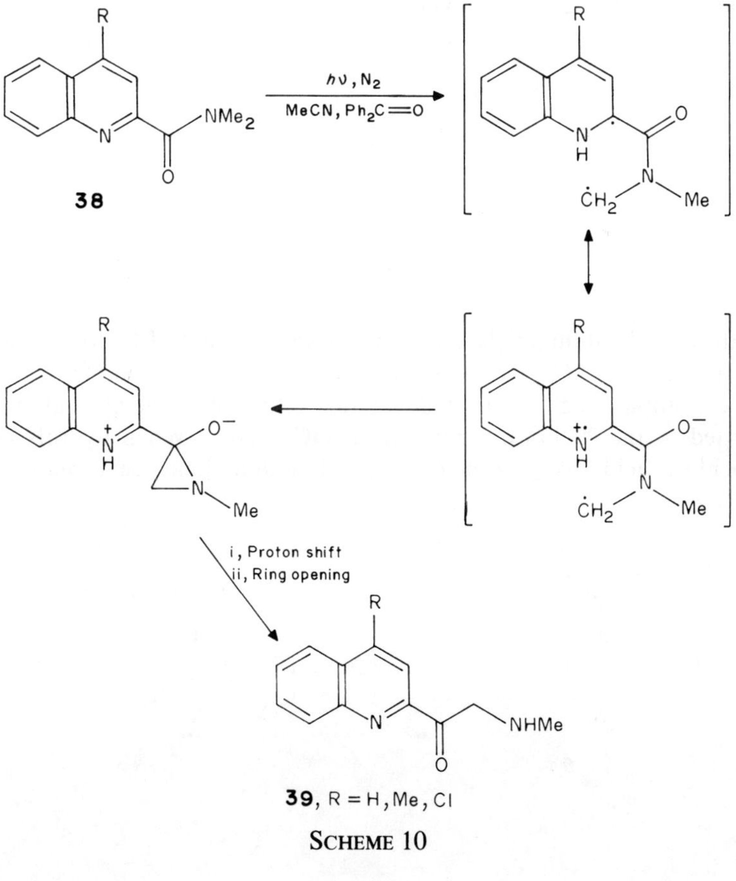

SCHEME 10

39, R = H, Me, Cl

F. Oxidations

The dione **40** was prepared by chromic acid oxidation of the corresponding diol.[34]

40, 93%, m.p. 283.5—284.5 °C

Di-2-quinolylmethane can be oxidized to the ketone by several reagents. An alcoholic solution of the hydrochloride oxidized on standing in air for two weeks or on treatment with dilute nitric acid.[35] The diquinolylmethane and 4-nitroso-N,N-dimethylaniline gave an intermediate, **41**, which was cleaved to the ketone by acid or alkali.[35,36] A solution of quinoline N-oxide and di-2-quinolylmethane in refluxing acetic acid gave a mixture of products, which included 8.4% of di-2-quinolyl ketone.[37]

41

Oxidation of the diquinolylacetonitrile **42** with 1.4 mol of hydrogen peroxide in acetic acid at 75 °C gave the ketone **43**. With 3 mols of peroxide under the same conditions, the $N-1'$-oxide of the ketone **43** formed, and could be converted to the 2'-chloro compound (POCl$_3$, HCl). When hydrolysed with hydrochloric acid this gave the corresponding quinolone; see Table 1.[38]

42   **43**

A 1904 paper claimed the preparation of di-(2-quinolyl) ketone from quinoline-2-carboxylic acid and acetic anhydride, but a later publication by the same author cast doubt on the structural assignment[39,40] and indeed the m.p. quoted (130–135 °C) does not correspond to that of the ketone prepared as described above.

The cyclobutenoquinoline 44 was oxidized by acid permanganate and assumed to give 2,3-dibenzoylquinoline (16.5%, m.p. 172–173°). Alternatively, an acetone solution kept in sunlight produced a 20% yield of the same compound.[41] It is noted that the melting points and the i.r. and n.m.r. spectra of these samples and of one prepared by another group[19] are different.

44

Ozonolysis of the olefine 45a gave the ketone 46[42], and the enols 45b gave the same ketone by air oxidation on standing in solution.[43]

45 a, R^1=H, R^2= Ph
   b, R^1=OH, R^2= Me,Ph,4-Py

46, 64%, m.p. 194–195 °C

G. Other Methods

Reduction (Al/Hg, $HgCl_2$) of a mixture of quinoline and a benzoic acid ester or amide gave a moderate yield of 2-benzoylquinoline mixed with other products.[44]

Quinoline N-oxide 47 and dimethylsulphonium acetyl(methoxycarbonyl)-methylide 48 in the presence of benzoyl chloride gave a mixture of the pyrroloquinolines 49a, b along with the benzoyl enol ester of the ketone 49a.

47   **48**

49 a, R = H
   b, R = 2–quinolyl

Compound **49a** could be converted to the ketone **49b** with benzoyl chloride in DMF. Di-2-quinolyl ketone was obtained from compound **49a** by oxidation (MCPBA) or from compound **49b** by careful hydrolysis.[45]

Treatment of the sulphonic ester **50** with the ylide **51** was assumed to give the intermediate **52**. Addition of an aldehyde or ketone gave an epoxide such as **53**. The epoxide could be isolated or rearranged *in situ* to a ketone **54**, Scheme 11.[46]

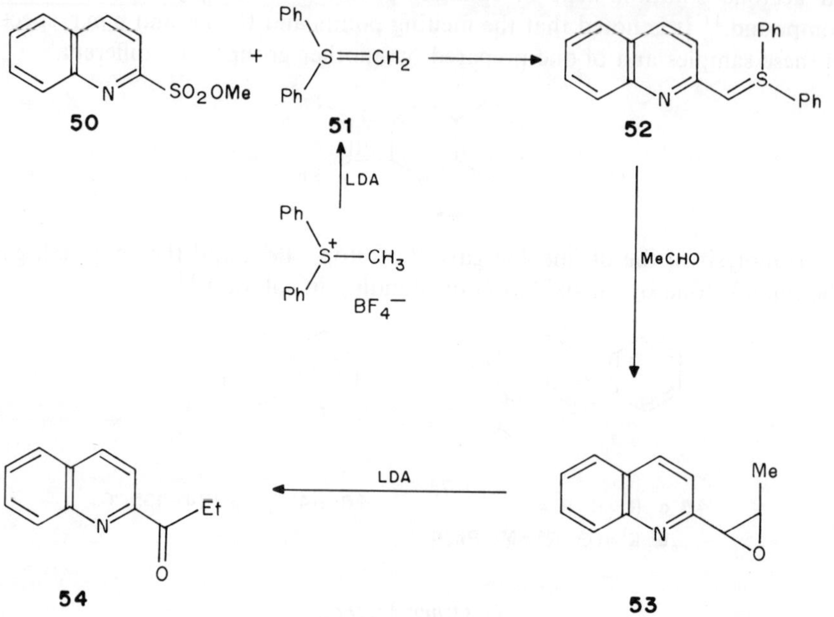

SCHEME 11

Quinoline *N*-oxide and esters of aryl glycolonitriles reacted in acetic anhydride to give the derivatives **55** in yields of 11–75%. Base treatment converted the nitriles into the ketones **56**.[47]

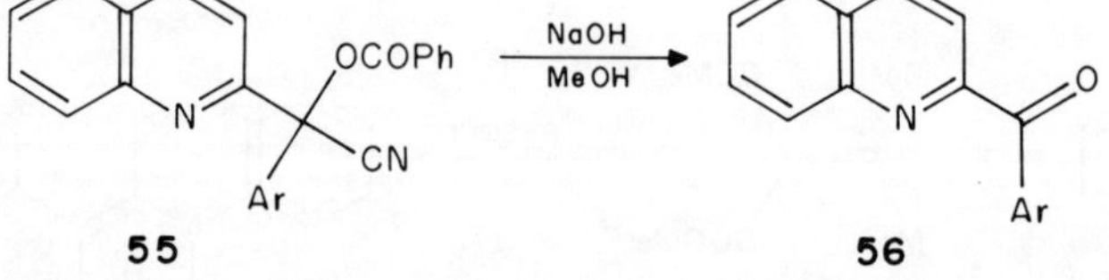

Quinoline-2,3-dicarboxylic acid anhydride and 1-ethyl-2-methylindole reacted in refluxing xylene to give a mixture of the ketones **57** and **58** (51.7%). The mixture (m.p. 196–198 °C) was used without separation.[48]

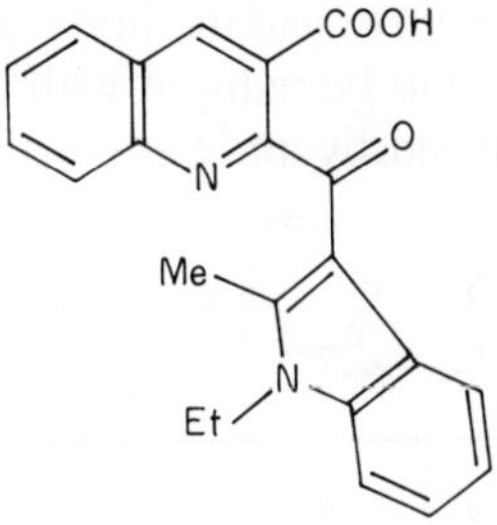

57 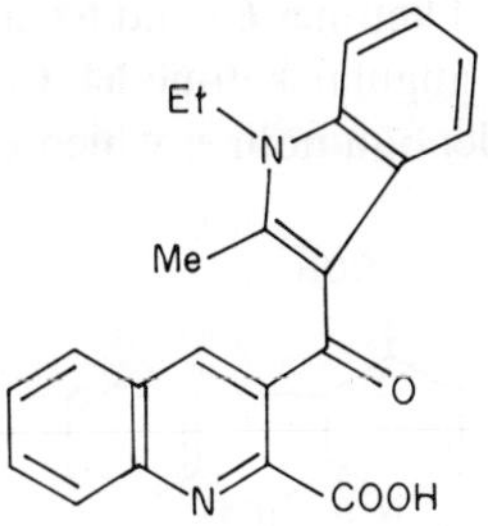   **58**

The hemiacetal **59** was treated with cyclohexyl isonitrile in methanol at room temperature to give the keto-amide **60**, but no yield or characteristics of the product were reported.[49]

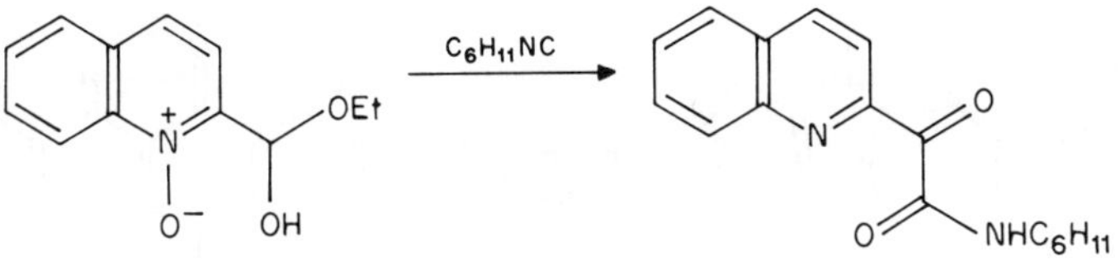

59   **60**

The 2-ketoquinolines **61** were prepared from the 3-phenylquinoline-2-carboxylic acids via the acid chlorides and Friedel–Crafts reactions.[50,51] The appropriate quinoline-4-carboxylic acids gave ketones **62a, b** either via the acid chloride as above or directly on heating in concentrated sulphuric acid.[51] Ring closing Friedel–Crafts reactions also gave rise to ketones **62c, e**[50,52] and **62d**.[53] Compound **62e** was converted into compound **62f** with sodium methoxide.[52] Use of the above procedures on 3-Phenylquinoline-2,4-dicarboxylic acid gave

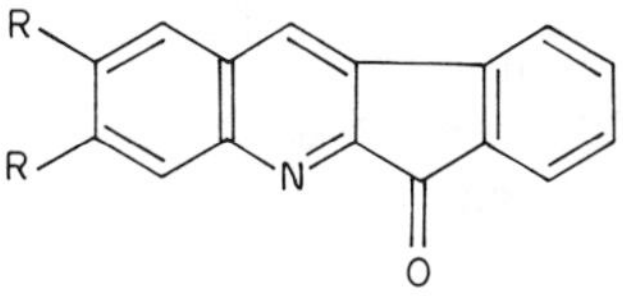

61

R = H, 50%, m.p. 190.5 °C

R, R = OCH$_2$O, m.p. 276 − 277 °C

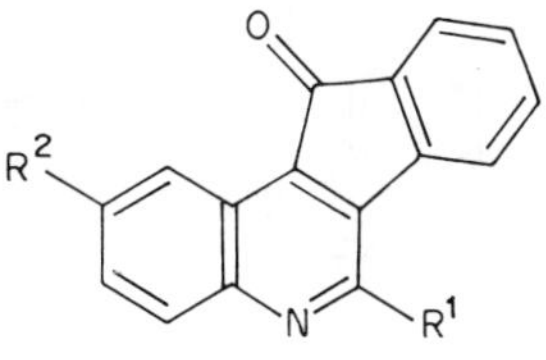

	R^1	R^2	%	M.p. (°C)
62 a,	H	H		238
b,	H	Me		237
c,	iPr	H	94	184
d,	Ph	H	62	259 − 261
e,	Cl	H		214.5
f,	OMe	H		173

mixtures of ketones **63** and **64**, although the sulphuric acid method gave mainly (80%) the angular ketone **64**. Compound **62a** could be reduced with hydrazine to the indenoquinoline, which could be re-oxidized by air.[51]

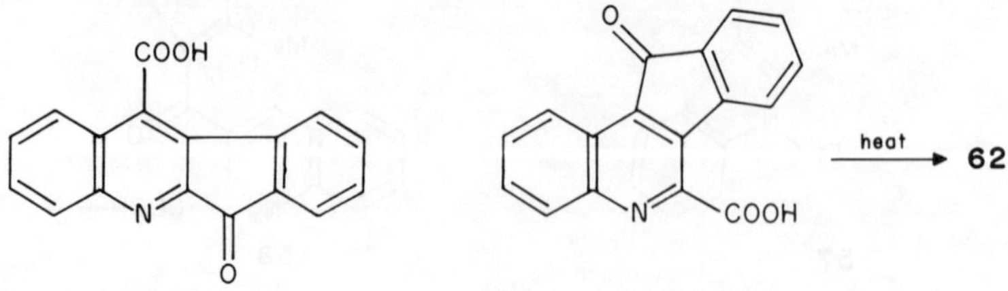

63, m.p. 313 °C (d)   **64**, m.p. 185 °C (d)
                                 Me ester, m.p. 206–207 °C

When air was was passed through a boiling solution of the fused oxepinone **65** in methanol, the dione **66** was formed.[54]

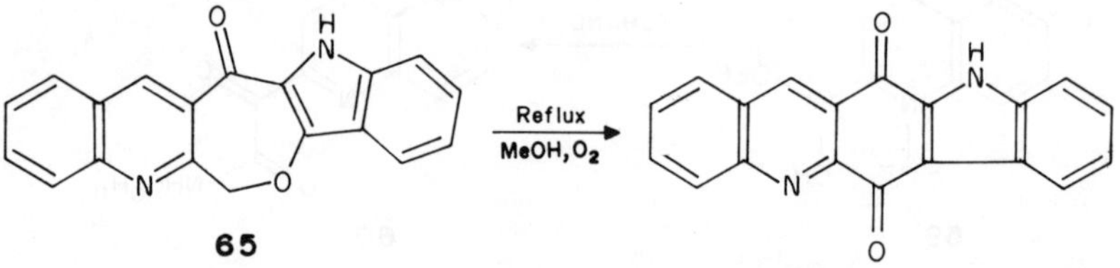

65

66, 75%, m.p. 153–158 °C

3. 3-Ketoquinolines

Friedländer or Pfitzinger procedures are usually used to prepare 3-ketoquinolines. These are summarized for the pentane-2,4-dione reactions in Scheme 12. The unstable 2-aminobenzaldehyde **67**, R = H was prepared *in situ* from the isocyanate **69** and treated with pentane-2,4-dione to give the ketone

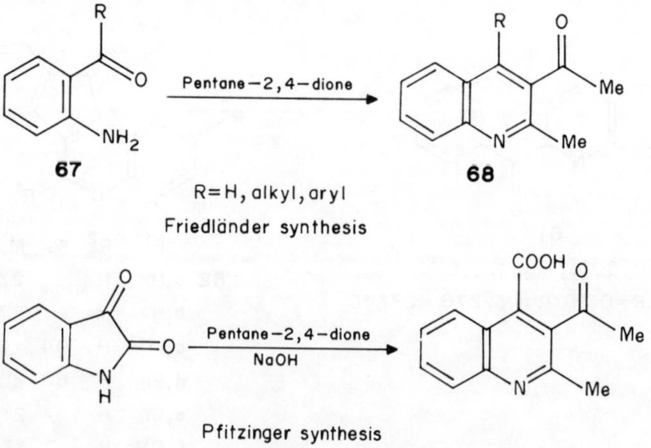

67   **68**

R=H, alkyl, aryl

Friedländer synthesis

Pfitzinger synthesis

SCHEME 12

68, R = H.[55] Direct reaction of the aminoaldehyde **67**, R = H with pentane-2,4-dione was observed[56] to give a coloured by-product. A recent reinvestigation showed that, in addition to the ketone **68**, R = H, the reaction gave 2-methylquinoline and the yellow derivative **70**; Scheme 13.[57]

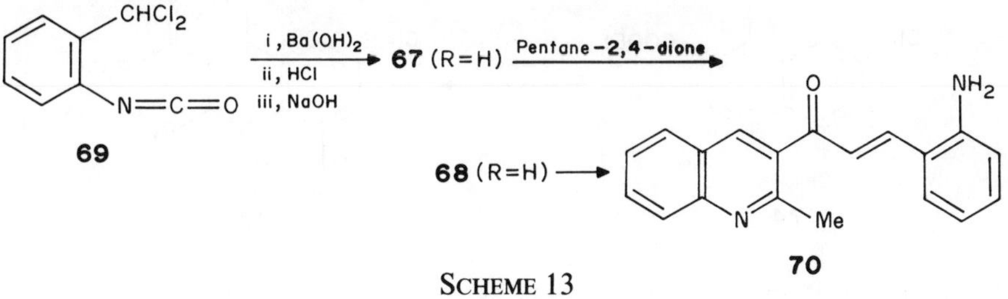

SCHEME 13

The anil **71** reacted in a similar way to the parent aldehyde with 1,3-diones to give ketones **72**.[50]

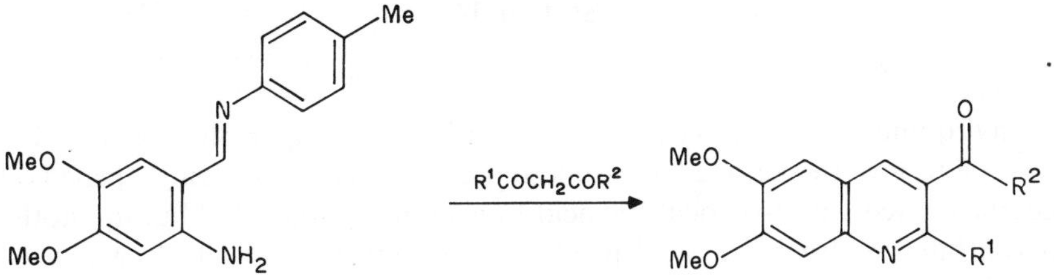

Improved yields from the Pfitzinger reaction were obtained when the preformed potassium salt of isatin was used instead of a potassium hydroxide solution. The acids produced can be decarboxylated if necessary; see Table 2.[58] Cyclohexane-1, 3-diones in the Friedländer synthesis gave the ketones **73a, b, c**[18] and, with potassium isatinate, the ketones **73d, e**.[58] In the latter two examples 3,3-disubstituted indolinones were by-products.

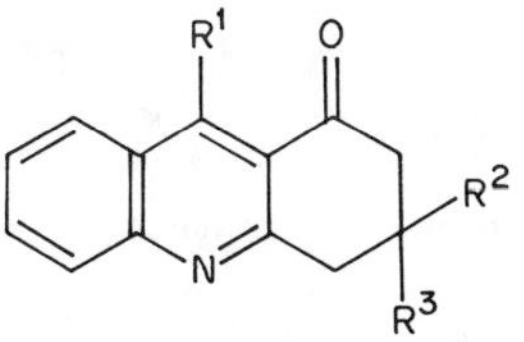

73

	R¹	R²	R³	%	M.p. (°C)
a,	Me	H	H	90	68
b,	Ph	H	H	85	156–157
c,	Ph	Me	Me	86	190–191
d,	COOH	Me	H	15.7	258–259
e,	COOH	Me	Me	26	292

Several derivatives of ethyl quinoline-3-carboxylate reacted with one molecule of 2-pyridyllithium to give the derived ketones, but the ester **74** accepted three molecules of the reagent to give the ketone **75**, as shown in Scheme 14.[59]

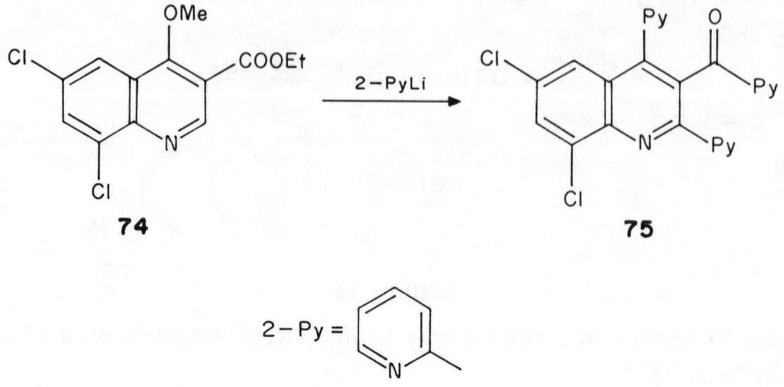

SCHEME 14

Ethyl quinoline-3-carboxylate reacted with *N*-methylpyrrolidone in sodium ethoxide to give the ketone **76**. In one report this ketone was ring opened and decarboxylated with hydrochloric acid to give the ketone **77**,[60] but in another, concentrated hydrochloric acid produced the pyrrolinium salt **78**, which was hydrolysed to the ketone **77** by sodium bicarbonate.[61]

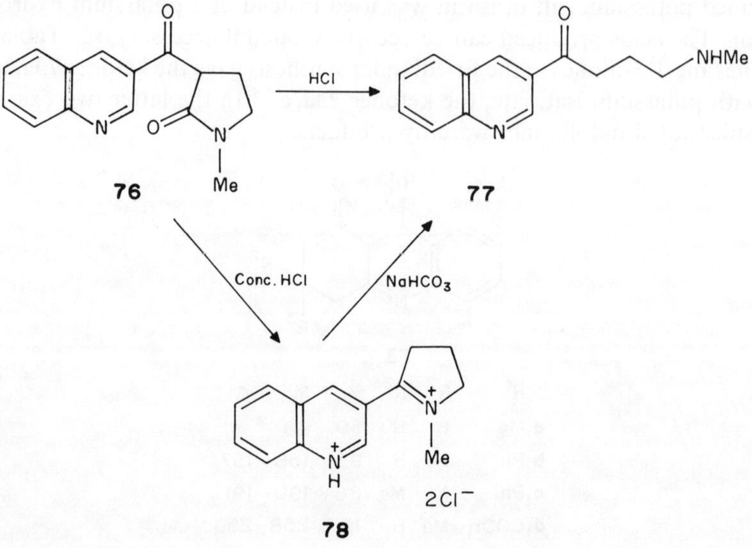

Thionyl chloride converted the quinolinone acid **79** to the 2-chloroquinoline acid chloride which cyclized to the angular ketone **80a** under Friedel–Crafts conditions.[50] When the mesitylene derivative **81** was heated strongly in polyphosphoric acid it gave the ketone **80b**.[62]

80 a, R = Cl, m.p. 215 – 217 °C
b, R = H, 55.7%, m.p. 223 – 224 °C

The indanedione derivative **82** is reported to react with urea in the presence of copper powder and cupric chloride at 240–250 °C to give the linear ketone **83**.[63]

Treatment of the *N*-oxide **84** with acetic anhydride gave the expected rearrangement to compound **85a**. Careful hydrolysis (KOH) produced the 2-hydroxymethyl ketone **85b**, which was oxidized to the keto-aldehyde **86**. An internal Cannizzaro reaction was assumed to occur whereby the keto-aldehyde **86** was converted to the lactone **87** in methanolic potassium hydroxide. The *N*-oxide **84** was converted directly to the lactone **87** by base. It was suggested that this reaction also went via the intermediate **86**.[64]

84

85 a, R = Ac
   b, R = H

87   86

Reduction (SnCl$_2$ or H$_2$, Pd) of the nitro compounds 88, R = Me, Ph gave mixtures of the quinolyl ketones 89 and their N-oxides. The N-oxides were reduced to the required ketones with phosphorus trichloride.[65]

88   89

Treatment of isatin with base followed by acid and 1-methoxybut-1-en-3-one gave the enaminone 90, which with acid or base cyclized to the ketonic acid 91. Further reactions of this keto-acid to give keto-amides and a keto-ester via pseudoanhydrides are shown in Scheme 15.[66–68]

The enaminediones 92 cyclized in hot polyphosphoric acid to the ketones 93.[69]

When 3-acetyl-1,2-dihydroquinoline-2-carboxylic acid was pyrolysed at 200 °C it gave 3-acetylquinoline (36%) along with 3-acetyl-1,4-dihydroquinoline (8%) and 3-acetyl-1,2,3,4-tetrahydroquinoline (13%).[70]

Some 3-ketoquinolines have been methylated at position 4 by the route shown in Scheme 16.[71] Recently, 2-chloro-3-cyanoquinolines 94 were shown to react with aryl Grignard reagents as in Scheme 17 to give 4-arylquinoline imines 95a, which were hydrolysed to ketones 95b. Yields were low, as several non-ketonic products were always obtained; see Scheme 17, Table 2.[72]

The ketone 96 was prepared from the parent heterocycle with lead tetraacetate.[73] It was methylated to give the 5,12-dihydro-12-methyl derivative (see Section X.4), which was aromatized to ketone 97.[71]

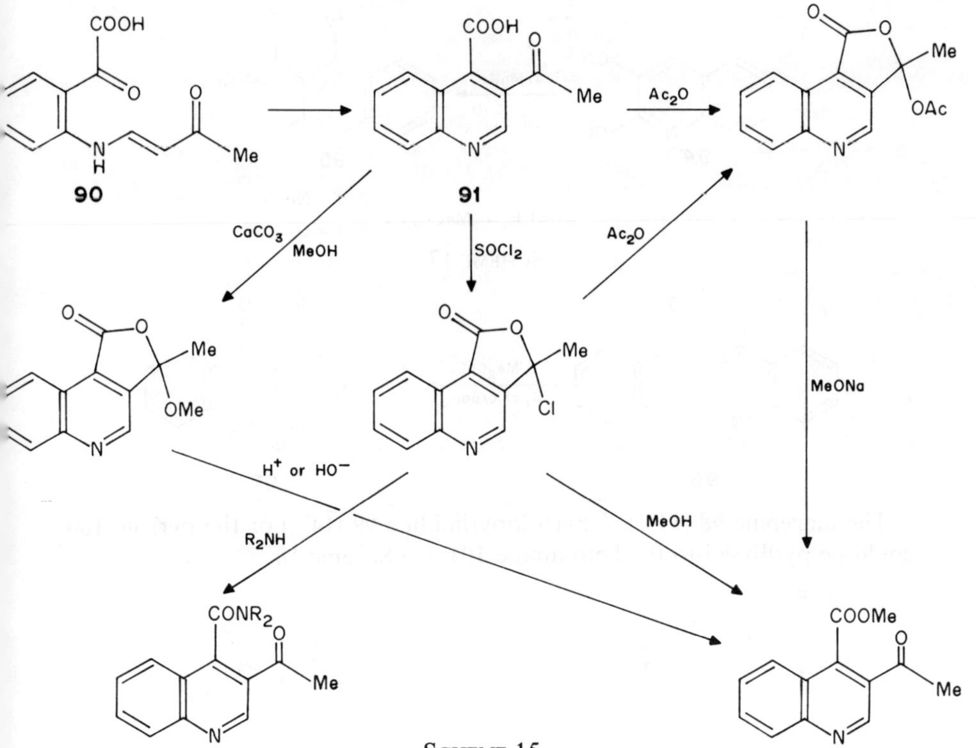

SCHEME 15

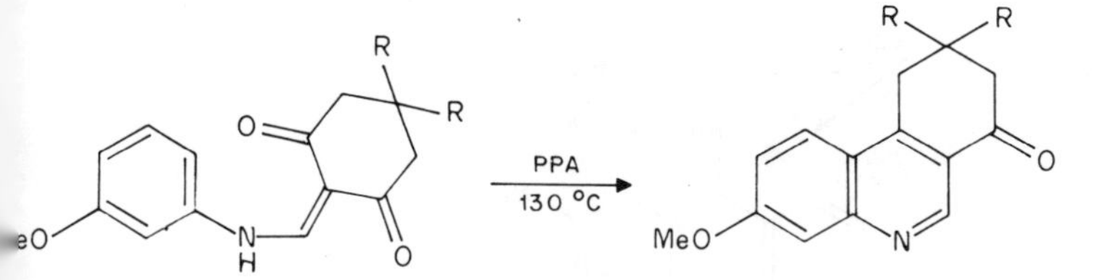

92

93

R = H, 70%, m.p. 147 °C

R = Me, 76%, m.p. 153 °C

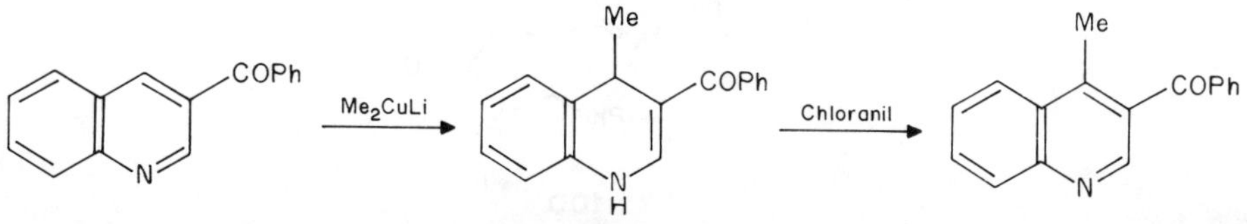

SCHEME 16

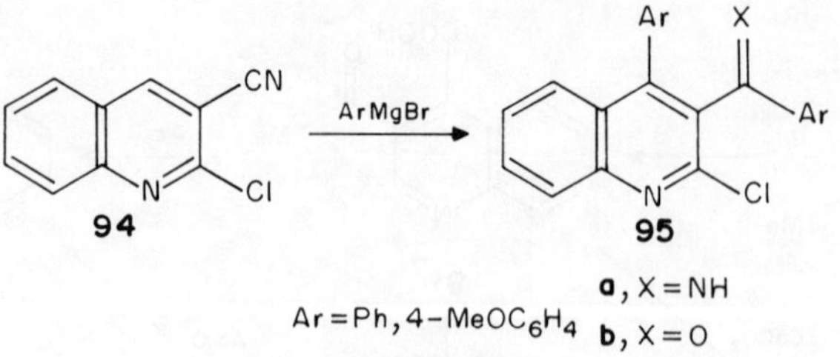

SCHEME 17

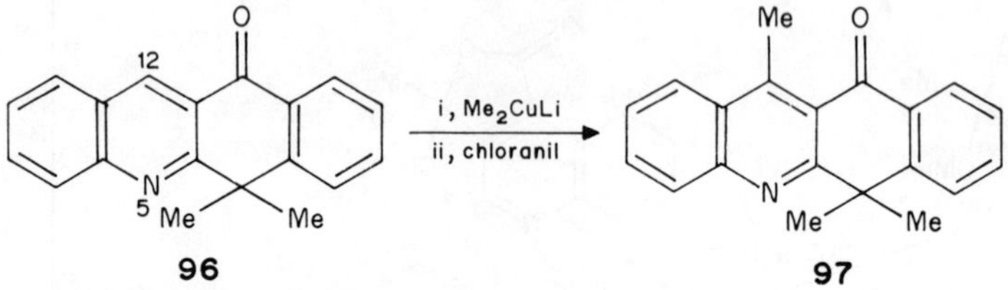

The diazepine **98** (11%),[75] pyrrolopyrimidine **99** (14%) or the pyrrole **100**[76] could be pyrolysed to the keto-amide **101**; see Scheme 18.[75]

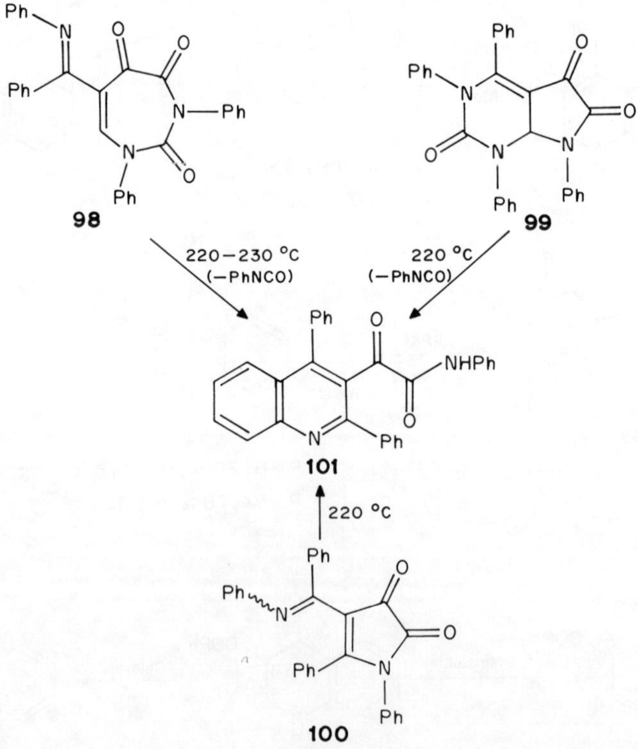

SCHEME 18

When the 2-chloro ketone **102**, R = Me was treated with methyl thioglycolate it gave the expected thio ether **103**. Under the same conditions, the ketone **102**, R = Ph gave the thienoquinoline **104**.[77]

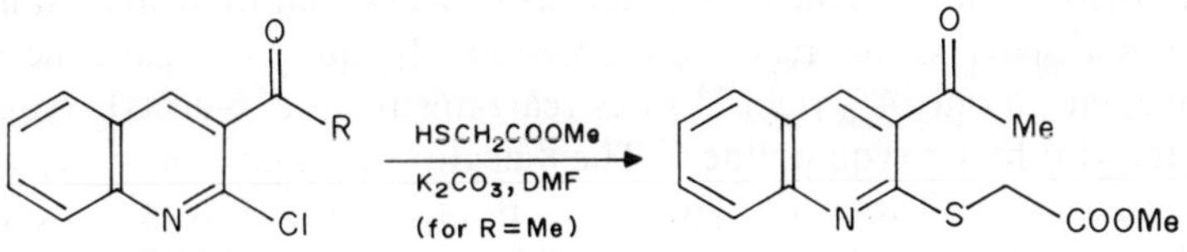

102 $\xrightarrow[\text{(for R = Me)}]{\substack{\text{HSCH}_2\text{COOMe} \\ \text{K}_2\text{CO}_3,\,\text{DMF}}}$ **103**

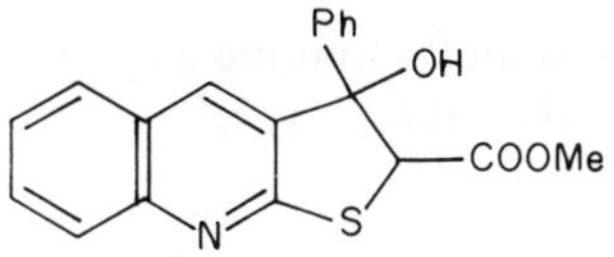

104

4. 5-Ketoquinolines

Many 5-ketoquinolines have been made by Friedel–Crafts reactions, particuarly on 8-hydroxyquinoline. The related Fries rearrangement of esters has also been employed; see Table 12. Generally it has been shown—or assumed—that the 7-ketone is not formed unless the 5-position is blocked, e.g. 5-ethyl--8-hydroxyquinoline reacts with diacid chlorides (AlCl$_3$, CCl$_4$, 70 °C) to give the diketones **105**.[78]

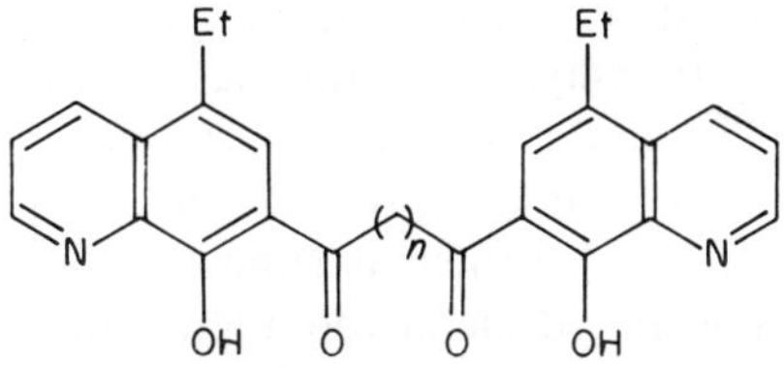

105

$n = 4$, 69%, m.p. 219–219.5 °C

$n = 6$, 69.4%, m.p. 184.6–185.2 °C

$n = 8$, 70%, m.p. 136–136.5 °C

A few papers have reported the isolation of small amounts of 7-keto-8-hydroxyquinolines. Pelargonoyl (nonanoyl) chloride gave both the 5- and 7-ketones with aluminium chloride in nitrobenzene at 70–80 °C.[79] Osadchii and co-workers[80] claimed to get only the 7-ketone (20%) from 8-hydroxyquinoline and heptanoyl chloride, although in another paper[81] they report the 5-ketone. Manecke and Aurich[82] treated 2-methyl-8-hydroxyquinoline with acetyl chloride and aluminium chloride to obtain a mixture of the 5-acetyl (70%) and 7-acetyl (5%) derivatives. It is reported[83] that 6-methyl-, 8-methyl- and 6,8-dimethyl-quinolines failed to react with acetyl chloride and aluminium chloride. A Friedel–Crafts reaction on 5,7-dimethylquinoline gave the 6-acetyl derivative, but in only 4% yield.[83] Fries rearrangement of 6-acetyloxyquinoline gave 5-acetyl-6-hydroxyquinoline.[84] The 8-acyloxyquinolines also rearrange to the 5-ketones; see Table 12. However, Borsche and Groth[83] noted that 8-acetyloxyquinoline with aluminium chloride in nitrobenzene gave only 8-hydroxyquinoline. A photo-Fries rearrangement of 8-benzoyloxyquinoline gave a mixture of 5-benzoyl-8-hydroxyquinoline (15%) and 7-benzoyl-8-hydroxyquinoline (20%).[85]

A Skraup reaction on 2-aminofluorenone gave a single product, m.p. 188 °C, which was formulated as either 106 or 107.[86]

106

107

Mannich reactions on 8-hydroxy-5-ketoquinolines introduce aminomethyl substituents at position 7; see Table 12. The reaction with formaldehyde and an unusual Mannich reaction are shown in Scheme 19.[87,88]

Condensation of 5-acetyl-8-hydroxyquinoline or its benzoate with formalin gave diketones 108.[89] The first of these condensations was repeated—39 years later by the same author!—in the presence of sodium hydroxide to give compound 109, which reacted further as shown in Scheme 20. The acetates were prepared with acetic anhydride at room temperature.[90]

Recently, 5-chloroacetyl-8-hydroxyquinoline was shown to undergo a Darzens reaction with benzaldehyde in sodium hydroxide to give the trans-oxirane 110.[91] This corrected a structure previously given in patents.[88] The oxirane could also be produced from the chalcone 111. Compound 110 was reduced back to the chalcone 111 by 3-methyl-2-selenoxobenzothiazole in trifluoroacetic acid.[91] Attempts to prepare hydrochlorides with hydrogen chloride in dichloromethane invariably gave ring opened derivatives such as the chloroketone 112a. When ethanol was used as a solvent, a mixture of the chloroketones 112a (55%) and 112b (45%) was obtained.[92] For this reason Mannich bases were prepared in the absence of acid; see Table 12.[91] Eschenmoser

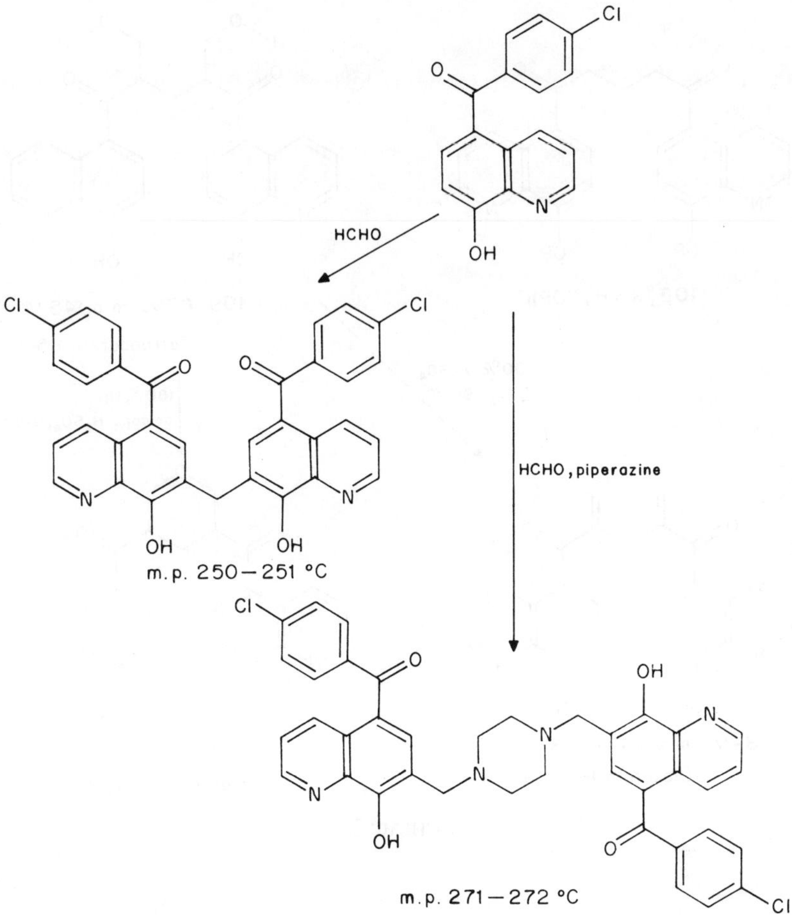

SCHEME 19

salts also gave Mannich bases without epoxide ring opening.[92] Both the oxirane
110 and its Mannich bases 113 gave complexes 114 (Scheme 21).[91]

Irradiation of the oxirane 110 gave the diketone 115, which was identical
with the compound prepared by Claisen condensation of methyl benzoate with
5-acetyl-8-hydroxyquinoline. The dione 115 gave complex 116 or 117 depending
on reactant proportions. Treatment with Eschenmoser salt gave a mixture of
the Mannich base 118 and its derived alkene 119 (Scheme 22). Only with 2 mol
of Eschenmoser salt was the 7-position attacked to give a di-Mannich base,
again accompanied by its vinyl derivative; see Table 12. All attempts to carry
out conventional Mannich reactions on the dione 115 gave the methylenebis
derivative 120.[93]

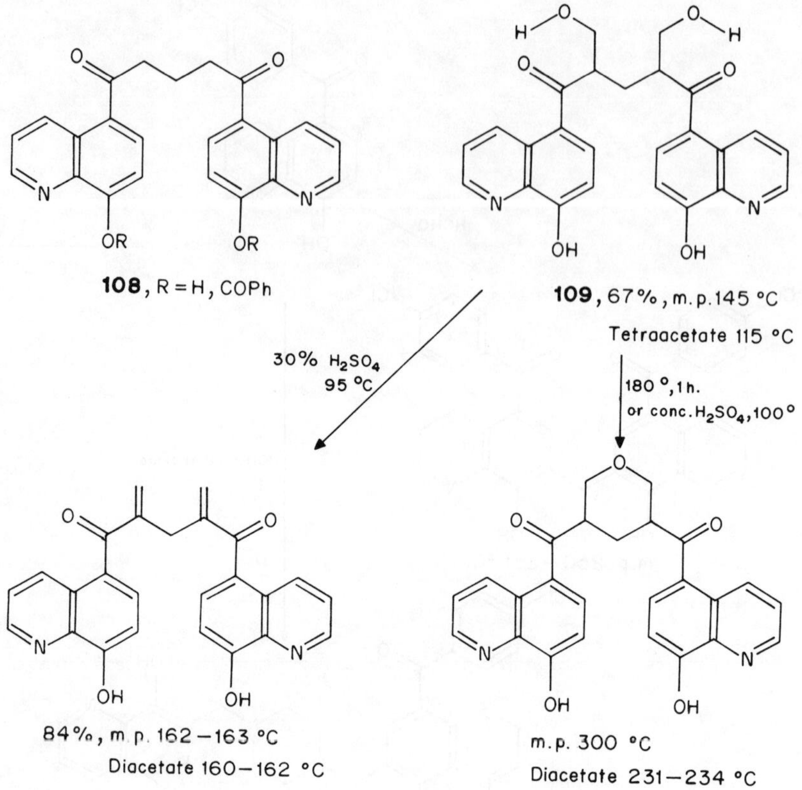
108, R = H, COPh
109, 67%, m.p. 145 °C
Tetraacetate 115 °C
30% H_2SO_4
95 °C
180 °, 1 h.
or conc. H_2SO_4, 100°
84%, m.p. 162–163 °C
Diacetate 160–162 °C
m.p. 300 °C
Diacetate 231–234 °C

SCHEME 20

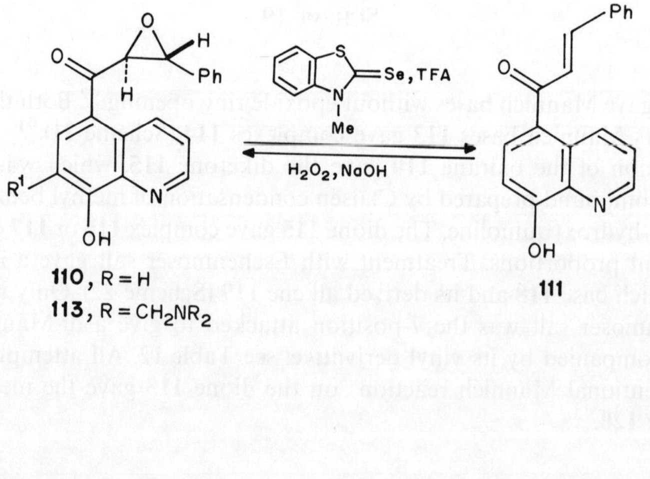
Se, TFA
Me
H_2O_2, NaOH
110, R = H
113, R = CH_2NR_2
111

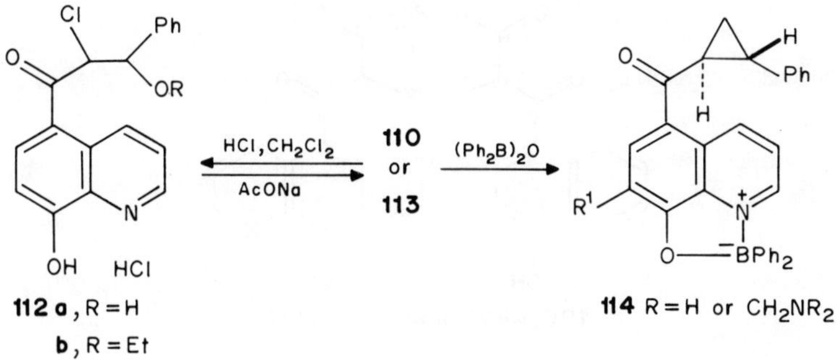

SCHEME 21

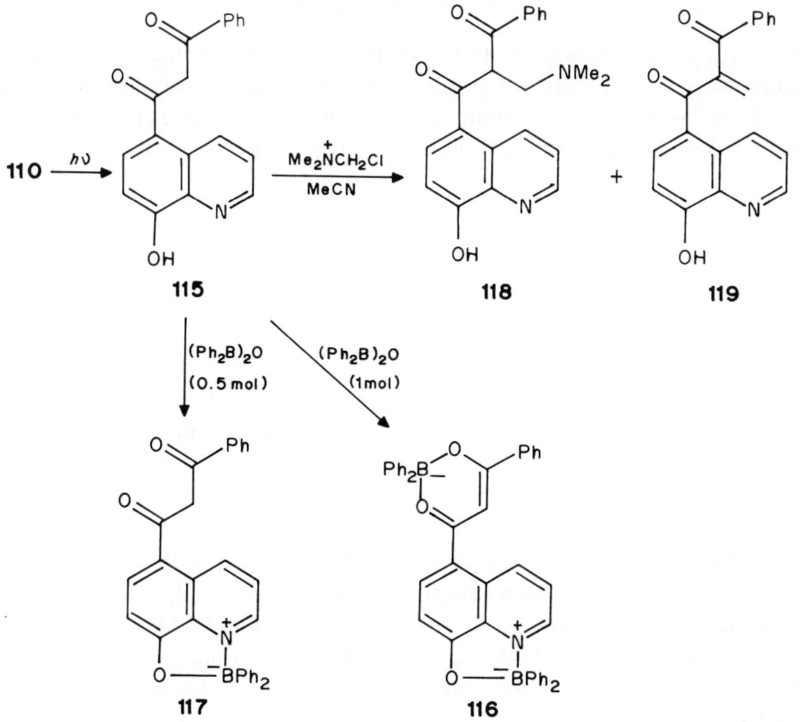

SCHEME 22

120, 48%, m.p. 192–197 °C

5. 6-Ketoquinolines

Ethyl groups at positions 6 and 8 of the quinoline nucleus are oxidized in preference to pyridine ring alkyl substituents by a limited amount of chromium trioxide or potassium dichromate in 3M-sulphuric acid, e.g. **121** to **122**. In some cases the ketone was obtained mixed with the carboxylic acid.[94]

The diazonium salt from 6-aminoquinoline reacted with propionaldehyde oxime in aqueous sodium sulphite and copper sulphate to give 6-propionylquinoline in 60% yield.[95] Compounds **123** and **124** were prepared by Skraup reactions from the appropriate diamino-diketones.[96]

With aldehydes and pyruvic acid in refluxing ethanol, 4-aminoacetophenones react to give 6-acetylquinoline-4-carboxylic acids, e.g. Scheme 23.[97]

The sulphoxide **125** reacted with chalcone under basic catalysis to give the ketone **127a**. The presumed intermediate was compound **126**. When the sulphoxide was replaced by its equivalent sulphone the isolated product was the sulphone derived from **126**; see Section XI. However, the same sulphone starting material and 4-nitrochalcone gave the aromatic product **127b**; see Table 16.[98]

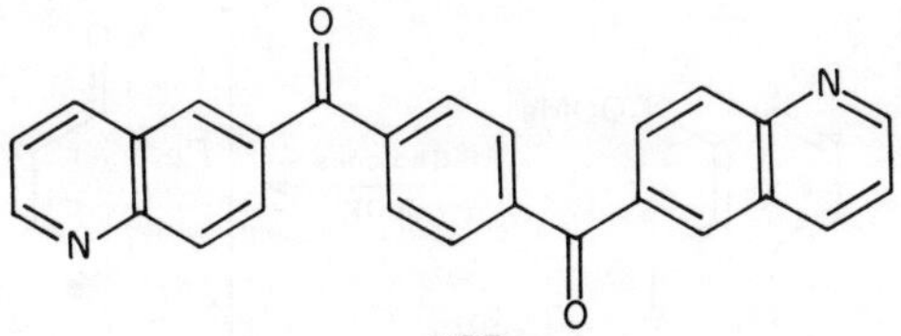

123

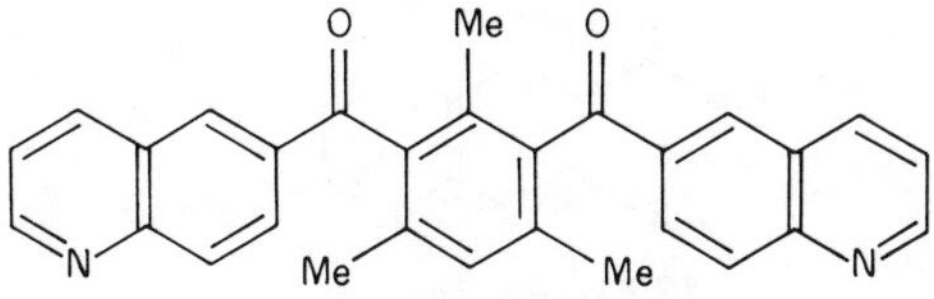

124

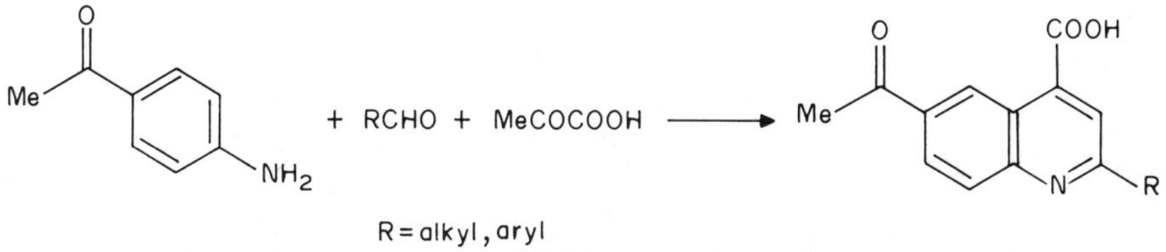

R = alkyl, aryl

SCHEME 23

6. 7-Ketoquinolines

In the Skraup reaction shown in Scheme 24 the anilide was deacetylated by the acid conditions. Reduction of the nitro group was achieved by two routes, as shown.[99]

The benzisoxazole **128** reacted with ethyl polyphosphate to give the quinolyl ketone **129**.[100]

Note also compounds **105** and **107**, Section II.4.

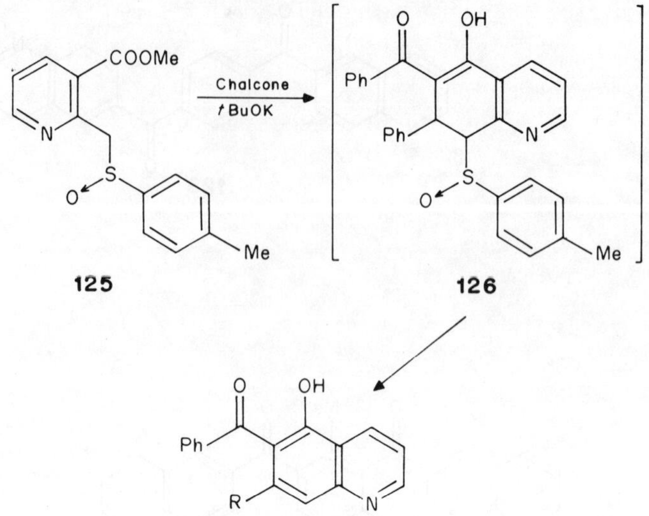

SCHEME 24

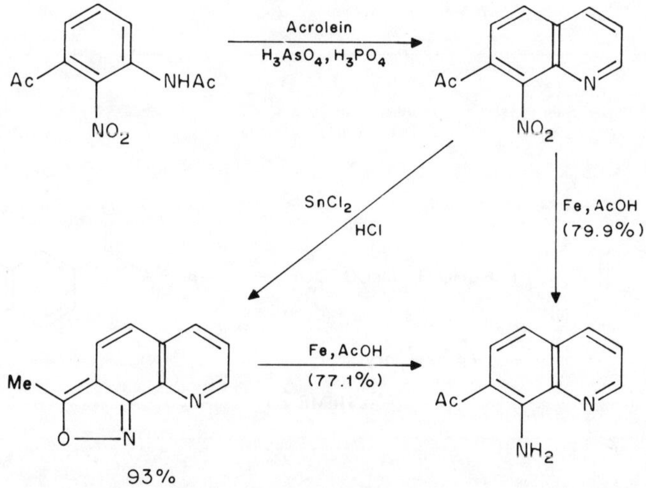

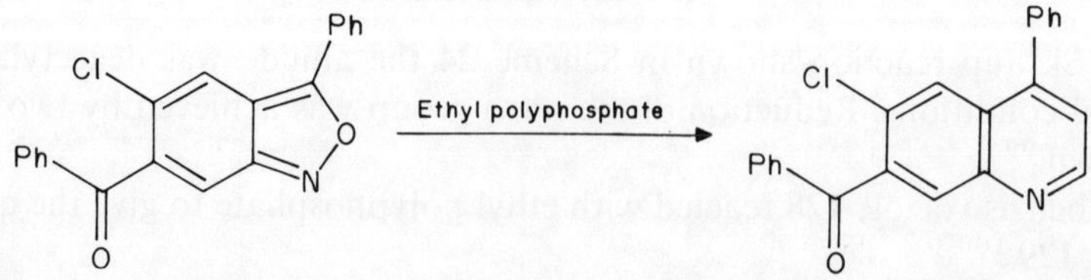

7. 8-Ketoquinolines

Quinoline-8-carboxaldehyde was converted into the complexes **130** and **131**, both of which were decarbonylated to quinoline on heating. However, when the complex **131** was warmed with 1-octene it gave a useful yield (55%) of the octyl ketone **132**.[101]

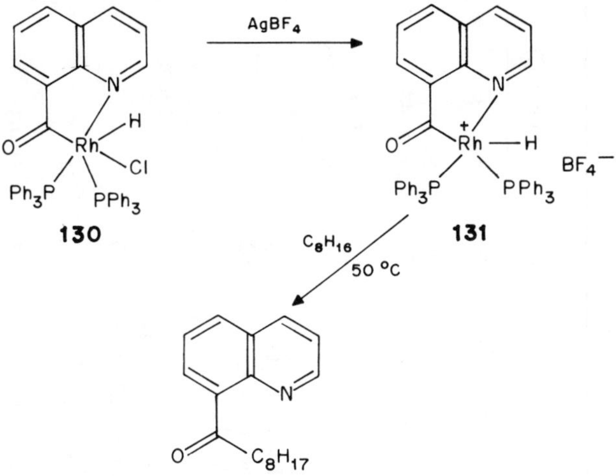

Two pinacol rearrangements run under the same conditions (AcCl, AcOH, C_6H_6, reflux) showed, in the first example, a quinoline ring migration and in the second, a phenyl group migration; Scheme 25.[102]

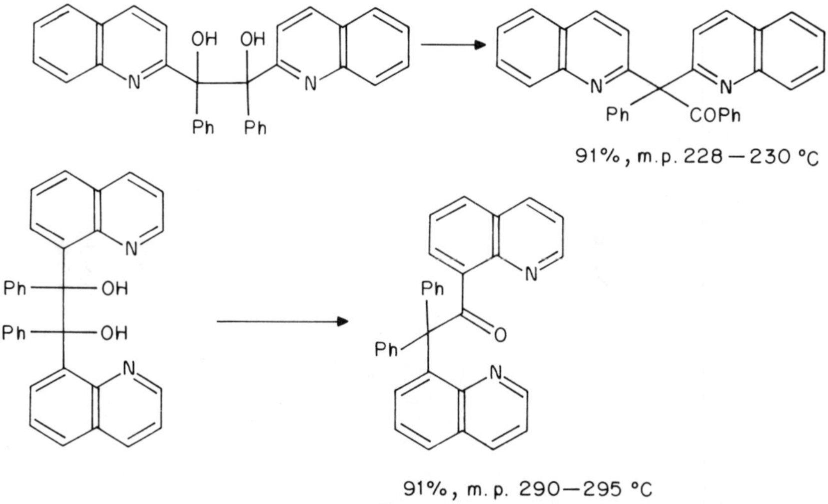

91%, m.p. 228−230 °C

91%, m.p. 290−295 °C

SCHEME 25

Table 1. 2-Quinolyl Ketones

R	Quinoline substituent(s)	Preparation	Yield (%)	M.p. (°C)	References
Me	None	Q-2-C(NH$_2$)=CHCONH$_2$, H$_2$SO$_4$	84.5	46	102
		Q-2-CN, MeMgI	79	52	104*, 105*
		Q-2-COCH$_2$COOEt, aq. H$_2$SO$_4$	76	47.5–48	106*, 107
		Q-2-C≡CH, HgSO$_4$, H$_2$SO$_4$	75	49–51 b.p. 117–119/1 mm	108
		Q-2-CHO, CH$_2$N$_2$, Et$_2$O	73	53	109
		2-Et-Q, Br$_2$, aq. H$_2$SO$_4$	62	46	9*
		Q-2-SnMe$_3$, AcCl	39	50–52 b.p. 93–95/0.45 mm	1, 2
		1-Ac-2-CN-1,2-diH-Q, NaH	31		25
		Q, MeCOCOMe, AgNO$_3$, Na$_2$S$_2$O$_8$, aq. H$_2$SO$_4$	30		14
		Q-2-C(CN)(COOEt)Me, KOH, 25°C		45	111
Me	3-Me	3-Me-Q-2-CHOHMe, CrO$_3$, HOAc	77	61–63	112
Me	3,8-diMe	3,8-diMe-Q-2-CHO, CH$_2$N$_2$	17	90	113*
Me	4-Me	4-Me-Q, MeCOCOMe or MeCHOHCOMe, AgNO$_3$, Na$_2$S$_2$O$_8$, H$_2$SO$_4$	93		14, 114

Me	6-Me	2-Et-4-Me-Q N-oxide, KCN, MeSO$_2$Cl	7		115
Me	4-Ph	4-Ph-Q-2-COCH$_2$COOEt, aq. H$_2$SO$_4$	50	340–342(d)	116, 117*
Me	4-(5-nitro-2-furyl)	2-Et-4-(5-NO$_2$-2-furyl)-Q, SeO$_2$, dioxane, reflux		205–211	118, 119
Me	4-Cl	4-Cl-Q, MeCHO, (tBuO)$_2$	71	99	16
Me	7-Cl	7-Cl-Q-2-COOEt, EtOAc, EtONa; aq. H$_2$SO$_4$	27	87–87.5	106*
Me	3-OH	See text, Section II.2.D	83	118	22
Me	3-OH-6-Me	See text, Section II.2.D	71	126	22
Me	3-OH-6-Cl	See text, Section II.2.D	55	149	22
Me	3-OAc				22
Me	4-OMe	4-MeO-Q, MeCHO, (tBuO)$_2$, FeSO$_4$		96	15
Me	6-OMe	6-MeO-Q-2-COCH$_2$COOEt, aq. H$_2$SO$_4$	98	97.5–98.5	106*
Me	4-COMe	Q, MeCHO, (tBuO)$_2$, FeSO$_4$	41	69	16
Me	3-COOMe-4-OMe	2-acetyl-3-methoxycarbonyl-4-quinolone, CH$_2$N$_2$	6	70	120
Me	4-COOEt	Q-4-COOEt, MeCHO, (tBuO)$_2$, FeSO$_4$		83	15
Me	4-CN	Q-4-CN, MeCHO, (tBuO)$_2$	93	168	16
Ch$_2$Ph		Q-2-COCH(CN)Ph, aq. H$_2$SO$_4$	100	78[a]	104
Me		Q-2-COCH(CN)Ph, H$_2$SO$_4$, AcOH, H$_2$O, 120 °C	90	78[a]	105*
		Q-2-CH=SPh$_2$, PhCHO, LDA, see text, Section II.2.G	65		46

Table 1. (*Contd.*)

R	Quinoline substituent(s)	Preparation	Yield (%)	M.p. (°C)	References
CH_2Ph		Q-2-CN, $PhCH_2CN$, Bu_2NMgBr, aq. H_2SO_4	31.9	81–83[a]	5
		Q-2-CN, $PhCH_2MgCl$	Trace		104
		Q-2-COOH, PhCOOMe, boil			8*
		Q-2-COOEt, $PhCH_2COOEt$, EtONa; HCl		110–112[a]	107
	4-COOH	2-Ph-C≡C-Q, 65% H_2SO_4		116[a]	121, 122
		2-Ph-C≡C-Q-4-COOH, 65% H_2SO_4		313	121, 122
$CH_2C_6H_4$-4-Cl		2-(4-ClC_6H_4)-C≡C-Q, 65% H_2SO_4		140.2–141	123
$CH(Br)C_6H_4$-4-Cl		Q-2-$COCH_2C_6H_4$-4-Cl, Br_2		HBr 103–105	123
$CH(Ph)NC_5H_{10}$		Q-2-$COCH_2Ph$, Br_2, piperidine		HCl 223–223.5	124
CH(CN)Ph		Q-2-COOEt, $PhCH_2CN$, EtONa		100	105
CH_2-2-Q		Q-2-COOMe, Q-2-Me, PhLi	87	212–213	4
		(Q-2-CHOH)$_2$, Py, HCl	85	221	125*
		Q-2-C≡C-2-Q, 65% H_2SO_4	83	213.5–215	126*
CH_2-2-(Q-3-Me)		(3-Me-Q-2-CHOH)$_2$, Py, HCl	47	235	449
CH_2-1-isoquinolyl		1-Me-isoquinoline, Q-2-COOMe, PhLi	53	169–170	4
CH_2Cl	5,7-diCl-8-OH	5,7-diCl-8-$PhCH_2$-O-Q-2-$COCH_2Br$, conc. HCl	35	242–243	127
CH_2Cl	7-Cl	7-Cl-Q-2-COOH, $SOCl_2$; CH_2N_2; HCl	45	HCl 127–129	106

	Ring substituents	Conditions	Yield (%)	mp (°C)	Ref.
CH$_2$Br		Q-2-Ac, Br$_2$, aq. HBr	85	81–82	106*
CH$_2$Br	4-(4-ClC$_6$H$_4$)-6-Ph	The diazoketone, 45% HBr	73	142–143	128
CH$_2$Br	4-(4-ClC$_6$H$_4$)-6-(4-FC$_6$H$_4$)	The diazoketone, 45% HBr	74	180–182	128
CH$_2$Br	4-(4-ClC$_6$H$_4$)-6,8-diCl	The diazoketone, 45% HBr	85	182–183	128
CH$_2$Br	4-(4-ClC$_6$H$_4$)-6-MeO-7-Cl	The diazoketone, 45% HBr	71	178–179	128
CH$_2$Br	5,7-diCl-8-OCH$_2$Ph	5,7-diCl-8-PhCH$_2$O-Q-2-COCHN$_2$, 47% HBr	50	157	127
CH$_2$Br	6,8-diCl	The diazoketone, 45% HBr	72	175–176	128
CH$_2$Br	7-Cl	7-Cl-Q-2-COOH, SOCl$_2$; CH$_2$N$_2$; HBr	65	117–118	106*
CH$_2$Br	4-OMe-6,8-diCl	The diazoketone, 45% HBr	75	204–205	128
CH$_2$Br	4,6-diOMe-7-Cl	The diazoketone, 45% HBr	70	218–220	128
CH$_2$Br	6-OMe	6-MeO-Q-2-COMe, Br$_2$, aq. HBr	Trace	122	106
CH$_2$Br	6-OMe-7-Cl	The diazoketone, 45% HBr	81	178–179	128
CH$_2$N$_2$	4-(4-ClC$_6$H$_4$)-6-Ph	The acid chloride, CH$_2$N$_2$	75	159–160	128
CH$_2$N$_2$	4-(4-ClC$_6$H$_4$)-6-(4-FC$_6$H$_4$)	The acid chloride, CH$_2$N$_2$	84	174–176	128
CH$_2$N$_2$	4-(4-ClC$_6$H$_4$)-6,8-diCl	The acid chloride, CH$_2$N$_2$	82	192–194	128
CH$_2$N$_2$	4-(4-ClC$_6$H$_4$)-6-MeO-7-Cl	The acid chloride, CH$_2$N$_2$	81	203–205	128
CH$_2$N$_2$	5,7-diCl-8-OCH$_2$Ph	The acid chloride, CH$_2$N$_2$	80	139	127
CH$_2$N$_2$	6,8-diCl	The acid chloride, CH$_2$N$_2$	77	183–185	106, 128
CH$_2$N$_2$	4-OMe-6,8-diCl	The acid chloride, CH$_2$N$_2$	86	198–199	128
CH$_2$N$_2$	4,6-diOMe-7-Cl	The acid chloride, CH$_2$N$_2$	83	163–165	128
CH$_2$N$_2$	6-OMe-7-Cl	The acid chloride, CH$_2$N$_2$	76	174–176	128
CH$_2$NH$_2$		Q-2-C(Me)=N-OSO$_2$C$_6$H$_4$-4-Me, EtOK	41	HCl 216(d)	129
CH$_2$NHMe		Q-2-CONMe$_2$, hv	70^b	Oil	33
CH$_2$NHMe	4-Me	4-Me-Q-2-CONMe$_2$, hv	61^b	80–81	33
CH$_2$NHMe	4-Cl	4-Cl-Q-2-CONMe$_2$, hv	6^b	Oil	33
CH$_2$NEt$_2$		Q-2-COCH$_2$Br, Et$_2$NH			106
CH$_2$N(nPr)$_2$		Q-2-COCH$_2$Br, nPr$_2$NH			106
CH$_2$N(iBu)$_2$		Q-2-COCH$_2$Br, iBu$_2$NH			106
CH$_2$N(Et)nBu		Q-2-COCH$_2$Br, nBuNHEt			106

Table 1. (*Contd.*)

R	Quinoline substituent(s)	Preparation	Yield (%)	M.p. (°C)	References
CH$_2$CN	4-COCH$_2$CN-6-NHCOC$_{17}$H$_{35}$	6-C$_{17}$H$_{35}$CONH-Q-2,4-diCOOMe, MeCN, KOH, H$_2$O, 90°C			130
COPh		Q-2-CHBr-COPh, DMSO	85	115–116	131
C(NNH$_2$)Ph		Q-2-COCOPh, N$_2$H$_4$·H$_2$O	61^c	174	131
COOsBu		2-I-Q, sBuOH, Et$_3$N, CO, PdCl$_2$[P(cycloC$_6$H$_{11}$)$_3$]$_2$	22^d		132
CONHcycloC$_6$H$_{11}$		2-CH(OH)OEt-Q N-oxide, cycloC$_6$H$_{11}$NC See text, Section II.2.G			49
		Q-1-oxide-2-CHO, EtOH; cycloC$_6$H$_{11}$NC See text, Section II.2.G			49
Et		Q-2-CN, EtMgBr or EtMgI	68	59–60	104*, 105* 133
		Q-2-SnMe$_3$, EtCOCl	64	b.p. 120–121/ 1 mm	2
		Q-2-CHO, MeCHN$_2$	52	59–60	134
		Q-2-COCl, MeCH(COOH)$_2$, dihydropyran, H$_2$SO$_4$, Py	80	60	7
		Q-2-CH=SPh$_2$, MeCHO, LDA	57		46
Et	3,8-diMe	3,8-diMe-Q-2-CHO, MeCHN$_2$		80	113
Et	3-OH-4-Me	See text, Section II.2.D			
Et	4-COEt	Q, EtCHO, (tBuO)$_2$, FeSO$_4$	82	82	22
2-(3-Piperidyl)ethyl		Q-2-COOEt, ethyl 3-(1-benzoyl-3-piperidyl)-propionate, KH; HCl	40	HCl 198	16 135

Side chain		Reagents; conditions	Yield (%)	mp (°C)	Ref.
2-(4-Piperidyl)ethyl		Q-2-COOEt, ethyl 3-(1-benzoyl-4-piperidyl)-propionate, EtOK; 8M-HCl	71.5	105–107	136
2-(1-Bromo-4-piperidyl)ethyl		Above, Br$_2$, NaOH		234–236	136
2-(2-Quinolyl)ethyl		Q-2-CHO, CH$_2$N$_2$		195	109
2-HO-2-(2-quinolyl)-ethyl^f		Q-2-CHO, Q-2-Ac, NaOH		164	137
CH=CHPh	4-Me	4-Me-Q-2-CHO, PhC≡CMgBr^e			110
CH=CH-2-Q^f		Q-2-CHO, Q-2-Ac, NaOH	94	138	137
CH=CH-2-furyl		2-Ac-Q, C$_4$H$_3$O-2-CHO, KOH		110–112	444
CH$_2$COC$_6$H$_4$-2-OH		Q-2-COCl, 2-HOC$_6$H$_4$COMe, LDA	80	147–148	32
		Q-2-COOC$_6$H$_4$-2-COMe, KOH, Py	33	148–149	29
CH$_2$COC$_6$H$_3$-2-OH-3-F		Q-2-COOC$_6$H$_3$-2-COMe-3-F, acid	32	164–165	30
CH$_2$COC$_6$H$_3$-2-OH-3-Br		Q-2-COOC$_6$H$_3$-2-COMe-3-Br, acid	40	164–165	30
CH$_2$COC$_6$H$_3$-2-OH-4-F		Q-2-COOC$_6$H$_3$-2-COMe-4-F, acid	33	156–157	30
CH$_2$COC$_6$H$_3$-2-OH-4-Cl		Q-2-COOC$_6$H$_3$-2-COMe-4-Cl, acid	23	148–149	30
CH$_2$COC$_6$H$_3$-2-OH-5-Cl		Q-2-COOC$_6$H$_3$-2-COMe-5-Cl, Py, 25°C	54	160–162	31
CH$_2$COC$_6$H$_3$-2-OH-4-Br		Q-2-COOC$_6$H$_3$-2-COMe-4-Br, acid	28	156–158	30
CH$_2$COC$_6$H$_2$-2-OH-3,4-diCl		Q-2-COOC$_6$H$_2$-2-COMe-3,4-diCl, acid	41	189–191	30
CH$_2$COC$_6$H$_2$-2-OH-3,4-diOMe		Q-2-COOC$_6$H$_2$-2-COMe-3,4-diOMe, Py, 25°C	60	123–124	31

Quinoline Ketones

Table 1. (Contd.)

R	Quinoline substituent(s)	Preparation	Yield (%)	M.p. (°C)	References
CH₂COOEt		Q-2-COOEt, EtOAc, EtONa	90	63–64	106
CH₂COOEt	4-Ph	4-Ph-Q-2-COOEt, EtOAc, EtONa			117
CH₂COOEt	6-OMe	6-MeO-Q-2-COOEt, EtOAc, EtONa	80	64–66	106
Pr		Q-2-CH=SPh₂, EtCHO, LDA, see text, Section II.2.G	45		46
		Q-2-CN, nPrMgBr	24.8	41–42 b.p. 146–150/3 mm	5
(CH₂)₃Cl		Q-2-COOMe, γ-butyrolactone, MeONa; HCl	48.2		138
(CH₂)₃OEt		Q-2-CN, EtO(CH₂)₃MgBr	86	b.p. 144–146/0.05 mm	139, 140*, 141*
(CH₂)₃-{1-[4-(4-chlorophenyl)-4-hydroxypiperidinyl]}		Q-2-CO(CH₂)₃-Cl, 4-(4-chlorophenyl)-4-hydroxypiperidine	33.1	125.5–127	138
(CH₂)₃-[1-(4-phenyl piperazinyl)]		Q-2-CO(CH₂)₃-Cl, 4-phenylpiperazine	40.6	142–143	138
CH₂CH₂COOH		See text, Section II.2.B.	72	145	7
		Q-2-CH(COOEt)CH₂COOEt, H₂O	34	144–145	142
CH₂CH₂COOMe		Q-2-CH₂CH₂COOMe, H₂O		147	143
		Q-2-CHO, N₂CHCH₂COOMe	70	71	143*

R	4-Sub	Reagents / Conditions	Yield (%)	b.p. / m.p.	Ref.
iPr		Q-2-SnMe₃, iPrCOCl	83	b.p. 105–108/0.55 mm	2
		Q-2-CH=SPh₂, MeCOMe, LDA; see text, Section II.2.G	51		46
iPr	4-Me	See text, Section II.2.B	81	b.p. 129–130/0.02 mm	144
Bu		Q-2-CN, nBuMgBr	35	42–44	7
				b.p. 178–182/10 mm	5
		Q-2-C(CN)(COOEt)nBu, KOH, 25 °C		35	111
Bu	4-Me	4-Me-Q-2-COiPr, (BuCOO)₂ CuOAc, H₂SO₄, AcOH	30		145
CH=CHCH=CH-C₆H₄-4-NMe₂	4-Me	4-Me-Q, 3-methylcyclopentane-1,2-dione AgNO₃, Na₂S₂O₈, H₂SO₄	60		13, 146
CH₂CH₂COMe		See text, Section II.2.B			14
(CH₂)₃COOH		See text, Section II.2.B	80	135	7
(CH₂)₃COOMe		Q-2-CO(CH₂)₃COOH, H₂O	68	133	143
(CH₂)₃COOEt		Q-2-CHO, N₂CH(CH₂)₂COOMe	75	57	143
		Q-2-COCl, CH(COOtBu)₂-CH₂CH₂COOEt, Na; 4-MeC₆H₄SO₃H		48–49	147
CH(Me)CH₂Me	4-Me	Q-2-CO(CH₂)₃OEt, MeI	85	b.p. 129–131/0.1 mm	144, 148
CH(Me)CH₂CH₂OEt		Q-2-CN, EtOCH₂CH₂CH(Me)MgBr	37	b.p. 136–138/0.03 mm	149

Table 1. (*Contd.*)

R	Quinoline substituent(s)	Preparation	Yield (%)	M.p. (°C)	References
$CH(COOEt)COMe$		Q-2-COCl, EAA, Na, C_6H_6		61	150
$CH(COOEt)CH_2$-$COOEt$		Q-2-COOEt, $(CH_2COOEt)_2$, NaH	63.5		142
tBu		Q-2-$SnMe_3$, tBuCOCl	95	b.p. 97–99/0.25 mm	1, 2
$(CH_2)_4Me$	4-Me	Q-2-COOEt,	45		144
$(CH_2)_5NH$		PhCONH$(CH_2)_5$COOEt, $NaNH_2$; conc. HCl		2HBr 240–241	151
$CHBr(CH_2)_4NH$		Above, Br_2, 48% HBr	97	2HBr 180–200	151
$(CH_2)_4COOH$		Q-2-cycloC_6H_9O, $MeCO_3H$ or $PhCO_3H$	100	107–109	152
		See text, Section II.2.B	74	109	7
$(CH_2)_4COOMe$		Q-2-CO$(CH_2)_4$COOMe, H_2O		110	143
		Q-2-CHO, $N_2CH(CH_2)_3COOMe$	71	61	143*
$CH(Me)CH_2CH_2Me$	4-Me	4-Me-Q-2-COiPr,[MeCH$_2$-CH$_2$CH(Me)COO]$_2$, CuOAc, H_2SO_4, AcOH	35		145
$C(Me)_2Et$		Q-2-COOMe, Et$(Me)_2$CMgCl; CrO_3, AcOH		b.p. 140/0.2 mm	153
$(CH_2)_5COOH$		Q-2-CO$(CH_2)_5$COOMe, H_2O		87	143
$(CH_2)_5COOMe$		Q-2-CHO, $N_2CH(CH_2)_4COOMe$	68	46	143*
$CH=CHCH=CH$-$CH=CHPh$					154

			Yield %	m.p. (°C)	Ref.
nC$_8$H$_{17}$	4-Me-5,8-diOMe	4-Me-5,8-diMeO-Q-2-CN, nC$_8$H$_{17}$MgBr	65	76–79	155
cycloC$_6$H$_{11}$		Q-2-SnMe$_3$, cycloC$_6$H$_{11}$COCl	76	88–90	1
				b.p. 154–156/0.5 mm	46
		Q-2-CH=SPh$_2$, cycloC$_6$H$_{10}$O, LDA, see text, Section II.2.G	52		
2-Quinuclidinyl		2-[3-(1-Bromo-4-)piperidinyl]-propionyl-Q, EtONa	77.5	125	136
Ph		Q-2-COCl, C$_6$H$_6$, AlCl$_3$	88	110–111	40
		Reissert comp., NaH, DMF	85	107–109	156, 157
			54	109–110	25*
		QSnMe$_3$, PhCOCl	74	109–111	1, 2
				b.p. 168–170/0.35 mm	
		Q-2-CHO, PhMgBr; CrO$_3$	64	107–108	102
		See text, Section II.2.E	54	110–111	28
		Q, PhCOOEt; Al/Hg/HgCl$_2$	29	111	44
		Q, PhCONMe$_2$, Al/Hg/HgCl$_2$	26.1		44
		Q, Cu(I)benzoate, heat	6		11
		Q N-oxide, PhCH$_2$COOH, Ac$_2$O	0.1ᵃ	106–107	158
		QCOOH, PhCOOMe	Trace	107–108	8
		Q-2-CHOHPh, CrO$_3$		110	27*
				Pic. 140	
Ph	4-Me	Q-2-CN, PhMgBr		111	104, 105*
		QCOOH, PhCHO, heat		107	10
		Reissert comp., NaH	62	107–108	157
			14	112–113	26
Ph	4-Me	4-Me-Q, PhCHOHCOPh, AgNO$_3$, Na$_2$S$_2$O$_8$, H$_2$SO$_4$	21		14

Table 1. (*Contd.*)

R	Quinoline substituent(s)	Preparation	Yield (%)	M.p. (°C)	References
Ph	4-Me	4-Me-Q-2-MgBr, PhCHO		109	159*
		4-Me-Q, PhCHO, $(tBuO)_2$		109	15
Ph	4-Ph	2-NH$_2$C$_6$H$_4$COPh, PhCOCOMe	57	121–122	160
Ph	4-Cl	4-Cl-Q, PhCHO, $(tBuO)_2$		122	15
Ph	3-COPh	2-CHOC$_6$H$_4$NHC(COPh)=CHCOPh, MeONa	58	118–120	19
Ph	3-COPh-4-Me	2-PhCH$_2$-3-PhCO-Q, CrO$_3$			19
		2-AcC$_6$H$_4$NHC(COPh)=CHCOPh, MeONa	79	205–206.5	20, 21
Ph	3-COPh-4-Ph	2-PhCOC$_6$H$_4$NHC(COPh)=CHCOPh, MeONa	81	203–205	20
Ph	3-COPh-4-NH$_2$	2-CNC$_6$H$_4$NHC(COPh)=CHCOPh, MeONa	63	194.5–195.5	20
Ph	3-OH	See text, Section II.2.D	15	86	22
Ph	6-OMe	Reissert comp., NaH	11	116.3–116.7	209
Ph	4-COPh	Q, PhCHO, $(tBuO)_2$	50	128	16
Ph	3-COOMe-4-OMe	2-PhCO-3-COOMe-4-quinolone, CH$_2$N$_2$	28	82	120
Ph	4-COOH-6,8-di-Cl	Pfitzinger synthesis or Q-2-PhCH$_2$-4-COOH-6,8-diCl, SeO$_2$ or Br$_2$, AcOH	69	248–250	17
Ph	4-COCl-6,8-di-Cl	Above, SOCl$_2$	97	176–178.5	17
Ph	4-COCH$_2$Br-6,8-di-Cl	Above, CH$_2$N$_2$, HBr	66	150–152	17
Ph	4-CHOHCH$_2$NBu$_2$-6,8-diCl	Above, Bu$_2$NH; NaBH$_4$; HCl	5	HCl 167–168.5	17
Ph	4-COOEt	Q-4-COOEt, PhCHO, $(tBuO)_2$		87	15

Ar	Quinolyl substituent	Reactants	Yield (%)	M.p. (°C)	Refs.
Ph	4-CN	Q-4-CN, PhCHO, (tBuO)$_2$	86	142	16
2-Me-3,4-di-HOC$_6$H$_2$		2-Me-3-MeO-4-HOC$_6$H$_2$CO-2-Q, 48% HBr			162, 163
2-Me-3-MeO-4-HOC$_6$H$_2$		2-Me-3,4-diMeOC$_6$H$_2$CO-2-Q, 48% HBr			162, 163
2-Me-3,4-di-MeOC$_6$H$_2$		2-Me-3,4-diMeOC$_6$H$_2$CHOH-2-Q, KMnO$_4$			162
3-MeC$_6$H$_4$	4-Cl	4-Cl-Q, 3-MeC$_6$H$_4$CHO, (tBuO)$_2$	64	107	16
3-MeC$_6$H$_4$	4-CN	Q-4-CN, 3-MeC$_6$H$_4$CHO, (tBuO)$_2$	87	156	16
4-MeC$_6$H$_4$		Q-2-CHOHC$_6$H$_4$-4-Me, CrO$_3$		68; Pic. 135	27*
4-MeC$_6$H$_4$	4-Cl	4-Cl-Q, 4-MeC$_6$H$_4$CHO, (tBuO)$_2$	67	104	16
4-MeC$_6$H$_4$	4-(COC$_6$H$_4$-4-Me)	Q, 4-MeC$_6$H$_4$CHO, (tBuO)$_2$	25	134	16
4-MeC$_6$H$_4$	4-CN	Q-4-CN, 4-MeC$_6$H$_4$CHO, (tBuO)$_2$	83	149	16
3-CF$_3$C$_6$H$_4$	4-COOH-6-Cl-8-CF$_3$	Doebner reaction	25	226–228	17
3-CF$_3$C$_6$H$_4$	4-COCl6-Cl-8-CF$_3$	Above, SOCl$_2$	96	110–111	17
3-CF$_3$C$_6$H$_4$	4-COCH$_2$Br-6-Cl-8-CF$_3$	Above, CH$_2$N$_2$; HBr	54		17
3-CF$_3$C$_6$H$_4$	4-CHOHCH$_2$NBu$_2$-6-Cl-8-CF$_3$	Above, Bu$_2$NH; NaBH$_4$; HCl	13	HCl 165–166	17
3,5-diCF$_3$C$_6$H$_3$	4-COOH-6,8-diCl	Doebner reaction	33	296–297	17
3,5-diCF$_3$C$_6$H$_3$	4-COCl-6,8-diCl	Above, SOCl$_2$	100	181–183	17
3,5-diCF$_3$C$_6$H$_3$	4-COCH$_2$Br-6,8-diCl	Above, CH$_2$N$_2$; HBr	55	145–146	17
3,5-diCF$_3$C$_6$H$_3$	4-CHOHCH$_2$NBu$_2$-6,8-diCl	Above, Bu$_2$NH; NaBH$_4$; HCl	11	HCl 194–195	17
4-CF$_3$C$_6$H$_4$	4-COOH-6,8-diCl	Doebner reaction	23	284–285	17
4-CF$_3$C$_6$H$_4$	4-COCl-6,8-diCl	Above, SOCl$_2$	96	151–154	17
4-CF$_3$C$_6$H$_4$	4-COCH$_2$Br-6,8-diCl	Above, CH$_2$N$_2$; HBr	47	155–156	17
4-CF$_3$C$_6$H$_4$	4-CHOHCH$_2$NBu$_2$-6,8-diCl	Above, Bu$_2$NH; NaBH$_4$; HCl	14	HCl 157–159	17

Table 1. (*Contd.*)

R	Quinoline substituent(s)	Preparation	Yield (%)	M.p. (°C)	References
$2\text{-}(C_6H_4\text{-}2\text{-}COOH)\text{-}C_6H_4$		Q-2-COOH, diphenic anhyd., $PhNO_2$, 160–165 °C	24	185–186 HCl 156–157 °C	12, 164
$2\text{-}(C_6H_4\text{-}2\text{-}CONHMe)\text{-}C_6H_4$		$Q\text{-}2\text{-}COC_6H_4\text{-}2\text{-}C_6H_4\text{-}2\text{-}COOH$, $SOCl_2$; $MeNH_2$			164
$4\text{-}FC_6H_4$		Reissert comp., NaH	99	129–130	156
$3\text{-}ClC_6H_4$	4-Cl	4-Cl-Q, $3\text{-}ClC_6H_4COCOOH$, $AgNO_3$, $(NH_4)_2S_2O_8$		135	15
$3\text{-}ClC_6H_4$	4-CN	Q-4-CN, $3\text{-}ClC_6H_4CHO$, $(tBuO)_2$	69	187	16
$3,4\text{-}diClC_6H_3$	$4\text{-}COOH\text{-}6\text{-}Cl\text{-}8\text{-}CF_3$	Doebner reaction	38	276.5–278.5	17
$3,4\text{-}diClC_6H_3$	$4\text{-}COCl\text{-}6\text{-}Cl\text{-}8\text{-}CF_3$	Above, $SOCl_2$	89	118.5–120.5	17
$3,4\text{-}diClC_6H_3$	$4\text{-}COCH_2Br\text{-}6\text{-}Cl\text{-}8\text{-}CF_3$	Above, CH_2N_2; HBr	88	133–134	17
$3,4\text{-}diClC_6H_3$	$4\text{-}CHOHCH_2NBu_2\text{-}6\text{-}Cl\text{-}8\text{-}CF_3$	Above, Bu_2NH; $NaBH_4$; HCl	15	HCl 185–188	17
$3,5\text{-}diClC_6H_3$	$4\text{-}COOH\text{-}6\text{-}Cl\text{-}8\text{-}CF_3$	Doebner reaction	39	285–286.5	17
$3,5\text{-}diClC_6H_3$	$4\text{-}COCl\text{-}6\text{-}Cl\text{-}8\text{-}CF_3$	Above, $SOCl_2$	100	164–166	17
$3,5\text{-}diClC_6H_3$	$4\text{-}COCH_2Br\text{-}6\text{-}Cl\text{-}8\text{-}CF_3$	Above, CH_2N_2; HBr	57	159–161	17
$3,5\text{-}diClC_6H_3$	$4\text{-}CHOHCH_2NBu_2\text{-}6\text{-}Cl\text{-}8\text{-}CF_3$	Above, Bu_2NH; $NaBH_4$; HCl	10	HCl 173–175	17
$4\text{-}ClC_6H_4$		$Q\text{-}2\text{-}CHOH\text{-}C_6H_4\text{-}4\text{-}Cl$, CrO_3		128 Pic. 150	27*
$4\text{-}ClC_6H_4$	4-Cl	4-Cl-Q, $4\text{-}ClC_6H_4CHO$, $(tBuO)_2$		158	15
$4\text{-}ClC_6H_4$	$4\text{-}(COC_6H_4\text{-}4\text{-}Cl)$	Q, $4\text{-}ClC_6H_4CHO$, $(tBuO)_2$	69	160	16
$4\text{-}ClC_6H_4$	$4\text{-}COOH\text{-}6\text{-}Cl\text{-}8\text{-}CF_3$	Doebner reaction	24	275–276	17
$4\text{-}ClC_6H_4$	$4\text{-}COCl\text{-}6\text{-}Cl\text{-}8\text{-}CF_3$	Above, $SOCl_2$	100	160–161	17
$4\text{-}ClC_6H_4$	$4\text{-}COCH_2Br\text{-}6\text{-}Cl\text{-}8\text{-}CF_3$	Above, CH_2N_2; HBr	60	160–161	17

4-ClC$_6$H$_4$	4-CHOHCH$_2$NBu$_2$-6-Cl-8-CF$_3$	Above, Bu$_2$NH; NaBH$_4$; HCl	HCl 165–169.5	12	291
4-ClC$_6$H$_4$	4-COOH-6,8-diCl	Pfitzinger synthesis or Doebner reaction	325–326(d)	40	17
4-ClC$_6$H$_4$	4-COCl-6,8-diCl	Above, SOCl$_2$	280–282	100	17
4-ClC$_6$H$_4$	4-COCH$_2$Br-6,8-diCl	Above, CH$_2$N$_2$; HBr	185–186	53	17
4-ClC$_6$H$_4$	4-CHOHCH$_2$NBu$_2$-6,8-diCl	Above, Bu$_2$NH; NaBH$_4$; HCl	HCl 159–161	59	17
4-ClC$_6$H$_4$	4-CN	Q-4-CN, 4-ClC$_6$H$_4$CHO, (tBuO)$_2$	193	81	16
2-HOC$_6$H$_4$	4-(2-COC$_6$H$_4$-2-OH)	Q, 2-HOC$_6$H$_4$CHO, (tBuO)$_2$	107	24	16
3,4-di-HOC$_6$H$_3$		3,4-diMeOC$_6$H$_3$CO-2-Q, 48% HBr	201–202 / HCl 219–227 / HBr 228–234		162, 163
3,5-diHOC$_6$H$_3$		3,5-diMeOC$_6$H$_3$CO-2-Q, 48% HBr			162, 163
3-MeOC$_6$H$_4$		Q-2-COOH, 3-MeOC$_6$H$_4$CHO	70–71	38	9*
3-MeOC$_6$H$_4$	4-CN	Q-4-CN, 3-MeOC$_6$H$_4$CHO, (tBuO)$_2$	188	88	16
3,4-diMeOC$_6$H$_3$		Q-2-COOH, 3,4-diMeOC$_6$H$_3$CHO	104	24	9*, 162, 163
3,5-diMeOC$_6$H$_3$		3,5-diMeOC$_6$H$_3$CHOH-2-Q, KMnO$_4$	201–202		162
3,4,5-triMeOC$_6$H$_2$		Reissert comp., NaH	159–160	53	156
4-MeOC$_6$H$_4$		Q-2-COOH, 4-MeOC$_6$H$_4$CHO	78	30	10*
4-MeOC$_6$H$_4$		Reissert comp., NaH	Pic. 165	83	156
4-MeOC$_6$H$_4$		Q-2-CHOHC$_6$H$_4$-4-OMe, CrO$_3$	82–83 / 82 / Pic. 167		27*
4-Me		Reissert comp., NaH	121–122		26
4-Cl		4-Cl-Q, 4-MeOC$_6$H$_4$CHO, (tBuO)$_2$	105	50	15

Table 1. (*Contd.*)

R	Quinoline substituent(s)	Preparation	Yield (%)	M.p. (°C)	References
4-MeOC$_6$H$_4$	4-(COC$_6$H$_4$-4-OMe)	Q, 4-MeOC$_6$H$_4$CHO, (tBuO)$_2$	25	147	16
4-MeOC$_6$H$_4$	4-CN	Q-4-CN, 4-MeOC$_6$H$_4$CHO, (tBuO)$_2$	73	138	16
2-EtO-4-Et$_2$NC$_6$H$_3$	3-COOH	Q-2-COOH, phthalic anhydride, PhNO$_2$, 100°C	43	225–226(d)	165
C$_6$H$_4$-2-COOH				HCl 182–183	12, 166
C$_6$H$_4$-4-COOH		See text, Section II.2.E	81	237–238	27
2-NO$_2$C$_6$H$_4$		See text, Section II.2.G	94.5	125–126	47
4-NO$_2$C$_6$H$_4$		See text, Section II.2.G	76.5	177–178	47
1-(C$_{10}$H$_6$-8-COOH)		Q-2-COOH, 1,8-naphthalic anhyd., PhNO$_2$, 160–165°C	43	179–180	12
3-Bromophenanthren-10-yl		Q-2-Li, 3-BrC$_{14}$H$_8$-9-COOH	47	239–241	167
3,6-DiCF$_3$-phenanthren-9-yl		Q-2-Li, 3,6-diCF$_3$C$_{14}$H$_7$-9-COOH	73	227–228	167
4-Hydroxy-3-coumarinyl		Q-2-COOH, 4-Hydroxycoumarin, POCl$_3$		96–98	168*
2-Furyl		Q-2-COOH, 2-furaldehyde	16	88–89	9*
2-Pyridyl		See text, Section II.2.G	54	b.p. 202–207/0.35 mm	47
2-Quinolyl		Q$_2$CH$_2$, KMnO$_4$	60	165–166 Pic. 179	169
		Q$_2$CH$_2$, HCl, O$_2$ or HNO$_3$		164	35*
		See text, Section II.2.G	50	164	45, 170
		See text, Section II.2.F	8.4	163–165	37

		Reagents	Yield (%)	m.p. (°C)	Ref.
2-Quinolyl		Q N-oxide, PhCH$_2$COOH, Ac$_2$O.	0.2[g]	163–165	158
4-Quinolyl		Q-2-CH(CN)-4-Q, H$_2$O$_2$	72	170–171	38
		Q$_2$CH$_2$, Me$_2$NC$_6$H$_4$N=O		167	36*
2-Chloro-4-quinolyl		Q-2-CO-(1-oxido-4-quinolyl), HCl, POCl$_3$	93	212.5	38
4-Quinolyl-N-oxide		Q-2-CH(CN)-4-Q, H$_2$O$_2$	70	196–197	38
4-Carbostyryl			95	270–272(d)	38
1-Isoquinolyl		Isoquinoline red, CrO$_3$		HCl 68–69(d)	171
1-Isoquinolylmethyl		1-Methylisoquinoline, Q-2-COOMe, PhLi	53	169–170	4
1-Et-2-Me-3-indolyl	3-COOH	Mixture, see text, Section II.2.G			48, 165

* These references include carbonyl derivatives.
a Inconsistent literature melting points.
b Formed in admixture with other products.
c Obtained as a mixture with the isomeric hydrazone (23%; m.p. 141 °C).
d Obtained as a mixture with Q-2-COOsBu (78%).
e See text, Section VI.5.
f Obtained as a mixture with the corresponding chalcone.
g Obtained in a mixture with several other products.

Table 2. 3-Quinolyl Ketones

R	Quinoline substituent(s)	Preparation	Yield (%)	M.p. (°C)	References
Me	None	Q-3-COCH$_2$COOEt, 25% H$_2$SO$_4$, 100 °C	95	97.5–98.5	172, 173*
		Q-3-C≡CH, HgSO$_4$, H$_2$SO$_4$	94	101–102	108
		1,2-diH-3-Ac-Q-COONa, KMnO$_4$, Py or K$_2$Cr$_2$O$_7$, AcOH	85	96–98	174, 175
		Q-3-SnMe$_3$, MeCOCl	70	100–102 b.p. 139–140/ 1.0 mm	1, 2
		1,4-diH-3-Ac-Q, KMnO$_4$	50	100–101	1, 2
		1,2-diH-3-Ac-Q-2-COOH, 200 °C	36		70
		Cl$_3$C-CHOH-CH$_2$-2-Q, NaOH; KMnO$_4$	17	99–100.5	176*
		3-Ac-Q-2-COOH, heat		100–101 b.p. 182/ 12 mm	174,177*, 178*
Me	2-Me	2-NO$_2$C$_6$H$_4$CH=C(Ac)$_2$, H$_2$/Pd; PCl$_3$	95	78–79	65
		2-Cl$_2$CHC$_6$H$_4$N=C=O, Ba(OH)$_2$; HCl; NaOH; AA	94	54–56	55
		2-NH$_2$C$_6$H$_4$CHO, AA, Py	92	78–79	179*
		2-NH$_2$C$_6$H$_4$CHO, AA, Dowex-2, EtOH, reflux	69.3	79–80	56*, 57, 180*
		Anthranil, AA, HgSO$_4$, xylene	8	74	181
		Quinazoline 3-oxide, AA	3.2	75–76	182
		2-Me-3-Ac-Q-4-COOH, 240 °C		74–75	58

Me	2,4-diMe	Friedländer synthesis	60	b.p. 163–164/1.3 mm	18
Me	2,4,6-triMe	Friedländer synthesis		74	183
Me	2,4-diMe-6-F	Friedländer synthesis		64	183
Me	2,4-diMe-6-Cl	Friedländer synthesis		76	183
Me	2,4-diMe-6-OH	Friedländer synthesis		245	183
Me	2,4-diMe-6-OAc	Friedländer synthesis		113	183
Me	2,4-diMe-6-OMe	Friedländer synthesis or 2-EtCONH-5-MeOC$_6$H$_3$COMe, AA, HCl	90 / 91	128–129	183
Me	2,4-diMe-6,7-diOMe	Friedländer synthesis		173	183
Me	2,4-diMe-6,7,8-triOMe	Friedländer synthesis		141	183
Me	2,4-diMe-7-OMe	Friedländer synthesis		92	183
Me	2,4-diMe-8-OMe	Friedländer synthesis		134	183
Me	2,4-diMe-6-OEt	Friedländer synthesis		120	183
Me	2,4-diMe-6-OBu	Friedländer synthesis		63	183
Me	2,4-diMe-6-OcycloC$_5$H$_9$	Friedländer synthesis		60	183
Me	2,4-diMe-6-OCH$_2$CH$_2$NEt$_2$	Friedländer synthesis		50–51	183
Me	2,4-diMe-6-OCH$_2$O-7	Friedländer synthesis		124	183
Me	2,4-diMe-7-Ac	Friedländer synthesis		99	183
Me	2,4-diMe-7-C(Me)=NOH	Friedländer synthesis		247	183
Me	2,4-diMe-6-NO$_2$	Friedländer synthesis		132	183
Me	2,4-diMe-6-NH$_2$	Friedländer synthesis		113	183
Me	2,4-diMe-6-NHAc	Friedländer synthesis		188	183
Me	2-Me-4-PhCH$_2$-6-Cl	Friedländer synthesis		119	183
Me	2-Me-4-Et-6-Cl	Friedländer synthesis		74	183
Me	2-Me-4-Pr-6-Cl	Friedländer synthesis		57	183
Me	2-Me-4-CH$_2$CH$_2$COOH-6,7-diOMe	Friedländer synthesis		245(d)	184
Me	2-Me-4-cycloC$_6$H$_{11}$-6-Cl	Friedländer synthesis		124	183

Table 2. (*Contd.*)

R	Quinoline substituent(s)	Preparation	Yield (%)	M.p. (°C)	References
Me	6,8-diBr	3-Ac-Q, KBrO₃, AcOH, 48% HBr, 100°C	20.1	181–183	185
Me	2-Me-4-Ph	Friedländer synthesis	84	113–114	186
Me	2-Me-4-Ph-6-Cl	Friedländer synthesis	96	153–155	187
Me	2-Me-4-(2-FC₆H₄)-6-Cl	Friedländer synthesis	67	121–123	188
Me	2-Me-4-(5-MeS-1,2,4-triazol-3-yl)				189
Me	2-Me-5,6-diOH	2-Me-3-Ac-5,6-diMeO-Q, HI, AcOH		210(d)	190
Me	2-Me-5-OH-6-OMe	Friedländer synthesis		132–133	190
Me	2-Me-5,6-diOMe	Friedländer synthesis	90	69	190*
Me	2-Me-7,8-diOH	2-Me-3-Ac-7,8-diMeO-Q, Py. HCl	55	184 HCl 237(d) Pic. 224	74
Me	2-Me-7,8-diOAc	2-Me-3-Ac-7,8-diHO-Q, Ac₂O		86	74
Me	2-Me-7-OH-8-OMe	Friedländer synthesis	65	120	74
Me	2-Me-6-OMe	Friedländer synthesis		Pic. 188 / 73	183
Me	2-Me-6,7-diOMe	Friedländer synthesis		148	183
Me	2-Me-7,8-diOMe	Friedländer synthesis	60	94	74
Me	2-Me-4-COOH	Pfitzinger reaction	72.7	Pic. 195 / 198	58
Me	2-Me-4-COOH	2-Me-3-Ac-Q-4-COOH N-oxide, H₂/Pd		199(d) HCl 200(d)	191
Me	2-Me-4-COOMe	2-Me-3-Ac-Q-4-COOH, CH₂N₂		55–56	58
Me	2-Me-4-NH₂	MeCOCH=C(Me)NHC₆H₄-2-CN, MeONa	75	151–153	192
Me	2-CH₂OH	2-Acetyloxymethyl-3-Ac-Q, KOH		104–105	193
Me	2-CH₂OAc	2-Me-3-Ac-Q N-oxide, Ac₂O	62	113–114	193
Me	2-CH₂OAc-6-OCH₂O-7	2-Me-3-Ac-6-OCH₂O-7-Q N-oxide, Ac₂O	81	176–177	193

	Substituents	Reactants / conditions	Yield (%)	m.p. (°C)	Ref.
Me	2-CH$_2$OCOPh	2-CH$_2$OH-3-Ac-Q, PhCOCl, Py	59	122–123	193
Me	4-Me	2-(MeCOCH=CHNH)C$_6$H$_4$-COMe, MeONa	87	93–94	20
Me	4-Me-6-OMe	2-NH$_2$-5-MeOC$_6$H$_3$COMe, MeCOCH$_2$CHO, HCl		111–112	183
Me	4-Me-7-OMe	3-MeOC$_6$H$_4$NHCH=C(Ac)$_2$, PPA	56	98	69
Me	2-Ph-4-Cl	2-Ph-3-Ac-4-quinolone, POCl$_3$	100	136	194, 195
Me	2-(4-MeOC$_6$H$_4$)-4-Cl	2-(4-MeOC$_6$H$_4$)-3-Ac-4-quinolone, POCl$_3$	90	138–140 / 146–148	194
Me	2-Ph-4,5(or 4,7)-diCl	2-Ph-3-Ac-5(or 7)-Cl-4-quinolone, POCl$_3$	90	125	196
Me	2-Ph-4-OMe	2-Ph-3-Ac-4-Cl-Q, MeONa		200 / 214–215 / HCl 199–202	195, 197 / 197 / 197
Me	2-Ph-4-OMe-5(or 7)-Me[a]			HCl 137	197
Me	2-Ph-4-OMe-6-Me			HCl 94–95	197
Me	2-Ph-4-OMe-8-Me			HCl 205–210[b]	197
Me	2-Ph-4-OMe-5-Cl			HCl 247–250[b]	197
Me	2-Ph-4-OMe-7-Cl			HCl 225	197
Me	2-Ph-4,6-diOMe			HCl 101	197
Me	2-Ph-4-OMe-6-OEt			HCl 203	197
Me	2-Ph-4-NHPh	2-Ph-3-Ac-4-Cl-Q, PhNH$_2$	69	114–116	194
Me	2-(4-MeOC$_6$H$_4$)-4-NH$_2$[c]	2-(4-MeOC$_6$H$_4$)-3-Ac-4-Cl-Q, NH$_3$	50	164–165	198
Me	2-(4-MeOC$_6$H$_4$)-4-NH$_2$-8-Me[c]	2-(4-MeOC$_6$H$_4$)-3-Ac-4-Cl-8-Me-Q, NH$_3$	50	167–168	198
Me	2-(4-MeOC$_6$H$_4$)-4-NH$_2$-6-OMe[c]	2-(4-MeOC$_6$H$_4$)-3-Ac-4-Cl-6-MeO-Q, NH$_3$	10	193–194	198

Table 2. (*Contd.*)

R	Quinoline substituent(s)	Preparation	Yield (%)	M.p. (°C)	References
Me	2-(4-MeOC$_6$H$_4$)-4-NH$_2$-6-OEt[c]	2-(4-MeOC$_6$H$_4$)-3-Ac-6-OEt-4-quinolone, POCl$_3$; NH$_3$	10	197–198	198
Me	2-(4-MeOC$_6$H$_4$)-4-NHPh	2-(4-MeOC$_6$H$_4$)-3-Ac-4-Cl-Q, PhNH$_2$	60	121–123	194
Me	2-(4-MeOC$_6$H$_4$)-4-NHPh[c]	2-(4-MeOC$_6$H$_4$)-3-Ac-4-quinolone, POCl$_3$; PhNH$_2$	50	213–214	198
Me	2-(4-MeOC$_6$H$_4$)-4-NHPh-8-Me[c]	2-(4-MeOC$_6$H$_4$)-3-Ac-8-Me-4-quinolone, POCl$_3$; PhNH$_2$	25	143–144	198
Me	2-(4-MeOC$_6$H$_4$)-4-NHPh-6-OMe[c]	2-(4-MeOC$_6$H$_4$)-3-Ac-6-MeO-4-quinolone, POCl$_3$; PhNH$_2$	20	163–164	198
Me	2-(4-MeOC$_6$H$_4$)-4-NHPh-6-OEt[c]	2-(4-MeOC$_6$H$_4$)-3-Ac-6-EtO-4-quinolone, POCl$_3$; PhNH$_2$		Crude	198
Me	4-Ph	2-(MeCOCH=CHNH)-C$_6$H$_4$COPh, MeONa	78	69–71	20
Me		4-Ph-Q-3-COCH(COOEt)$_2$, aq. H$_2$SO$_4$		76–78	199
Me	4-(2-MeC$_6$H$_4$)	4-(2-MeC$_6$H$_4$)-Q-3-COCl, CH$_2$(COOEt)$_2$; aq. H$_2$SO$_4$		92–94	199
Me	4-Ph-6-Cl	1,2-diH-3-Ac-4-Ph-6-Cl-Q, MnO$_2$		129–131	200
Me	2-Cl	2-Cl-Q-3-CHO, MeMgI; PyClCrO$_3$	97	75–76	77
Me	2-Cl-7-Me	2-Cl-7-Me-Q-3-CHO, MeMgBr; CrO$_3$		81–82[d]	201
Me	2-Cl-6-OMe	2-Cl-3-CN-7-Me-Q, MeMgI; HCl	22.5	58[d]	72
Me		2-Cl-3-MeCHOH-6-MeO-Q, PyClCrO$_3$	85	100	202*
Me	2-Cl-6,7-diOMe	2-Cl-6,7-diMeO-Q-3-CHO, MeMgI; PyClCrO$_3$	96	172	77
Me		2-Cl-3-MeCHOH-6,7-diMeO-Q, PyClCrO$_3$	90	172	202*

			Yield (%)	mp	Ref.
Me	2-Cl-6,7-diOEt	2-Cl-3-MeCHOH-6,7-diEtO-Q, PyClCrO$_3$	90	126–128	202*
Me	4,7-diCl	3-Ac-7-Cl-4-quinolone, POCl$_3$	55	99–100	175
Me	4-Cl-6-OMe-8-NO$_2$	3-Ac-6-MeO-8-NO$_2$-4-quinolone, POCl$_3$	15	118–119	203
Me	7-Cl	3-Ac-4,7-diCl-Q, H$_2$/Pd, AcOH	Trace	130–133	175
Me	6,8-diBr	3-Ac-Q, KBrO$_3$, AcOH, 48% HBr	20.1	181–183	185
Me	4-OCOPh-7-Cl	3-Ac-7-Cl-4-quinolone, (PhCO)$_2$O, Et$_3$N		245–246	204
Me	2-OMe	2-MeO-Q-3-COOEt, EtOAc, EtONa; aq. H$_2$SO$_4$	30	110–112	173*
Me	2-OCH$_2$CH$_2$NMe$_2$-4-Ph-6-Cl^e	3-Ac-4-Ph-6-Cl-2-quinolone, Me$_2$NCH$_2$CH$_2$Cl, NaH, DMF	38	156–157	205
Me	6-OCH$_2$O-7	2-NO$_2$-4-OCH$_2$O-5-C$_6$H$_2$CH=C(Ac)$_2$, H$_2$/Pd; PCl$_3$	94		65
Me	2-SCH$_2$COOMe	2-Cl-3-Ac-Q, HSCH$_2$COOMe, K$_2$CO$_3$, DMF	91	140–141	77
Me	2-SCH$_2$COOMe-6,7-diOMe	2-Cl-3-Ac-6,7-diMeO-Q, HSCH$_2$COOMe, K$_2$CO$_3$, DMF	95	170	77
Me	2-COOH	3-Ac-Q-2-COOEt, NaOH; 1,2-diH-3-Ac-Q-2-COONa, KMnO$_4$, Py		141; 141–142(d)	174, 177; 174
Me	2-COOEt	1,2-diH-3-Ac-Q-2-COOEt, KMnO$_4$, Py	50	93–94	174
Me	4-COOH	Friedländer synthesis; MeCOCH=CHNHC$_6$H$_4$-2-COCOOH, NaOH or HCl	80.7	92–93; 228–230(d)	174, 177; 66–68
Me	4-COOH-6-Me	5-Me-2-MeCOCH=CHNHC$_6$H$_3$-COOH, NaOH or HCl	72	239.5–241	66–68
Me	4-COOH-7-Me	4-Me-2-MeCOCH=CHNHC$_6$H$_3$-COOH, NaOH or HCl	85	245–246	66–68
Me	4-COOH-6-Cl	5-Cl-2-MeCOCH=CHNHC$_6$H$_3$-COOH, NaOH or HCl	88	228–229.5	66–68
Me	4-COOH-7-Cl	4-Cl-2-MeCOCH=CHNHC$_6$H$_3$-COOH, NaOH or HCl	73	231–235(d)	66–68

Table 2. (*Contd.*)

R	Quinoline substituent(s)	Preparation	Yield (%)	M.p. (°C)	References
Me	4-COOCOMe	3-Ac-Q-4-COOH, Ac$_2$O		116.5	66–68
Me	4-COOCOMe-6-Me	3-Ac-6-Me-Q-4-COOH, Ac$_2$O		110–112	66
Me	4-COOCOMe-7-Me	3-Ac-7-Me-Q-4-COOH, Ac$_2$O		121–123	66
Me	4-COOCOMe-6-Cl	3-Ac-6-Cl-Q-4-COOH, Ac$_2$O		168–172	66
Me	4-COOCOMe-7-Cl	3-Ac-7-Cl-Q-4-COOH, Ac$_2$O		144–145	66
Me	4-COOMe	3-Ac-Q-4-COOH, CH$_2$N$_2$, MeOH or MeOH, HCl	95	110–111	68
Me	4-CONMe$_2$	See text, Scheme 15	60	136–138	68
Me	4-CON(CH$_2$)$_5$	3-Ac-Q-4-COOCOMe, Py	35	118–121	66, 67
CH$_2$Ph	2-OMe	Q-2-SO$_2$CPh=C(OH)-3, MeONa or 2-(PhCH$_2$SO$_2$)-Q-3-COOMe, MeONa	87.5 92 92	88–89.5	206 206
CH$_2$Br		3-Ac-Q, Br$_2$, 45% HBr	96.5	120	173
		3-Ac-Q, Br$_2$, AcOH, 40% HBr		HBr 220(d)	185
CH$_2$Br	2-Me	2-Me-3-Ac-Q, Br$_2$, HBr		102–103	207
CH$_2$Br	8-CF$_3$	8-CF$_3$-Q-3-COCl, CH$_2$N$_2$; HBr	66	142–143	59
CH$_2$Br	4-Cl-6,8-diMe	4-Cl-6,8-diMe-Q-3-COCl, CH$_2$N$_2$; HBr	58	76.5–78	59
CH$_2$Br	4-Cl-8-CF$_3$	3-Cl-8-CF$_3$-Q-3-COCl, CH$_2$N$_2$; HBr	79	98–99	59
CH$_2$Br	4-Cl-8-Ph	4-Cl-8-Ph-Q-3-COCl, CH$_2$N$_2$; HBr	98	132–133	59
CH$_2$Br	4,6,8-triCl	4,6,8-triCl-Q-3-COCl, CH$_2$N$_2$; HBr	87	136–137.5	59
CH$_2$Br	4,7-diCl	4,7-diCl-Q-3-COCl, CH$_2$N$_2$; HBr	83	104–106	59
CH$_2$Br	6,8-diCl	6,8-diCl-Q-3-COCl, CH$_2$N$_2$; HBr	81	197–199(d)	59
CH$_2$Br	2-OMe	2-MeO-3-Ac-Q, Br$_2$, 45% HBr		126–127	173

R	Quinoline subst.	Preparation	Yield (%)	M.p./b.p. and derivatives	Ref.
CH_2NMe_2		Q-3-$COCH_2Br$, Me_2NH		HCl 157–158; diPic. 147–149	173
CH_2NMe_2	2-OMe	2-MeO-Q-3-$COCH_2Br$, Me_2NH		HCl 177	173
CH_2NEt_2		Q-3-$COCH_2Br$, Et_2NH		HBr 142–145; diPic. 150–151	173
CH_2NEt_2	2-OMe	2-MeO-Q-3-$COCH_2Br$, Et_2NH		HBr 134–136	173
$CH_2N(CH_2)_5$		Q-3-$COCH_2Br$, $C_5H_{11}N$		b.p. 165–168/15 mm; HBr 245–246; diPic. 139–141	173
$CH_2N(CH_2)_5$	2-OMe	2-MeO-Q-3-$COCH_2Br$, $C_5H_{11}N$		69–71; HBr 251–256(d)	173
Et		Q-3-$SnMe_3$, EtCOCl	76	83–85; b.p. 140–142/0.8 mm	2
Et		Q-3-CN, EtMgBr	70	76	133
Et		3-EtCO-Q-2-COOH, heat		79–80	178*
Et	2-Me-4-COOH	Pfitzinger reaction	66.7	132	58
Et	2-Et-4-Me-6-OMe	Friedländer synthesis		77	183
Et	2-Cl-7-Me	2-Cl-3-CN-7-Me-Q, EtMgBr; HCl	27	62	72
Et	2-COOH	3-EtCO-Q-2-COOEt, aq. KOH	95	153(d)	178
Et	2-COOEt	Friedländer synthesis	53	107–108	178
CHBrMe		3-EtCO-Q, Br_2		HBr 231–232	133
$CH_2CH_2NHCH(Me)$-CHOHPh		3-Ac-Q, norephedrine, Mannich reaction	25	205–206	208
2-(3-Piperidyl)ethyl		Q-3-COOEt, 1-PhCO-C_5H_9N-3-CH_2CH_2COOEt, KH; HCl		Crude	135
2-(3-Et-4-piperidyl)-ethyl		1-PhCO-3-Et-C_5H_8N-4-CH_2CH_2COOEt, Q-3-COOEt, EtONa; HCl	70	b.p. 225/9 mm	173*

Table 2. (*Contd.*)

R	Quinoline substituent(s)	Preparation	Yield (%)	M.p. (°C)	References
2-(1-Br-3-Et-4-piperidyl)-ethyl		Above, HOBr		137–139	173
2-(3-Et-4-piperidyl)ethyl	2-OMe	1-PhCO-3-Et-C_5H_8N-4-CH_2CH_2COOEt, 2-MeO-Q-3-COOEt, EtONa; HCl		b.p. 197–200/5 mm	173*
2-(1-Br-3-Et-4-piperidyl)-ethyl	2-OMe	Above, HOBr		158–162	173
CH=CHPh		3-Ac-Q, PhCHO, KOH or $ZnCl_2$		123^d / 223–224^d	209 / 177*
CH=CHC$_6$H$_4$-4-OH	4-COOH-6-OMe	Natural product			210
CH=CHC$_6$H$_3$-3-OMe-4-OH	4-COOH-6-OMe	Natural product			210
CH=CH-C$_6$H$_3$-3-OCH$_2$O-4	2-Me-4-CH$_2$CH$_2$COOH-6,7-diOMe	2-Me-3-Ac-4-CH$_2$CH$_2$COOH-6,7-diMeO-Q, piperonal		240(d)	211
CH=CHC$_6$H$_4$-2-NH$_2$	2-Me	2-Me-3-Ac-Q, 2-NH$_2$C$_6$H$_4$CHO, NaOH	39	164	57
CH$_2$COOEt		Q-3-COOEt, AcOEt, EtONa	80	84	172, 173
(CH$_2$)$_3$NHMe		Q-3-CO-3'-(1'-methyl-2'-pyrrolidinone), HCl		b.p. 165–175/0.01 mm	60
(CH$_2$)$_3$N(Me)-CO-C$_6$H$_4$-4-NO$_2$		See text, Section II.3	75	139–140	61
CH$_2$COMe	2-OMe	2-OMe-Q-3-COOMe, MeCOMe, Na	25	86–87	212
CH$_2$CH$_2$COOH		Q-3-CH(COOEt)CH$_2$COOEt, acid	52	208–208.5	142
CH$_2$COCOOEt iPr	2-COOEt	Friedländer synthesis / Q-3-SnMe$_3$, iPrCOCl	82	124–126 / b.p. 131–132/1.0 mm	209 / 2
CH(COOEt)$_2$	4-Ph	4-Ph-Q-3-COCl, CH$_2$(COOEt)$_2$		106–107	199

Substituent	Ring substituent	Reagents	M.p. / b.p.	Yield (%)	Reference
CH(COOEt)CH$_2$COOEt tBu		Q-3-COOEt, (CH$_2$COOEt)$_2$, NaH	b.p. 141–143/ 0.9 mm	74	142
		Q-3-SnMe$_3$, tBuCOCl		73	1
(CH$_2$)$_5$NH$_2$		Q-3-COOEt, EtOOC(CH$_2$)$_5$NHCOPh, NaNH$_2$; conc. HCl	2HBr 251–253	13	151
CHBr(CH$_2$)$_4$NH$_2$		Q-3-CO(CH$_2$)$_5$NH$_2$, Br$_2$, 48% HBr		96	151
CycloC$_6$H$_{11}$		Q-3-SnMe$_3$, C$_6$H$_{11}$COCl	72–73 / b.p. 179–180/ 0.5 mm	80	1, 2
3-(1-methyl-2-oxo)-pyrrolidinyl		Q-3-COOEt, 1-methyl-pyrrolidinone, EtONa	113–116	58.2	61
5-Et-2-quinuclidinyl		3-(1-Br-3-Et-4-piperidyl)-CH$_2$CH$_2$CO-Q, EtONa	120; 122–124; Pic. 167–168		60; 173
5-Et-2-quinuclidinyl	2-OMe	2-MeO-3-(1-Br-3-Et-4-piperidyl)-CH$_2$CH$_2$CO-Q, EtONa	155–156		173
CHO		3-Ac-Q, SeO$_2$, dioxane	134–136		213*
COOEt		Q-3-MgBr, (COOEt)$_2$			214
COOEt	2-Me-4-OEt	2-Me-3-COCOOEt-4-quinolone, EtOTl; EtI		79[f]	215
COOPr		Q-3-MgBr, (COOPr)$_2$	240–242	11	214
CONHPh		See text, Scheme 18	74–76	71	75
Ph	2,4-diPh	QSnMe$_3$, PhCOCl	b.p. 165–167/ 0.25 mm		1, 2
		QCOCl, C$_6$H$_6$, AlCl$_3$	76–77 / Pic. 216–218	65	62
Ph	2-Me	QLi, PhCN	75.6	41	216
		2-NO$_2$C$_6$H$_4$CH=CH(COMe)-COPh, H$_2$/Pd	68–69	88	65
		2-NH$_2$C$_6$H$_4$CHO, PhCOCH$_2$COMe	61–62	79	217*

Table 2. (*Contd.*)

R	Quinoline substituent(s)	Preparation	Yield (%)	M.p. (°C)	References
Ph	2-Me	2-Me-3-PhCO-Q-4-COOH, 310–320 °C		61–62	58*
Ph	2,4-diMe	2-NH$_2$C$_6$H$_4$COMe, PhCOCH$_2$COMe	80	94.5–95	18
Ph	2,4-diMe-6-Cl	2-NH$_2$-5-ClC$_6$H$_3$COMe, PhCOCH$_2$COMe		211	183
Ph	2-Me-4-Ph	2-NH$_2$C$_6$H$_4$COPh, PhCOCH$_2$COMe	86	138–139	160
Ph	2-Me-4-Cl	2-Me-3-PhCO-4-quinolone, POCl$_3$	90	123–124	194, 195
Ph	2-Me-5,6-diOH	2-Me-3-PhCO-5,6-diMeO-Q, Py. HCl	39	246–248 Pic 213–215	190
Ph	2-Me-7,8-diOH	2-Me-3-PhCO-7,8-diMeO-Q, Py. HCl or 2-Me-7-HO-8-MeO-Q, Py. HI	55 63	225 HCl 69 Pic. 219	74
Ph	2-Me-7-OH-8-OMe	2-NH$_2$-3-MeO-4-HOC$_6$H$_2$CHO, PhCOCH$_2$COMe	50	164	74
Ph	2-Me-4-OMe	2-Me-3-PhCO-4-Cl-Q, NaOMe		219–220	195
Ph	2-Me-5,6-diOMe	2-NH$_2$-5,6-diMeOC$_6$H$_2$CHO, PhCOCH$_2$COMe	58	157 Pic. 177–179	190
Ph	2-Me-6,7-diOMe	2-NH$_2$-4,5-diMeOC$_6$H$_2$CHO anil, MeCOCH$_2$COMe		158	50
Ph	2-Me-7,8-diOMe	2-NH$_2$-3,4-diMeOC$_6$H$_2$CHO, PhCOCH$_2$COMe	65	120 Pic. 175	74
Ph	2-Me-4-COOH	Isatin, PhCOCH$_2$COMe	41.8	280	58
Ph	2-Me-4-COOMe	2-Me-3-PhCO-Q-4-COOH, CH$_2$N$_2$		119–120	58
Ph	2-CH$_2$Ph	Friedländer condensation	43	79–80	19
Ph	2-CH$_2$Ph-4-Me	2-NH$_2$C$_6$H$_4$COMe, PhCOCH$_2$COCH$_2$Ph	70	170.5–171.5	18

			Yield (%)	mp	Ref.
Ph	2-CH₂Ph-4-OMe	2-PhCH₂-3-PhCO-4-quinolone, CH₂N₂	27	93–94	218
Ph	2-CH₂OH	2-CH₂OAc-3-PhCO-Q, KOH		75	64
Ph	2-CH₂OAc	2-Me-3-PhCO-Q N-oxide, Ac₂O		97–98	64
Ph	4-Me	1,4-diH-3-PhCO-4-Me-Q, chloranil	69	94–96	71
Ph	2-Ph	2-NH₂C₆H₄CHO, PhCOCH₂COPh	78	134–135	217*
		2-Ph-3-PhCO-Q-4-COOH, 290 °C		133–135	58
Ph	2-Ph-4-Me	2-NH₂C₆H₄COMe, PhCOCH₂COPh	80	127	18
Ph	2-Ph-4-Me-6-Cl	2-NH₂-5-ClC₆H₃COMe, PhCOCH₂COPh		142	183
Ph	2-Ph-4-CH₂CH₂COOH-6,7-diOMe	2-NH₂-4,5-diMeOC₆H₂COCH₂-CH₂COOH, PhCOCH₂COPh		229	184
Ph	2,4-diPh	2-NH₂C₆H₄COPh, PhCOCH₂COPh	59	146–147	160
		2-NH₂C₆H₄COPh, PhC≡C-COPh or PhCOCH₂COPh	52 / 43	146–147.5(d)	20
Ph	2-Ph-4-OAc	2-Ph-3-PhCO-4-quinolone, Ac₂O	82	126	219
Ph	2-Ph-4-OAc-6-Me	2-Ph-3-PhCO-6-Me-4-quinolone, Ac₂O	89	144	219
Ph	2-Ph-4-COOH	Isatin, PhCOCH₂COPh	63.2	257	58
		2-Ph-3-PhCO-4-COOH-Q N-oxide, H₂/Pd		266(d)	191
Ph	2-Ph-4-COOMe	2-Ph-3-PhCO-Q-4-COOH, CH₂N₂		159	58
Ph	4-Ph	1,4-diH-3-PhCO-4-Ph-Q, chloranil	75	115–116	62
		3-PhCO-4-Ph-Q-2-COOH, 175–195 °C	57	113–114	160
Ph or 2-Furyl or	2-(2-Furyl)-4-Me 2-Ph-4-Me	2-NH₂C₆H₄COMe, C₄H₄OCOCH₂COPh	46	150.5–152	18

Table 2. (*Contd.*)

R	Quinoline substituent(s)	Preparation	Yield (%)	M.p. (°C)	References
Ph	2-Cl	2-Cl-Q-3-CHO, PhMgBr; PyClCrO₃	92	82–83	77
Ph	2-Cl-7-Me	2-Cl-7-Me-Q-3-CHO, PhMgBr; CrO₃		115–117	201
Ph	2-Cl-4-Ph	2-Cl-3-CN-Q, PhMgBr; HCl	7	158	72
Ph	2-Cl-6-OMe	2-Cl-3-CHOHPh-6-MeO-Q, PyClCrO₃	90	127–128	202*
Ph	2-Cl-6,7-diOMe	2-Cl-3-CHOHPh-6,7-diMeO-Q, PyClCrO₃	98	142–144	202*
		2-Cl-6,7-diMeO-Q-3-CHO, PhMgBr; PyClCrO₃	97	142–144	77
Ph	6,7-diOMe	3-PhCO-6,7-diOMe-Q-2-COOH, heat		156–157	50
Ph	2-CHO	2-AcOCH₂-3-PhCO-Q N-oxide, KOH		132–133	64
Ph	2-COPh	See Table 9			
Ph	2-COPh-4-Me	See Table 9			
Ph	2-COPh-4-Ph	See Table 9			
Ph	2-COPh-4-NH₂	See Table 9			
Ph	2-COOH	Q-2,3-diCOOH anhydride, C₆H₆, AlCl₃	50	161–162	220
Ph	2-COOH-4-Ph	3-PhCO-4-Ph-Q-2-COOEt, KOH or 2-Me-3-PhCO-4-Ph-Q, SeO₂; H₂O₂	100 74	182–183	160
Ph	2-COOH-6,7-diOMe	3-PhCO-6,7-diMeO-Q-2-COOEt, KOH		206–207	50
Ph	2-COOEt	2-NH₂C₆H₄CH=NC₆H₄-4-Me, PhCOCH₂COCOOEt		89	50
Ph	2-COOEt-4-Ph	2-NH₂C₆H₄COPh, PhCOCH₂COCOOEt	65	155–156	160, 273

R	Quinolyl substituent	Reagents/Conditions	Yield (%)	m.p. (°C)	Ref.
Ph	2-COOEt-4-OMe	Not given		131–135	221
Ph	2-COOEt-6,7-diOMe	2-NH$_2$-4,5-diMeOC$_6$H$_2$CH=NC$_6$H$_4$-4-Me, PhCOCH$_2$COOEt		196–197	50
Ph	2-COOEt-6-OCH$_2$O-7	2-NH$_2$,-4-OCH$_2$O-5-C$_6$H$_2$CH=NC$_6$H$_4$-4-Me, PhCOCH$_2$COOEt		247–248	50
Ph	4-COOH	Isatin, (MeO)$_2$CHCH$_2$COPh	60	236–237(d)	67, 68
Ph	4-COOH	Isatin, PhCOCH=CHOMe		236–237	66
Ph	4-COOCOMe	3-PhCO-Q-4-COOH, Ac$_2$O		136–138.5	66
Ph	4-NH$_2$-6-Me	4-MeC$_6$H$_4$NHCH=C(CN)COPh, AlCl$_3$	53	218–221	222
2-MeC$_6$H$_4$		QCN, 2-MeC$_6$H$_4$MgBr	41.6	80.5–81.4	223
2,4,6-triMeC$_6$H$_2$		QCOCl, mesitylene	37	90.5–92	62
2,4,6-triMeC$_6$H$_2$-4-Ph		Dihydro derivative, chloranil	92.5	98.5–100	62
4-MeOC$_6$H$_4$	2-Me-4-Cl	The quinolone, POCl$_3$	90	138–140	194
4-MeOC$_6$H$_4$	2-Cl	2-Cl-Q-3-CHO, 4-MeOC$_6$H$_4$MgBr; PyClCrO$_3$			72
4-MeOC$_6$H$_4$	2-Cl-4-(C$_6$H$_4$-4-OMe)	2-Cl-3-CN-Q, 4-MeOC$_6$H$_4$MgBr; HCl	95	212	72
4-MeOC$_6$H$_4$	2-Cl-4-(C$_6$H$_4$-4-OMe)-6-OMe	2-Cl-3-CN-6-MeO-Q, 4-MeOC$_6$H$_4$MgBr; HCl	18	157	72
4-MeOC$_6$H$_4$	2-Cl-6,7-diOMe	2-Cl-3-CHOH-C$_6$H$_4$-4-OMe-6,7-diMeO-Q, PyClCrO$_3$	80	168–170	202*
2-Furyl	2-COOH	3-C$_4$H$_3$O-CO-Q-2-COOEt, HCl		169	224
2-Furyl	2-COOEt	2-NH$_2$C$_6$H$_4$CHO, C$_4$H$_3$O-COCH$_2$COOEt, HCl		133	224
2-Thienyl	2,4-diMe	2-NH$_2$C$_6$H$_4$COMe, C$_4$H$_4$SCOCH$_2$COMe	70	112–113	18
2-Pyridyl	6,8-diMe	6,8-diMe-Q-3-COOEt, PyLi	27	97.5–98	59
2-Pyridyl	8-CF$_3$	8-CF$_3$-Q-3-COOEt, PyLi	58	99–99.5	59
2-Pyridyl	8-Ph	8-Ph-Q-3-COOEt, PyLi	66	118–118.5	59
2-Pyridyl	2,4-di-(2-Pyridyl)-6,8-diCl	4-MeO-6,8-diCl-Q-3-COOEt, PyLi	25	239–241	59

Table 2. (*Contd.*)

R	Quinoline substituent(s)	Preparation	Yield (%)	M.p. (°C)	References
2-Pyridyl	4-Cl-6,8-diMe	4-Cl-6,8-diMe-Q-3-COOEt, PyLi	63	148(d)	59
2-Pyridyl	4-Cl-8-CF$_3$	4-Cl-8-CF$_3$-Q-3-COOEt, PyLi	63	155	59
2-Pyridyl	4-Cl-8-Ph	4-Cl-8-Ph-Q-3-COOEt, PyLi	27	102–103	59
2-Pyridyl	4-OMe-8-CF$_3$	4-Cl-8-CF$_3$-Q-3-COPy, MeONa	37	172.5–174	59
2-Pyridyl	6-OMe	6-OMe-Q-3-COOEt, PyLi	66	129–131.5	59
2-Pyridyl	4-Et$_2$N-8-CF$_3$	4-Cl-3-COPy-8-CF$_3$-Q, Et$_2$NH	60	130.5–131	59
4-Pyridyl		QLi, PyCN	44	146.7–147	216
1-PhSO$_2$-2-indolyl	2-CH$_2$OH	1-PhSO$_2$-indole, BuLi;2-CH$_2$OH-Q-3-COOH lactone	57	181–184	54
1-Et-2-Me-3-indolyl	2-COOH	Mixture, see text, Section II.2.G	51.7	196–198	48

* These references include carbonyl derivatives.

[a] A mixture was obtained and one of the isomers purified, but not identified. The ketones of Ref. 197 were prepared by methylation of the appropriate 4-quinolones, Table 52, but the reagent was not given.

[b] The starting material was a mixture (Table 52); the product was separated, but the isomeric structures were not identified.

[c] The starting materials for these reactions were probably wrongly formulated; see Section XIV.9 and Table 52.

[d] Inconsistent literature melting points.

[e] The reaction product was reported to be either *O*-alkylated, as shown, or *N*-alkylated, see Table 47.

[f] Formed as a mixture with 1-Et-2-Me-3-COCOOEt-4-quinolone (Table 52).

Table 3. Indeno[1,2-*b*]quinolinones

Entry	Substituent(s)	Preparation	Yield(%)	M.p.(°C)	References
1	None	2-NH$_2$C$_6$H$_4$CHO, 1,3-indanedione	100	175.5	225*
2	None	Parent heterocycle, Na$_2$Cr$_2$O$_7$, AcOH	82.6	173–174	226
3	None	Decarboxylation of the 10-carboxylic acid at 300–320 °C	75	175–176	227, 228
4	None	2-Ph-Q-3-COOH, SOCl$_2$, AlCl$_3$		175.5 Pic. 198.5	52*
5	None	See text, Section II.3		172–173	63
6	2-Me	Parent heterocycle, Na$_2$Cr$_2$O$_7$, AcOH	70.6	196–196.5	226
7	2,3-diMe	Parent heterocycle, Na$_2$Cr$_2$O$_7$, AcOH	52	198–200	226
8	2,4-diMe	Parent heterocycle, Na$_2$Cr$_2$O$_7$, AcOH	41	192–194	226
9	2,4,6-triMe	Parent heterocycle, Na$_2$Cr$_2$O$_7$, AcOH		177–179	226
10	2,6-diMe	Parent heterocycle, Na$_2$Cr$_2$O$_7$, AcOH	41	210–211.5	226
11	3-Me	Parent heterocycle, Na$_2$Cr$_2$O$_7$, AcOH	66	170–171	226
12	3,6-diMe	Parent heterocycle, Na$_2$Cr$_2$O$_7$, AcOH	74	175–178	226
13	6-Me	Parent heterocycle, Na$_2$Cr$_2$O$_7$, AcOH		169–172	226
14	7-Me	Parent heterocycle, Na$_2$Cr$_2$O$_7$, AcOH		162–164	226
15	8-Me-10-Cl	2-Ph-4-Cl-6-Me-Q-3-COCl, PPA	73	209–210	229
16	8-Me-10-NHMe	Entry 15, MeNH$_2$	90	195–196	229
17	10-Ph	2-NH$_2$C$_6$H$_4$COPh, 1,3-indanedione	83	187–188	160
18	8-F-10-Cl	2-Ph-4-Cl-6-F-Q-3-COCl, PPA	52	205–206	229
19	8-F-10-NHMe	Entry 18, MeNH$_2$	85	212–213	229
20	2,10-diCl	2-(4-ClC$_6$H$_4$)-4-Cl-Q-3-COCl, PPA	29	231–233	229
21	2-Cl-10-NHMe	Entry 20, MeNH$_2$	74	229	229
22	6,10-diCl	2-Ph-4,8-diCl-Q-3-COCl, PPA	57	228–229	229
23	6-Cl-10-NHMe	Entry 22, MeNH$_2$	95	256–257	229
24	7,10-diCl	2-Ph-4,7-diCl-Q-3-COCl, PPA	73	204–205	229
25	7-Cl-10-NHMe	Entry 24, MeNH$_2$	93	221–222	229

Table 3. (*Contd.*)

Entry	Substituent(s)	Preparation	Yield(%)	M.p.(°C)	References
26	8,10-diCl	2-Ph-4,6-diCl-Q-3-COCl, PPA	66	225–226	229
27	8-Cl-10-NHMe	Entry 26, MeNH$_2$	71	224–226	229
28	10-Cl	2-Ph-4-Cl-Q-3-COCl, PPA, 130 °C	62	211–212	230
		or entry 33, SOCl$_2$	36		230
29	8-Br-10-Cl	2-Ph-4-Cl-6-Br-Q-3-COCl, PPA	50	223–225	229
30	8-Br-10-NHMe	Entry 29, MeNH$_2$	80	218–219	229
31	2-OH-10-NHMe	Entry 35, HBr, AcOH	60	322(d)	229
32	8-OH-10-NHMe	Entry 37, HBr, AcOH	75	182–183	229
33	10-OH	Ethyl 2-phenyl-4-quinolone-3-carboxylate, PPA, 160–170 °C or entry 28, KOH, EtOH	90	> 360	230
34	2-OMe-10-Cl	2-(4-MeOC$_6$H$_4$)-4-Cl-Q-3-COCl, PPA	65	221–223	229
35	2-OMe-10-NHMe	Entry 34, MeNH$_2$	72	273	229
36	8-OMe-10-Cl	2-Ph-6-MeO-Q-3-COCl, PPA	43	232–233	229
37	8-OMe-10-NHMe	Entry 36, MeNH$_2$	70	181–182	229
38	10-OMe	Entry 28, MeONa	80	152–154	230
39	7,8-methylenedioxy	2-Ph-6-O-CH$_2$-O-7-Q-3-COOH, SOCl$_2$, AlCl$_3$		245–246	52
40	10-OPh	Entry 28, PhOH, reflux (no solvent)	46	244–245	230
41	10-COOH	Isatin, 1,3-indanedione, NaOH		340	228
42	10-NH$_2$	Entry 28, NH$_3$/MeOH, 90–100 °C	45	230	230
43	10-NHMe	Entry 28, MeNH$_2$	86	177–178	230, 231
44	10-NHCH$_2$Ph	Entry 28, PhCH$_2$NH$_2$			231
45	10-NHCH$_2$CH$_2$OH	Entry 28, NH$_2$CH$_2$CH$_2$OH	79	208–209 HCl 221–223	230, 231
46	10-NMe$_2$	Entry 28, Me$_2$NH	71	156–157 HCl 185–186	230, 231
47	10-Piperidyl	Entry 28, piperidine	52	128–129 HCl 185–186	230, 231
48	10-Morpholinyl	Entry 28, morpholine			231
49	10-NHPh	Entry 28, PhNH$_2$	88	214–215	230, 231
50	10-N(Me)Ph	Entry 28, PhNHMe	59	188–189	230

*These references include carbonyl derivatives.

Table 4. 4-Quinolyl Ketones

Ketones derived from Claisen condensations of quinoline-4-carboxylic esters are listed in Table 5. Diazomethyl 4-quinolyl ketones and their derivatives are in Table 6 and aminomethyl 4-quinolyl ketones in Table 7.

R	Quinoline substituent(s)	Preparation	Yield (%)	M.p. (°C)	References
Me[a]	None	Q-4-C$\equiv$CH, $HgSO_4$, H_2SO_4	75	b.p. 115–120/1 mm	232
		2-Methylindol-3-ylacetaldehyde, O_2, hv	62.3	Pic. 165–170	233
		Q-4-CN, MeMgI, PhOMe	60	b.p. 138/2 mm; HCl 200–214; Pic. 165–170(d)	234
		Q-4-CN, MeMgI, C_6H_6	51	b.p. 111–112/0.2 mm	176, 235
		Q-4-SnMe$_3$, AcCl, $PdCl_2$, C_6H_6, reflux	24	b.p. 118–121/1 mm	1,2
Me	2,8-diCF$_3$	2-Methyltryptophan, NaOCl	20	b.p. 105/0.5 mm	236, 237
Me[a]	2-Ph	2,8-diCF$_3$-Q-4-Li; MeCHO; CrO_3; AcOH	87	85–86	238
		PhCH=NPh, MeCOC$\equiv$CH, DPDC[b]	70	75–76	239
Me	2-Ph-3-Cl	2-Ph-4-CN-Q, MeMgBr		HBr 240	240
Me[a]	2-Ph-6-OEt	2-Ph-3-NH_2-4-Ac-Q, HNO_2, Conc.HCl	50	100–101	241
Me	2-Ph-8-NO_2	2-Ph-4-CN-6-EtO-Q, MeMgI			242
		2-Ph-8-NO_2-Q-4-COCl, $CH_2(COOEt)_2$, Na; H_2SO_4		157–157.5	243*
Me	2-Ph-3-NH_2	2-Ph-3-NH_2-4-CN(or $CONH_2$)-Q, MeMgI; HCl, 70°C		93–94	241
Me	2-Ph-3-phthalimido	2-Ph-3-phthalimido-Q-4-COOH, $SOCl_2$; $NaCH(COOEt)_2$; aq. H_2SO_4		240–241	241*

Table 4. (Contd.)

R	Quinoline substituent(s)	Preparation	Yield (%)	M.p. (°C)	References
Me	2-Ph-8-NH$_2$	2-Ph-4-COCH$_2$Cl-8-NO$_2$-Q or 2-Ph-4-Ac-8-NO$_2$-Q, H$_2$/Pd		103–103.5	243
Me	2-(4-MeC$_6$H$_4$)-6-Me	2-(4-MeC$_6$H$_4$)-6-Me-Q-4-COOH, MeLi	86	117–118	3, 336
Me	2-(4-MeC$_6$H$_4$)-6,8-diMe	2-(4-MeC$_6$H$_4$)-6,8-diMe-Q-4-COOH, MeLi	72	123.5–124.5	3
Me	2-(4-FC$_6$H$_4$)-6-Me	2-(4-FC$_6$H$_4$)-6-Me-Q-4-COOH, MeLi	68	116–118	3
Me[a]	2-(4-ClC$_6$H$_4$)	4-ClC$_6$H$_4$CH=NPh, MeCOC≡CH, DPDC[b]	65	101–102	239
Me	2-(4-ClC$_6$H$_4$)-6-Me	2-(4-ClC$_6$H$_4$)-6-Me-Q-4-COOH, MeLi	85	133–134	3
Me	2-(4-MeOC$_6$H$_4$)	4-MeOC$_6$H$_4$CH=NPh, MeCOC≡CH, DPDC[b]	85	b.p. 205–208 /0.1 mm	239
Me	2-(4-MeOC$_6$H$_4$)-6-Me	2-(4-MeOC$_6$H$_4$)-6-Me-Q-4-COOH, MeLi	83	123–124	3
Me	2-(4-MeOC$_6$H$_4$)-6,8-diMe	2-(4-MeOC$_6$H$_4$)-6,8-diMe-Q-4-COOH,	81	101–102	3
Me	2-(4-MeOC$_6$H$_4$)-6-MeO-7-Cl	2-(4-MeOC$_6$H$_4$)-6-MeO-7-Cl-Q-COOH, MeLi			245
Me	2-(3-NO$_2$C$_6$H$_4$)	3-NO$_2$C$_6$H$_4$CH=NPh, MeCOC≡CH, DPDC[b]	30	141–143	239
Me	2-(4-NO$_2$C$_6$H$_4$)	4-NO$_2$C$_6$H$_4$CH=NPh, MeCOC≡CH, DPDC[b]	13	194–195	239
Me[a]	2-Cl	4-Ac-2-quinolone, POCl$_3$	47	99.5–100.5	246
		2-Cl-Q, MeCHO, (tBuO)$_2$, H$_2$SO$_4$, AcOH	100		16
Me	6-Cl	6-Cl-Q-4-COCH$_2$Br, H$_2$/Pd	60	HBr 248–258	247
Me	7-Cl				248
Me	6-Br-8-Ph	6-Br-8-Ph-Q-4-CHO, MeMgBr; P$_2$O$_5$, DMSO	80	109.5–110.5	249
Me	8-OH	4-Ac-8-HO-Q-2-COOH, 245°C		110–113	250, 251
Me	2-OMe	2-MeO-Q, MeCHO, (tBuO)$_2$, H$_2$SO$_4$, AcOH	75	75	16
Me[a]	6-OMe	6-MeO-Q-4-CN, MeMgI, PhOMe		92	234, 252
		6-MeO-Q-4-CHOHCH$_2$N(CH$_2$)$_5$, AcOH		90	253

R	Quinolyl substituent	Reagents / Method	Yield (%)	m.p. (°C)	Ref.
Me[a]	2-OEt	2-EtO-Q-4-CN, MeMgI	50	57; b.p. 195–200/20 mm; HCl 108; Pic. 101	254
Me	6-OEt	6-EtO-Q-4-CN, MeMgI	80	80–81	255
Me	2-COOH-8-OH	Thiostrepton, 1M-HCl; Thiopeptin B, or Sch 51640, 6M-HCl		199–200(d); 235–240	250; 256,257
Me	2-COOH-8-OMe	4-Ac-8-MeO-Q-2-COOMe, NaOH	62	128–130	250
Me	2-COOMe-8-OMe	4-Ac-8-OH-Q-2-COOH, CH_2N_2	58	165–167	250, 251, 256
Me	2-COOEt	2-CH_2CH_2OH-4-Et-8-MeO-Q, $KMnO_4$; CH_2N_2; Q-2-COOEt, MeCHO, $(tBuO)_2$, H_2SO_4, AcOH	79	157–162	16
Me	2-NHAc	2-NH_2-4-Ac-Q, Ac_2O	83	207.5–212.5	258
Me	2-N(Ac)CH(Me)$CH_2CH_2NEt_2$	2-NHCH(Me)$CH_2CH_2NEt_2$-4-Ac-Q, Ac_2O		120–160(d)[c]	259
Me	2-CN	Q-2-CN, MeCHO, $(tBuO)_2$, H_2SO_4, AcOH	90	160	16
Me[a]	2-OEt	2-EtO-Q-4-CN, MeMgI	50	57; b.p. 195–200/20 mm; HCl 108; Pic. 101	254
Me	6-OEt	6-EtO-Q-4-CN, MeMgI	80	80–81	255
Me	2-COOH-8-OH	Thiostrepton, M HCl; Thiopeptin B, or Sch 51640, 6M HCl		199–200(d); 235–240	250; 256,257
Me	2-COOH-8-OMe	4-Ac-8-MeO-Q-2-COOMe, NaOH	62	128–130	250
Me	2-COOMe-8-OMe	4-Ac-8-OH-Q-2-COOH, CH_2N_2	58	165–167	250, 251, 256
Me	2-COOEt	2-CH_2CH_2OH-4-Et-8-MeO-Q, $KMnO_4$; CH_2N_2; Q-2-COOEt, MeCHO, $(tBuO)_2$, H_2SO_4, AcOH	79	157–162	16
Me	2-NHAc	2-NH_2-4-Ac-Q, Ac_2O	79	207.5–212.5	258

Table 4. (*Contd.*)

R	Quinoline substituent(s)	Preparation	Yield (%)	M.p. (°C)	References
Me		$2\text{-}N(Ac)CH(Me)CH_2CH_2NEt_2$, $2\text{-}NHCH(Me)CH_2CH_2NEt_2\text{-}4\text{-}Ac\text{-}Q$, Ac_2O	83	120–160(d)[c]	259
Me	2-CN	$Q\text{-}2\text{-}CN$, $MeCHO$, $(tBuO)_2$, H_2SO_4, $AcOH$	90	160	16
$CH_2CH(CH_2CH_2)_2O$		$Q\text{-}4\text{-}CN$, $C_5H_9OCH_2CN$, Bu_2NMgBr; H_2SO_4, reflux		HBr 214 (d)	6
$CHCH(CH_2CH_2)_2O$ — $CONH_2$		$Q\text{-}4\text{-}CN$, $C_5H_9OCH_2CN$, Bu_2NMgBr; H_2SO_4, 25 °C		211(d)	6
CH_2Ph		$Q\text{-}4\text{-}COOEt$, $PhCH_2MgCl$	8	91	260
CH_2Ph		$Q\text{-}4\text{-}CN$, $PhCH_2CN$, Bu_2NMgBr; 50% H_2SO_4		89–89.5; Pic. 192	6
CH_2Ph		$Q\text{-}4\text{-}CN$, $PhCH_2MgCl$		104	261
CH_2Ph	2-Ph	$2\text{-}Ph\text{-}Q\text{-}4\text{-}CN$, $PhCH_2MgCl$		Pic. 178	262*
CH_2(2-pyridyl)		$Q\text{-}4\text{-}COOEt$, 2-MePy, LDA		crude[d]	135
CH_2(2-pyridyl)	2-Ph	$2\text{-}Ph\text{-}Q\text{-}4\text{-}COOEt$, 2-MePy, LDA		250	135
$CHPh$(2-pyridyl)	2-Ph-7-Cl	$2\text{-}Ph\text{-}7\text{-}Cl\text{-}Q\text{-}4\text{-}COOEt$, $2\text{-}PhCH_2Py$, PhLi	17	162.5–163.5	263
$CH_2\text{-}2\text{-}Q$		$Q\text{-}4\text{-}COOME$, 2-Me-Q, PhLi	30	170	4
CH_2Cl	2-Ph-8-OH	$2\text{-}Ph\text{-}4\text{-}COCH_2Cl\text{-}8\text{-}AcO\text{-}Q$, EtOH, AcCl		138–139	243
CH_2Cl	8-OH	$4\text{-}COCH_2Cl\text{-}8\text{-}AcO\text{-}Q$, HCl, MeOH		132–133	259
CH_2Br	$2,8\text{-}diCF_3$	$2,8\text{-}diCF_3\text{-}4\text{-}Ac\text{-}Q$, Br_2, 48% HBr			238
CH_2Br	$7\text{-}CF_3$	$4\text{-}Ac\text{-}7\text{-}CF_3\text{-}Q$, Br_2, AcOH	63	203–205(d)	3
CH_2Br	$2\text{-}(4\text{-}MeC_6H_4)\text{-}6\text{-}Me$	$2\text{-}(4\text{-}MeC_6H_4)\text{-}4\text{-}Ac\text{-}6\text{-}Me\text{-}Q$, Br_2, AcOH	71	132–135	3, 244
CH_2Br	$2\text{-}(4\text{-}MeC_6H_4)\text{-}6,8\text{-}diMe$	$2\text{-}(4\text{-}MeC_6H_4)\text{-}4\text{-}Ac\text{-}6,8\text{-}diMe\text{-}Q$, Br_2, AcOH	88	145–147	3
CH_2Br	$2\text{-}(4\text{-}FC_6H_4)\text{-}6\text{-}Me$	$2\text{-}(4\text{-}FC_6H_4)\text{-}4\text{-}Ac\text{-}6\text{-}Me\text{-}Q$, Br_2, AcOH	72	134–136	3
CH_2Br[e]	$2\text{-}(ClC_6H_4)\text{-}6,8\text{-}diMe$	$2\text{-}(ClC_6H_4)\text{-}6,8\text{-}diMe\text{-}Q\text{-}4\text{-}COCBr(COOEt)_2$, 48% HBr	74		264
CH_2Br[e]	$2\text{-}(4\text{-}ClC_6H_4)\text{-}6\text{-}Cl$	$2\text{-}(4\text{-}ClC_6H_4)\text{-}6\text{-}Cl\text{-}Q\text{-}4\text{-}COCBr(COOEt)_2$, 48% HBr	82	HBr 249–251	264

R	Substituent	Reagents	Yield (%)	mp / salt	Ref
CH_2Br[e]	$2\text{-}(4\text{-}ClC_6H_4)\text{-}7\text{-}Cl$	$2\text{-}(4\text{-}ClC_6H_4)\text{-}7\text{-}Cl\text{-}Q\text{-}4\text{-}COCBr(COOEt)_2$, 48% HBr	89.6	HBr 236–239	264
CH_2Br[e]	$2\text{-}(4\text{-}ClC_6H_4)\text{-}6\text{-}OMe\text{-}7\text{-}Cl$	$2\text{-}(4\text{-}ClC_6H_4)\text{-}6\text{-}MeO\text{-}7\text{-}Cl\text{-}Q\text{-}4\text{-}COCBr(COOEt)_2$, 48% HBr	85	HBr 185–188	264
CH_2Br	$2\text{-}(4\text{-}MeOC_6H_4)\text{-}6\text{-}Me$	$2\text{-}(4\text{-}MeOC_6H_4)\text{-}4\text{-}Ac\text{-}6\text{-}Me\text{-}Q$, Br_2, AcOH	75	106–108	3
CH_2Br	$2\text{-}(5\text{-}Br\text{-}2\text{-thienyl})$	$2\text{-}(2\text{-Thienyl})\text{-}4\text{-}Ac\text{-}Q$, Br_2, AcOH	44.2	123.5–125	265
CH_2Br[e]	$2\text{-}OEt$	$2\text{-}EtO\text{-}4\text{-}Ac\text{-}Q$, Br_2		HBr 80	254
CH_2Br	$6\text{-}OEt$	$4\text{-}Ac\text{-}6\text{-}EtO\text{-}Q$, Br_2, 48% HBr	90	99–100(d) HCl 190 HBr 207	266
CH_2OAc	$2\text{-}Ph$	$2\text{-}Ph\text{-}Q\text{-}4\text{-}COCHN_2$, AcOH, $Cu(CrO_4)_2$	45	108–109	267
CH_2OCOEt	$2\text{-}Ph$	$2\text{-}Ph\text{-}Q\text{-}4\text{-}COCHN_2$, EtCOOH, $Cu(CrO_4)_2$	40	47–48	267
CH_2OCOPh	$2\text{-}Ph$	$2\text{-}Ph\text{-}Q\text{-}4\text{-}COCHN_2$, PhCOOH	60	121–122	267
$CH_2OCOC_6H_4\text{-}2\text{-}OAc$	$2\text{-}Ph$	$2\text{-}Ph\text{-}Q\text{-}4\text{-}COCHN_2$, $2\text{-}AcOC_6H_4COOH$	60	116–117	267
$CH_2OCOC_6H_4\text{-}3\text{-}OAc$	$2\text{-}Ph$	$2\text{-}Ph\text{-}Q\text{-}4\text{-}COCHN_2$, $3\text{-}AcOC_6H_4COOH$	71	136–137	267
$CH_2OCOC_6H_4\text{-}4\text{-}OAc$	$2\text{-}Ph$	$2\text{-}Ph\text{-}Q\text{-}4\text{-}COCHN_2$, $4\text{-}AcOC_6H_4COOH$	83	190–191	267
$CH_2OCO\text{-}2\text{-furanyl}$	$2\text{-}Ph$	$2\text{-}Ph\text{-}Q\text{-}4\text{-}COCHN_2$, 2-furoic acid	67	159–160	267
$CH_2OCO\text{-}3\text{-}Py$	$2\text{-}Ph$	$2\text{-}Ph\text{-}Q\text{-}4\text{-}COCHN_2$, Py-3-COOH	68	169–170	267
CHN_2	$2\text{-}Ph$	$Q\text{-}4\text{-}COCl$, CH_2N_2	85	104–105	267
$CH{=}NOH$	$2\text{-}Ph$	$2\text{-}Ph\text{-}4\text{-}Ac\text{-}Q$, $C_5H_{11}NO_2$, EtONa		182	240
CH_2NH_2	$2\text{-}Ph$	$2\text{-}Ph\text{-}Q\text{-}4\text{-}COCH{=}NOH$, Zn, HCOOH		2HCl 145	240
CH_2CN	$2\text{-}Me$	$2\text{-}Me\text{-}Q\text{-}4\text{-}COOEt$, MeCN, EtONa		154–155	268
$2\text{-}COCH_2CN\text{-}6\text{-}NHCOC_{17}H_{35}$		See Table 1			
CH_2CN	$2\text{-}C_{17}H_{35}$	$2\text{-}C_{17}H_{35}\text{-}Q\text{-}4\text{-}COOMe$, MeCN, MeONa			130
CH_2CN	$2\text{-}C_6H_4\text{-}4\text{-}C_{18}H_{37}$	$2\text{-}(4\text{-}C_{18}H_{37}\text{-}C_6H_4)\text{-}Q\text{-}4\text{-}COOMe$, MeCN, MeONa			130
CH_2CN	$2\text{-}C_6H_4\text{-}4\text{-}Ph$	$2\text{-}(4\text{-}PhC_6H_4)\text{-}Q\text{-}4\text{-}COOMe$, MeCN, MeONa	90		130
CH_2CN	$2\text{-}C_6H_4\text{-}4\text{-}OC_{18}H_{37}$	$2\text{-}(4\text{-}C_{18}H_{37}OC_6H_4)\text{-}Q\text{-}4\text{-}COOMe$, MeCN, MeONa			130
CH_2CN	$2\text{-}C_6H_4\text{-}4\text{-}NHCOC_{17}H_{35}$	$2\text{-}(4\text{-}C_{17}H_{35}CONHC_6H_4)\text{-}Q\text{-}4\text{-}COOMe$, MeCN, MeONa			130

Table 4. (*Contd.*)

R	Quinoline substituent(s)	Preparation	Yield (%)	M.p. (°C)	References
Et[a]		Q-4-SnMe₃, EtCOCl, PdCl₂, C₆H₆, reflux	55	b.p. 128–130 /1 mm	2
Et		Q-4-CN, EtMgI		b.p. 163–166 /8–9 mm	261
Et	2-Ph-3-NH₂	2-Ph-3-NH₂-4-CONH₂-Q, EtMgI	40.5	78	269
Et	2-Ph	2-Ph-Q-4-CN, EtMgBr		114 Pic. 182	262*
Et	2-(4-ClC₆H₄)-6,8-diMe	2-(4-ClC₆H₄)-6,8-diMe-Q-4-COOH, EtLi	65	120–121	3
Et	2-Cl	2-Cl-Q, EtCHO, (tBuO)₂, H₂SO₄, AcOH	52		16
Et	2-OEt	2-EtO-Q-4-CN, EtMgI	56	b.p. 198–203 /20 mm HCl 90 Pic. 158	254
Et	6-OEt	6-EtO-Q-4-CN, EtMgI	92		255
CH₂CH₂-(4-Me-3-piperidyl)	6-OMe	6-MeO-Q-4-COOEt, 1-PhCO-4-Me-3-piperidyl-CH₂COOEt, EtONa	34.9	diPic. 177.5–179.5	279
CHBrMe	2-OEt	2-EtO-Q-4-COEt, Br₂		HBr 114	254
CH(COPh)CN	2-NH₂-3-CN	See text, Section II.2.E, Scheme 7		240	270
Pr	2-OEt	2-EtO-Q-4-CN, PrMgI	62	b.p. 190–200 /20 mm Pic. 150	254, 266
CCH₂CH(CH₂CH₂)₂O =NOH		Q-4-COCH₂CH₂CH(CH₂CH₂)₂O, C₅H₁₁NO₂, EtONa	76	158.5–159.5	271
CCH₂CH(CH₂CH₂)₂O =NOH	6-OMe	6-MeO-Q-4-COCH₂CH₂CH(CH₂CH₂)₂O, C₅H₁₁NO₂, EtONa	82	167.5–168	271
CH₂CH₂CH(CH₂CH₂)₂- ...CH Ph	2-tBu	2-tBu-Q-4-COCH₂CH₂CH(CH₂CH₂)₂NH, PhCH₂CH·Br, K₂CO₃,		2HCl 130	272

R	Substituent	Reagents/Conditions	Yield (%)	Physical properties	Ref.
CHBrEt		2-EtO-Q-4-COPr, Br_2		HBr 123 64–65	254
CH_2COMe	2-OEt	Q-4-COOEt, MeCOMe, EtONa		b.p. 205–207/17mm HCl 180–181	273*
CH_2CH_2COOH	2-Me	2-Me-Q-4-COCH(COOEt)CH_2COOEt, H_3O^+	59.5	160–161	142
iPr		Q-4-SnMe$_3$, iPrCOCl, PdCl$_2$, C$_6$H$_6$, reflux	28.5	b.p. 128–130 /1 mm	2
iPr	2,8-diCF$_3$	2,8-diCF$_3$-Q-4-Li, iPrCHO; K$_2$Cr$_2$O$_7$, AcOH	90	Oil	238
CBrMe$_2$	2,8-diCF$_3$	2,8-diCF$_3$-Q-4-COiPr, Br$_2$	66	108–110	238
CH(COOEt)$_2$	2-(4-ClC$_6$H$_4$)-6,8-diMe	2-(4-ClC$_6$H$_4$)-6,8-diMe-Q-4-COCl, CH$_2$(COOEt)$_2$; (EtO)$_2$Mg	98		264
CH(COOEt)$_2$	2-(4-ClC$_6$H$_4$)-6-Cl	2-(4-ClC$_6$H$_4$)-6-Cl-Q-4-COCl, CH$_2$(COOEt)$_2$; (EtO)$_2$Mg			264
CH(COOEt)$_2$	2-(4-ClC$_6$H$_4$)-7-Cl	2-(4-ClC$_6$H$_4$)-7-Cl-Q-4-COCl, CH$_2$(COOEt)$_2$; (EtO)$_2$Mg			264
CH(COOEt)$_2$	2-(4-ClC$_6$H$_4$)-6-OMe-7-Cl	2-(4-ClC$_6$H$_4$)-6-MeO-7-Cl-Q-4-COCl, CH$_2$(COOEt)$_2$; (EtO)$_2$Mg			264
CH(COOEt)$_2$	2-Cl-6-OMe	2-Cl-6-MeO-Q-4-COCl, CH$_2$(COOEt)$_2$, EtOMg			274,275
CBr(COOEt)$_2$	2-(4-ClC$_6$H$_4$)-6,8-diMe	2-(4-ClC$_6$H$_4$)-6,8-diMe-Q-4-COCH(COOEt)$_2$, Br$_2$	87.5		264
CBr(COOEt)$_2$	2-(4-ClC$_6$H$_4$)-6-Cl	2-(4-ClC$_6$H$_4$)-6-Cl-Q-4-COCH(COOEt)$_2$, Br$_2$			264
CBr(COOEt)$_2$	2-(4-ClC$_6$H$_4$)-6-OMe-7-Cl	2-(4-ClC$_6$H$_4$)-6-MeO-7-Cl-Q-4-COCH(COOEt)$_2$, Br$_2$			264
CH=C(Me)NH$_2$		Q-4-COCH$_2$COMe, NH$_3$		184	273
tBu		4-Me$_3$Sn-Q, tBuCOCl, PdCl$_2$, C$_6$H$_6$, reflux	70	b.p. 130–132/mm	1,2
		Q-4-COOMe, tBuMgCl; CrO$_3$, AcOH		74–75 Pic. 198	153
tBu	2-Me	2-Me-Q-4-COOMe, tBuMgCl; CrO$_3$, AcOH		86 Pic. 180	153

Table 4. (*Contd.*)

R	Quinoline substituent(s)	Preparation	Yield (%)	M.p. (°C)	References
tBu	2,6,8-triMe	2,6,8-triMe-Q-4-COOMe, tBuMgCl; CrO$_3$, AcOH		56–57 b.p. 130/0.15 mm	153
tBu	2-Ph	2-Ph-Q-4-COOMe, tBuMgCl; CrO$_3$, AcOH		107 Pic. 178–180	153
CH(Ac)COOEt	2-NH$_2$-3-CN	See text, Section II.2.E, Scheme 7	61	270	270
CBr$_2$(CH$_2$)$_4$NH$_2$	2-(4-ClC$_6$H$_4$)	2-(4-ClC$_6$H$_4$)-Q-CO(CH$_2$)$_5$NH$_2$, Br$_2$, 48% HBr; boil		HBr 167(d)	276
CBr$_2$(CH$_2$)$_4$NH$_2$	6-OMe	6-MeO-Q-4-COCHBr(CH$_2$)$_4$NH$_2$, Br$_2$, 48% HBr	60	2HBr 170–172	277
CH$_2$CO(CH$_2$)$_3$NEt$_2$	2-Ph-7-Cl	2-Ph-7-Cl-Q-4-COOEt, MeCO(CH$_2$)$_3$NEt$_2$, NaNH$_2$	15	88	278
CMe$_2$Et		Q-4-COOMe, EtCMe$_2$MgCl; CrO$_3$, AcOH		b.p. 140/0.3 mm Pic. 199	153
CMe$_2$Et	6,8-diMe	6,8-diMe-Q-4-COOMe, EtCMe$_2$MgCl	57	b.p. 125/0.1 mm Pic. 126–127	24
CMe$_2$Et	6,8-diCH$_2$CH$_2$$i$Pr	4-COMe$_2$Et-6,8-diCH$_2$CH$_2$CHMe$_2$-Q-2-COOH, 200 °C		31–34 b.p. 210–215 /0.04 mm	24
CMe$_2$Et	2-COOH-6,8-diCH$_2$CH$_2$$i$Pr	See text, Scheme 8		134	24
CMe$_2$Et	2-COOMe-6,8-diCH$_2$-CH$_2$$i$Pr	Above, CH$_2$N$_2$	34	41–45	24
CHAc$_2$	2-NH$_2$-3-COOEt	See text, Section II.2.E, Scheme 7	74	b.p. 180/0.03 mm	270
CH$_2$CH$_2$CH-(CH$_2$CH$_2$Br)$_2$		Q-4-COCH$_2$CH$_2$CH(CH$_2$CH$_2$)$_2$O, 67% HBr, 100 °C	97	255 HBr 142–143	271
CHBrCH$_2$CH-(CH$_2$CH$_2$Br)$_2$		Q-4-COCH$_2$CH$_2$CH(CH$_2$CH$_2$Br)$_2$, Br$_2$, 48% HBr	95	HBr 136–137	271
CH$_2$CH$_2$CH$_2$CH(CH=CH$_2$)$_2$		Q-4-COCH$_2$CH$_2$CH(CH$_2$CH$_2$Br)$_2$, KOH, EtOH	59	HBr 207 Pic. 204–208	271

R	Quinoline substituent	Reagents, conditions	Yield (%)	m.p. (°C) / b.p.	Ref.
CycloC$_6$H$_{11}$		4-Me$_3$Sn-Q, cycloC$_6$H$_{11}$COCl, PdCl$_2$, C$_6$H$_6$, reflux	50	75–77, b.p. 170–171/1 mm	1,2
2-Piperidyl	2,8-diCF$_3$	2,8-diCF$_3$-4-Ac-Q, MMC[f], DMF, Tripiperideine	50	219–222	238
1-Azabicyclo[3.2.1]-oct-7-yl	6-OMe	6-MeO-Q-4-COCH$_2$CH$_2$(3-piperidyl), Br$_2$; Na$_2$CO$_3$	74.9	65 diPic. 140	279
4-Me-1-azabicyclo-[3.2.1]oct-7-yl	6-OMe	6-MeO-Q-4-COCH$_2$CH$_2$-(4-Me-3-piperidyl), Br$_2$; Na$_2$CO$_3$	82.5	60–66 diPic. 124–126	279
Ph		Q-4-CH$_2$Ph, CrO$_3$	75	60 Pic. 220	261
Ph		Q-4-SnMe$_3$, PhCOCl	47	57–61, b.p. 163–165/0.25 mm	1,2
Ph		Q-4-COOEt, PhMgBr	45	60	261
Ph		Q-4-CN, PhMgBr		58–59	234*, 252*
Ph				b.p. 155/0.5 mm HCl 204 Pic. 214(d)	237
Ph	2-Me	Q-4-COOEt (1 eq), PhMgBr (2 eq)		294[g]	280
Ph	2-Ph	2-Me-Q-4-CN, PhMgBr		118	281
Ph	2,3-diPh	2-Ph-Q-4-COCl, C$_6$H$_6$, AlCl$_3$		114 Pic. 213–214	52*
Ph	2-Ph-6-OMe	2-Ph-Q-4-CN, PhMgBr		116	282*
Ph	2-Cl	Dibenzoylstilbene, NH$_3$, H$_2$SO$_4$	53	130–132 Pic. 190–192	53
Ph	2-Cl-6-Me	6-MeO-Q-4-COOEt, PhLi	Trace		283
Ph	2-Cl-6-OMe	2-Cl-Q, PhCHO, (tBuO)$_2$	80	115	16
Ph	2-OMe	2-Cl-Q-4-COCl, C$_6$H$_6$, AlCl$_3$		105–107	52*
Ph	6-OMe	4-PhCO-6-Me-Q N-oxide, POCl$_3$	58	72.5–73	284
Ph	2-COPh	4-PhCO-6-MeO-Q N-oxide, POCl$_3$, HCl	94	128–129	285
Ph		2-MeO-Q, PhCHO, (tBuO)$_2$	35	140	16
Ph		6-MeO-Q-4-COOEt, PhLi	Trace		283
Ph		See Table 1			

Table 4. (*Contd.*)

R	Quinoline substituent(s)	Preparation	Yield (%)	M.p. (°C)	References
Ph	2-COOEt	Q-2-COOEt, PhCHO, (tBuO)$_2$	70	60	16
Ph	2-CN	Q-2-CN, PhCHO, (tBuO)$_2$	70	123	16
2-MeC$_6$H$_4$	2-CN	Q-4-CN, 2-MeC$_6$H$_4$MgBr	62	63–64.5	223
				Pic. 191–192	
4-MeC$_6$H$_4$	2-(COC$_6$H$_4$-4-Me)	See Table 1			
4-ClC$_6$H$_4$	2-(COC$_6$H$_4$-4-Cl)	See Table 1			
2-HOC$_6$H$_4$	2-(COC$_6$H$_4$-2-OH)	See Table 1			
4-MeOC$_6$H$_4$	2-(COC$_6$H$_4$-4-OMe)	See Table 1			
2-Pyrrolyl		Q-4-COCl, C$_4$H$_4$NMgI		153	286
2-Pyrrolyl	2-Ph	2-Ph-Q-4-CN, C$_4$H$_4$NLi		177	262
				Pic. 238	
2-Pyrrolyl	6-OMe	6-MeO-Q-4-COCl, C$_4$H$_4$NMgI		153	286
2-Pyrrolyl	6-EtO	6-EtO-Q-4-COCl, C$_4$H$_4$NMgI		139	286
3,5-dimethyl-2-pyrrolyl	6-OMe	6-MeO-Q-4-COCl, C$_6$H$_8$NMgI		165	286
4-Hydroxy-3-coumarinyl		Q-4-COOH, 4-hydroxycoumarin, POCl$_3$		98–100	168*
3-Pyridyl	2,8-diCF$_3$	2,8-diCF$_3$-Q-4-CHOH-Py, CrO$_3$	81	78–81	287
2-Quinolyl	2-Cl	See Table 1			
2-Quinolyl	2-Cl	See Table 1			
5,6,7,8-Tetrahydro-2-quinolyl	2-Cl	Q-4-CO-C$_9$H$_{10}$NN-oxide, POCl$_3$, HCl	98	148–149	38

*These references include carbonyl derivatives.
Py = pyridine.
[a] See also Table 7.
[b] Diisopropylperoxydicarbonate.
[c] Salt with methylenebis-(2-hydroxy-3-naphthoic acid).
[d] Reduced before characterization.
[e] See also Table 6.
[f] Magnesium methyl carbonate.
[g] Reported to be mixed with diphenyl 4-quinolylmethanol.
[h] Inconsistent literature melting points.

Table 5. 4-Quinolyl Ketones Derived from Base Catalysed Ester Condensation

TABLE 5A. 3-Keto-esters and their Derivatives

R	Quinoline substituent(s)	Compound	Reagent or method	Yield (%)	M.p. (°C)	References
H	None	A	EtONa NaNH$_2$	60–70	65 Hbr 166–168 H$_2$SO$_4$ 150(d) Pic. 160–163	288–291
		B[a]	H$_2$SO$_4$		b.p. 108–110/0.2 mm 153–155/8 mm Pic. 165–167	291–293
		C	Br$_2$, method 2		95–96 HBr 127–128	293
		D	Br$_2$, method 2	73	74–75 HBr 225–227 (d)	293 288 289

TABLE 5A. (Contd.)

R	Quinoline substituent(s)	Compound	Reagent or method	Yield (%)	M.p. (°C)	References
H	2-Me	A	EtONa	52.8	Oil	294
		B			68–69; HCl 153; Pic. 177–178	294
H	6,7-diMe	A	$NaNH_2$	58	HBr 162	295
		B		60	93	295
		D	Br_2, method 1 or 2	90	HBr 236–238	295
H	2-Ph	A	EtONa	57	52–54	240
		B[a]	HCl		77–78; HBr 240	240, 296, 297, 298*
		D	Br_2, method 1		91; HBr 225 (d)	298
H	2-Ph-6,7-diMe	A	$NaNH_2$	68	108	295
		B	NaOH	50	127–128	295
		D	Br_2, method 1		HBr 218–220	295
H	2-Ph-6-OEt	A	H_2SO_4		98–99	242
		B[a]			107; HCl 220(d)	242
		D	Br_2, method 1		129; HCl 207(d)	242
H	2-(3-$CF_3C_6H_4$)-6,8-diCl	A	EtONa	28	129–130	299
		B		36	145–146.5	299
		D	Br_2. method 1		168–169	299
H	2-(4-FC_6H_4)-6,8-diCl	A	EtONa	17	97.5–98.5	299
		B		12	137–138	299
H	2-(4-ClC_6H_4)	A	EtONa	76	99.8–100.3	258
		B[a]		72	102–103	258
		D	Br_2, method 1	95.3	HBr 250(d)	258

		Method		Yield (%)	m.p. (°C)	Refs.
H	2-(4-ClC$_6$H$_4$)-7-CF$_3$	B	Br$_2$, method 1	54	123.5–124	299
		D[a]	EtONa	95	169–171	299
H	2-(4-ClC$_6$H$_4$)-8-CF$_3$	A		51	119–120	299
		B	Br$_2$, method 1	80	178–179	299
H	2-(4-ClC$_6$H$_4$)-6-Cl-8-CF$_3$	D	Br$_2$, method 1	33	162–164	299
H	2-(4-ClC$_6$H$_4$)-6,8-diCl	B	Br$_2$, method 1	41.7	209–209.5	299
		D	EtONa; H$_2$SO$_4$	87	195–196	258
H	2-(1-C$_{10}$H$_7$)	B	Br$_2$, method 1	99	179.2–181	258
		D	EtONa; H$_2$SO$_4$	90	181–190	258
H	2-(2-C$_{10}$H$_7$)	B	Br$_2$, method 1	95	106–108	258
		D	EtONa	96	91–93	258
		A	Br$_2$, method 1	92	97–98	258
		B	EtONa	41	184–185	258
H	2-(2-Thienyl)	D	H$_2$SO$_4$	94	72–74	265
		A	EtONa; H$_2$SO$_4$	Trace	105–107	265
H	2-(3-Pyridyl)-6,8-diCl	B	Br$_2$, method 1	72	189–190	258
H	2-(4-Pyridyl)-6,8-diCl	B	EtONa; H$_2$SO$_4$	34	202–204	258
		D	EtONa		58–60	258
H	6-Cl	B	NaNH$_2$		101–103	289, 295
		A	15% H$_2$SO$_4$	83	101–103	289, 295
		B[a]	Br$_2$, method 1		HBr 236	289, 295
H	6-OMe	A	NaNH$_2$	57	84–85; HBr 159–160	288, 293
		B[a]	25% H$_2$SO$_4$	100	89–90	293
		C[a]	Br$_2$, method 2		81–82; HBr 129	289, 293
		D	Br$_2$, method 2	85	66–67; HBr 197–198	288, 293
H	2-OEt	A	EtONa	81.5	56–57	258
		B[a]	40% H$_2$SO$_4$	55	61–63	300
		C[b]	Br$_2$, method 2		161–162	258

TABLE 5A. (Contd.)

R	Quinoline substituent(s)	Compound	Reagent or method	Yield (%)	M.p. (°C)	References
H	2-OEt-6-OMe	A	EtONa	80	80	301
H		B	40% H_2SO_4	67	95–96	301
H	2-SPh	B	EtONa; H_2SO_4	40	82–82.7	258
H		D	Br_2, method 2	63	132–134	258
H	2-NH_2	B	EtONa; H_2SO_4	30.8	194–195	258
H	2-NBu_2	B	EtONa; H_2SO_4	48	b.p. 191–193/1 mm	258
H	2-$NHCH(Me)CH_2CH_2NEt_2$	B	EtONa; H_2SO_4	75	153–166[c]	258
H	2-NHAc	B			207.5–212.5	258
		D	Br_2, method 1		146.5–149.5	258, 259
Me		A	A, R = H, MeONa; MeI	88	Crude[d]	293
		B[a]	$NaNH_2$; HCl		AcOH 86–87 b.p. 170–174/12–13 mm	293, 302
Me	6-OMe	A	EtONa or $NaNH_2$		Pic. 137–138	291
		B	25% H_2SO_4	100	57–58	291, 302
		B	EtONa; HCl	60.4	46 b.p. 187/0.02 mm HCl 166–166.5 Pic. 154.5–155	271*
$CH_2CH(CH_2CH_2)_2O$		B	EtONa; HCl	40	54.5–55.5 b.p. 195–205/0.2 mm HCl 204–205 Pic. 173–173.5	271
$CH_2CH(CH_2CH_2)_2O$6-OMe						
CH_2-3-piperidyl	2-Ph	B	KH; HCl		HCl 163	135
CH_2-3-piperidyl	2-Ph	B	KH; HCl[e]		HCl 224	135
CH_2-3-piperidyl	6-OMe	D	EtONa or KH; HCl[e]		HCl 169	135
					diPic. 64–68	279

Side chain	Ring substituent	Method	Reagents	Yield (%)	Derivative, m.p.	Ref.
CH₂(4-Me-3-piperidyl)	6-OMe	D	EtONa; HCl[e]		diPic. 177.5–179.5	279
CH₂-4-piperidyl	2-tBu	B	NaH		diHCl 200	272
CH₂-4-piperidyl	2-cyclo-C₆H₁₁	B				272
CH₂-4-piperidyl	2-cyclo C₆H₁₁	B				272
Et	6-OMe	B	NaNH₂; 25% H₂SO₄	50	HCl 180	302
CH₂CH₂NMe₂	2-Ph	B	tBuOK; H₃O⁺	55	HCl 118–120	303
CH₂CH₂NMe₂	2-(4-ClC₆H₄)-7-Cl-8-Me	B	tBuOK; H₃O⁺	57	Pic. 188–190	303
CH₂CH₂NMe₂	6-OMe	B	tBuOK; H₃O⁺		b.p. 170/0.06 mm; Pic. 183–185	303
CH₂CH₂NMe₂	2-(N-piperidyl)	B	tBuOK; H₃O⁺	47	2Pic. 161–163	303
CH₂COOEt	2-Me	A	NaH	61.5		142
C₄H₉	6-OMe	B	NaNH₂; 25% H₂SO₄	17	Pic. 134–135	302
(CH₂)₄NHCOPh		A	PhCONH(CH₂)₅-COOEt		Pic. 153–154	304
(CH₂)₄NH₂	2-Ph	B	A, above, 20% HCl	17	2HCl 206	304
(CH₂)₄NH₂	6-OMe	B	tBuOK; H₃O⁺	33	HBr 230–232	303
(CH₂)₄N(Me)COPh		B	Caprolactam[f]	5		303
		A	PhCONH(CH₂)₅-COOEt	45	Pic. 105(d)	304
(CH₂)₄NHMe		B	A, above, 20% HCl		2HCl 179–180	304
		D	Br₂, method 1		diHBr 136	304
(CH₂)₄NHMe	6-OMe	B	N-Methylcaprolactam[f]		2Pic 145	304
(CH₂)₄NMe₂	2-Ph	B	tBuOK; H₃O⁺	60	Pic. 173–173.5	303
(CH₂)₄N(Et)COPh		A	PhCONEt(CH₂)₅-COOEt			304
CH(CH₂CH₂)₂-CHNHAc	6-OMe	B	A, above, 20% HCl	58	Pic. 90–91	304
		A	EtONa	30	Oil	294*
3-Pyrrolidinyl	2-Ph	B	25% H₂SO₄	30	Pic. 170	135
3-Piperidyl	2-Ph	B	KH; HCl[e]		Crude[d]	135
3-Piperidyl	2-Ph	B	KH; HCl[e]		HCl 218	135
3-Piperidyl	2-Ph-6-Cl	B	KH; HCl[e]		HCl 260	135

TABLE 5A. (Contd.)

R	Quinoline substituent(s)	Compound	Reagent or method	Yield (%)	M.p. (°C)	References
3-Piperidyl	2-(3-CF$_3$C$_6$H$_4$)	B	KH; HCl[e]		Crude[e]	135
3-Piperidyl	2-(4-ClC$_6$H$_4$)	B	KH; HCl[e]		Crude	135
3-Piperidyl	2-(2-Pyridyl)	B	KH; HCl[e]		Crude[d]	135
3-Piperidyl	2-(3-Pyridyl)	B	KH; HCl[e]		Crude[d]	135
3-Piperidyl	6-OMe	B	KH; HCl[e]		HCl 203	135
4-Piperidyl	2-Ph	B	tBuOK; HCl[e]		diHCl 259	272
SOMe	2-Ph	B	tBuOK; H$_3$O$^+$	64		303
NEt$_2$	tBu	A	NaH	96.9	153–154	265
NEt$_2$	2-(2-Thienyl)	A	NaH	93	159–160	265
					172–173	

[a] See also Table 4.
[b] Hydrolysed during bromination to a stable β-keto acid.
[c] Salt with methylene bis-(2-hydroxy-3-naphthoic acid).
[d] Crude ketone reduced to alcohol before characterization.
[e] Secondary amine group initially protected by benzoylation.
[f] Caprolactam or N-methylcaprolactam was used instead of the aliphatic ester.

TABLE 5B. Piperidyl Quinolyl Ketones

R^1		R^2	R^3	R^4	Compound	Yield (%)	Salt	M.p. (°C)	References
H		H	H	H	A, B, C[a]				305
H		Cl	H	H	A		2HBr	233–235	306
					B	56.5	2HBr	185–186	
					C			Crude[a]	
H		H	Cl	H	A		2HBr	Crude[a]	305
					B	47	2HBr	184–186(d)	
					C			Crude[a]	
H		H	H	Cl	A	45.1	2HCl	173(d)	306
							2HBr	189–190	
					B	78	2HBr	187–188	
					C			Crude[a]	
H		Cl	H	Cl	A	27	2HBr	205–206	306
					B	75.5	2HBr	209–210	
					C			Crude[a]	
H		MeO	H	H	A		2HBr	210–210.5	277
					B	34.5	2HBr	192(d)	
					C		2HCl	233(d)	
							2HBr	183–185(d)	
Me		MeO	H	H	B		2Hbr	207(d)	307
iPr		H	H	H	B		2HBr	196–197(d)	307
cyclo C_6H_{11}		H	H	H	B		2HBr	145–147(d)	307
CF_3		H	H	CF_3	C			Crude[a]	308
Ph		H	H	H	A		2HBr	225–227(d)	309
					B	52	2HBr	197–198(d)	
								Crude[a]	
Ph		Me	H	H	A	48.3	2HBr	244–245(d)	310
					B	67.7	2HBr	187.7–188.1	
					C			Crude[a]	
Ph		H	H	Me	A		2HBr	136–137	310
					B		HBr	175.8–176	
					C			Crude[a]	

TABLE 5B. (*Contd.*)

R^1	R^2	R^3	R^4	Compound	Yield (%)	Salt	M.p. (°C)	References
Ph	H	H	Ph	A	43	2HBr	224–226	310
				B		HBr	177.1–177.4	
				C			Crude[a]	
Ph	Cl	H	H	A	33.8	2HBr	Crude	276
				B	67	2HBr	211–212(d)	
				C			Crude[a]	
Ph	H	Cl	H	A	45	2HBr	262–263	309
				B	91	2HBr	245–248(d)	
				C			Crude[a]	
Ph	H	H	Cl	A		2HBr	Crude	276
				B	67	HBr	171(d)	
				C			Crude	
Ph	Cl	Cl	H	A		HBr	168–181	309
				B	90	2HBr	227–228(d)	
				C			Crude[a]	
Ph	Cl	H	Cl	A	48.8	2HBr	236–237(d)	276
						Base	172.6–173.1	
				B			Crude	
				C			Crude[a]	
Ph	MeO	H	H	A			Crude[a]	309
				B	23	2HBr	174–175(d)	
				C			Crude[a]	
Ph	H	H	MeO	A	44	2HBr	103–111	309
				B	70	2HBr	146.5–147(d)	
				C			Crude[a]	
4-MeC$_6$H$_4$	H	H	H	A, B, C			Crude[a]	311
2,5-diMeC$_6$H$_3$	H	H	H	A, B, C			Crude[a]	311
2-ClC$_6$H$_4$	H	H	H	A, B, C			Crude[a]	311
2,4-diClC$_6$H$_3$	H	H	H	A, B, C			Crude[a]	311
2,5-diClC$_6$H$_3$	H	H	H	A, B, C			Crude[a]	311
3-ClC$_6$H$_4$	H	H	H	A, B, C			Crude[a]	311
3,4-diClC$_6$H$_3$	H	H	H	A, B, C			Crude[a]	311
4-ClC$_6$H$_4$	H	H	H	A, B, C			Crude[a]	311
4-ClC$_6$H$_4$	Cl	H	Cl	A	38.9	2HBr	155.6–156	276
				B	63.7	HBr	197(d)	
				C			Crude[a]	
4-BrC$_6$H$_4$	H	H	H	A, B, C			Crude[a]	311
4-HOC$_6$H$_4$	H	H	H	A, B, C			Crude[a]	311
4-MeOC$_6$H$_4$	H	H	H	A, B, C			Crude[a]	311
4-Et$_2$NC$_6$H$_4$	H	H	H	A, B, C			Crude[a]	311
4-Xenyl	H	H	H	A, B, C			Crude[a]	311
1-NBu$_2$	H	H	H	A			Crude[a]	312
				C			Crude[a]	
1-Piperidyl	H	H	H	A		2HBr	Crude	312
				C		2HBr	Crude[a]	
1-Morpholinyl	H	H	H	A		2HBr	Crude	312
				C			Crude[a]	
3-Pyridyl	H	H	H	A, B, C			Crude[a]	311

[a] Crude ketone reduced to alcohol before characterization.

TABLE 5C. Piperidylmethyl Quinolyl Ketones

R^1	R^2	R^3	R^4	Base	M.p. (°C)	References
H	H	H	H	KH	Oil	313
H	MeO	H	H	tBuOK	2HCl 160	313
H	MeO	H	Me	NaH	(cis) Oil	313
H	MeO	H	Me	NaH	(trans) Oil	313
H	MeO	H	CH=CH$_2$	tBuOK	2HCl 200	313
tBu	H	H	H	KH	Oil	313
Ph	H	H	H	tBuOK	2HCl 259	272, 313
Ph	H	H	CH=CH$_2$	tBuOK	2HCl 210[a]	313
Ph	H	Cl	H	NaH	MeSO$_3$H 220	313
Ph	Me	H	H	NaH	HCl > 260	313
Ph	Cl	H	H	KH	HCl > 260	313
4-MeC$_6$H$_4$	H	H	H	tBuOK	2HCl > 260	313
N-CH$_2$CH$_2$Ph derivative from above,						
PhCH$_2$CH$_2$Br, K$_2$CO$_3$					2HCl 195	313
3-CF$_3$C$_6$H$_4$	H	H	H	KH	Oil	313
4-FC$_6$H$_4$	H	H	H	NaH	Fumara 206	313
2-ClC$_6$H$_4$	H	H	H	NaH	HCl 246	313
3-ClC$_6$H$_4$	H	H	H	NaH	MeSO$_3$H 210	313
4-ClC$_6$H$_4$	H	H	H	NaH	MeSO$_3$H 170	313
4-MeOC$_6$H$_4$	H	H	H	tBuOK	2HCl > 260	313
2-Thienyl	H	H	H	KH	Oil	313
2-Pyridyl	H	H	H	KH	Crude[b]	313
4-Pyridyl	H	H	H	KH	2HCl > 250	313

[a] Crude ketone reduced to alcohol before characterization.
[b] For 3-R, 4-S isomer.

TABLE 5D. Piperidonyl and Pyrrolidinyl Quinolyl Ketones and their Derivatives

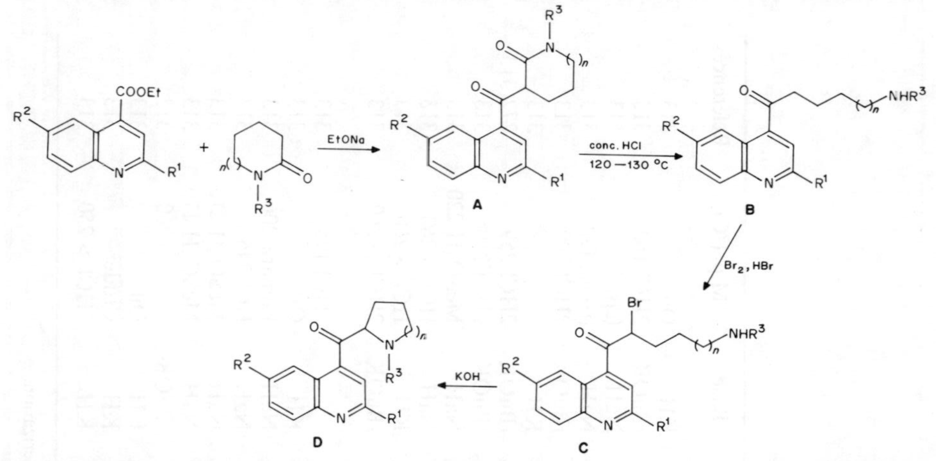

R¹	R²	R³	n	Compound A Yield (%)	Compound A M.p. (°C)	Compound B Yield (%)	Compound B M.p. (°C)	Compound C[a] M.p. (°C)	Compound D M.p. (°C)	References
H	H	H	2	17	Prl 190		2PrI 206–207 2HCl 206	2HBr 185	Crude [b]	304, 314
H	H	Me	1	60	95 Pic. 207(d)	100	HCl 171–172	2HBr 149	HCl 180	304, 314
H	H	Me	2		Pic. 222–225		2HCl 179–180 2PrI 195	2HBr 136	HCl 159–160 2Pic. 145	304, 314
H	H	Et	1	63			2HCl 166–167	2HBr 161–162	2PrI 170–172	304, 314
H	MeO	H	2			5	2PrI 229–230			304
H	MeO	Me	1	50	125–126 Pic. 211–212		2HCl 167–168	2HBr 134–135	2PrI 155(d)	304, 314
Ph	H	Me	1		135 Pic. 174		2HCl 110–120			304

Prl = picrolonate.
[a] Shown as the N-bromamine in Ref. 522, but the formula shown here is probably correct.
[b] Crude ketone reduced to alcohol before characterization.

Table 6. Diazomethyl 4-Quinolyl Ketones and their Derivatives

Quinoline substituent(s)	Compound A M.p. (°C)	X	Compound B Yield(%)	M.p. (°C)	References
None	83–84	Cl	50	101	288
7-Me	152–153(d)	Br	61	204–205	315
8-Me		Br	44	101–102	315
2-CH_2Ph-6,8-diMe		Br	66	89–91	17
2-CF_2Ph-6,8-diMe		Br	58	133–134	17
2-tBu-6-Me		Br	79		316
2-tBu-6-Cl		Br	29		316
2-tBu-6-NO_2		Br	95	HBr 125–129	316
2-(1-Adamantyl)-8-CF_3		Br	95	139–143	316
2-(1-Adamantyl)-6-Cl		Br	41		316
2-(1-Adamantyl)-6,8-diCl		Br	86	114–117	316
2-Ph	170(d)[a](70%)	Br	70	100–101 HBr 233(d)	317, 318
		OH		118	319
	98[a]	OAc[b]	71	112	267, 317, 319
		OCOEt[c]	40	47–48	267
		OCOPh[c]	60	121–122	267
		OCOC_6H_4-2-COOMe	60	116–117	267
		OCOC_6H_4-3-COOMe	71	136–137	267
		OCOC_6H_4-4-COOMe	83	190–191	267
		OCO-3-Py	86	169–170	267
		OCO-2-furyl	67	159–160	267
2-Ph-6-Me	135 (50%)	Br	80	115	317
		OAc	70	118	317

Table 6. (*Contd.*)

Quinoline substituent(s)	Compound A M.p. (°C)	X	Compound B Yield(%)	M.p. (°C)	References
2,8-diPh		Br	70	133–134(d)	315
2-Ph-5-Cl		Br	37	130–132 HBr 220–226	315
2-Ph-5,8-diCl	155(d)	Br	14	181–182	315
2-Ph-6-Cl	128–130(d)	Br	42	140–141 HBr 210–212(d)	315
2-Ph-6,8-diCl		Br	78	154–155(d)	315
2-Ph-7-Cl	141–142(d)	Br	44	106–107 HBr 210–213(d)	315
2-Ph-7-Cl-8-Me		Br	63	128–129	315
2-Ph-8-Cl		Br	63	114–115	315
2-Ph-3-Br		Br	60	95–96	320
2-Ph-6-I	150 (66%)	Br	80	137	317
		OAc	50	180	317
2-Ph-8-OAc	167–168 (96%)	Cl	87	188–188.5	243
2-Ph-6-OMe	149–150(d)	Br	70	120–121 HBr 205–206(d)	315
		OH		140	319
		OAc		123	319
2-Ph-6-OMe-7-Cl	157–159(d)	Br	56	213–215(d)	315
2-Ph-7-OMe	142	OH		162	319
		OAc		123	319
2-Ph-8-OMe		OH		74	319
		OAc		63	319
2-Ph-8-NO$_2$	167–168 (96%)	Cl	94	187–187.5	243
2-(2-Me-4,5-diClC$_6$H$_2$)-7-Cl	179–180 (d)	Br	72	138–139 HBr 214–215	315
2-(3-CF$_3$C$_6$H$_4$)-6,8-diCl					299
2-(4-MeC$_6$H$_4$)-6,8-diMe	159–160 (78%)	Br	95	168–169	3

Substituents	mp (°C)	X	mp (°C)	Ref.
2-(4-FC_6H_4)-7-F		Br	118–121	321
2-(4-FC_6H_4)-6,8-diCl	124–126(d)	Br	173–174	299
2-(4-FC_6H_4)-7-Cl	145(d)	Cl	121–124	315
2-(3,4-diClC_6H_3)		Br	114–116	315
2-(3,4-diClC_6H_3)-6,8-diCl	179–180(d)	Br	166–168	315
2-(3,4-diClC_6H_3)-6-OMe	147.5–148.5	Br	227–230	322
2-(3,4-diClC_6H_3)-7-OMe	161–163	Br	153–156	322
2-(4-ClC_6H_4)	123–125	Cl	121–123	315
			117–119	315
			HBr 259–261	
2-(4-ClC_6H_4)-6,8-diMe[b]	160–162(d)	Br	147–148	315
			HBr 224–227(d)	
2-(4-ClC_6H_4)-7-Me	163–165(d)	Br	108–109	315
2-(4-ClC_6H_4)-8-Me		Br	162–164	315
2-(4-ClC_6H_4)-8-CF$_3$		Br	156–158	299
2-(4-ClC_6H_4)-8-Ph	160(d)	Br	178–180	315
2-(4-ClC_6H_4)-3-F-6,8-diCl		Br	245–246(d)	320
2-(4-ClC_6H_4)-6-F		Br	239–241(d)	299
2-(4-ClC_6H_4)-7-F		Br	193–193.5	299
2-(4-ClC_6H_4)-3,6,8-triCl	172–173(d)	Br	170–171	320
2-(4-ClC_6H_4)6-Cl[b]		Br	HBr 250–253(d)	315
2-(4-ClC_6H_4)-6-Cl-8-CF$_3$	150–154(d)	Br	195–196	299
2-(4-ClC_6H_4)-6,8-diCl	162–165	Br	195	315
2-(4-ClC_6H_4)-6-Cl-7-OMe	173	Br	216–217	322
2-(4-ClC_6H_4)-7-Cl[b]	160–162(d)	Br	HBr 238–240	315
2-(4-ClC_6H_4)-7-Cl-8-Me		Br	150–152	315
2-(4-ClC_6H_4)-8-Cl		Br	144–147	315
2-(4-ClC_6H_4)-7-I		Br	158–160	323
2-(4-ClC_6H_4)-6-OMe	151–157(d)	Br	HBr 233–235(d)	315
2-(4-ClC_6H_4)-6-OMe-7-Cl[b]	165–167(d)	Br	166–168[a]	315
			HBr 233–235(d)	
2-(4-ClC_6H_4)-6,7-diOMe	162–163	Br	188–190[a]	322
	160–161	Br	231–232	322

Table 6. (*Contd.*)

Quinoline substituent(s)	Compound A M.p. (°C)	X	Compound B Yield(%)	M.p. (°C)	References
2-(3-IC$_6$H$_4$)-6,8-diCl		Br	88	177–178	323
2-(4-IC$_6$H$_4$)-6,8-diCl		Br	86	221–223	323
2-(4-IC$_6$H$_4$)-6-I		Br	79	HBr 258–263	323
2-(3-AcOC$_6$H$_4$)-6,8-diCl		Br	59	185–187	323
2-(4-AcOC$_6$H$_4$)-6,8-diCl		Br	85	208–210	323
2-(3,4-diMeOC$_6$H$_3$)-6-Cl	148–149	Br	82	236–238	322
2-(3,4-diMeOC$_6$H$_3$)-7-Cl	124–129	Br	88	207–209	322
2-(4-MeOC$_6$H$_4$)-6,7-diCl	170–173	Br	93.8	213–215	322
2-(4-MeOC$_6$H$_4$)-6-Cl-7-OMe	155–157	Br	91.1	196–197	322
2-(4-MeOC$_6$H$_4$)-7-Cl	169–170(d)	Br	43	233–237(d)	315
2-(4-MeOC$_6$H$_4$)-6-OMe-7-Cl	155–157	Br	94.5	222–223	322
2-(4-N$_2$CHCOC$_6$H$_4$)-6-OMe[d]	123				319
2-(4-HOCH$_2$COC$_6$H$_4$)-6-OMe[d]		OH		102–106	319
2-(4-AcOCH$_2$COC$_6$H$_4$)-6-OMe[d]		OAc		70	319
2-(2-Pyridyl)-6,8-diCl		Br	95	HBr 265–268	324
8-Ph	165(d)	Br	70	133–134(d)	315
2-Cl		Br	86	101–102	325
2-Cl-6-Me		Br	80	97–98	325
2-Cl-6,8-diMe		Br	73	71–72.5	325
2,6,8-triCl		Br	77	98–99	325
6-Cl		Br	69	HCl 205–210(d) / HBr 224–227	247
7-Cl		Br	80	120(d) / HBr 242–243	247
8-OAc	121.5–122(d) (90%)	Cl	86	140–140.5	259
6-OMe		Br	27	HBr 196.5–197	318
6-OMe-8-NO$_2$	170–171(d) (97%)	Cl	80	193.5–194.5	259
2-OEt		Br	73	100–100.5	259

2-SPh	Cl	40	130–132	259
2-COPh-6,8-diCl	Br	See Table 1		
2-CO(C_6H_4-3-CF_3)-6-Cl-8-CF_3	Br	See Table 1		
2-CO(C_6H_3,5-diCF_3)-6,8-diCl	Br	See Table 1		
2-CO(C_6H_4-4-CF_3)-6,8-diCl	Br	See Table 1		
2-CO(C_6H_3-3,4-diCl)-6-Cl-8-CF_3	Br	See Table 1		
2-CO(C_6H_3-3,5-diCl)-6-Cl-8-CF_3	Br	See Table 1		
2-CO(C_6H_4-Cl)-6-Cl-8-CF_3	Br	See Table 1		
2-CO(C_6H_3-4-Cl)-6,8-diCl	Br	See Table 1		

[a] Inconsistent literature melting points.
[b] See also Table 4.
[c] Prepared in the presence of $Cu(CrO_4)_2$.
[d] Phenyl substituents reacted simultaneously with the side chain.

Table 7. Aminomethyl 4-Quinolyl Ketones

R¹	Quinoline substituent(s)	R²	R³	Yield (%)	M.p. (°C)	References
H	None	Et	Et		Crude^a	288
H		Pr	Pr		Crude^a	288
H		C_5H_{11}	C_5H_{11}		Crude^a	288
H			$-(CH_2)_5-$	80	HBr 178 / diPic. 133–134	288, 293
H			$-(CH_2)CH(Et)CH(CH_2COOEt)-CH_2CH_2-$		diPrl 146–147	293
H			$-(CH_2)_2-O-(CH_2)_2-$		Crude^a	289
H			$-(CH_2)_2CH-[CH(CH_2CH_2)_2NH]-(CH_2)_2-$		Crude^a	288
H	6,7-diMe		$-(CH_2)_5-$		Crude^a	295
H	6,7-diMe		$-(CH_2)_2-O-(CH_2)_2-$		Crude^a	295
H	2-Ph	Ph	H		123–125	326
H	2-Ph	Me	Me		HCl 208(d) / HBr 206	240
H	2-Ph	Et	Et		HBr 164(d)^b / HBr 188(d)^b	240, 326
H	2-Ph		$-(CH_2)_5-$		HCl 235 / HBr 241(d)	240
H	2-Ph-6,7-diMe		$-(CH_2)_5-$		Crude^a	295
H	2-Ph-6,7-diMe		$-(CH_2)_2-O-(CH_2)_2-$		Crude^a	295
H	2-Ph-6-OEt	Me	Me		HBr 230(d)	242
H	2-Ph-6-OEt	Et	Et		HBr 210(d)	242

R^2	Substituent	R (ketone group)	% Yield	M.p. (°C) / derivative	Ref.
H	2-Ph-6-OEt	$-(CH_2)_5-$		82; HCl 260(d); HBr 235–240(d)	242
H	6-Cl	Et		Crude[a]	289
H	6-Cl	Bu		Crude[a]	289
H	6-OMe	Et		Crude[a]	288, 289
H	6-OMe	Bu		Crude[a]	289
H	6-OMe	Et		Crude[a]	289
H	6-OMe	Bu		Crude[a]	289
H	6-OMe	Bu		Crude[a]	288
H	6-OMe	iBu		Crude[a]	288
H	6-OMe	Bu		Crude[a]	288
H	6-OMe	$CH_2CH_2CHMe_2$		Crude[a]	288
H	6-OMe	C_5H_{11}		Crude[a]	288
H	6-OMe	C_6H_{13}		Crude[a]	288
H	6-OMe	C_7H_{15}	70	HBr 182–185	288, 293
H	6-OMe	$-(CH_2)_2CH-[CH(CH_2CH_2)_2NH]-(CH_2)_2-$		Crude[a]	288
H	2-OEt	Et	60	b.p. 175/18 mm; 2HCl 65	254
H	6-OEt	Me	66	132	266
H	6-OEt	Et		131	266
H	6-OEt	$-(CH_2)_5-$	75	2HBr 193–194(d); 158; HBr 189–190	266
Me	2-OEt	Et		b.p. 178–180/18 mm; 2HCl 89	254
Et	2-OEt	Et		b.p. 180/20 mm; 2HCl 100	254

Prl = picrolonate.
[a] Crude ketone reduced to alcohol before characterization.
[b] Inconsistent literature melting points.

Table 8. 2-Pyridyl 4-Quinolyl Ketones

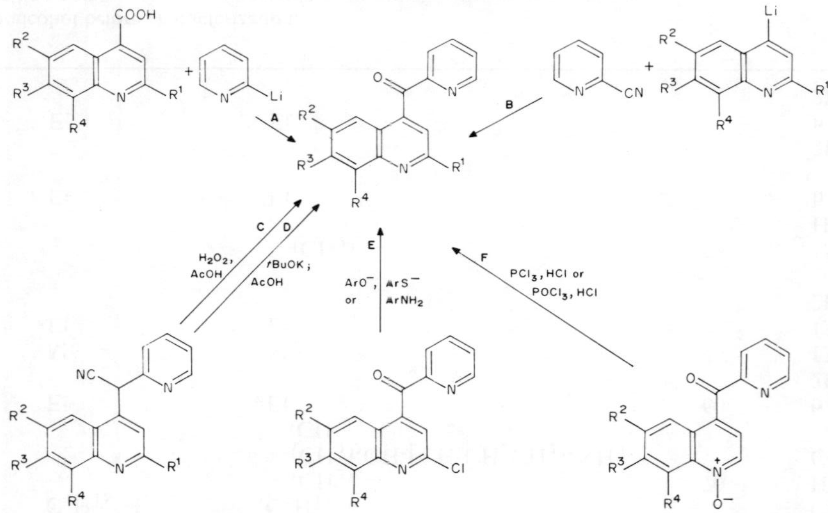

R^1	R^2	R^3	R^4	Method	Yield (%)	M.p. (°C)	References
H	H	H	H	C or F	60	116–117	327
H	Me	H	H	F	84	125–126	328
H	MeO	H	H	C	85	140–142	329
H	H	CF_3	H	A	16	118–119.5	3
H	H	CF_3	H	A on ester	67	118–119.5	3
CF_3	H	H	CF_3	A	76	141–141.5	3
CF_3	H	H	H	A	63	130–132	330
CF_3	Me	H	H	A	60	125–126[a]	330
CF_3	Me	H	H	B	76	155[a]	3
CF_3	Me	H	Me	A	73	119–120	330
CF_3	CF_3	H	H	A	13	138.5–140	331

			R	Method	Yield	m.p. (°C)	References
CF₃	H	H	CF₃	A on ester	55	138.5–140	331
H	CF₃	H	CF₃	A	27	124.5–125.5	331
H	H	Me	CF₃	A	64	98–99	330
H	H	CF₃	CF₃	D	97.8	128–129	332
H	H	CF₃	CF₃	A	61	128–129.5	331, 333
F	H	CF₃	CF₃	A for ¹⁴C=O			334
H	H	H	CF₃	Not given	90	121–122	3
Cl	H	F	CF₃	Not given	62	130–132	3
Cl	H	H	CF₃	B	48	152–153	330
Cl	H	Cl	CF₃	D	43		3
Cl	H	Cl	CF₃	A	80	193–194ª	332
H	H	Cl	CF₃	A	34	138–139ª	330
MeO	H	H	CF₃	A	31	116–117	330
MeO	H	CF₃	CF₃	A	67	132–133	330
H	H	H	CF₃	A	90	164–165	331
Cl	H	Cl	tBu	A	71	128–129	265
Cl	H	H	tBu	A	65	123–124.5	265
Me	H	Me	1-Adamantyl	A	84	138–140	335
Me	H	Me	Ph	A	45	140.5–142	336
H	H	Me	Ph	A	76	143–145	336
H	H	CF₃	Ph	A	84	130.5–132.5	336
Me	H	Me	Ph	A	72	145–146.5	336
Me	H	Me	4-MeC₆H₄	A	60	142–143	244, 336
H	H	CF₃	4-MeC₆H₄	A	81	144–145	336
F	H	H	4-MeC₆H₄	A	59	142.5–144	336
MeO	H	MeO	4-MeC₆H₄	A	74	162.5–163.5	336
Me	H	Me	4-MeC₆H₄	A	49	172–174	3, 336
Me	H	Me	4-MeC₆H₄	A	45	166–167	336
H	H	CF₃	4-FC₆H₄	A	49	155–156.5	336
H	H	H	4-FC₆H₄	A	62	140.5–142	336
Me	H	Me	4-FC₆H₄	A	75	141.5–142.5	336
Me	H	Me	4-FC₆H₄	A	60	206–207	336
H	H	H	4-ClC₆H₄	A	50	192.5–193	336
H	H	CF₃	4-ClC₆H₄	A	65	175–176	336
Me	H	Me	4-ClC₆H₄	A	70	144–146	336
Me	H	CF₃	4-ClC₆H₄	A	85	192–193	336

Table 8. (*Contd.*)

R¹	R²	R³	R⁴	Method	Yield (%)	M.p. (°C)	References
4-MeOC₆H₄	Me	H	H	A	47	147–148	336
4-MeOC₆H₄	Me	H	Me	A	68	146–147	336
4-MeOC₆H₄	H	H	Me	A	66	143–144	336
4-MeOC₆H₄	H	H	CF₃	A	66	162–163	336
2-Pyridyl	H	CF₃	H	A	12	170–172	3
2-Thienyl	H	H	H	A	70	178–180	265
Cl	H	H	H	A	69	149–150	325
Cl	H	H	H	F	84	150–151	327
Cl	Me	H	H	F	95	159–160	328
Cl	Me	H	Me	A	68	154.5–155.5	325
Cl	Cl	H	H	A	74	168–169	325
Cl	Cl	H	Cl	A	54	203–204.5	325
Cl	MeO	H	H	F	68	212–214	325
Cl	H	MeO	H	F	89	189–191	329
Cl	H	H	H	F	88	138–139	329
4-MeC₆H₄O	Me	H	H	E	71	136–137.5	325
4-MeC₆H₄O	Cl	H	H	E	82	111–112.5	325
4-MeC₆H₄O	H	H	H	E	35	87–89	325
4-ClC₆H₄O	Me	H	Me	E	40	151–153	325
4-ClC₆H₄O	Cl	H	H	E	73	134–135	325
4-ClC₆H₄O	Cl	H	Cl	E	71	163.5–165	325
4-ClC₆H₄O	Cl	H	Cl	E	80	207–208	325
3,4-diClC₆H₃O	Cl	H	H	E	52	222–223	325
4-MeSC₆H₄O	H	H	H	E	61	174.5–176	325
4-ClC₆H₄S	H	H	H	E	83	149.5–151	325
4-ClC₆H₄NH	Me	H	H	E	56	182–183	325
4-ClC₆H₄NH	Me	H	Me	E	79	180–182	325
4-ClC₆H₄NH	Me	H	H	E	45	208–209.4	325
4-ClC₆H₄NH	Cl	H	H	E	78	212–213	325
4-ClC₆H₄NH	Cl	H	Cl	E		236–237	325

[a] Inconsistent literature melting points.

Table 9. 5-Quinolyl Ketones

Carbonates and esters from 5-keto-8-hydroxyquinolines are in Tables 10 and 11 respectively. Chalcones from 5-acetyl-8-hydroxyquinolines are in Table 12.

R	Quinoline substituent(s)	Preparation	Yield (%)	M.p. (°C)	References
Me	None	Q-5-CN, MeMgI	40.5	b.p. 124–127/0.4 mm	176*
Me	2-Me-7-Br-8-OH	2-Me-5-Ac-8-HO-Q, Br$_2$	78	192	82
Me	2-Me-7-Br-8-OEt	2-Me-5-Ac-7-Br-8-HO-Q, Et$_2$SO$_4$	47	109.5	82
Me	2-Me-7-Br-8-OCH$_2$Ph	2-Me-5-Ac-7-Br-8-HO-Q, PhCH$_2$Br	53	106	82
Me	2-Me-8-OH	Friedel–Crafts reaction	70[a]	83.5	82,337
Me	2-Me-8-OAc	Friedel–Crafts reaction		114–116	337
Me	2-Me-8-OCONHMe	2-Me-5-Ac-8-HO-Q, MeNHCOCl, MeN=C=O	50	129–131	338
Me	2-Me-8-OEt	2-Me-5-Ac-8-HO-Q, EtI, KOH	77	120	82
Me	2-Me-8-OCH$_2$Ph	2-Me-5-Ac-8-HO-Q, PhCH$_2$Br	65	153	82
Me	7-CH$_2$N(CH$_2$)$_5$-8-OH	Mannich reaction	67	208	339
Me	7-CH$_2$N(CH$_2$CH$_2$CN)-COC$_{11}$H$_{23}$	Friedel–Crafts reaction			340
Me	3-Et-8-OCH$_2$COOMe	3-ET-5-Ac-8-HO-Q, ClCH$_2$COOMe, NaI, K$_2$CO$_3$		94–95	341
Me	2-C$_{12}$H$_{25}$				342
Me	2-C$_{14}$H$_{29}$				342

Table 9. (Contd.)

R	Quinoline substituent(s)	Preparation	Yield (%)	M.p. (°C)	References
Me	2-C$_{16}$H$_{33}$				342
Me	2-C$_{18}$H$_{37}$				342
Me	2-Et-8-OCH$_2$COOMe	3-Et-5-Ac-8-HO-Q, ClCH$_2$COOMe, KOH		94–95	343
Me	2-Ph-8-Cl	2-Ph-8-Cl-Q-5-COOH, SOCl$_2$; CH$_2$(COOEt)$_2$; 20% H$_2$SO$_4$	44	144.5–146	344
Me	3-Ph-6,8-diMe	3-Ph-6,8-diMe-Q-5-COCl, MeMgBr	75	133.5–134.5	345
Me	3-(4-ClC$_6$H$_4$)-6,8-diMe	3-(4-ClC$_6$H$_4$)-6,8-diMe-Q-5-COCl, MeMgBr	55	131–136	345
Me	7-Cl-8-OH	5-Ac-8-HO-Q, NaOCl	61	163–166	346
Me	7—Cl—8—OCO—[isoxazole]				347
Me	7—Cl—8—OCO—[thiazole]				347
Me	8-Cl	8-Cl-Q-5-COOH, SOCl$_2$; CH$_2$(COOEt)$_2$; 20% H$_2$SO$_4$	48	86–87.5	344
Me	7-Br-8-OH	5-Ac-8-HO-Q, KBrO$_3$, KBr, HCl	89	180 176	82, 90
Me	7-Br-8-OEt	5-Ac-7-Br-8-HO-Q, Et$_2$SO$_4$	8.5	85	82
Me	7-I-8-OH	5-Ac-8-HO-Q, I$_2$, KI	90	183(d)	348
Me	6-OH	6-AcO-Q, AlCl$_3$, PhNO$_2$	30	142–143	84*
Me	7,8-diOH	5-Ac-7-NO-8-HO-Q, Fe, conc.HCl		241–242(d) HCl 280–305(d)	349*

			Yield	mp (°C) / props	Ref.
Me	8-OH	8-AcO-Q, AlCl$_3$, PhNO$_2$, 95 °C	62–66		90,350–352
		8-AcO-Q, AlCl$_3$, 120–125 °C	53	112–113 HCl 284–285	353*
		Friedel–Crafts reaction 5-ClCH$_2$CO-8-HO-Q, Fe, 40% HCl	45	112–112.5 HCl 284–285 H$_2$SO$_4$ 248(d) Pic. 188–189	354*,355 355,356*,357
Me	7,8-diOAc	5-Ac-7,8-diHO-Q, Ac$_2$O, AcONa		121–122	349
Me	8-OAc	Friedel–Crafts reaction		134–136	337
Me	8-OCOCH$_2$CH$_2$Cl	Friedel–Crafts reaction		Oil	337
Me	8-OCOCH=CMe$_2$	Friedel–Crafts reaction		146.5–147.5	337
Me	8-OCOtBu	Friedel–Crafts reaction		100–101	337
Me	8-OCO(CH$_2$)$_8$CH=CH$_2$	Friedel–Crafts reaction		Oil	337
Me	8-OCOCH$_2$Ph	Friedel–Crafts reaction		93–94	337
Me	8-OCOCH$_2$C$_6$H$_4$-4-NO$_2$	Friedel–Crafts reaction		165–167	337
Me	8-OCOCH$_2$OPh	Friedel–Crafts reaction		118–119	337
Me	8-OCOCH$_2$OC$_6$H$_4$-2-Cl	Friedel–Crafts reaction		143–144	337
Me	8-OCOCH$_2$OC$_6$H$_4$-4-Cl	Friedel–Crafts reaction		106–107	337
Me	8-OCOCH$_2$OC$_6$H$_3$-2,4-diCl	Friedel–Crafts reaction		156–157	337
Me	8-OCOCH$_2$OC$_6$H$_2$-2,4,5-triCl	Friedel–Crafts reaction		162–163	337
Me	8-OCOCH$_2$SPh	Friedel–Crafts reaction		80.5–81.5	337
Me	8-OCOCH$_2$SC$_6$H$_3$-3,4-diCl	Friedel–Crafts reaction		84–85	337
Me	8-OCOCH$_2$CMe$_2$Ph	Friedel–Crafts reaction		101–102	337
Me	8-OCOCH=CHPh	Friedel–Crafts reaction		160–161	337
Me	8-OCOCH=CHC$_6$H$_4$-4-Cl	Friedel–Crafts reaction		157–158	337
Me	8-(Phenoxyacetyl-6-APA)[b]	5-Ac-8-HO-Q, phenoxyacetyl-6-APA, DCCI	38	78–81	358

Table 9. (Contd.)

R	Quinoline substituent(s)	Preparation	Yield (%)	M.p. (°C)	References
Me	8-OCOC$_6$H$_4$-2-Cl	Friedel–Crafts reaction		129.5–130.5	337
Me	8-OCOC$_6$H$_3$-3,4-diCl	Friedel–Crafts reaction		185–186	337
Me	8-OCOC$_6$H$_3$-3-NO$_2$-4-Cl	Friedel–Crafts reaction		168–169	337
Me	8-OCOC$_6$H$_4$-4-NO$_2$	Friedel–Crafts reaction		170–171	337
Me	8-OCONHMe	5-Ac-8-HO-Q, MeNHCOCl, MeN=C=O	43	119–121	338
Me	8-OMe	Friedel–Crafts reaction	33	125–126 b.p. 210–212/30 mm Pic. 175	83*
Me	8-OCH$_2$Ph	5-Ac-8-HO-Q, PhCH$_2$Cl, K$_2$CO$_3$, DMF	76.1	131–133	359
Me		5-Ac-8-HO-Q, PhCH$_2$Br		137	82
Me	8-OEt	5-Ac-8-HO-Q, EtI, NaOH	47	156–157	355*
Me	8-OCH$_2$COOH	5-Ac-Q-8-OCH$_2$COOEt, 15% HCl		236–238	341, 343
Me		5-Ac-8-HO-Q, ClCH$_2$COOH, NaOH			341, 343
Me	8-OCH$_2$COOMe	5-Ac-8-HO-Q, ClCH$_2$COOMe, KOH		165–167	341, 343
Me	8-OCH$_2$COOEt	5-Ac-8-HO-Q, ClCH$_2$COOEt, KOH		HCl 96–98	341, 343
Me	8-OCH$_2$COOBu	Friedel–Crafts reaction		89–90	360
Me		5-Ac-8-HO-Q, ClCH$_2$COOBu, KOH			341, 343
Me	8-OCH$_2$CONH$_2$	5-Ac-8-OCH$_2$COOEt, NH$_3$		218–220	360
Me	8-OCH$_2$CONEt$_2$	5-Ac-8-OCH$_2$COOEt, Et$_2$NH		144–145	360
Me	8-OCH$_2$CONMeBu	5-Ac-8-OCH$_2$COOEt, MeBuNH		104–106	360
Me	8-OCH$_2$CON-(CH$_2$CH$_2$)$_2$O	5-Ac-8-OCH$_2$COOEt, morpholine		167–169	360

R	Substituent	Reagents / Conditions	Yield (%)	mp	Ref.
Me	8-OSO$_3$H				361
Me	8-OPO(OEt)(SPr)	5-Ac-8-HO-Q, EtO(PrS)POCl, Et$_3$N			362
Me	8-OSbClMe$_2$Ph				363
Me	8-OSbBrMe$_2$Ph				363
Me	8-OHgMe	5-Ac-8-HO-Q, MeCOOHgMe, KOH		163	364
Me	7-NO-8-OH	5-Ac-8-HO-Q, HNO$_2$	84	190–195(d)	349
Me	6-NO$_2$-8-OMe				365
Me	7-NO$_2$-8-OH	5-Ac-8-HO-Q, NaNO$_2$, HNO$_3$, AcOH	62	280–290(d)	346
Me	7-NO$_2$-8-OCO-3-isoxazolyl				347
Me	7-NO$_2$-8-OCO-4-(1,2,3-thiadiazolyl)			165(d)	347
Me	6-NH$_2$-8-OMe			148–149(d)	365
Me	7-NH$_2$-8-OH	5-Ac-7-NO-8-HO-Q, SnCl$_2$, HCl	30	HCl 145(d)	349
Me	6-NHCOCOOMe-8-OMe	5-Ac-6-NH$_2$-8-MeO-Q, ClCOCOOMe, Py		Crude	365
CH$_2$-2-furyl	8-OH	Friedel–Crafts reaction		156	340
CF$_3$	6-OMe-8-NHCH(Me)-(CH$_2$)$_3$NHCOCF$_3$	Primaquine, TFAA, Py	24	124–126; 121–122	366, 367
CF$_3$	6-OMe-8-N(COCF$_3$)Me-(CH$_2$)$_3$NHCOCF$_3$	Primaquine, TFAA, CHCl$_3$, reflux	16.1	89–92	367
CF$_3$	6-OMe-8-NHMe(CH$_2$)$_3$NH$_2$	Above, KOH, EtOH, 25 °C	28	Oxalate 190–192	367
CF$_3$	6-OMe-8-NHCHMe(CH$_2$)$_3$NH$_2$				368
CH$_2$Cl	7-I-8-OH	5-ClCH$_2$CO-8-HO-Q, I$_2$, KI	95	227(d)	348
CH$_2$Cl	8-OH	Friedel–Crafts reaction	45.6	157–158; HCl 287(d); H$_2$SO$_4$ 263(d); Pic. 179	350, 356, 369

Table 9. (*Contd.*)

R	Quinoline substituent(s)	Preparation	Yield (%)	M.p. (°C)	References
CH₂Cl^c	8-OMe	Friedel–Crafts reaction	80	58	370
CH₂Cl^d	8-OMe	Friedel–Crafts reaction		b.p. 152/22 mm	83
CH₂CL	2-SH-8-OH			159 Pic. 183	371
CH₂Br	3-Ph-8-Me	3-Ph-8-Me-Q-5-COCl, CH₂N₂; HBr	95	HBr 237–240	345
CH₂Br	3-(4-ClC₆H₄)-8-Me	3-(4-ClC₆H₄)-8-Me-Q-5-COCl, CH₂N₂; HBr	91	94–96 HBr 220–224	345
CH₂Br	6,8-diCl	6,8-diCl-Q-5-COOH, SOCl₂, AcCl, CH₂N₂; HBr	61	HBr 244–249	323
CH₂Br	8-Cl	5-Ac-8-Cl-Q, KBrO₃, 48% HBr, AcOH	57	129–130.5	344
CHBr₂	2-Ph-8-Cl	2-Ph-5-Ac-8-Cl-Q, KBrO₃, 48% HBr, AcOH	22	181–182.5	344
CH₂I	8-OH	5-ClCH₂CO-8-HO-Q, NaI, MeCOMe	96	135(d)	348
Phthalimidomethyl	8-OMe	5-ClCH₂CO-8-MeO-Q, K phthalimide Above, conc.HCl, 190–200 °C		219	370
CH₂NH₂	8-OMe			198	370
CH₂NHiPr	2-SH-8-OH				371
CH₂N(CH₂)₅	8-OH	5-ClCH₂CO-8-HO-Q, C₅H₁₁N		HCl 262(d) 2HCl 246–249(d)	369
CH₂N(CH₂CH₂)₂O	8-OH	5-ClCH₂CO-8-HO-Q, morpholine		HCl 250(d)	369
Et	2-Me-8-OH	Friedel–Crafts reaction		97–98	337
Et	6-OH	2,3-diMe-pyrano[3,2-f]-quinolin-1-one, KOH		145–147	372

R	Substituent	Reaction	Yield (%)	m.p. (°C) / derivative	Ref.
Et	8-OH	Friedel–Crafts reaction		91^e; 124–125^e; 127–128^e; HCl 279–280^e; HCl 224^e; H_2SO_4 230–231	373*, 337, 355*, 373, 355, 373
CH_2CH_2Ph	7-CH_2NMe_2-8-OH			Pic. 192	373
CH_2CH_2Ph	8-OH	5-PhCH=CHCO-8-HO-Q, H_2/Pd/$BaSO_4$		88–90	88
CH_2CH_2Ph	8-OEt	5-PhCH=CHCO-8-EtO-Q, H_2/Pd		103–104	355
$CH_2CH_2C_6H_4$-4-NH_2	8-OH	5-(4-$NO_2C_6H_4$)CH=CHCO-8-HO-Q, H_2/Pd		144	355
$CH_2CH_2C_6H_4$-3-NH_2	8-OH	5-(3-$NO_2C_6H_4$)CH=CHCO-8-HO-Q, H_2/Pd		143–144	355
$CH_2CH_2C_6H_4$-3-NHAc	8-OH	5-(3-$NH_2C_6H_4$)CH_2CH_2CO-8-HO-Q, Ac_2O		127	355
CHClCHOHPh	7-CH_2NMe_2-8-OH	5-(2-Ph-3-oxiranyl)CO-7-(Me_2NCH_2)-8-HO-Q, HCl	58	AcOH 155–156	355
CHClCHOHPh	8-OH	5-(2-Ph-3-oxiranyl)CO-8-HO-Q, HCl	68	HCl 163–165	92
CH=CHPh^f	8-OH	Friedel–Crafts reaction		HCl >200(d)	92
CH=CHPh	8-OEt	5-Ac-8-EtO-Q, PhCHO, NaOH		143–144; HCl 252–254	340, 350
CH=CHC$_6$H$_3$-2,4-diCl	6-OH	5-Ac-6-HO-Q, 2,4-diClC_6H_3CHO, NaOH	78	115	355
CH=CHC$_6$H$_4$-4-Br	6-OH	5-Ac-6-HO-Q, 4-BrC_6H_4CHO, NaOH	53	153–154	84
CH=CHC$_6$H$_4$-4-OH	4-COOH-6-OMe	Natural product		164	84
CH=CHC$_6$H$_3$-3-OMe-4-OH	4-COOH-6-OMe	Natural product			210, 210

Table 9. (*Contd.*)

R	Quinoline substituent(s)	Preparation	Yield (%)	M.p. (°C)	References
CH=CHC$_6$H$_4$-4-OMe	6-OH	5-Ac-6-HO-Q, 4-MeOC$_6$H$_4$CHO, NaOH	76	182–183	84
CH=CHC$_6$H$_4$-3-NO$_2$[f]	8-OH	Friedel–Crafts reaction		226	350
CH=CHC$_6$H$_4$-4-NO$_2$[f]	8-OH	Friedel–Crafts reaction		222	350
CH=CH-2-furanyl	8-OH	Friedel–Crafts reaction		156	350
7-CH$_2$CH$_2$N-(CH$_2$CH$_2$)$_2$O	8-OH	5-Ac-8-HO-Q, (HCHO)$_3$, morpholine		HCl 205(d)	374
CH$_2$COPh	8-OH	5-Ac-8-HO-Q, PhCOOMe, NaNH$_2$	39	172	93
		5-(2-Ph-3-oxiranyl)CO-8-HO-Q, hv	23		93
COCOOH[g]	8-OH	8-HO-Q-5-CH$_2$CSCOOH, NH$_2$OH, EtONa	45	Ox >300	375
Pr	2-Me-8-OH	Friedel–Crafts reaction		79–81	337
Pr	8-OH	Friedel–Crafts reaction	25	96–97 HCl247	337, 355*
Pr	8-OAc	Friedel–Crafts reaction		Oil	337
Pr	8-OCONHMe	5-PrCO-8-HO-Q, MeNHCOCl, MeN=C=O	62	105–107	338
Pr	8-OCH$_2$COOMe	5-PrCO-8-HO-Q, ClCH$_2$COOMe, KOH		114–115	343
Pr	8-OCH$_2$CONH$_2$	5-PrCO-8-OCH$_2$COOEt-Q, NH$_3$		200–201	360
CF$_2$CF$_2$CF$_3$	6-OMe-8-NHCHMe-(CH$_2$)$_3$NH$_2$				368
CF$_2$CF$_2$CF$_3$	6-OMe-8-NHCHMe-(CH$_2$)$_3$NHCOCF$_2$CF$_2$CF$_3$				368
CH$_2$CH=CHPh	8-OH	Friedel–Crafts reaction		226	340

R	Quinoline substituent	Reaction	Yield (%)	M.p. (°C)	Ref.
$CH_2COCOOEt$	$6\text{-}NO_2\text{-}8\text{-}OMe$	$5\text{-}Ac\text{-}6\text{-}NO_2\text{-}8\text{-}MeO\text{-}Q$, $(COOEt)_2$, EtONa		Crude	365
iPr	8-OH	Friedel–Crafts reaction		75–76	337
iPr	8-OAc	Friedel–Crafts reaction		Oil	337
$CH(COPh)CH_2NMe_2$[h]	$7\text{-}CH_2NMe_2\text{-}8\text{-}OH$	$5\text{-}PhCOCH_2CO\text{-}8\text{-}HO\text{-}Q$, $Me_2N{=}CH_2^+$ Cl^- 2 mols	35		93
$CH(COPh){=}CH_2$[h]	$7\text{-}CH_2NMe_2\text{-}8\text{-}OH$	$5\text{-}PhCOCH_2CO\text{-}8\text{-}HO\text{-}Q$, $Me_2N{=}CH_2^+$ Cl^- 1 mol	40		93
$CH(COPh)CH_2NMe_2$[h]	8-OH				
$CH(COPh){=}CH_2$[h]	8-OH				
Bu	$8\text{-}OCH_2CONH_2$	$5\text{-}BuCO\text{-}8\text{-}OCH_2COOEt\text{-}Q$, NH_3		206–208	360
$CH{=}CH\text{-}CH{=}CHPh$	8-OH	$5\text{-}Ac\text{-}8\text{-}HO\text{-}Q$, $PhCH{=}CHCHO$, NaOH	70	147; HCl 257(d)	376
$CH{=}CH\text{-}CH{=}CHPh$	8-OH	Friedel–Crafts reaction		226	350
$CH{=}CHCH{=}CHPh$	$7\text{-}CH_2NMe_2\text{-}8\text{-}OH$	Mannich reaction		73–74; HCl 140	377
CH_2iPr	2-Me-8-OH	Friedel–Crafts reaction		63–64	337
CH_2iPr	2-Me-8-OAc	Friedel–Crafts reaction		Oil	337
CH_2iPr	2-Me-8-OCONHMe	$2\text{-}Me\text{-}5\text{-}iPrCH_2CO\text{-}8\text{-}HO\text{-}Q$, MeNHCOCl, MeN=C=O		108–109	338
CH_2iPr	8-OH	Friedel–Crafts reaction	37	87–88; HCl 202–203	355*, 337
CH_2iPr	8-OAc	$5\text{-}iPrCH_2CO\text{-}8\text{-}HO\text{-}Q$, MeNHCOCl, MeN=C=O		Oil	337
CH_2iPr	8-OCONHMe	MeNHCOCl, MeN=C=O	59	104–105	338
$(CH_2)_5NH_2$		$Q\text{-}5\text{-}COOEt$, $EtOOC(CH_2)_5NHCOPh$, $NaNH_2$; conc.HCl	28	2HBr 235–237	151
$(CH_2)_5NH_2$	8-Cl	$8\text{-}Cl\text{-}Q\text{-}5\text{-}COOEt$, $EtOOC(CH_2)_5NHCOPh$, $NaNH_2$, conc.HCl	11	2HBr 272(d)	151

Table 9. (*Contd.*)

R	Quinoline substituent(s)	Preparation	Yield (%)	M.p. (°C)	References
$CHBr(CH_2)_4NH_2$		Q-5-$CO(CH_2)_5NH_2$, Br_2, 48% HBr	100	2HBr 127–132	151
$CHBr(CH_2)_4NH_2$	8-Cl	5-$NH_2(CH_2)_5CO$-8-Cl-Q, Br_2, 48% HBr	64		151
C_6H_{13}	8-OH	Friedel–Crafts reaction		69–70 b.p. 180–195/2 mm	81, 340
C_8H_{17}[i]	8-OH	Friedel–Crafts reaction	17	62–63 HCl 163–164 HBr 229–231(d)	79, 339
C_9H_{19}	8-OH	Friedel–Crafts reaction	20–25	60–61	378
$C_{11}H_{23}$	2-Me-8-OH	Friedel–Crafts reaction		52–54	337
$C_{11}H_{23}$	2-Me-8-OCONHMe	2-Me-5-$C_{11}H_{23}CO$-8-HO-Q, MeNHCOCl, MeN=C=O	36	87–90	338
$C_{11}H_{23}$	8-OH	Friedel–Crafts reaction		45–46	337, 340, 350
$C_{15}H_{31}$	7-$CH_2N(CH_2)_5$-8-OH	Mannich reaction	54	167	339
$C_{15}H_{31}$	8-OH	Friedel–Crafts reaction	11	HBr 223	339
$C_{21}H_{43}$	8-OH	Friedel–Crafts reaction		43	340
CH-CHPh (*trans*) (epoxide)	7-CH_2NMe_2-8-OH[k]	Mannich reaction	81	172 HCl 175	91
CH-CHPh (*trans*) (epoxide)	7-CH_2NEt_2-8-OH	Mannich reaction	85	157	91

Ketone	Substituent	Method/Reagents	Yield (%)	M.p.	Ref.
CH-CHPh (trans) (epoxide)	7-CH$_2$N(CH$_2$)$_4$-8-OH	Mannich reaction	90	181	91
CH-CHPh (trans) (epoxide)	7-CH$_2$N(CH$_2$)$_5$-8-OH	Mannich reaction	80	147	91
CH-CHPh (trans) (epoxide)	7-CH$_2$N(CH$_2$)6-8-OH	Mannich reaction	87	133	91
CH-CHPh (trans) (epoxide)	7-CH$_2$N(CH$_2$CH$_2$)$_2$O-8-OH	Mannich reaction	67	88	91
CH-CHPh (trans) (epoxide)	8-OH^k	5-ClCH$_2$CO-8-HO-Q, PhCHO, NaOH	55	123	91
		5-PhCH=CHCO-8-HO-Q, H$_2$O$_2$, NaOH	72		91
		5-PHCHOHCHClCO-8-HO-Q, AcONa			92
Ph		Q-5-CN, PhMgBr		118–119 b.p. 218–221/3 mm	379
Ph	2-Me-8-OH	Friedel–Crafts reaction		110–111	337
Ph	2-Me-8-OAc	2-Me-5-PhCO-8-HO-Q, Ac$_2$O		177.5–178.5	337
Ph	2-Me-8-OCH$_2$COOEt	2-Me-5-PhCO-8-HO-Q, ClCH$_2$COOEt, K$_2$CO$_3$		125–126	341, 343

Table 9. (*Contd.*)

R	Quinoline substituent(s)	Preparation	Yield (%)	M.p. (°C)	References
Ph	2-Me-8-OCONHMe	2-Me-5-PhCO-8-HO-Q, MeNHCOCl or MeN=C=O, Et$_3$N	33	152–154	338
Ph	2-Me-8-OCH$_2$CONH$_2$	2-Me-5-PhCO-8-OCH$_2$COOMe-Q, NH$_3$		208–209	360
Ph	2-Me-8-OCH$_2$CONHNH$_2$	2-Me-5-PhCO-8-OCH$_2$-COOMe-Q, NH$_2$NH$_2$		168–169	360
Ph	7-CH$_2$NMe$_2$-8-OH	Mannich reaction		113 HCl 135–140	88
Ph	7-CH$_2$N(Et)CH$_2$CH$_2$NEt$_2$-8-OH	Mannich reaction		54–56	88, 377
Ph	7-CH$_2$N(CH$_2$)$_5$-8-OH	Mannich reaction	81	251	339
Ph	7-I-8-OH	5-PhCO-8-HO-Q, I$_2$, KI	90	209–210	348
Ph	8-OH	Friedel–Crafts reaction	77	118–119^e HCl 225–226 BHSO$_4$ 219–220 Pic. 143–145^e	337, 355, 356*
		5-(4-NH$_2$C$_6$H$_4$CO)-8-HO-Q, HNO$_2$	50	112^e Pic. 123^e	380*
		8-HO-Q copper chelate, PhCOCl, AlCl$_3$,	36	118.5–119.5	352, 381
		8-PhCOO-Q, $h\nu$[l],	151	73	85
		8-PhCOO-Q, AlCl$_3$		120 HCl 249–251	351, 353*
Ph	8-OAc	5-PhCO-8-HO-Q, Ac$_2$O		103–104	337
Ph	8-O-CO-NHMe	5-PhCO-8-HO-Q, MeNHCOCl or MeN=C=O, Et$_3$N	61	128–129	338

Ar	Quinolyl substituent	Method	Yield (%)	M.p. (°C) / derivatives	Ref.
Ph	8-OMe[d]	Friedel–Crafts reaction		115	83
Ph	8-O-CH$_2$-CO-NH$_2$	5-PhCO-8-O-CH$_2$-COOMe-Q, NH$_3$		198–199	360
2-MeC$_6$H$_4$		Q-5-CN, 2-MeC$_6$H$_4$MgBr	33	91.7–92.2 b.p. 190–200/2 mm	382
2-ClC$_6$H$_4$	2-Me-8-OH	Friedel–Crafts reaction		125–126	337
2-ClC$_6$H$_4$	7-CH$_2$NMe$_2$-8-OH	Mannich reaction		207	88
2-ClC$_6$H$_4$	8-OH	Friedel–Crafts reaction		127–128 HCl 225(d) BHSO$_4$ 208–210 Pic. 178–180	88, 337, 383
2-ClC$_6$H$_4$	8-O-CO-NHMe	5-ArCO-8-HO-Q, MeNHCOCl or MeN=C=O, Et$_3$N	49	113–115	338
2,4-diClC$_6$H$_3$	7-CH$_2$NMe$_2$-8-OH	Mannich reaction		HCl 222(d) 192–194	88
2,4-diClC$_6$H$_3$	8-OH	Friedel–Crafts reaction		HCl 164 BHSO$_4$ 230 Pic. 175	88, 383
2,4-diClC$_6$H$_4$	8-O-CO-NHMe	5-ArCO-8-HO-Q, MeNHCOCl or MeN=C=O, Et$_3$N	42	125(d)	338
3,4-diClC$_6$H$_3$	8-OH	Friedel–Crafts reaction		182–183 HCl 258 BHSO$_4$ 235 Pic. 148–150	383
4-ClC$_6$H$_4$	2-Me-8-OCONHMe	2-Me-5-PhCO-8-HO-Q, MeNHCOCl or MeN=C=O, Et$_3$N	23	134–135	338
4-ClC$_6$H$_4$	7-CH$_2$NMe$_2$-8-OH	Mannich reaction		92–93 HCl 140	87, 88, 377
4-ClC$_6$H$_4$	7-CH$_2$NEt$_2$-8-OH	Mannich reaction		146–147	

Table 9. (*Contd.*)

R	Quinoline substituent(s)	Preparation	Yield (%)	M.p. (°C)	References
4-ClC$_6$H$_4$	7-Piperidylmethyl-8-OH	Mannich reaction		142–143	87
4-ClC$_6$H$_4$	7-Morpholinylmethyl-8-OH	Mannich reaction		154–155	87
4-ClC$_6$H$_4$	7-I-8-OH	5-(4-ClC$_6$H$_4$CO)-8-HO-Q, I$_2$, KI		188–189	87
4-ClC$_6$H$_4$	8-OH	Friedel–Crafts reaction	25.6	218–219 HCl 250 BHSO$_4$ 220 Pic. 166–168	87,88,366 337,383
4-ClC$_6$H$_4$	8-O-CO-NHMe	5-ArCO-8-HO-Q, MeNHCOCl or MeN=C=O, Et$_3$N	45	216–217	338
2-BrC$_6$H$_4$	8-O-CO-NHMe	5-ArCO-8-HO-Q, MeNHCOCl or MeN=C=O, Et$_3$N	32	116–117	338
4-BrC$_6$H$_4$	8-O-CO-NHMe	5-ArCO-8-HO-Q, MeNHCOCl or MeN=C=O, Et$_3$N	47	213–215	338
4-NO$_2$C$_6$H$_4$	8-OH	Friedel–Crafts reaction		192–194 HCl 272 BHSO$_4$ 166 Pic. 205–208	383
4-NO$_2$C$_6$H$_4$	8-OMe	Friedel–Crafts reaction	6	201–202 HCl 133–136 Pic. 208–209(d)	380*
4-NH$_2$C$_6$H$_4$	8-OH	5-(4-NO$_2$C$_6$H$_4$CO)-8-MeO-Q,HI	86	255 Pic. 180(d)	380
7-Me-2,3-dihydro-4-indenyl		Q-5-CN, ArMgBr	17.5	135–135.5 b.p. 240–250/2 mm	382

| 8-HO-5-quinolyl | 8-OH | 8-HO-Q, CCl$_4$, KOH, EtOH, reflux | 3^m | 282–283 HCl 309–311 | 384 |
| 8-AcO-5-quinolyl | 8-OAc | | | 201–202 | 384 |

*These references include carbonyl derivatives.

BHSO$_4$ = acid sulphate

a Product contaminated with 5% of 7-acetyl-8-hydroxy-2-methylquinoline; see Table 14.

c Position of substitution in Friedel–Crafts reaction not determined, but presumably went to C-5 as shown.

d Burger[369] noted the ready cleavage of 8-methoxyquinolines during Friedel–Crafts reactions. The melting point given for this compound is close to that of the 8-hydroxy derivative.

e Inconsistent literature melting points.

f See also Table 12.

g Only oxime isolated.

h Mixtures formed were not separated.

i Obtained as a mixture with the 7-ketone; see Table 14.

k Original structure[88] corrected by Möhrle[91].

l Obtained as a mixture with 7-benzoyl-8-hydroxyquinoline; see Table 14.

m The main products were 8-hydroxyquinoline-5-carboxylic; acid (22%) and its ethyl ester (6%).

Table 10. Carbonates from 8-Hydroxy-5-quinolyl Ketones[385]
See also Table 11 for 7-substituted derivatives

R[1]	R[2]	R[3]	M.p. (°C)	B.p. (°C)
Me	H	Me	127–129	
Me	H	CH₂—(tetrahydrofuran-2-yl)		190–195/0.002 mm
Me	H	CH₂—(2,2-dimethyl-1,3-dioxolan-4-yl)		Crude
Me	H	CH_2Ph	75–76	
Me	H	$CH_2C_6H_4\text{-}4\text{-}Cl$	83.5–84.5	
Me	H	Et	80–81	
Me	H	CH_2CH_2Cl	94–96	
Me	H	CH_2CH_2OMe	60–61	
Me	H	$CH_2CH_2OCH_2Ph$		242–245/0.03 mm
Me	H	CH_2CH_2OEt		179–183/0.005 mm
Me	H	CH_2CH_2OBu		190–195/0.003 mm
Me	H	$CH_2CH_2OC_6H_{13}$		191–196/0.004 mm
Me	H	CH_2CH_2OPh	103–104	
Me	H	$CH_2CH_2OC_6H_4\text{-}2\text{-}Cl$	79–81	
Me	H	CH_2CH_2SPh	63–64	
Me	H	Pr	49–50	
Me	H	$CH_2CH{=}CH_2$		185–186/0.12 mm
Me	H	iPr	81–82	
Me	H	iBu	51–53	
Me	H	sBu	57–59	
Me	H	C_5H_{11}		193–196/0.13 mm
Me	H	C_6H_{13}		196–199/0.1 mm
Me	H	$CH_2CHEtC_4H_9$		185–187/0.005 mm
Me	H	$C_{12}H_{25}$	47–49	
Me	H	cycloC_6H_{11}	85–86	
Me	H	Ph	93–95	
Me	H	$C_6H_4\text{-}2\text{-}Me$	105–106	
Me	H	$C_6H_4\text{-}4\text{-}Me$	93–94	
Me	H	$C_6H_4\text{-}4\text{-}i$Bu	119–120	
Me	H	$C_6H_4\text{-}2\text{-}Cl$	63–65	
Me	H	$C_6H_3\text{-}2,4\text{-diCl}$	141–143	
Me	H	$C_6H_3\text{-}3,4\text{-diCl}$	122–123	
Me	H	$C_6H_4\text{-}4\text{-}Cl$	104–105	
Me	H	$C_6H_4\text{-}2\text{-OMe}$	109–110	
Me	H	$C_6H_4\text{-}4\text{-}NO_2$	144.5–145.5	

Table 10. (*Contd.*)

R^1	R^2	R^3	M.p. (°C)	B.p. (°C)
Me	Me	$CH_2C_6H_4$-4-Cl	124–125	
Me	Me	Et	76–78	
Me	Me	CH_2CH_2SPh	77–78	
Me	Me	Ph	112.5–113.5	
Me	Me	C_6H_4-2-OMe	131.5–132.5	
Me	Me	C_6H_4-4-NO_2	86–88	
$ClCH_2$	H	Me	129.5–130.5	
Pr	H	Me	68–69	
iPrCH$_2$	H	Me	85–86	
iPrCH$_2$	H	Ph	107.5–108	
Ph	H	Me	138.5–139.5	
Ph	H	CH_2CH_2SPh	102–103	
Ph	Me	Me	178.5–180	
2-ClC$_6$H$_4$	H	Me	128.5–129.5	

Table 11. Esters and Carbonates from 7-Substituted 5-Acetyl-8-hydroxyquinolines[346]

R^1	R^2	M.p. (°C)
Cl	Et	86–88
Cl	$CH=CH_2$	120–122
Cl	tBu	82–86
Cl	$(CH_2)_8CH=CH_2$	Oil
Cl	3-Methylcyclohexyl	Oil
Cl	Ph	
Cl	OMe	138–140
Cl	$OCH_2CH=CH_2$	87–88
Cl	OcycloC$_6$H$_{11}$	
Cl	SEt	
Br	Me	142–143
Br	$CH=CH_2$	129–130

Table 11. (*Contd.*)

R^1	R^2	M.p. (°C)
Br	C(Br)=CH$_2$	132–134
Br	CH=CH-Me	144–147
Br	C(Me)=CH$_2$	124–126
Br	CH=CMe$_2$	119–123
Br	tBu	90–94
Br	CH=CH-CH=CH-Me	94–96
Br	cycloC$_6$H$_{11}$	107–109
Br	3-Methylcyclohexyl	Oil
Br	Cyclohex-3-enyl	91–93
Br	C$_6$H$_4$-4-Cl	135–137
Br	C$_6$H$_4$-4-NO$_2$	
Br	OMe	142–144
Br	OEt	107–109
Br	SBu	79–80
I	Me	135–138
I	CH$_2$Cl	120–123
I	CH=CH$_2$	118–120
I	CH=CH-Me	127–128
I	C(Me)=CH$_2$	112–114
I	CH=CMe$_2$	136–139
I	CH=CH-CH=CH-Me	133–136
I	cycloC$_6$H$_{11}$	106–108
I	3-Methylcyclohexyl	Oil
I	Ph	192–194
I	OMe	106–108
NO$_2$	Me	
NO$_2$	CH$_2$Cl	
NO$_2$	CH=CH$_2$	112–114
NO$_2$	C(Me)=CH$_2$	114–116
NO$_2$	CH$_2$$i$Pr	
NO$_2$	tBu	
NO$_2$	(CH$_2$)$_8$CH=CH$_2$	
NO$_2$	cycloC$_6$H$_{11}$	
NO$_2$	Ph	127–130
NO$_2$	OMe	126–128
NO$_2$	OCH$_2$-CH=CH$_2$	76–78
NO$_2$	SEt	

Table 12. Chalcones from 5-Acetyl-8-hydroxyquinoline and their Mannich derivatives. Other chalcones are in Table 9

Ar	Compound A			R¹	R²	Compound B		References
	Yield(%)	M.p. (°C)	HCl M.p. (°C)			M.p. (°C)	HCl M.p. (°C)	
Ph	73	143–144	252–254					340, 376, 386
	39	142		Me	Me	193		88, 355, 377, 387
				Et	Et	85–90	180–185	88, 377
				HOCH$_2$CH$_2$	HOCH$_2$CH$_2$	136		88, 377
				Et$_2$NCH$_2$CH$_2$	Et	85–90		88, 377
				Et$_2$NCH$_2$CH$_2$	Et$_2$NCH$_2$CH$_2$	85–90		88, 377
				Bu	Bu	51–53		88, 377
				C$_{12}$H$_{25}$	Me	76		88, 377
				PhCH$_2$	Me	124	227(d)	88, 377
				—(CH$_2$)$_4$—		178	232(d)	88, 377
				—(CH$_2$)$_5$—		71–72	220(d)	88, 377
				—(CH$_2$)$_2$-NMe-(CH$_2$)$_2$—		181	203–204(d)	88, 377

Table 12. (*Contd.*)

Ar	Compound A			Compound B				References
	Yield(%)	M.p. (°C)	HCl M.p. (°C)	R¹	R²	M.p. (°C)	HCl M.p. (°C)	
2-MeC₆H₄	53	140						388
3-MeC₆H₄	65.6	127–128						389
4-MeC₆H₄	76.4	182–183						389
2-ClC₆H₄	72.5	176–177						389
2,6-diClC₆H₃	77	116						388
4-ClC₆H₄	76	197		Me	Me	183	215–216(d)	88, 377
	82	193–195						376
2-BrC₆H₄	72	197						390, 391
2-Br-4,5-diMeOC₆H₂	74	214						388
2-Br-3,4-OCH₂OC₆H₂	58	234						388
3-Br-4-HO-5-MeOC₆H₂	67	174						388
3-Br-4,5-diMeOC₆H₂	85	132						388
4-BrC₆H₄	36	197						388
2-HOC₆H₄	46	203						388
2-HO-5-BrC₆H₃	32	97						388
2,4-diHOC₆H₃	58	183(d)						388
2-HO-4-MeOC₆H₃	37	274(d)						388
3-HOC₆H₄	35	235(d)						388
4-HOC₆H₄	69	230						388
2-MeOC₆H₄	68	104						388
2,3-diMeOC₆H₃	64	143						388
3-MeOC₆H₄	70	137						390, 391
3-MeO-4-HOC₆H₃	53	181–182	268(d)					389
3-MeO-4-HO-5-NO₂C₆H₂		192						376
3,4-diMeOC₆H₃	86.8	185		Me	Me	132	232	88, 377, 389

R	%	mp	mp (deriv.)	R'	mp	mp (deriv.)	References
4-MeOC$_6$H$_4$	80	193–194	271(d)				376
3,4-OCH$_2$O-C$_6$H$_3$	94	180–181	256				89, 376
2-NO$_2$C$_6$H$_4$	65	164					388
2-NO$_2$-4,5-diMeOC$_6$H$_2$	68	147					388
3-NO$_2$C$_6$H$_4$[a]	63.5	229–230					355, 389
4-NO$_2$C$_6$H$_4$[a]	89	224	280				340, 355, 376
4-Me$_2$NC$_6$H$_4$	74	191–192	263–265(d)				376
2-AcNHC$_6$H$_4$		214–215					355
Furan-2-yl	89	155–155.5		Me	190	220	88, 90, 340, 377
5-NO$_2$-furan-2-yl		190		Me		254(d)	88, 377
4-Pyridyl		191					392
8-Hydroxyquinolin-5-yl	80	271	267(d)				376
Quinolin-6-yl		210		Me	220	263–265	88, 377

[a] See also Table 9.

Table 13. 6-Quinolyl Ketones

R	Quinoline substituent(s)	Preparation	Yield(%)	M.p. (°C)	References
Me	None	Q-6-COOEt, EtOAc, EtONa; 25% H_2SO_4	90	76	172, 176, 393, 394
		1,2,3,4-tetraH-6-Ac-Q, $Hg(OAc)_2$, aq. AcOH, 188 °C, 5 h	50	75–76	281
		Q-6-CN, MeMgI	42	Pic. 242	395
		Q-6-CHOHMe, CrO_3, AcOH	39.5	73–73.5	396*
				75–76 Pic. 236	
Me	2-Me	2-Me-6-Et-Q, CrO_3, $3M$-H_2SO_4	75	92	94*
		4-$AcC_6H_4NH_2$, $(MeCHO)_3$, HCl		b.p. 318–320 Pic. 208–211	397*
Me	2,4-diMe	2,4-diMe-6-Et-Q, CrO_3, $3M$-H_2SO_4	30		94*
Me	2-Me-4-Cl	2-Me-6-Ac-4-quinolone, $POCl_3$	77	137–137.5	398*
Me	4-Me				399
Me	5,7-diMe	5,7-diMe-Q, AcCl, $AlCl_3$, CS_2	4	74–78	83*
Me	2-Et-3-Me	2,6-diEt-3-Me-Q, CrO_3, $3M$-H_2SO_4	85		94*
Me	2-CH=CHPh-4-COOH	4-$AcC_6H_4NH_2$, 2-HOC_6H_4CHO, MeCOCOOH	80	131	400
Me	2-CH=CHPh-4-CONHNHCOPh	2-CH=CHPh-6-Ac-Q-4-COOH, $SOCl_2$; $PhCONHNH_2$		273	97
Me	2-CH=CHPh-4-CONHNHCOCH$_2$Ph	2-CH=CHPh-6-Ac-Q-4-COOH, $SOCl_2$; $PhCH_2CONHNH_2$		209	97

Me	2-Ph-4-COOH	4-AcC$_6$H$_4$NH$_2$, PhCHO, MeCOCOOH	92	173–175	401, 402
Me	2-Ph-4-CONHNHCOPh	2-Ph-6-Ac-Q-4-COOH, SOCl$_2$; PhCONHNH$_2$		205	97
Me	2-Ph-4-CONHNHCOCH$_2$Ph	2-Ph-6-Ac-Q-4-COOH, SOCl$_2$; PhCH$_2$CONHNH$_2$		215	97
Me	2-(2-ClC$_6$H$_4$)-4-COOH	4-AcC$_6$H$_4$NH$_2$, 2-ClC$_6$H$_4$CHO, MeCOCOOH	73	138–140	401
Me	2-(4-ClC$_6$H$_4$)-4-COOH	4-AcC$_6$H$_4$NH$_2$, 4-ClC$_6$H$_4$CHO, MeCOCOOH	78	188–190	401
Me	2-(2-HO-3-BrC$_6$H$_3$)-4-COOH	4-AcC$_6$H$_4$NH$_2$, 2-HO-3-BrC$_6$H$_3$CHO, MeCOCOOH	78	190–192	401
Me	2-(2-HO-3,5-diBr-C$_6$H$_2$)-4-COOH	4-AcC$_6$H$_4$NH$_2$, 2-HO-3,5-diBrC$_6$H$_2$CHO, MeCOCOOH	89	175–177	401
Me	2-(2-HOC$_6$H$_4$)-4-COOH	4-AcC$_6$H$_4$NH$_2$, 2-HOC$_6$H$_4$CHO, MeCOCOOH	80	195	400, 401
Me	2-(2-HOC$_6$H$_4$)-4-CONHNHCOPh	2-(2-HOC$_6$H$_4$)-6-Ac-Q-4-COOH, SOCl$_2$; PhCONHNH$_2$		195	97
Me	2-(2-HOC$_6$H$_4$)-4-CONHNHCOCH$_2$Ph	2-(2-HOC$_6$H$_4$)-6-Ac-Q-4-COOH, SOCl$_2$; PhCH$_2$CONHNH$_2$		178	97
Me	2-(3-HOC$_6$H$_4$)-4-COOH	4-AcC$_6$H$_4$NH$_2$, 3-HOC$_6$H$_4$CHO, MeCOCOOH	86	164–166	401
Me	2-(4-HOC$_6$H$_4$)-4-COOH	4-AcC$_6$H$_4$NH$_2$, 4-HOC$_6$H$_4$CHO, MeCOCOOH	81	157–159	401
Me	2-(3-MeO-4-HOC$_6$H$_3$)-4-COOH	4-AcC$_6$H$_4$NH$_2$, 3-MeO-4-HOC$_6$H$_3$CHO, MeCOCOOH	80	144	400
Me	2-(3-MeO-4-HOC$_6$H$_3$)-4-CONHNHCOPh[a]	2-(3-MeO-4-HOC$_6$H$_3$)-Q-4-COOH, SOCl$_2$; PhCONHNH$_2$		222	97

Table 13. (*Contd.*)

R	Quinoline substituent(s)	Preparation	Yield(%)	M.p. (°C)	References
Me	2-(3-MeO-4-HOC$_6$H$_3$)-4-CONHNHCOCH$_2$Ph[a]	2-(3-MeO-4-HOC$_6$H$_3$)-Q-4-COOH, SOCl$_2$; PhCH$_2$CONHNH$_2$		228	97
Me	2-(4-MeOC$_6$H$_4$)-4-COOH	4-AcC$_6$H$_4$NH$_2$, 4-MeOC$_6$H$_4$CHO, MeCOCOOH	82	125–127	401
Me	2-(4-NO$_2$C$_6$H$_4$)-4-COOH	4-AcC$_6$H$_4$NH$_2$, 4-NO$_2$C$_6$H$_4$CHO, MeCOCOOH	80	158	97, 400, 401
Me	2-(4-NO$_2$C$_6$H$_4$)-4-CONHNHCOPh	2-(4-NO$_2$C$_6$H$_4$)-6-Ac-Q-4-COOH, SOCl$_2$; PhCONHNH$_2$		245	97
Me	2-(4-NO$_2$C$_6$H$_4$)-4-CONHNHCOCH$_2$Ph	2-(4-NO$_2$C$_6$H$_4$)-6-Ac-Q-4-COOH, SOCl$_2$; PhCH$_2$CONHNH$_2$		192	97
Me	4-Cl	6-Ac-4-quinolone, POCl$_3$	20	47–48	403
Me	4-Cl-7-OH-8-Pr				404
Me	5,8-diOH-7-COMe	Py-2,3-diCOOEt, (CH$_2$Ac)$_2$, Na, xylene	36	243(d)	405
Me	2-OEt-4-OMe	2-EtO-6-Ac-4-quinolone, MeI, K$_2$CO$_3$	62	138.5–139.5	410
Me	2-COOH-4-Me-7-OH-8-Pr	1-Ac-2-HO-3-Pr-4-NHCH(COOH)CH$_2$Ac-C$_6$H$_2$, PPA, 100 °C	30	125–127	406,407
Me	2-COOH-4-NHMe-7-OH-8-Pr	2-COOMe-4-Cl-6-Ac-7-HO-8-Pr-Q, MeNH$_2$; 70% H$_2$SO$_4$			406, 407
Me	2-COOH-4-NMe$_2$-7-OH-8-Pr	2-COOMe-4-Cl-6-Ac-7-HO-8-Pr-Q, Me$_2$NH; 70% H$_2$SO$_4$	20		406
Me	2-COOH-4-NHEt-7-OH-8-Pr	2-COOMe-4-Cl-6-Ac-7-HO-8-Pr-Q, EtNH$_2$; 70% H$_2$SO$_4$	76		406
Me	2-COOH-4-N(CH$_2$)$_4$-7-OH-8-Pr	2-COOMe-4-Cl-6-Ac-7-HO-8-Pr-Q·(CH$_2$)$_4$NH; 70% H$_2$SO$_4$			406, 407

	Product	Reagents/Conditions	Yield (%)	m.p. (°C)	Ref.
Me	2-COOMe-4-Cl-7-OH-8-Pr	2-COOMe-6-Ac-7-HO-8-Pr-4-quinolone, POCl$_3$	90	163–164	406–408
Me	2-COOMe-7-OH-8-Pr	2-COOMe-4-SEt-6-Ac-7-HO-8-Pr-Q, Raney Ni	73	110–111	406, 407
Me	2-COOMe-4-OPh-7-OH-8-Pr	2-COOMe-4-Cl-6-Ac-7-HO-8-Pr-Q, PhONa	69	195–196	406, 407
Me	2-COOMe-4-SEt-7-OH-8-Pr	2-COOMe-4-Cl-6-Ac-7-HO-8-Pr-Q, EtSNa, DMF	74	193–195	406, 407
Me	2-COOMe-4-SPh-7-OH-8-Pr	2-COOMe-4-Cl-6-Ac-7-HO-8-Pr-Q, PhSNa	33	171–172	406
Me	2-COOEt-4-Me-7-OH-8-Pr	4-Me-6-Ac-7-HO-8-Pr-Q-2-COOH, EtOH, HCl	15	150–151	404, 406, 407
Me	2-COOEt-4-NHMe-7-OH-8-Pr	4-MeNH-6-Ac-7-HO-8-Pr-Q-2-COOH, EtOH, HCl		219–220	406
Me	2-COOEt-4-NMe$_2$-7-OH-8-Pr	4-Me$_2$N-6-Ac-7-HO-8-Pr-Q-2-COOH, EtOH, HCl	76		406
Me	2-COOEt-4-NHEt-7-OH-8-Pr	4-EtNH-6-Ac-7-HO-8-Pr-Q-2-COOH, EtOH, HCl	82		406, 407
Me	2-COOEt-4-N(CH$_2$)$_4$-7-OH-8-Pr	4-(CH$_2$)$_4$N-6-Ac-7-HO-8-Pr-Q-2-COOH, EtOH, HCl			406
Me	2-COOEt-4-NHPh-7-OH-8-Pr	2-COOMe-4-Cl-6-Ac-7-HO-8-Pr-Q, PhNH$_2$, TsOH, 175 °C; 70% H$_2$SO$_4$; EtOH, HCl	20		406
Me	2-CONH$_2$-4-Cl-7-OH-8-Pr	2-COOMe-4-Cl-6-Ac-7-HO-8-Pr-Q, NH$_3$, MeOH, 100 °C			406
CH$_2$Br		6-Ac-Q, Br$_2$, 48% HBr	91	115 / HBr 236–237(d)	393, 409
CH$_2$Br	2-OEt-4-OMe	2-EtO-4-MeO-6-Ac-Q, Br$_2$, HBr		Crude	410
CH$_2$N$_3$	2-OEt-4-OMe	2-EtO-4-MeO-Q-6-COCH$_2$Br, NaN$_3$, DMF	85	133.5–134	410
CH$_2$NH$_2$		Q-6-COCH$_2$Br, (CH$_2$)$_6$N$_4$; Conc. HCl		2HCl 229–230(d)	409
CH$_2$NH$_2$	2-OEt-4-OMe	2-EtO-4-MeO-Q-6-COCH$_2$N$_3$, SnCl$_2$, EtOH		Crude	410
CH$_2$NHAc		Q-6-COCH$_2$NH$_2$, Ac$_2$O, AcONa		120–122	393, 409

Table 13. (Contd.)

R	Quinoline substituent(s)	Preparation	Yield(%)	M.p. (°C)	References
CH$_2$NHCOCHCl$_2$		Q-6-COCH$_2$NH$_2$, Cl$_2$CHCOCl	57	156–157(d)	393, 409
CHO		6-Ac-Q, SeO$_2$, AcOH, 90 °C		145	411, 412
Et		6-NH$_2$-Q, HNO$_2$, EtCH= NOH, CuSO$_4$, Na$_2$SO$_3$	60	75.5–76.5	95*
		Q-6-CHOHEt, CrO$_3$, AcOH		76–77 Pic. 209	396*
		Q-6-CHO, EtNO$_2$; H$_2$, Ni			413*
CH=CHPh	2-Ph-4-COOH	2-Ph-6-Ac-Q-4-COOH, PhCHO, NAOH	64	158–160	401
CH=CHPh	2-(2-ClC$_6$H$_4$)-4-COOH	2-(2-ClC$_6$H$_4$)-6-Ac-Q-4-COOH, PhCHO, NaOH	50	153–155	401
CH=CHPh	2-(4-ClC$_6$H$_4$)-4-COOH	2-(4-ClC$_6$H$_4$)-6-Ac-Q-4-COOH, PhCHO, NaOH	51	198–200	401
CH=CHPh	2-(2-HOC$_6$H$_4$)-4-COOH	2-(2-HOC$_6$H$_4$)-6-Ac-Q-4-COOH, PhCHO, NaOH	54	110	401
CH=CHPh	2-(2-HO-3-BrC$_6$H$_3$)-4-COOH	2-(2-HO-3-BrC$_6$H$_3$)-6-Ac-Q-4-COOH, PhCHO, NaOH	57	132–134	401
CH=CHPh	2-(2-HO-3,5-diBrC$_6$H$_2$)-4-COOH	2-(2-HO-3,5-diBrC$_6$H$_2$)-6-Ac-Q-4-COOH, PhCHO, NaOH	53	105	401
CH=CHPh	2-(3-HOC$_6$H$_4$)-4-COOH	2-(3-HOC$_6$H$_4$)-6-Ac-Q-4-COOH, PhCHO, NaOH	60	195–197	401
CH=CHPh	2-(4-HOC$_6$H$_4$)-4-COOH	2-(4-HOC$_6$H$_4$)-6-Ac-Q-4-COOH, PhCHO, NaOH	61	147–148	401
CH=CHPh	2-(4-MeOC$_6$H$_4$)-4-COOH	2-(4-MeOC$_6$H$_4$)-6-Ac-Q-4-COOH, PhCHO, NaOH	56	95	401
CH=CHPh	2-(4-NO$_2$C$_6$H$_4$)-4-COOH	2-(4-NO$_2$C$_6$H$_4$)-6-Ac-Q-4-COOH, PhCHO, NaOH	54	300	401
CH=CH-(2-furyl)		6-Ac-Q, C$_4$H$_3$O-CHO, KOH	90	100–102	444
CH(CH$_2$OH)NHAc		Q-6-COCH$_2$NHAc, (HCHO)$_3$, K$_2$CO$_3$	74	185–187	393, 409
CH(CH$_2$OAc)NHAc		Q-6-COCH(CH$_2$OH)NHAc, Ac$_2$O, 100 °C	60	109–110	393

R	Quinoline substituents	Method / Reagents	Yield (%)	m.p. (°C)	References
CH(CH₂OCPh₃)NHAc		Q-6-COCH(CH₂OH)NHAc, Ph₃CCl, Py		119–120[b]	393, 409
CH(CH₂OH)NHCOCHCl₂		Q-6-COCH₂NHCOCHCl₂, (HCHO)₃, K₂CO₃	75	154–155[b]	393, 409
CH₂CHO		6-Ac-Q, SeO₂, AcOH		190–191	414
CH(N=O)COOEt		Q-6-COOEt, EtOAc; NaNO₂, AcOH	67	144.5(d)	393
Pr	2-COOH-4-OEt-7-Me	4-EtO-6-PrCO-7-Me-Q-2-COOEt, NaOH, EtOH		203–204	415
Pr	2-COOEt-4-OEt-7-Me	2-COOEt-6-PrCO-7-Me-4-quinolone, NaH, DMF; EtI		123–124	415
(CH₂)₅NH₂		Q-6-COOEt, EtOOC(CH₂)₅NHCOPh, NaNH₂; Conc. HCl	26	2HBr 267–269	151
CHBr(CH₂)₄NH₂		Q-6-CO(CH₂)₅NH₂, Br₂, 48% HBr	94	2HBr 128–130	151
		Q-6-COCH(CH₂OH)NHAc, Ac₂O, Py, 25°C	93	136–138	393
Ph		4-Cl-6-PhCO-Q, H₂/Pd	84	59–60	416
Ph		6-PhCH₂-Q, K₂Cr₂O₇, H₂SO₄	76	59–60	416*
Ph		Q-6-CN, PhMgBr	71.5	104–105; b.p. 215–217/3 mm	379, 395
Ph		Skraup reaction	50	60.5	281*, 398*
Ph		Skraup reaction	30	b.p. 240/16 mm	
Ph		Friedel–Crafts reaction	26	63	237, 416
Ph		Q-6-CHOH-Ph, CrO₃, AcOH		Pic. 219	396*
Ph	2-Me	4-PhC₆H₄NH₂, paraldehyde		67–68	417
Ph	8-Me	Skraup reaction		142; Pic. 199	281*
Ph	2-CH₂COOMe	2-CH₂COOMe-4-Cl-6-PhCO-Q, H₂, Pd	64	98–100	418

Table 13. *(Contd.)*

R	Quinoline substituent(s)	Preparation	Yield(%)	M.p. (°C)	References
Ph		2-CH_2COOMe-6-PhCO-4-quinolone, $POCl_3$	43	117–118	418
Ph	2-Ph	Q-6-CN, PhMgBr	2.2	163–166	395
Ph	4-Cl	6-PhCO-4-quinolone, $POCl_3$	86	116–117	416
Ph	5-HO-7-Ph	2-TsCH$_2$-Py-3-COOMe, PhCOCH=CHPh	72	134.5–135.5	98
Ph	5-HO-7-(4-$NO_2C_6H_4$)	2-TsCH$_2$-Py-3-COOMe, PhCOCH=CHC$_6$H$_4$-4-NO_2	20	165–165.5	98
2,4-diClC$_6$H$_3$		Skraup reaction		131–132	281
2,5-diClC$_6$H$_3$		Skraup reaction		134–135 Pic. 208–209	281
3,4-diClC$_6$H$_3$		Skraup reaction		139–140 Pic. 173–174	281
4-ClC$_6$H$_4$		Skraup reaction		127–128 Pic. 204	281
3-$NO_2C_6H_4$		6-PhCO-Q, HNO_3, H_2SO_4		160	281
3-$NO_2C_6H_4$	8-Me	6-PhCO-8-Me-Q, HNO_3, H_2SO_4		156	281
3-$NH_2C_6H_4$		Above, $SnCl_2$		142	281
3-$NH_2C_6H_4$	8-Me	6-(3-$NO_2C_6H_4CO$)-8-Me-Q, $SnCl_2$		187	281
6-Quinolyl		4,4'-diaminobenzophenone, Skraup reaction	50	193–194	96, 419
Q-6-COC$_6$H$_4$-4-		Skraup reaction, see text, Section II.5	22	240–241	96
Q-6-CO-2,4,6-triMeC$_6$H-3-		Skraup reaction, see text, Section II.5		120	96

*These references include carbonyl derivatives.
[a]There is a printing error in the paper[97]; this correct structure was kindly supplied by Dr. K. A. Thaker.
[b]Diamorphic forms.

Table 14. 7-Quinolyl Ketones

R	Quinoline substituent(s)	Preparation	Yield(%)	M.p. (°C)	References
Me	None	Q-7-COOEt, EtOAc, EtONa; aqHCl	51	76.5–77.5	176*
Me	2,4-diMe-3-Ac	See Table 2			420
Me	2,5-diMe-8-OH	Friedel–Crafts reaction		73–74	420
Me	2-Me-5-Cl-8-OH	Friedel–Crafts reaction		123–125	420
Me	2-Me-8-OH	Friedel–Crafts reaction	5[a]	100	82
Me	5-Me-8-OH	Friedel–Crafts reaction	60–65	140–142	420
Me	5-CH$_2$Ph-8-Oh	Friedel–Crafts reaction		141–143	420
Me	5-Et-8-OH	Friedel–Crafts reaction		122–124	420
Me	5-Bu-8-OH	Friedel–Crafts reaction		105–107	420
Me	5-CH$_2$CH$_2$iPr-8-OH	Friedel–Crafts reaction		101–102	420
Me	5-Ph-8-OH	Friedel–Crafts reaction		157–159	420
Me	5-Cl-8-OH	Friedel–Crafts reaction[b]	60–70	152–154	420
			40–50	160–161	421
Me	8-Cl	7-Ac-8-NH$_2$-Q, NaNO$_2$, 9M-HCl	53	49–50	99
Me	5-Br-8-OH	Friedel–Crafts reaction		175–177	420
Me	5,8-diOH-6-Ac	See Table 13			
Me	3-NO$_2$-4-Me	1,4-diAcC$_6$H$_3$NH$_2$, HON=CHCH$_2$NO$_2$, aq. HCl	38	150–151	422
Me	8-NO$_2$	Skraup reaction, see Section II.6, Scheme 24	46.1	210–211	99
Me	8-NH$_2$	7-Ac-8-NO$_2$-Q, Fe, AcOH or	79.9	108–109	99, 423
		7-Ac-8-AcNH-Q, aq. HCl		107–109	
Me	8-NHAc	2,3-dimethylpyrrolo[2,3-h]quinoline, O$_3$	35	157–158.5	424
CH$_2$Cl	5-Cl-8-OH	Friedel–Crafts reaction		167–169	420

Table 14. (*Contd.*)

R	Quinoline substituent(s)	Preparation	Yield(%)	M.p. (°C)	References
Et	2,5-diMe-8-OH	Friedel–Crafts reaction		83–85	420
Et	2-Me-5-Et-8-OH	Friedel–Crafts reaction		73–75	420
Et	2-Me-5-Cl-8-OH	Friedel–Crafts reaction		86–88	420
Et	5-Me-8-OH	Friedel–Crafts reaction		124–126	420
Et	5-Cl-8-OH	Friedel–Crafts reaction		123–125	420
COCOOH[c]	8-OH	8-OH-Q-7-CH$_2$CSCOOH, NH$_2$OH, EtONa	48		375*
Pr	2,5-diMe-8-OH	Friedel–Crafts reaction		86–88	420
Pr	2-Me-5-Cl-8-OH	Friedel–Crafts reaction		78–80	420
Pr	5-Me-8-OH	Friedel–Crafts reaction		117–119	420
Pr	5-Et-8-OH	Friedel–Crafts reaction		81–82	420
Pr	5-Cl-8-OH	Friedel–Crafts reaction		109–111	420
CH$_2$iPr	2,5-diMe-8-OH	Friedel–Crafts reaction		72–74	420
CH$_2$iPr	5-Me-8-OH	Friedel–Crafts reaction		79–81	420
CH$_2$iPr	5-Cl-8-OH	Friedel–Crafts reaction		92–94	420
(CH$_2$)$_5$NH$_2$		Q-7-COOEt, EtOOC(CH$_2$)$_5$NHCOPh, NaNH$_2$; conc. HCl	30	2HBr 223–224	151
CHBr(CH$_2$)$_4$NH$_2$	8-OH	Q-7-CO(CH$_2$)$_5$NH$_2$, Br$_2$, 48% HBr	78	2HBr 180–181	151
C$_6$H$_{13}$	8-OEt	Friedel–Crafts reaction	20	75–76	80
C$_6$H$_{13}$	8-OH	7-C$_6$H$_{13}$CO-8-HO-Q, EtI, KOH	71	b.p. 188–194/2 mm	81
C$_8$H$_{17}$	8-OH	Friedel–Crafts reaction[d]		63–64	79
C$_9$H$_{19}$	5-Cl-8-OH	Not given	10	68–69	378*
C$_9$H$_{19}$	8-OH	Not given	5–8	61–62	378*
Ph	8-OH	Q-7-CN, PhMgBr		129–130.5 b.p. 219–223/3 mm	379* 379*
Ph	2,5-diMe-8-OH	Friedel–Crafts reaction		135–137	420
Ph	2-Me-5-Cl-8-OH	Friedel–Crafts reaction		138–139	420
Ph	5-Me-8-OH	Friedel–Crafts reaction		136–138	420
Ph	4-Ph-6-Cl	See text, Section II.6	54	135–136	100
Ph	5-Cl-8-OH	Friedel–Crafts reaction		151–153	420

			Yield (%)	M.p. (°C)	Ref.
Ph	8-OH	8-PhCOO-Q, hv	20^e	111–113	85
Ph	5-(N=NC$_6$H$_3$-3-OH-4-COOH)-8-OH				425
2-ClC$_6$H$_4$	2,5-diMe-8-OH	Friedel–Crafts reaction	50–55	157–158	420
2-ClC$_6$H$_4$	2-Me-5-Cl-8-OH	Friedel–Crafts reaction	35–45	192–193	420
2-ClC$_6$H$_4$	5-Me-8-OH	Friedel–Crafts reaction		187–188	420
2-ClC$_6$H$_4$	5-Cl-8-Oh	Friedel–Crafts reaction		164–166	420
4-ClC$_6$H$_4$	2,5-diMe-8-OH	Friedel–Crafts reaction		171–172	420
4-ClC$_6$H$_4$	2-Me-5-Cl-8-OH	Friedel–Crafts reaction		185–187	420
4-ClC$_6$H$_4$	5-Me-8-OH	Friedel–Crafts reaction		160–161	420
4-ClC$_6$H$_4$	5-Cl-8-OH	Friedel–Crafts reaction		185–187	420
C$_6$H$_4$-2-COOH	5-Br-8-OH	7-(2-COC$_6$H$_4$COOH)-8-HO-Q, NaOBr		227–228(d)	373
C$_6$H$_4$-2-COOH	5-I-8-OH	7-(2-COC$_6$H$_4$COOH)-8-HO-Q, I$_2$, 0.2M-HCl, EtOH		224(d)	373
C$_6$H$_4$-2-COOH	8-OH	8-HO-Q, phthaloyl dichloride	52	226 HCl 224(d)	373
C$_6$H$_4$-2-COOMe	8-OH			143–144	373
C$_6$H$_4$-2-CONHNH$_2$	8-OH			326(d)	373
5-(N=N-C$_6$H$_3$-3-OH-4-COOH)-8-OH					425

*These references include carbonyl derivatives.

[a] By-product from the preparation of 2-Me-5-Ac-8-HO-Q; see Table 9.

[b] Also prepared from 5-Cl-8-MeO-Q, which underwent O-demethylation during the Friedel–Crafts reaction.

[c] Only the oxime was isolated.

[d] By-product from the preparation of 5-C$_8$H$_{17}$CO-8-HO-Q; see Table 9.

[e] Obtained as a mixture with 5-benzoyl-8-hydroxyquinoline; see Table 9.

Table 15. 8-Quinolyl Ketones

R	Quinoline substituent(s)	Preparation	Yield (%)	M.p. (°C)	References
Me	None	Q-8-COOEt, EtOAc, EtONa; 20% H_2SO_4	52	42–43.5; b.p. 114–116/0.7 mm; b.p. 176–180/12 mm	176, 237, 426, 427
		8-Et-Q, $K_2Cr_2O_7$, $3M\text{-}H_2SO_4$	40[a]		94, 428*
		Q-8-CHOH-Me, $K_2Cr_2O_7$, H_2SO_4		45	436
Me	2-Me	2-Me-8-Et-Q, $K_2Cr_2O_7$, $3M\text{-}H_2SO_4$	46	Pic. 182(d)	94, 428*
Me	2,3-diMe	2,3-diMe-8-Et-Q, CrO_3	12[a]		94*
Me	2,3,4-triMe	2,3,4-triMe-8-Et-Q, $K_2Cr_2O_7$, $3M\text{-}H_2SO_4$	50[a]		94*
Me	2,4-diMe	2,4-diMe-8-Et-Q, CrO_3	36		94*
Me	3-Me	3-Me-8-Et-Q, $K_2Cr_2O_7$, $3M\text{-}H_2SO_4$	80		94*
Me	2-Et-3-Me	2,8-diEt-3-Me-Q, $K_2Cr_2O_7$, $3M\text{-}H_2SO_4$	55[a]		94*
Me or CH_2Br	?-Br-6-OMe	6-MeO-8-Ac-Q, Br_2, $CHCl_3$[b]		98	429
Me	?-Br-6-OMe	6-MeO-8-Ac-Q, Br_2, conc.H_2SO_4			429
Me	6-OMe	6-MeO-8-CN-Q, MeMgBr	79	80; HBr 215	426*
CH_2Ph		6-MeO-Q-8-$COCH_2COOEt$, 6M-HCl			426
CD_2Ph		Q-8-COOEt; $PhCH_2COOEt$			430
					430
CH(Me)Ph		Q-8-CHO, PhCH(Me)MgBr; Swern oxidation			431
$C(Me)_2Ph$		Q-8-CHO, $PhC(Me)_2$MgBr; Swern oxidation			431

Group	Substituent	Method / Reagents	Yield (%)	Properties	References
CH$_2$Cl	6-OMe	6-MeO-Q-8-COCl, CH$_2$N$_2$; HCl		120 HCl 151–152	429
CH$_2$Br		8-Ac-Q, Br$_2$, HBr	82	HBr 176–177	409, 426
CH$_2$NH$_2$		8-BrCH$_2$CO-Q, (CH$_2$)$_6$N$_4$			409
CH$_2$NHAc		Q-8-COCH$_2$NH$_2$, Ac$_2$O, AcONa			409
CH(CH$_2$OH)NHAc		Q-8-COCH$_2$NHAc, (HCHO)$_3$, K$_2$CO$_3$			409
CH$_2$NEt$_2$		Q-8-COCH$_2$Br, Et$_2$NH		Crude^c	426
CH$_2$NEt$_2$	6-OMe	6-MeO-Q-8-COCH$_2$Cl, Et$_2$NH		b.p. 150–160/ 0.1 mm 2Pic. 142	429
CH$_2$NBu$_2$		Q-8COCH$_2$Br, Bu$_2$NH		Crude^c	426
CH$_2$NBu$_2$	6-OMe	6-MeO-Q-8-COCH$_2$Cl, Bu$_2$NH		b.p. 186–190/ 0.5 mm 2Pic. 141	429
COOH	2-Cl-4-Me	4-Me-8-COCOOMe-2-quinolone, SOCl$_2$		150	432
COOH	2-Ph-4-COOH-5-NH$_2$-6-Me	Pfitzinger synthesis		>360	433
COOEt	6-F	6-F-8-Br-Q, (COOEt)$_2$	32	114–117	434
COOEt	6-Cl	6-Cl-8-Br-Q, (COOEt)$_2$	35	107–110	434
Et		Q-8-COBu^d, [(C$_2$H$_4$)$_2$RhCl$_2$]$_2$; Py; PPh$_3$	90–100		431, 435
		Q-8-CHOH-Et, K$_2$C$_2$O$_7$, H$_2$SO$_4$			436*
		Q-8-COOEt, EtCOOEt, EtONa			430
		Q-8-CHOH-C≡C-Ph, MnO$_2$			437
C≡C-Ph				171–172	437
CH$_2$COOEt	6-OMe	6-MeO-Q-8-COOEt, EtOAc, EtONa	57	HBr 110 Pic. 122	426, 429
CH$_2$CH$_2$NMe$_2$		8-Ac-Q, Mannich reaction	20	Oil	438
(CH$_2$)$_3$OH		Q-8-CO-(2-oxotetrahydrofuran-3-yl), 5% HCl		72–73 b.p. 168–175/ 3 mm	439
(CH$_2$)$_3$NEt$_2$		Q-8-CO(CH$_2$)$_3$OH, 48% HBr; Et$_2$NH		b.p. 165–167/ 2 mm	439
Bu		Q-8-CHO, BuMgBr; Swern oxidation			431
(CH$_2$)$_5$NH$_2$		Q-8-COOEt, EtOOC(CH$_2$)$_5$NHCOPh, NaNH$_2$; HCl	22.9	2HBr 230–230.5	440

Table 15. (*Contd.*)

R	Quinoline substituent(s)	Preparation	Yield (%)	M.p. (°C)	References
$CHBr(CH_2)_4NH_2$		$Q\text{-}8\text{-}CO(CH_2)_5NH_2$, Br_2, 48% HBr		Crude	440
$C{\equiv}C\text{-}tBu$		$Q\text{-}8\text{-}CHOH\text{-}C{\equiv}C\text{-}tBu$, MnO_2			437
C_8H_{17}		$Q\text{-}8\text{-}CORhH(PPh_3)_2BF_4$, $C_6H_{13}CH{=}CH_2$	55		101
$C_{15}H_{31}$		Q, $C_{15}H_{31}COCl$, $AlCl_3$, DCM			441
$cycloC_6H_{11}$		$8\text{-}Q\text{-}CHO$, $cycloC_6H_{11}MgBr$; Swern oxidation			431
2-Piperidyl		$Q\text{-}8\text{-}CO\text{-}CHBr(CH_2)_4NH_2$, NH_2, NaOH	71	Crude[c]	440
2-Oxotetrahydrofuran-3-yl		$Q\text{-}8\text{-}COOEt$, butyrolactone, EtONa		47	439
Ph		$Q\text{-}8\text{-}COCl$, C_6H_6, $AlCl_3$	64	92–93	102
		$Q\text{-}8\text{-}CHOH\text{-}Ph$, CrO_3, AcOH		94	436*
		$Q\text{-}8\text{-}CN$, PhMgBr		92–94 b.p. 212–215/3 mm	379
Ph	2-Me	2-Aminobenzophenone, $(MeCHO)_3$, H_2SO_4		107–108	442
Ph	6-Cl	Skraup reaction	45	129–130	443
7-Me-2,3-diH-4-indenyl		$Q\text{-}8\text{-}CN$, ArMgBr	57	135–135.6 b.p. 240–245/2 mm	382

*These references include carbonyl derivatives.

a These ketones were formed in admixture with the corresponding acids; see text, Section II.5.

b Mono- and di-bromo derivatives not characterized.

c Crude ketone reduced to alcohol before characterization.

d The butyl group could be replaced by other alkyl groups or a phenyl group, with the same result.

III. Quinaldoins and Quinaldils

Reaction of quinoline-2-carboxaldehyde **133** with potassium cyanide under benzoin condensation conditions gave quinaldoin **134** in up to 90% yield.[125,445,446] There was always a second product, the diol **135**, which could be obtained in up to 40% yield by a careful choice of conditions. When oxygen was passed through a solution of compound **134** in dioxane, the diketone **136** (quinaldil) was produced. Quinaldil was reduced back to quinaldoin by hydrogenation over Adams' catalyst under conditions in which further reduction did not occur. The diol **135** has been dehydrated to ketone **137** (desoxyquinaldoin), which on treatment with oxygen[125,445] or selenium dioxide[447,448] gave quinaldil **136**.

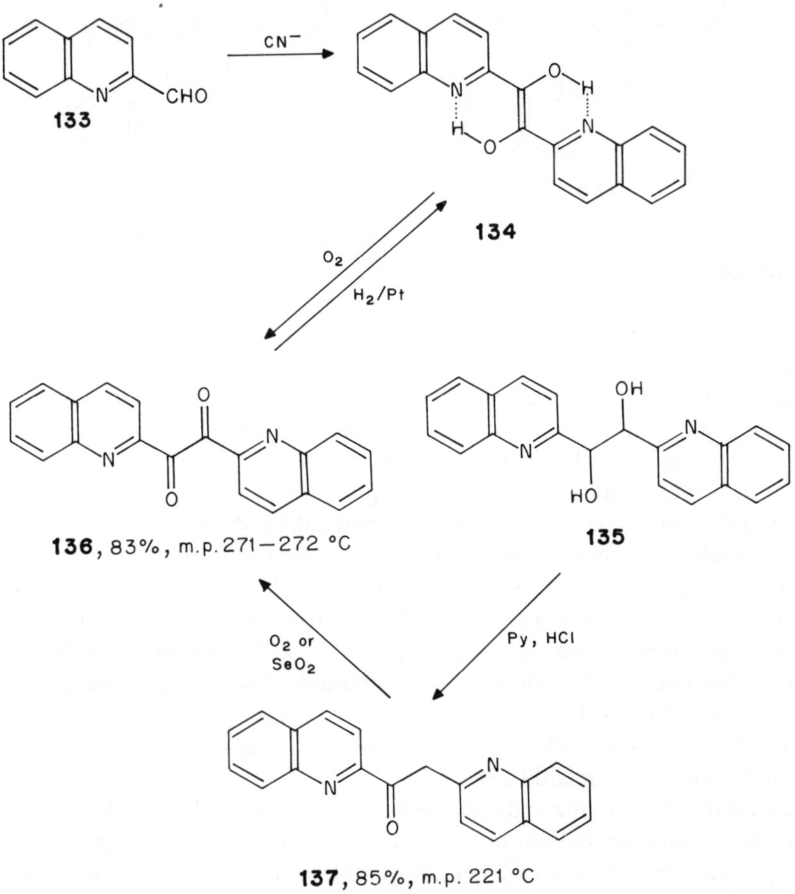
CN⁻
CHO
133
134
O₂
H₂/Pt
136, 83%, m.p. 271–272 °C
135
O₂ or
SeO₂
Py, HCl
137, 85%, m.p. 221 °C

Buehler and co-workers[125,449] suggested the ene-diol structure for quinaldoin **134** on the basis of its reducing properties and i.r. spectrum. Although they saw no carbonyl absorption, they were also unable to find the O–H stretching vibration. They explained this as being due to the strong intra-molecular hydrogen bonds which would be present. Gill and Morgan[450] confirmed these findings and noted that they could see no i.r. vibrations between $5000 \, cm^{-1}$ and $1625 \, cm^{-1}$.

Brown and Hammick[451] obtained the same product under benzoin condensation conditions as from the decarboxylation of picolinic acid or quinoline-2-carboxylic acid in the presence of quinoline-2-carboxaldehyde, but gave the structure wrongly as **138**. This compound (m.p. 266–267 °C) was named 'quinocoll'. It was shown by others to be quinaldil **136**.[125,448,452]

138   **139**

Many quinaldoins appear to be oxidized *in situ* unless the benzoin condensation is protected from atmospheric oxygen. The three 'quinocolls' claimed by Brown and Hammick are therefore assumed to be diketones in this report. Brown and Hammick[453] show 3-methylquinoline-2-carboxaldehyde as giving 'dihydroquinocoll' **139**, which must be 3,3'-dimethylquinaldoin, in a reaction which they ran under nitrogen. It was suggested that this compound did not readily undergo autoxidation because of the steric hindrance of the 3-methyl group, but when exposed to the air for several days it changed to the 'quinocoll', which must be 3,3'-dimethylquinaldil.

An early claim to have prepared compound **136** by selenium dioxide oxidation of 2-methylquinoline (dioxane, 45° C) to a product with m.p. 175 °C was later refuted[125]. Reduction of this product was claimed to give compound **134**, but again the melting point was wrong (135°). However, another group[454] later found conditions to oxidize 2-methylquinoline with selenium dioxide to give either quinoline-2-carboxaldehyde or quinaldoin **134** in major amount. They also achieved the oxidation to quinaldoin over a metal oxide mixture (V_2O_5, MoO_3, WO_3) at 450–480 °C and several oxidizing agents were investigated for the preparation of quinaldil **136**.

A mixture of quinoline and either ethyl benzoate or *N,N*-dimethylbenzamide was treated with aluminium amalgam and mercuric chloride to give 29% and 26.1% yields, respectively, of 2-benzoylquinoline, Table 1. In both cases a total of 12–13% of compounds **134** and **136** was claimed to be present as by-products.[44]

Air oxidation of the quinaldoin[449] gave 6,6'-dimethylquinaldil (83%, m.p.

278 °C). This is presumably the third of Brown and Hammick's 'quinocolls', but they gave[453] m.p. 250–251 °C. The 3,3'-dimethyl derivative of ketone **137** was made by dehydration of the appropriate saturated diol, Table 1. The X-ray crystal structure of quinaldil **136** has been determined.[452] The compound is not planar and is highly polar.

Little work has been reported on benzoin condensations of other quinoline aldehydes. Only the 6-carboxaldehyde has given a benzoin—named quinoloin (5%, m.p. 166–167 °C)—on treatment with potassium cyanide.[455]

Quinoline-3-carboxaldehyde was reacted with potassium cyanide and presumably gave the benzoin,[456] but it was treated *in situ* with warm nitric acid to give the diketone (75%, m.p. 244–246 °C). One report of the reaction of quinoline-4-carboxaldehyde with potassium cyanide in aqueous methanol showed no trace of a benzoin. The products, which were isolated in high yield, were the diol **140** and quinoline-4-carboxylic acid, the result of a modified

140

Cannizzaro reaction.[457] However, 6-methoxyquinoline-4-carboxaldehyde was converted into its cyanohydrin, which was hydrolysed (EtOH/HCl) to the ester **141**. Treatment with acid or cupric acetate then gave the benzoin **142**. A patent notes that the ketoalcohol **142** could also be obtained by benzoin condensation.[458]

141

142, m.p. 66 °C; HCl, m.p. 250–252 °C

Table 16. Quinaldoins

Quinoline substituents (both rings)	Yield (%)	M.p. (°C)	References
H	90	232–233(d)	125, 445, 446, 459
3-Me		154–155	451, 459
3-Me-6-OMe			459
3-Me-6-NO$_2$			459
6-Me	60	258	449, 459
4-Cl			459
6-Cl			459
4-Br			459
6-Br			459
4-OMe	50	277–280	459, 459A
4-OPh	40	247–249	459, 459A
4-OC$_6$H$_4$-4-Me	10	220–222	459, 459A
4-SC$_6$H$_4$-4-Me			459
6-NO$_2$			459
4-CN	35	274–278	459, 459A
4-N$_3$			459

The compounds reported in Ref. 459 were used in a u.v. study; no characteristics or references to their preparation were given.

IV. Cinchona Alkaloids

1. Ketonic Alkaloids and Synthetic Intermediates

The use of cinchona alkaloids as antimalarial drugs over more than two centuries has been reflected in an intense study of the structure and synthesis of the active compounds. This work, culminating in the total synthesis of quinine, was reviewed in 1953.[460] After a quiet period, a shortage of quinine from natural sources led to renewed interest in the synthesis in the 1970s, which has also largely been covered in reviews.[461,462]

In the context of the present summary, the various ketones that have been prepared fall into the classes of alkyl 4-quinolyl ketones and 4-ketomethylquinolines, but it is convenient to discuss and tabulate them here, as the methods involved are, in the main, different from those used for the compounds of Tables 4 to 7.

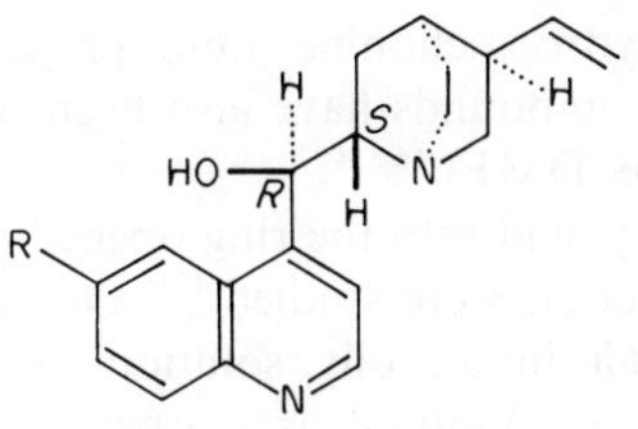

143 a, R=H, cinchonidine
  b, R=OMe, quinine

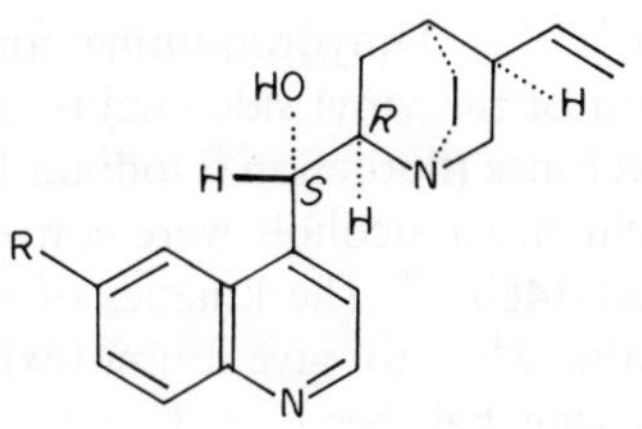

144 a, R=H, cinchonine
  b, R=OMe, quinidine

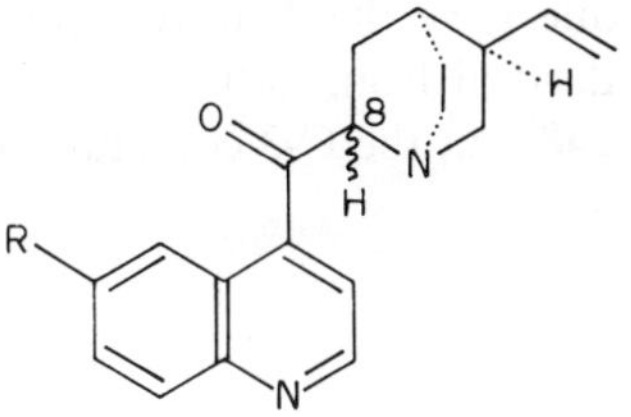

145

a, R=H 8α, cinchoninone
b, R=H 8β, cinchonidinone
c, R=OMe 8α, quinidinone
d, R=OMe 8β, quininone

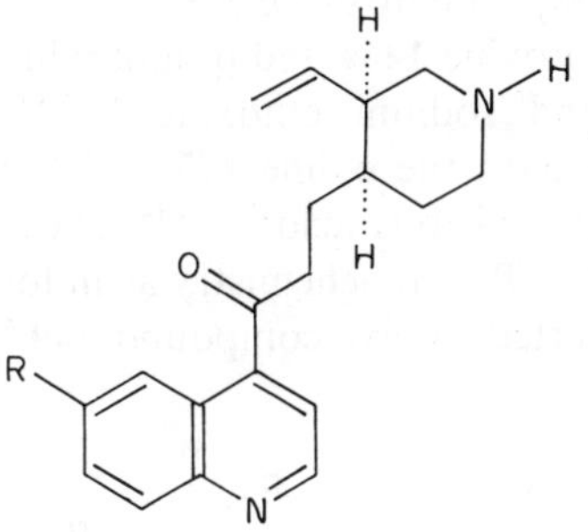

146 a, R=H, cinchotoxine
  b, R=OMe, quinotoxine

SCHEME 26

The most important cinchona alkaloids are shown in formulae **143** and **144**. Early oxidations produced the ketones **145**. Fluorenone was suggested as being superior to benzophenone in the Oppenauer procedure[463], but recent work has concentrated on the use of its sodium ketyl.[464-467] In all reports it must be assumed, unless otherwise stated, that the products are equilibrium mixtures of the position 8 epimers, i.e. **145a**/**145b** and **145c**/**145d**,[468] as confirmed by optical rotatory dispersion (o.r.d.) and circular dichroism (c.d.) studies.[469] Careful recrystallization has produced pure samples of cinchonidinone **145b** and quinidinone **145c**, the less soluble members of each pair. In one report, 80% of quinidinone was claimed to have crystallized from the reaction mixture.[470] These ketones readily form the common enols in solution and show rapid mutarotation.[471,472] The details of an h.p.l.c. separation of cinchona alkaloids, including the isolation of quinidinone **145c**, recently appeared[473], and cinchoninone **145a** has been detected in *Cinchona ledgeriana*.[474] In a biosynthetic study, cinchonidinone **145b** was also shown to be a natural product.[475,476] Dihydroquinine and dihydrocinchonine were prepared by reduction of the vinyl side chains. These compounds have also been oxidized to the ketones (fluorenone, sodium hydride, DMF).[465,477]

The cinchona alcohols were converted by acid into the ring opened ketones **146a** and **146b**.[478] The kinetics of the reaction were studied.[479] Quinine was also claimed[480] to give quinotoxine **146b** in an oily solution at 140 °C. Quinotoxine has been used under the name Viquidil as a vasodilator and antiarrhythmic drug.[481] It has been prepared with a ^{14}C label in the vinyl group.[482] A nomenclature note; 4-quinolyl 2-quinuclidinyl ketone **145a** without the vinyl group has been called 9-rubanone, and compound **146a**, without the vinyl group, rubatoxanone-9.[499]

Cinchotoxine **146a** and quinotoxine **146b** gave the oximes **147a, b** with amyl nitrite and sodium ethoxide.[483,484] The appropriate dihydrocinchotoxine similarly gave the oxime **147c**.[485] Quinotoxine gave the N-nitrosamine (m.p. 94 °C) with nitrous acid.[483,484] The toxins derived from natural products all have the R, R stereochemistry as in formula **148**, but the C-3 configuration has been inverted to give compound **149**.[486]

147

a, R¹ = H, R² = CH=CH₂, R³ = H, m.p. 169–170 °C; HCl, m.p. 268 °C
b, R¹ = OMe, R² = CH=CH₂, R³ = H, m.p. 168–170 °C; HCl, m.p. 244 °C
c, R¹ = H, R² = Et, R³ = COOEt, 64%, m.p. 155–156 °C

pH 3.4, 140 °C, 48 h

148   **149**

2. The Preparation of Cinchotoxine and its Derivatives

The schemes below illustrate the preparative methods used for the ketones listed in Table 17 (p. 235). In Method 1, hydrolysis after Claisen condensation gives the ketone **151**, R = H.

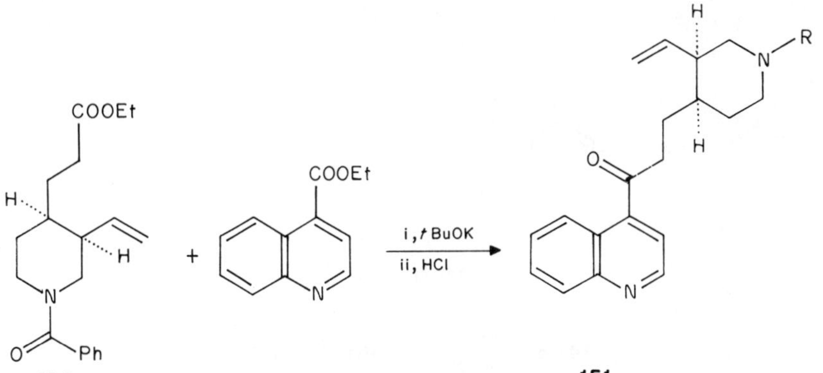

METHOD 1

The secondary base **151**, R = H was treated with an alkyl chloride in the presence of triethylamine or potassium carbonate/sodium iodide to give the tertiary amine **151**, R = alkyl, or with an acid chloride and triethylamine to give the amide **151**, R = acyl (Method 2). Sometimes such amides were reduced (LAH) to the amines. In one example[487] propane sultone was used to give compound **151**, R = $(CH_2)_3SO_3H$. When an epoxide was employed as the alkylating agent, a 2-aminoethanol such as compound **152** resulted.[511]

$$\textbf{151}, R=H \xrightarrow{R^1Cl} \textbf{151}, R=R^1$$

METHOD 2

$$\textbf{144a} \xrightarrow{50\%,\ H_2SO_4} \textbf{146a}$$

METHOD 3

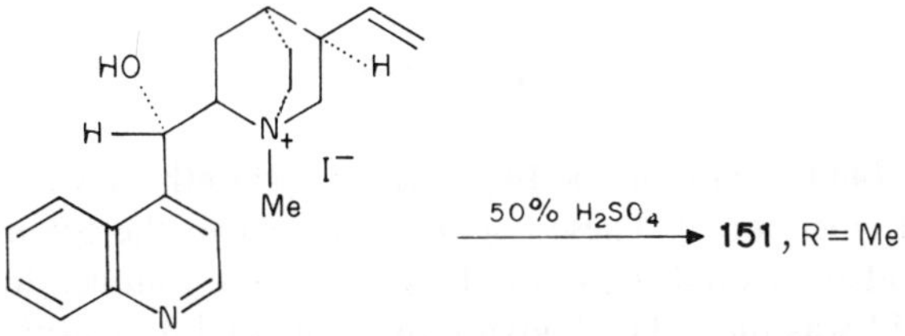

$$\xrightarrow{50\%\ H_2SO_4} \textbf{151}, R = Me$$

METHOD 3a

In Method 4, the alcohol **152**, R = H was esterified (EtCOCl, Et$_3$N) to give ester **152**, R = COEt.[490]

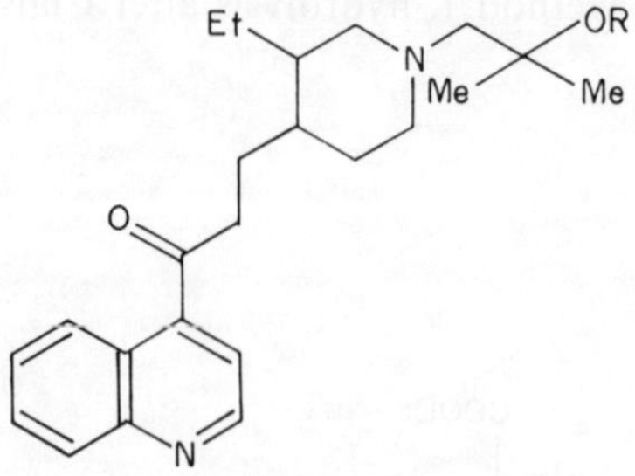

152

METHOD 4

146a $\xrightarrow[\text{Br}_2,\ \text{NaOH}]{\text{NaOCl}}$ **151**, R = Cl, Br

METHOD 5

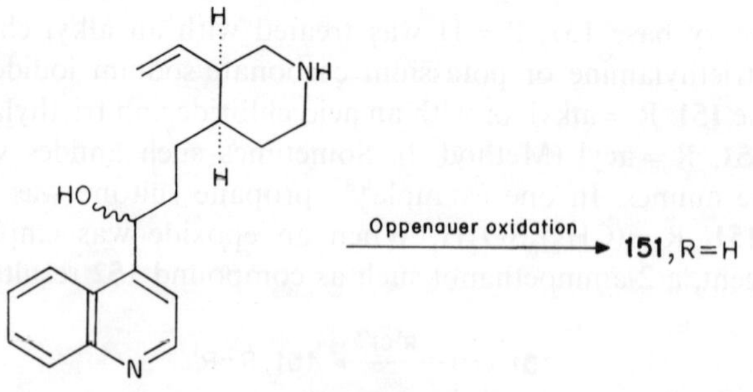

$\xrightarrow{\text{Oppenauer oxidation}}$ **151**, R = H

METHOD 6

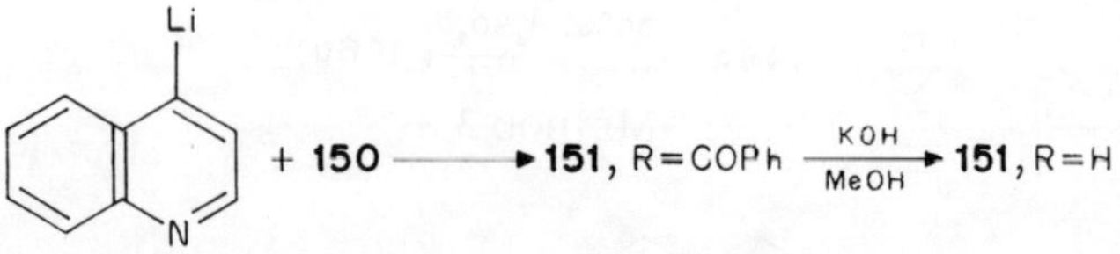

+ **150** $\longrightarrow$ **151**, R = COPh $\xrightarrow[\text{MeOH}]{\text{KOH}}$ **151**, R = H

METHOD 7

Cinchonine **144a** and quinidine **144b** formed cyclic ethers with acidic reagents (e.g. **153** from cinchonine). These could be opened to the toxins **154**.[488] This ring opening also succeeded (Method 3a) when the quinuclidine nitrogen of cyclic ether **153** was methylated, giving the N-methyl derivative of compound **154**.[489]

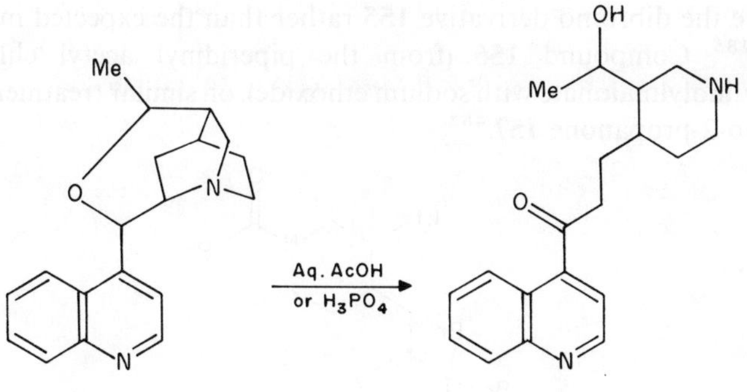
Me
O
N
N
Aq. AcOH
or H3PO4
OH
Me
NH
O
N

153 154

METHOD 8

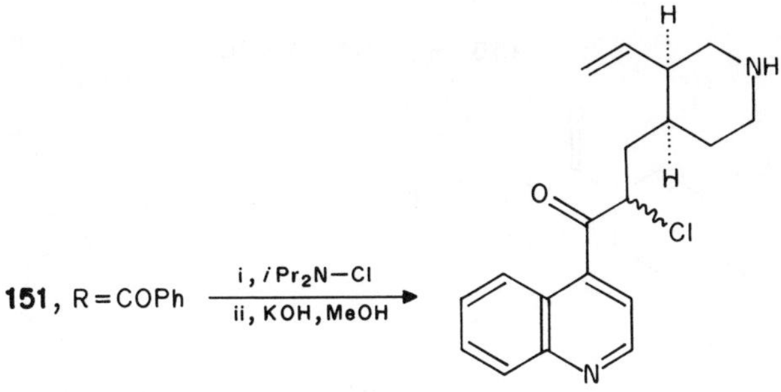
151, R=COPh
i, iPr2N—Cl
ii, KOH, MeOH
H
NH
H
O
Cl
N

METHOD 9

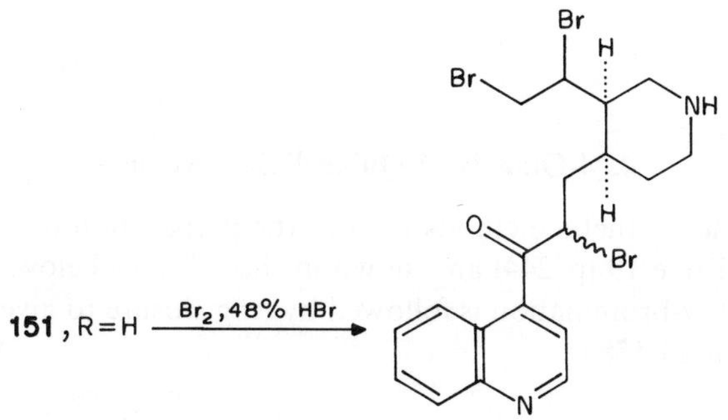
151, R=H
Br2, 48% HBr
Br
Br
H
NH
H
O
Br
N

METHOD 10

3. 1-(4-Piperidyl)-3-(4-quinolyl)-2-propanones

Most of the 2-propanones in Table 18 (p. 241) were prepared by standard methods. The compounds are intermediates to some of the ketones in the next Section. The appropriate 2-propanone, Table 18, reacted with bromine in 48%

HBr to give the dibromo derivative **155** rather than the expected monobromo derivative.[485] Compound **156** (from the piperidinyl acetyl chloride and diethyl 4-quinolylmalonate with sodium ethoxide), on similar treatment, gave the 1,3-dibromo-2-propanone **157**.[485]

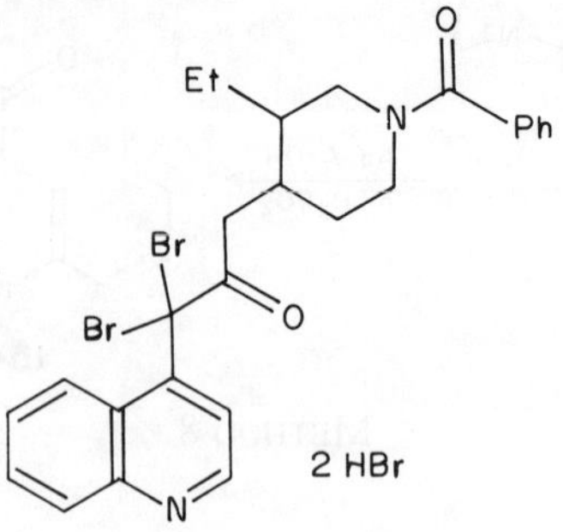

155, m.p. 194−196 °C

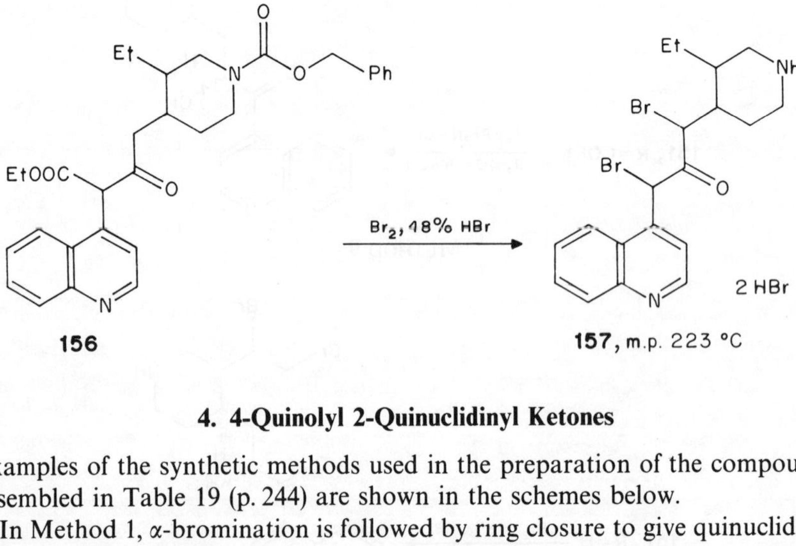

Br₂, 18% HBr

156   **157**, m.p. 223 °C

4. 4-Quinolyl 2-Quinuclidinyl Ketones

Examples of the synthetic methods used in the preparation of the compounds assembled in Table 19 (p. 244) are shown in the schemes below.

In Method 1, α-bromination is followed by ring closure to give quinuclidines such as compound **158**.

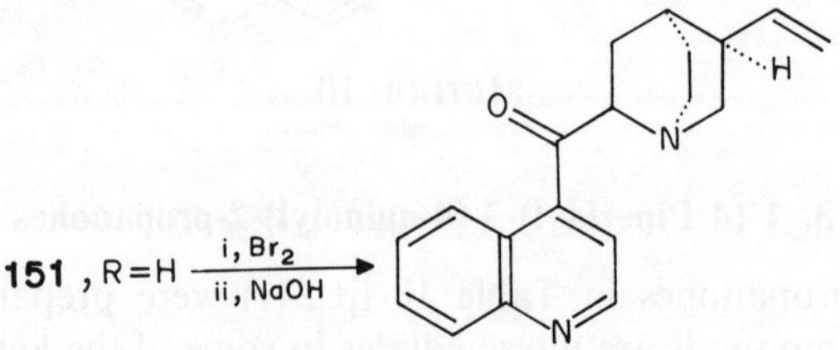

151, R=H $\xrightarrow{\begin{array}{c} \text{i, Br}_2 \\ \text{ii, NaOH} \end{array}}$

158

METHOD 1

The condensation of quinuclidine *N*-oxide with ethyl quininate using t-butyllithium gave the ketone **159**, which was deoxygenated with hexachlorodisilane (Method 2) to give 6-methoxy-9-rubanone **160** in good yield.[496]

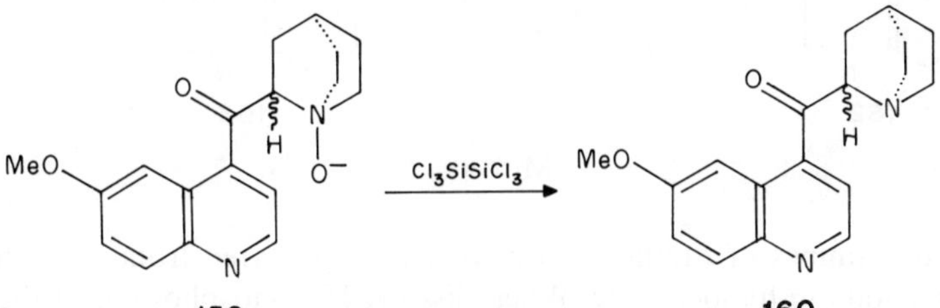

159   **160**

METHOD 2

143a   **145**

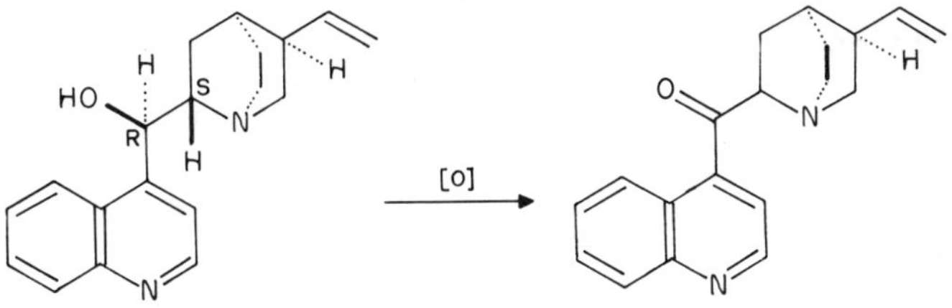

METHOD 3

Method 3 is alcohol oxidation. Early workers used chromic acid. Recently, Oppenauer oxidation with fluorenone and potassium *t*-butoxide or, better, oxidation with fluorenone ketyl have given improved yields.[470,494]

In Method 4 reaction between a 4-quinolyllithium and a quinuclidine ester gives a compound such as the ketone **161**.

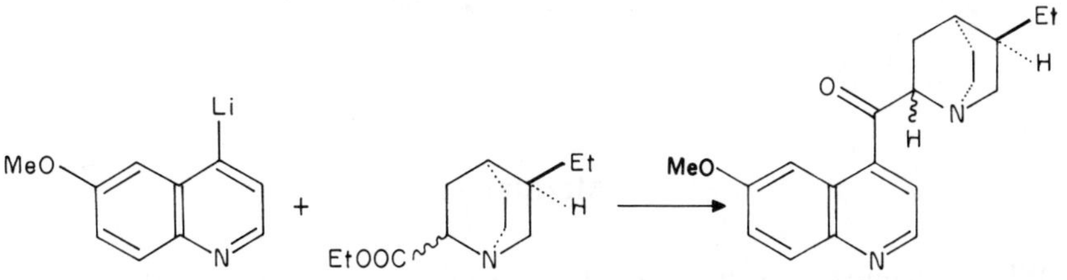

161

METHOD 4

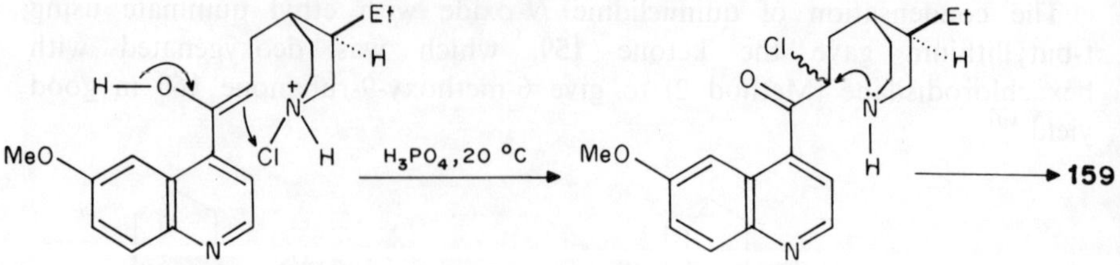

162   **163**

METHOD 5

The chloramines of Method 5, e.g. **162**, were prepared from the secondary bases and sodium hypochlorite. When dissolved in phosphoric acid they were considered to change to the α-chloroketones **163**, which then cyclized. The use of an external chloramine, *N*-chlorodiisopropylamine, in phosphoric acid also gave the quinuclidines.[491,492]

In one example, the by-product dichloroketone **164** was isolated by preparative t.l.c.[470,492,493]

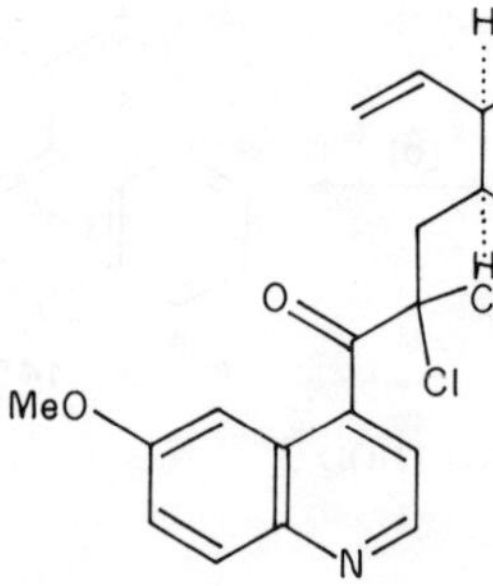

164 , 6%

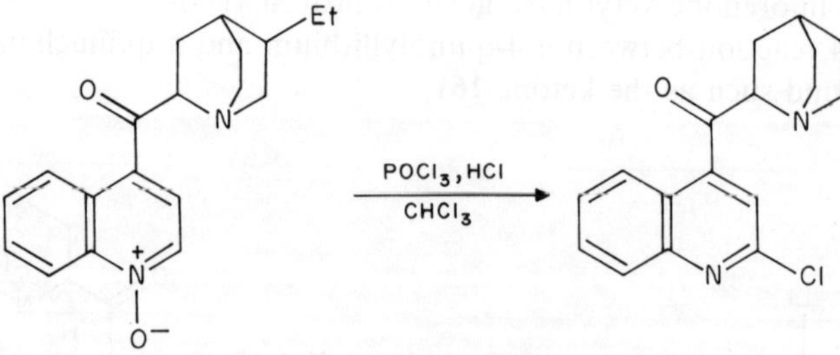

METHOD 6

When attempts were made to run Chichibabin reactions on the cinchona alkaloids, it was discovered (Method 7) that sodamide in boiling xylene acted instead as an oxidizing agent.[477] Addition of hydrogen bromide (Method 8) or bromine (Method 9) to compounds **145c, d**, gave, for example, the dibromo derivative **165**.

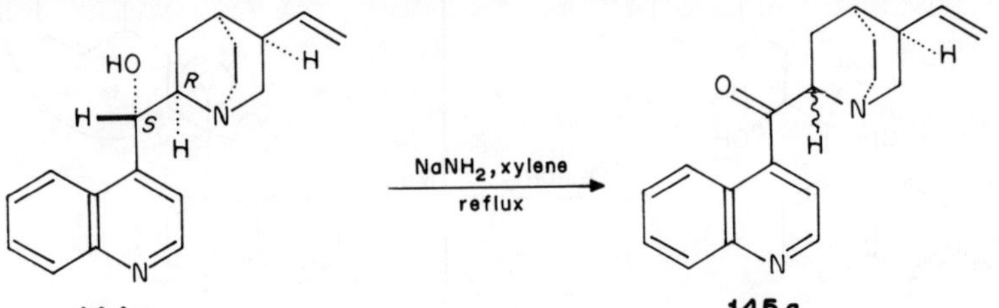

144 a

NaNH₂, xylene
reflux

145 a

METHOD 7

145 c,d

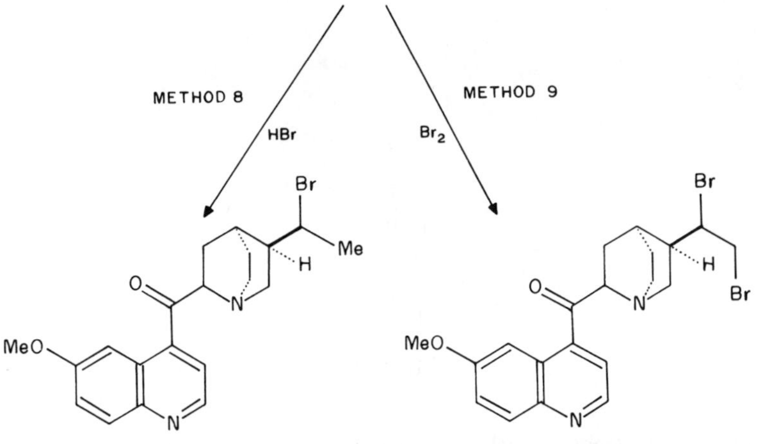

METHOD 8

HBr

METHOD 9

Br₂

165

165 — i, NaI, acetone / ii, 25% KOH →

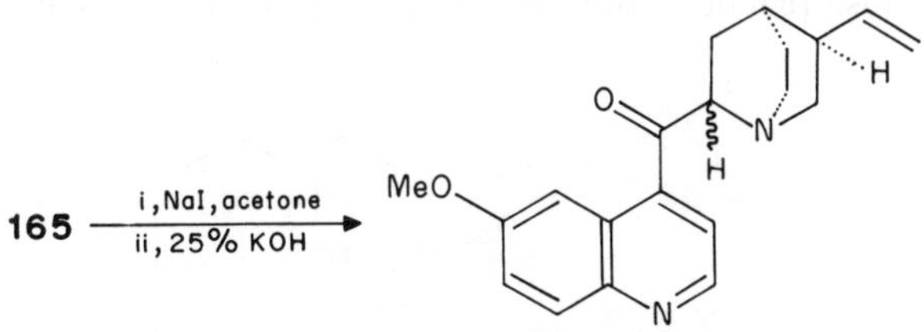

145 c,d

METHOD 10

Compound **166** was prepared (Method 11) from the appropriate 2-propanone, Table 18, and excess N-chlorodiisopropylamine followed by sodium borohydride reduction. Barium hydroxide then gave the ketone as shown.[470]

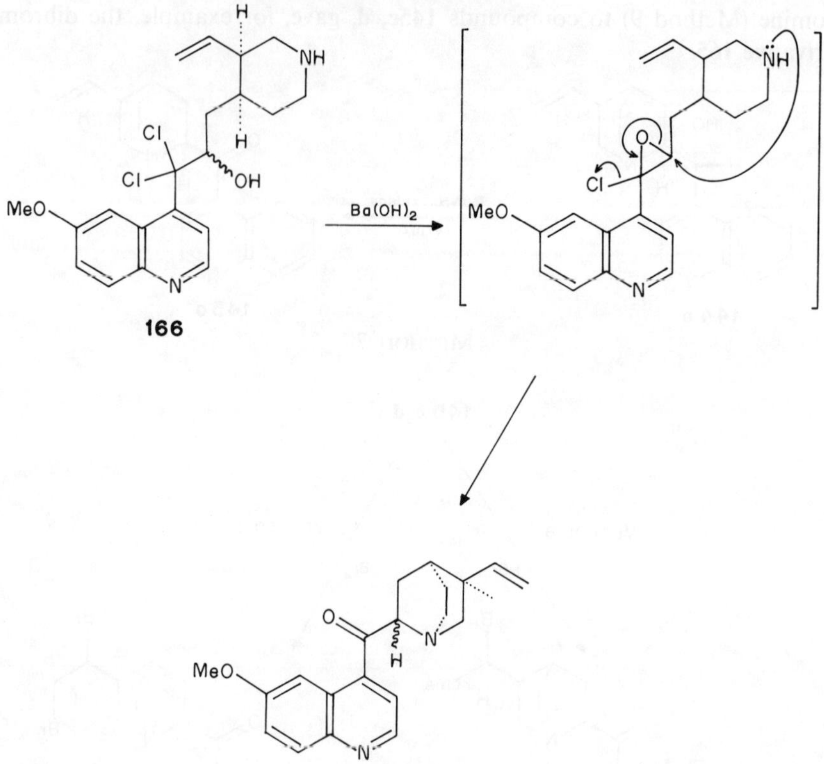

METHOD 11

Quinine was treated with the selective oxidizing agents triaryl bismuth carbonates. The expected product, quininone, was never obtained in better than 34% yield because reaction continued to give up to 90% of ketones **167**.[495]

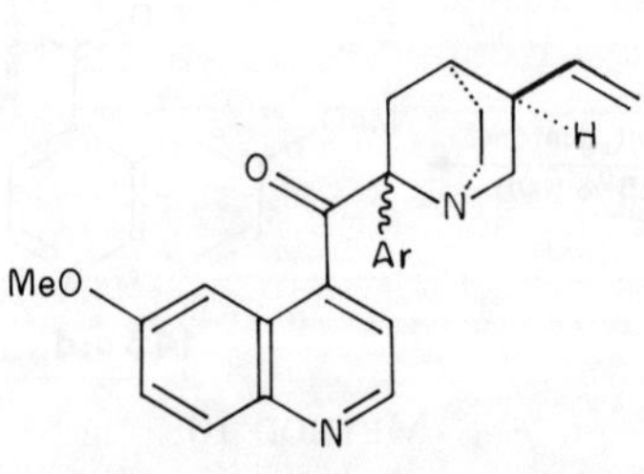

167 , Ar = Ph, *p*-CH₃C₆H₄

METHOD 12

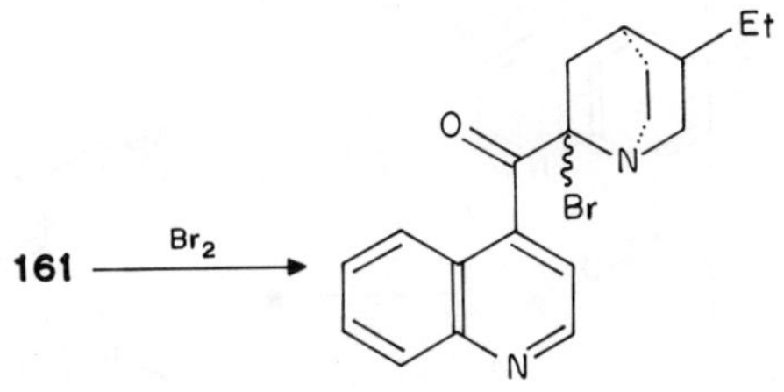

161 $\xrightarrow{Br_2}$

METHOD 13

5. Other Related Ketones

A Claisen condensation gave the 3-piperidyl propanones **168** which were cyclized under the conditions of Method 1, Table 19 to ketones **169**.[279]

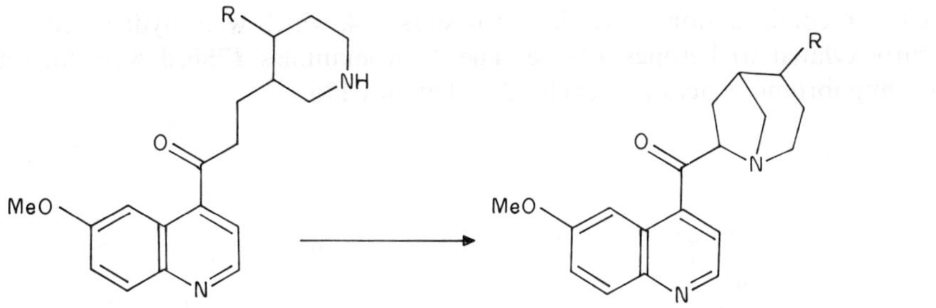

168, R = H, 53.6%
  2 Pic. m.p. 64−68 °C
  R = Me, 34.9%
  2 Pic. m.p. 177.5−179.5 °C

169, R = H, 74.9%, m.p. 65 °C
  2 Pic. 140 °C
  R = Me, 82.5%, m.p. 60−66 °C
  2 Pic. 124−126 °C

Similar procedures gave propanones **170** and **172b** (via isolated **172a**) which were cyclized to quinuclidines **171** and **173**.[497,498]

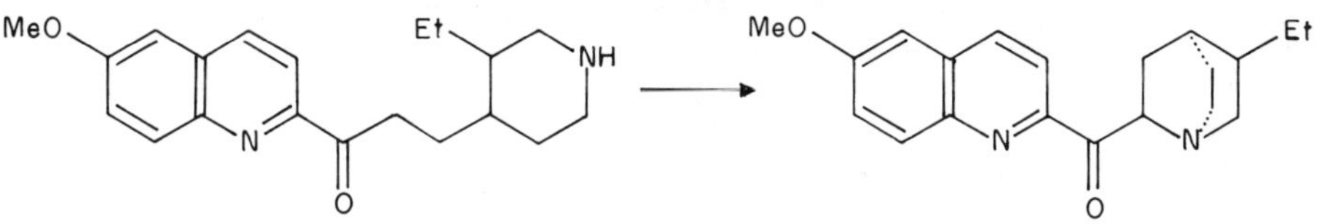

170, m.p. 195−196 °C
  1 HCl 226.5−228 °C
  2 HBr 192−193 °C
  2 Pic. 167−168 °C

171, m.p. 125−126 °C

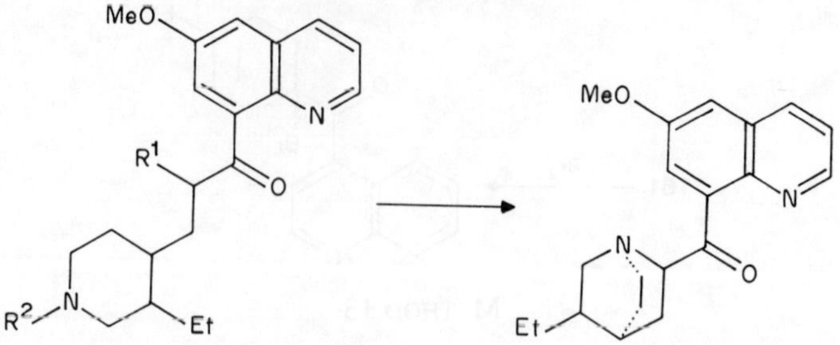

172 a, R¹ = COOEt, R² = COPh
m.p. 54−56 °C
b, R¹, R² = H
2HBr m.p. 193−194 °C

173, m.p. 152−153 °C

Claisen condensations gave the keto-esters **174**, which were hydrolysed and decarboxylated to ketones **175a, c**. The *N*-bromamines **175b, d** were formed with hypobromous acid and cyclized to ketones **176**.[173]

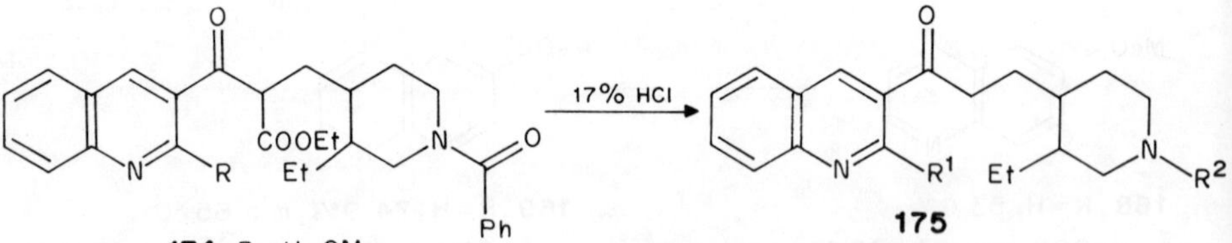

174, R = H, OMe

17% HCl →

175

a, R¹, R² = H, 70%, b.p. 225 °C/9 mm
b, R¹ = H, R² = Br, m.p. 137−139 °C
c, R¹ = OMe, R² = H, b.p. 197−200 °C/5 mm
d, R¹ = OMe, R² = Br, m.p. 158−162 °C

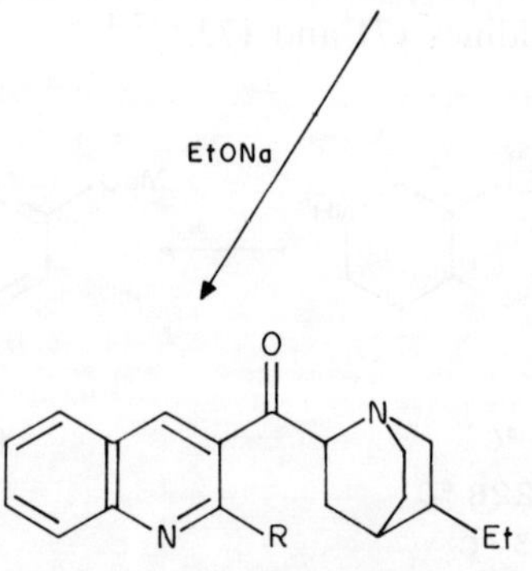

EtONa

176, R = H, m.p. 122−124 °C
R = OMe, m.p. 155−156 °C

Table 17. Cinchotoxine and its Derivatives

R¹	R²	R³	Quinoline substituent(s)	Method	Yield (%)	M.p. (°C)	References
H	H	H	None	1	56		499
H	H	H	2-Me-6-OMe	1		2HCl 210	500
H	H	H	6-Bu	1			501, 502
H	H	H	2-tBu	1		2HCl 200	272
H	H	H	2-cycloC$_6$H$_{11}$	1		1HCl 190–191	272
H	H	H	2-Ph				272
H	H	H	6-OMe[a]	1	71	Oil	499, 503, 504
H	H	CH$_2$CH$_2$Ph	2-tBu	2		2HCl 130	272
							505
H	Et	H	2-Me	3		BF[b] 159–160	506, 507
H	Et	H	2-Me-6-OMe	3		Oil	506, 507
				3		1HCl 239	500
H	Et	H	2-Me-6-OCH$_2$CH$_2$iPr	3		BO[b] 154–155	506, 507
H	Et	H	6-Me	1	68		508
H	Et	H	2-Pr	3		TOx[b] 148–152	506, 507
H	Et	H	2-Pr-6-OMe	3			506, 507
H	Et	H	2-iPr	3	51	BO 97	506, 507
H	Et	H	6-Cl	1		1HCl 157	508
H	Et	H	6-OMe	1		1HBr 156–157	508, 509
H	Et	H	6,7-diOMe	1	40	DBT[b] 161.5–163.5	508

Table 17. (*Contd.*)

R^1	R^2	R^3	Quinoline substituent(s)	Method	Yield (%)	M.p. (°C)	References
H	Et	H	7-OMe	1	54	DBT 174–175.5	508
H	Et	H	6-OEt				505
H	Et	H	6-OPr				505
H	Et	H	6-OCH$_2$CH$_2$iPr				505
H	Et	H	6-OCH$_2$CH$_2$OMe				505
H	Et	Me					510
H	Et	Me	2-OMe	3a		BO 187	507
H	Et	Me	2-OPr	3a		BO 138–141	507
H	Et	CH$_2$cycloC$_3$H$_5$^{c}		2		BO 201	490, 511
H	Et	CH$_2$cycloC$_3$H$_4$-1-OH		2		BO 172–174	490, 511
H	Et	CH$_2$cycloC$_5$H$_8$-1-OH		2		BO 82–85	511
H	Et	CH$_2$CH$_2$C$_6$H$_3$-3,4-diOM^c		2		BF 170–172	490, 511
H	Et	CH$_2$CH$_2$C$_6$H$_2$-3,4,5-triOMe^c		2		BO 155–160(d)	490, 511
H	Et	CH$_2$CH$_2$OH		2		BO 157–160	490
H	Et	CH$_2$CHOHPh		2		BO 163–165	490, 511
H	Et	CH$_2$CH$_2$NMe$_2$^{c}		2		Oil	490, 511
H	Et	CH$_2$CH$_2$NEt$_2$		2		BF 175–176	490
H	Et	CH$_2$CH$_2$OMe^c		2		BF 140–142	490, 511
H	Et	CH$_2$COPh		2		BO 158–160	490
H	Et	Pr^c		2		BO 187–189	490, 511
H	Et	Pr	2-Me	2		BF 147–149	507
H	Et	Pr	2-Me-6-OMe	2		HCl 224	507
H	Et	Pr	2-Pr	2		MO^b 130	507
H	Et	Pr	6-OH	2		HCl 158–159	511
H	Et	Pr	6-OMe	2		BO 202–204	511
						1MeI 180–185	
H	Et	Pr	6-OPr	2		BF 135–136	511
H	Et	Pr	6-OCH$_2$CH$_2$iPr	2		BO 190–192	490, 511
H	Et	(CH$_2$)$_3$OMe		2			511

H	Et	CH$_2$CHOHMe		2	BO 188–190	490
H	Et	CH$_2$COH(Me)Ph		2	HBr 228–229	511
H	Et	CH$_2$CHOHCH$_2$NEt$_2$		2	TO[b] 122–125	511
H	Et	CH$_2$CH$_2$COC$_6$H$_4$-4-F		2	BF 153–155	490, 511
H	Et	CH$_2$C≡CH		2	BO 203–205	490, 511
H	Et	Bu	2-Me	2	BF 160–162	490, 511
H	Et	Bu	2-Pr	2	BF 163–165	507
H	Et	Bu	6-OH	2	BO 140–143	507
H	Et	Bu	6-OMe	2	HCl 158–159	490
H	Et	Bu	6-OPr	2	BO 202–204	490
H	Et	(CH$_2$)$_3$CH(C$_6$H$_4$-4-F)$_2$		2	BF 135–136	490
H	Et	(CH$_2$)$_4$OH		2	HCl 170(d)	490, 511
H	Et	(CH$_2$)$_4$OMe		2	BO 140–143	511
H	Et	(CH$_2$)$_4$OMe	2-Pr	2	BF 108–110	490, 511
H	Et	(CH$_2$)$_4$CN		2	BO 156–158	507
H	Et	(CH$_2$)$_4$CN	2-Pr	2	BF 149–151	490, 511
H	Et	(CH$_2$)$_3$COPh		2	BO 138	507
H	Et	(CH$_2$)$_3$COPh		2	BF 127–129	490, 507, 511
H	Et	(CH$_2$)$_3$COPh	2-Me	2	BF 71–74	507
H	Et	(CH$_2$)$_3$COPh	2-Pr	2	TOx 127–131	507
H	Et	(CH$_2$)$_3$COC$_6$H$_4$-4-Me		2	BF 121–123	490, 511
H	Et	(CH$_2$)$_3$COC$_6$H$_4$-4-OH		2	Fu[b] 175–177	490, 511
H	Et	(CH$_2$)$_3$COC$_6$H$_4$-4-OMe		2	BF 139–141	490, 511
H	Et	(CH$_2$)$_3$COC$_6$H$_4$-4-Cl		2	BF 128–130	490, 511
H	Et	(CH$_2$)$_3$COC$_6$H$_4$-4-F		2	BF 143–146	490, 511
H	Et	(CH$_2$)$_3$COC$_6$H$_3$-3,4-diCl		2	BF 122–124	490, 511
H	Et	(CH$_2$)$_3$CO-2-thienyl		2	BF 157–160	490, 511
H	Et	CH$_2$iPr		2	BF 148–150	490, 511
H	Et	CH$_2$C(OH)Me$_2$	2-Me	2	BO 132–134	490, 511
H	Et	CH$_2$C(OH)Me$_2$	2-iPr	2		511
H	Et	CH$_2$C(OH)Me$_2$		2		511
H	Et	CH$_2$C(OCOEt)Me$_2$		4	BF 132–134	490, 511
H	Et	CH$_2$C(OH)Me$_2$	6-OMe	2	BO 132–134	490, 511
H	Et	C$_5$H$_{11}$		2	BF 102–104	490, 511

Table 17. (*Contd.*)

R¹	R²	R³	Quinoline substituent(s)	Method	Yield (%)	M.p. (°C)	References
H	Et	(CH₂)₄CHOHC₆H₄-4-F		2		BO 189–192	511
H	Et	(CH₂)₄COC₆H₄-4-F		2		BO 175–178	511
H	Et	CH₂CH₂iPr		2		BF 180–182	490, 511
H	Et	CH₂CH₂C(OH)Me₂		2		BO 142–145	511
H	Et	C₆H₁₃		2		BF 106–108	490, 511
H	Et	(CH₂)₃iPr		2		BF 138–140	490, 511
H	Et	CH₂C(OH)Et₂		2		BO 72–75	511
H	Et	C₇H₁₅		2		BF 90–93	490, 511
H	Et	C₉H₁₉		2		BF 105–107	490, 511
H	Et	cycloC₆H₁₁		2		BO 108–109	511
H	Et	CH₂-2-tetrahydrofuryl		2		BF 146(d)	490, 511
H	Et	CH₂Ph		3a		BO 130–134	512
H	Et	CH₂CH=CHPh		2		BF 153–155	490, 511
H	Et	Cl	6-OMe	5			508
H	Et	COOEt					485
H	CH=CH₂	H		3	98	Oil	513
H	CH=CH₂	H	2-Me	3		BF 166–168	506, 507
H	CH=CH₂	H	2-Me-6-OMe	3		1HCl 110	500
H	CH=CH₂	H	6-OMe	1	64	DBT 183	513
H	CH=CH₂	H		1		(+)-DBT 185.5–186 (−)-DBT 185–186	514, 515
H	CH=CH₂	H		6	15.7	DBT 184–185	516
H	CH=CH₂	H		7			517
H	CH=CH₂	H		3		BO 166–167	518
H	CH=CH₂[d]	H	6-OMe	1	97		519
H	CH=CH₂	Me	6-OCH₂O-7	3a	41	130–131	508
H	CH=CH₂						513, 520
H	CH=CH₂	CH₂CH₂OMe	6-OMe	2		BF 126–128	490, 511

					Yield	mp / deriv.	Ref.
H	CH=CH₂	CH₂CH₂NMe₂	6-OMe	2		2HCl 223–225	490
H	CH=CH₂	CH₂CH₂N-(CH₂CH₂)₂O	6-OMe	2	76	HCl 130–140	521
H	CH=CH₂	CH₂CH₂N-(CH₂CH₂)₂NCH₂-CH=CHPh	6-OMe	2	98	TO 219(d)	521
H	CH=CH₂	CH₂CH₂N-(CH₂CH₂)₂-NCOCH=CHPh	6-OMe	2	95	BF 204–205	521
H	CH=CH₂	CH₂CH₂N-(CH₂CH₂)₂-NCOCH=CH-C₆H₂-3,4,5-triOMe	6-OMe	2	99	HCl 130–140	521
H	CH=CH₂	CH₂COPh	6-OMe	2		BO 150–152	490, 511
H	CH=CH₂	Pr	6-OMe	2		BO 170–172	490
H	CH=CH₂	Pr	6-OMe	2		BF 115–117	490
H	CH=CH₂	(CH₂)₃OMe	6-OMe	2		Oil	511
H	CH=CH₂	CH₂CH=CH₂	6-OMe	2		BF 97–99	490, 511
H	CH=CH₂	(CH₂)₃SO₃H	6-OMe	2		HCl 217–218	487, 522
H	CH=CH₂	CHMeCH₂NMe₂	6-OMe	2		BO 158	521
H	CH=CH₂	Bu	6-OMe	2		BF 148	521
H	CH=CH₂	(CH₂)₃CH(C₆H₄-4-F)₂	6-OMe	2		HCl 185	490
H	CH=CH₂	(CH₂)₃COPh	6-OMe	2		BF 117–119	490, 511
H	CH=CH₂	(CH₂)₃COC₆H₄-4-F	2-Me	2		BO 168–170	507
H	CH=CH₂	C₅H₁₁	6-OMe	2		BF 142–144	490
H	CH=CH₂	(CH₂)₃iPr	6-OMe	2		BF 158–160	490
H	CH=CH₂	C₉H₁₉	6-OMe	2		BF 134–136	490, 511
H	CH=CH₂	CH₂CH₂cycloC₆H₁₁	6-OMe	2		BF 183–185	490, 511
H	CH=CH₂	Cl	6-OMe	5	54	153	508
H	CH=CH₂	Br	6-OMe	5		123	523
H	CH=CH₂	Br	6-OMe	5		Pic. 118–119	524
H	CH=CH₂	Ac	6-OMe	2	96		525
H	CH=CH₂	COCH=CHPh	6-OMe	2	98	HCl 105–110	521
H	CH=CH₂	COCH=CHC₆H₂-3,4,5-triOMe	6-OMe	2	92	145–146	521

Quinoline Ketones

Table 17. (*Contd.*)

R¹	R²	R³	Quinoline substituent(s)	Method	Yield (%)	M.p. (°C)	References
H	CH=CH$_2$	COPh		2			513
H	CH=CH$_2$	COPh	6-OMe	6		111–112	516
H	CH=CH$_2$	COPh	6-OCH$_2$O-7	7	34.5	HCl 145–146	517, 526
H	CH=CH$_2$	COC$_6$H$_4$-2-OAc	6-OMe	1	28	70	508
H	CH=CH$_2$	COC$_6$H$_4$-4-Cl	6-OMe	2	95	HCl 111	521
H	CH=CH$_2$	COOEt	6-OMe	2	83	107.5–108	521
H	CHOMe	Me		2	96	HCl 148–149; Pic. 139–140	525
H	CH=CH$_2$	COPh		8			489
H	CH=CH$_2$	H	6-OMe	8			489, 527
Cl	CH=CH$_2$	COPh	6-OMe	9			526
Cl	CH=CH$_2$	H	6-OMe	9			492
Br	H	H	6-OMe	10		2HBr 184	499
Br	H	H	6-OMe	10	50		499
Br[e]	Et	H	6-OMe	10		270(d)	503
Br	CHBrCH$_2$Br	H		10		1HBr 194–195	528
Br	CHBrCH$_2$Br	H	6-OMe	10		2HBr 158–159(d)	529
Br	CHBrCH$_2$Br	H	6-OMe	10		2HBr 152–155(d)	529
Br	CHOHMe	H	6-OMe	10		241	489
COOEt	H	COPh	6-OMe	1	88	Oil	504
COOEt	CH=CH$_2$	H	6-OMe	1	63.4		514, 515
COOEt	CH=CH$_2$	COPh	6-OMe	1	59		513

[a] 6-Methoxyrubatoxanone; the nomenclature is based on rubane, see Ref. 530.
[b] DBT = di-O-benzoyl-(+)-tartrate, BF = bifumarate (mol. ratio 1:1), BO = bioxalate (mol. ratio 1:1), MO = monoxalate (mol. ratio 2:1), TO = trioxalate (mol. ratio 2:3), TOx = tetraoxalate (mol. ratio 1:2), Fu = fumarate.
[c] Reference 511 has a further 48 ketones, all prepared by N-alkylation as for the examples listed; Ref. 507 likewise has 12 more ketones.
[d] The trans isomer prepared from quinotoxine with aqueous formaldehyde.
[e] By-product from 6-methoxyrubanone synthesis; described as 6-methoxy-5-(?)-bromorubatoxanone-9.

Table 18. 1-(4-Piperidyl)-3-(4-quinolyl)-2-propanones

(see Footnote a)

Entry	R¹	R²	R³	Quinoline substituent(s)	Preparation	Yield (%)	M.P. (°C)	References
1	H	H	Ac	6-OMe	No details given			531
2	H	Et	H	2-CF$_3$	As above	44.5	BF^b 140	507, 532
3	H	Et	H	2,7-diCF$_3$	As above			533
4	H	Et	H	2,8-diCF$_3$	As above			533
5	H	Et	H	2-CF$_3$-7-Cl	As above			533
6	H	Et	H	2-CF$_3$-8-Cl	As above			533
7	H	Et	H	7-CF$_3$	As above	66		534, 535
8	H	Et	H	7-Cl	As above	72	HCl 215–216	534, 535
9	H	Et	H	2-OMe	As above		Oil	507, 532
10	H	Et	H	6-OMe	As above			511
11	H	Et	H	6,8-diOMe	As above	75	Oil	535
12	H	Et	H	6-OCH$_2$O-7	As above	65		534, 535
13	H	Et	Pr	2-OMe	Entry 9, PrBr, K$_2$CO$_3$, DMF, 110–120 °C		BO^b 138–139	507
14	H	Et	Pr	6-OMe	Entry 10, PrI, DMF, 50–60 °C		Oil	511
15	H	Et	C$_5$H$_{11}$	2-CF$_3$	Entry 2, C$_5$H$_{11}$Br, K$_2$CO$_3$, DMF, 110–120 °C		BF 160	507

Table 18. (*Contd.*)

Entry	R¹	R²	R³	Quinoline substituent(s)	Preparation	Yield (%)	M.P. (°C)	References
16	H	Et	PhCO	2,7-diCF$_3$	As above			485
17	H	Et	PhCO	2,8-diCF$_3$	As above			533
18	H	Et	PhCO	2-CF$_3$-5-Cl	As above			533
19	H	Et	PhCO	2-CF$_3$-7-Cl	As above			533
20	H	Et	PhCO	2-CF$_3$-8-Cl	As above			533
21	H	Et	PhCO	7-Cl	As above			533
22	H	Et	PhCO	6-OMe	As above	69		535
23	H	Et	PhCO	2,8-diCF$_3$	As above	45	Oil	534, 535
24	H	CH=CH$_2$	H	7-CF$_3$	As above			533
25	H	CH=CH$_2$	H	7-Cl	As above	75		534, 535
26	H	CH=CH$_2$	H	6-OMe	As above		1HCl 236–237(d)	534, 535
27	H	CH=CH$_2$	H	6,8-diOMe	Entry 35, 1.5M-H$_2$SO$_4$	75	58–60	470
28	H	CH=CH$_2$	H	2,8-diCF$_3$	As above			534, 535
29	H	CH=CH$_2$	PhCO	5-CF$_3$	As above			533
30	H	CH=CH$_2$	PhCO	6-CF$_3$	As above			534
31	H	CH=CH$_2$	PhCO	6-Cl	As above			534
32	H	CH=CH$_2$	PhCO	6,8-diCl	As above	46[c]		534, 535
33	H	CH=CH$_2$	PhCO	7-Cl	As above	40		534
34	H	CH=CH$_2$	PhCO	6-OMe	As above		147–148	534, 535
35	H	CH=CH$_2$	PhCO	6,8-diOMe	As above	78	Oil	470, 534–537
36	H	CH=CH$_2$	PhCO	6,8-diOMe	As above			535
37	Br	Et	PhCO	2,7-diCF$_3$	Entry 17, NBS, (PhCOO)$_2$	61		533
38	Br	Et	PhCO	2,8-diCF$_3$	Entry 18, NBS, (PhCOO)$_2$			533

39	Br	Et	PhCO	2-CF$_3$-5-Cl	Entry 19, NBS, (PhCOO)$_2$	533
40	Br	Et	PhCO	2-CF$_3$-7-Cl	Entry 20, NBS, (PhCOO)$_2$	533
41	Br	Et	PhCO	2-CF$_3$-8-Cl	Entry 21, NBS, (PhCOO)$_2$	533
42	Br	CH=CH$_2$	PhCO	6-OMe	Entry 35, NBS, (PhCOO)$_2$	470, 534–536, 538
43	COOEt	Et	COOCH$_2$Ph		See text, Section IV.3	485
44	COOEt	Et	H		Above, HCl, EtOH Pic. 154–156	485

[a] The stereochemistry is not always specified. Where it is given, it is as shown.
[b] See Table 17, footnote b for definitions.
[c] Accompanied by 14% of 2-(6-chloro-4-quinolyl)-1-phenylethanone; see Table 25.

Table 19. 4-Quinolyl 2-Quinuclidinyl Ketones

R¹	R²	Quinoline substituent(s)	R¹ orientation	Method	Yield (%)	M.p. (°C)	References
H	H	None	a/b	1		1Pic. 170–180	499
H	H	6-OMe	a/b	1	50.5	90–91	504, 539
			a/b			1Pic. 211–211.5	499, 503, 540
				2	17.5		496
H	Et		a/b	1		130	528, 541
						HI 196	
				3	34	1HCl 265	542
						1Pic. 186	
H	Et	2,7-diCF$_3$	a/b	4			533
H	Et	2,8-diCF$_3$	a/b	4			533
H	Et	2-CF$_3$-5-Cl	a/b	4			533
H	Et	2-CF$_3$-7-Cl	a/b	4			533
H	Et	2-CF$_3$-8-Cl	a/b	4		HCl 245–247	533
H	Et	6-Me	a/b	4		105–108	543, 544
				5			508
H	Et	7-CF$_3$	a/b	5		106–111	508
H	Et	2-Cl	a/b	6	81	163–164	545
H	Et	2-Cl-6-OMe	a/b	6		163–164	545
H	Et	6-Cl	a/b	4		104–107	543, 544
				5	69	97.5–100.5	508
				3			508

R	R′	Ar subst.	a/b	No.	% Yield	mp (°C)	Refs
H	Et	6,8-diCl	a/b	4			546
H	Et	7-Cl	a/b	4		124–127	543, 544
				5			508, 534
				3			535
H	Et	6-OMe	a	5	73	102–104	508
H	Et	6-OMe	b	5		100–104	508
H	Et	6-OMe	a/b	1		98–99; Pic. 224	541
H	Et	6,7-diOMe	a/b	4	35	86–90	543, 544, 546, 547
H	Et	7-OMe	b	7	40	98–99	477
H	Et	7-OMe	a/b	5	65	115–118	508
H	Et	6-OCH$_2$O-7	a/b	4	82	111–117	508
H	CHBrMe	6-OMe	a/b	4	56	103–108	543, 544
H	CHBrCH$_2$Br	6-OMe	a/b	5	74	151–151.5	508
H	CHBrCH$_2$Br	6-OMe	a/b	3		171–172(d)	546
H	CHOHMe		a/b	8		177(d)	472
				9		172–173	529
				9		170	529
				5		Pic. 222–224	472
				3			527
H	CH=CH$_2$		b	3	85	133–134	494, 548
H	CH=CH$_2$		a/b	5	46	126–127	523
H	CH=CH$_2$	2,8-diCF$_3$	a/b	3	11.4	1HCl 245–247	466, 542, 549
H	CH=CH$_2$	6,8-diCl	a/b	10		120	529
H	CH=CH$_2$	7-Cl	a/b	7	21	125–127	477
H	CH=CH$_2$	2-OH-6-OMe[a]		4			533
H	CH=CH$_2$	6-OMe	a	4	62.3	100–102	546
				5	78	99–100.5	508
				11		113–113.5	550
				5			470
				3			492, 508, 524
							548

Table 19. (*Contd.*)

R¹	R²	Quinoline substituent(s)	R¹ orientation	Method	Yield (%)[b]	M.p. (°C)	References
H	CH=CH₂	6-OMe	*b*	5	Trace[b]		492
H	CH=CH₂	6-OMe	*a/b*	4			546, 547
				3	94	102	466
						1HCl 210–212	542
						1H₂SO₄ 106–108	
						1Pic. 232–233	
				11			470
				10	34	101–102	472, 479
				7			477
H	CH=CH₂	6-OCH₂O-7	*a/b*	5	84		508, 546
Ph	CH=CH₂	6-OMe	*a/b*	12	75		495
4-MeC₆H₄	CH=CH₂	6-OMe	*a/b*	12	90		495
Br	Et		*a/b*	13		161–162	528

[a] Metabolic of quinidine.
[b] Quininone, detected by t.l.c.

V. Ketomethyl Quinolines

The 4-ketomethylquinolines prepared for the synthesis of cinchona alkaloids and their derivatives are covered in Section IV.3 and Table 18. All other quinoline derivatives carrying ketone groups in alkyl side chains are dealt with here and in subsequent sections.

1. Structure of 2-Ketomethylquinolines

Several studies have shown that these ketomethylquinolines exist as equilibrium mixtures of the enaminone forms **177a** and the unconjugated forms **177b**.

177 a   **177 b**

For example, **177**, R = tBu was examined by u.v. and n.m.r. spectroscopy in 25 solvents. In most, **177a** was the main form, but a significant proportion (11.3%·to 58.8%) of **177b** was always present.[551] For a series of compounds **177**, R = various aryl groups, the ratio **177a**:**177b** ranged from 100:0 to 65:35.[552] In aqueous solution compound **177**, R = Ph was shown by potentiometric titration to favour form **177a** with $pK_T = 1.09$ [$pK_a (-H^+)$ 13.29; $pK_a (+H^+)$ 3.73].[553,554] The chemical shift of the N–H proton was $\delta 16$, and of the ring C3–H $\delta 6.9$. The same proton for the N-methyl derivative **178**, now in the deshielding zone of the carbonyl group, was $\delta 9.14$.[554] The corresponding figures for the diketone **179** were C3–H $\delta 9.21$ and N–H $\delta 18.1$.[555] The mechanism of the imine–enamine tautomerism for compound **177**, R = Ph has been studied.[553]

178   **179**

Compound **177**, R = Me, was investigated by u.v. and n.m.r. spectroscopy in carbon tetrachloride[556]. The data were interpreted to show an 85:15 preponderance of the enaminone form. Another n.m.r. investigation showed for compound **177**, R = Me in $CDCl_3$ a 73:27 ratio of **177a**:**177b**. For compound **179** in $CDCl_3$ only the enaminone form could be detected, and quinophthalone,

Section V.8, in the same solvent showed this as the major form.[555] The cyclopentanone derivative **180**, $n = 1$ had no i.r. carbonyl band above $1660 \, cm^{-1}$ (nujol). This was considered reasonable for the enaminone form of a five-membered ring ketone. The keto-ester **181** showed bands at 1690, 1632 and $1615 \, cm^{-1}$ (CCl$_4$).[557]

180   **181**

Other workers suggested for compound **177**, R = COOEt that the enol rather than the enaminone form was dominant. In one report, bromine titration in dry methanol and u.v. spectroscopy, λ_{max} (toluene) 460 nm (log ε3.79); λ_{max}. (H$_2$O) 432 nm (log ε4.39) were the methods used[558]. Another group studied the n.m.r. spectrum and assigned a signal at δ6.38 (DMSO-d$_6$) or δ6.60(CDCl$_3$) to a side chain =CH of the enol. As they could find no trace of a CH$_2$ signal they concluded that the compound was completely enolized. They reached a similar conclusion for the 4-ketomethyl compound **184**.[559] For compound **177**, R = Ph in CDCl$_3$, signals at δ4.62 (keto) and δ6.00—taken to represent enol—were integrated to give 94% of the enol form.[560] However, a signal in the region δ6.00 to 6.60 could be the vinyl proton of the enaminone form. The i.r. spectrum of compound **180**, n = 2 was reported to show no carbonyl band, but νC=C at $1613 \, cm^{-1}$.[561] In consequence the structure was drawn as an enol zwitterion; however, this band is not too low to be the carbonyl of an enaminone form.

The balance of evidence favours the enaminone form **177a**. Recent work in the reviewer's laboratory has confirmed this except for 3-(2-quinolyl)butanone, which was 100% in the keto form.[562] In spite of this single exception, all structures in this section are drawn in the enaminone form whether or not they are shown this way in the original reports.

2. Structure of 4-Ketomethylquinolines

182   **183**

184 a   184 b

By bromine titration, it was claimed that the 4-ketomethylquinoline **184** had 76% of the enol form **184b** after 20 min in absolute methanol, which dropped to 22% at equilibrium. Intermolecular hydrogen bonded stabilization of the enol was suggested.[558] A similar conclusion was reached[563] for compound **182** (R = iPr). A partial separation of tautomers was achieved by recrystallization. A solid of m.p. 45.4–46.6° had only $v_{C=O}$ at 1700 cm^{-1} in the carbonyl region, while a second sample m.p. 160–165 °C, was a mixture with additional bands at 1625 and 1540 cm^{-1}. A band at 2450–2300 cm^{-1} was assigned to a hydrogen bonded enol. However, others[556] claimed to detect a small proportion of the dienaminone form of compound **184**. One 4-acetonylquinoline, which had $v_{C=O}$ 1680 cm^{-1}, was concluded from this and other evidence to be exclusively in the dienaminone form.[182] Compounds with the ketomethyl group at position 4 of quinoline lack the possibility of intramolecular hydrogen bonding to stabilize a dienaminone form comparable with **177a**. Thus compound **182** (R = Ph) has pK_a 5.24, while its N-methyl derivative **183** has pK_a 7.02, i.e. log K_T 1.78, in favour of the keto form. This conclusion was confirmed by u.v. spectroscopy (pK_T 2.31). Also recorded for compound **182** (R = Ph) were pK_a (−H$^+$) 12.37 and various u.v. spectral data.[553,554]

Clearly more work needs to be done on this tautomerism, but in this Chapter the 4-ketomethylquinolines are shown in the keto forms.

3. Preparation of 2- and 4-Ketomethylquinolines

The following are examples of the synthetic methods used. Details and full references are in Tables 20 (p. 266) and 25 (p. 285).

Method 1. Base catalysed condensation of 2-methylquinoline **185** or 4-methylquinoline with an ester or anhydride.

185

METHOD 1

Method 2. Nucleophilic attack by a deprotonated methylene ketone on 2-chloroquinoline **186** (R = Cl). The kinetics of this reaction have been studied. Reaction is slow in the presence of lithamide (liquid NH_3, $-33\,°C$), but is faster under u.v. irradiation or in the presence of dilithiobenzoylacetone.[564–566] Some derivatives of 4-chloroquinolines have been similarly prepared.[567] In one example a sulphonic ester group underwent a similar nucleophilic displacement. Compound **186** (R = SO_3Me) reacted with acetophenone to give ketone **187**, $R^1 = Ph$, $R^2 = H$.[568]

186   **187**

METHOD 2

Method 3. Hydration of 2-, 3- and 4-quinolylacetylenes. Some reports stress that only single products, the ketomethylquinolines, are obtained, e.g. **188**.[124,569,570]

188

METHOD 3

Method 4. Base catalysed reaction of a 2-alkylquinoline with a nitrile.

METHOD 4

Method 5. Active methylene compounds add to quinoline *N*-oxides in acetic anhydride. The mechanism proposed for the pyruvonitrile reaction is shown.[571] The product, **189**, was converted (EtOH, HCl) into the ester **189** (in which CN was replaced by COOEt), identical with an authentic sample. When benzoylacetonitrile was used, a high percentage of the ylid **190** was obtained.[572]

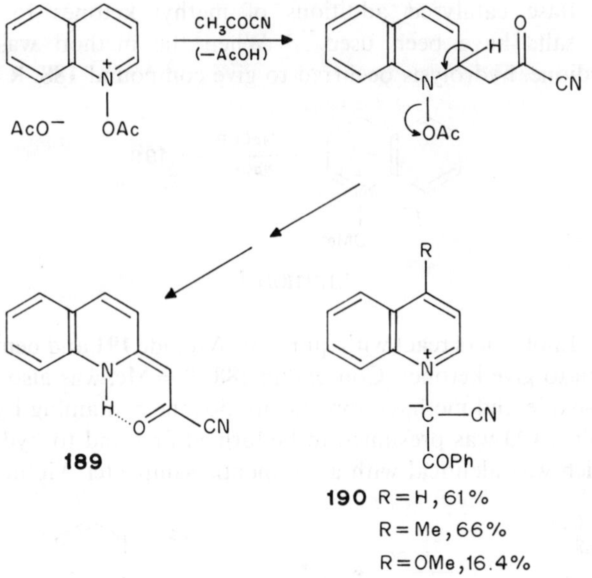

189

190 R = H, 61%
R = Me, 66%
R = OMe, 16.4%

METHOD 5

Method 6. Under the conditions of Method 5, *N*-acylmethylpyridinium salts react to give pyridinium ketones **192** which can be reduced with zinc and acetic acid to the simple acylquinolines **188**. Warm sodium carbonate solution hydrolysed the ketone groups in a few minutes to give the quinolylmethyl-pyridinium salts.[573] The identity of compound **192**, R = Ph was confirmed by its preparation from 2-phenacylquinoline **188**, R = Ph by bromination and treatment with pyridine.[552,573,574]

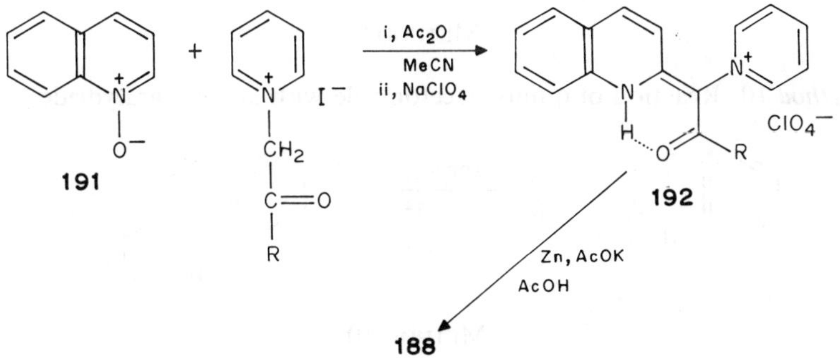

191

192

188

METHOD 6

Method 7. Base catalysed additions of methyl ketones to *N*-methoxy-quinolinium salts have been used.[575] When the method was applied to pentane-2,4-dione, hydrolysis occurred to give compound **188**, R = Me.[576]

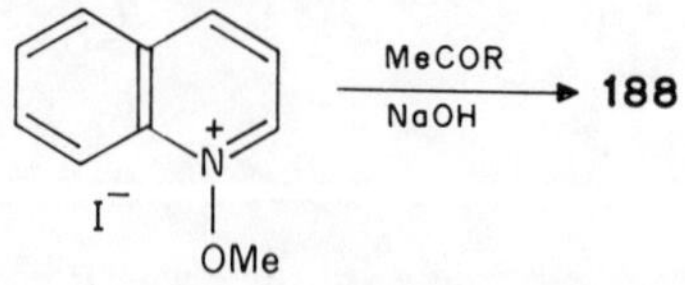

METHOD 7

Method 8. Enol ethers react with quinoline *N*-oxide **191** and benzoyl chloride in chloroform to give ketones. Compound **188**, R = Me, was also formed from quinoline *N*-oxide and methacrylonitrile in dioxane containing hydroquinone. The cyanhydrin **193** was presumed to be formed first and to hydrolyse to the product, which was identical with an authentic sample (cf. Method 15).[577,578]

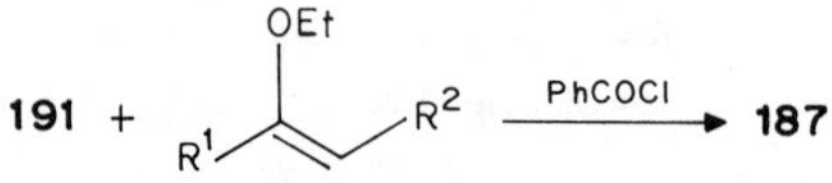

191 +

METHOD 8

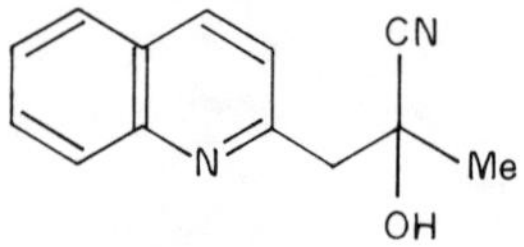

193

Method 9. Reductive ring closure of *o*-nitroketones. The 2-acetonylquinoline **188**, R = Me produced could be hydrolysed to 2-methylquinoline in a sulphuric/hydrochloric acid mixture at 160–170 °C.[579,580]

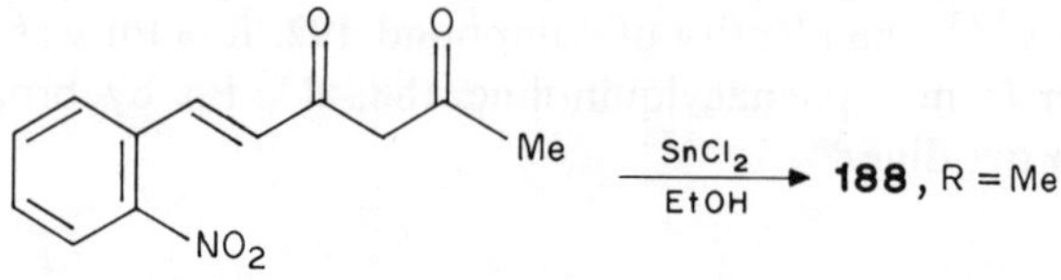

METHOD 9

Method 10. Reaction of quinolylacetonitrile with an acid anhydride.

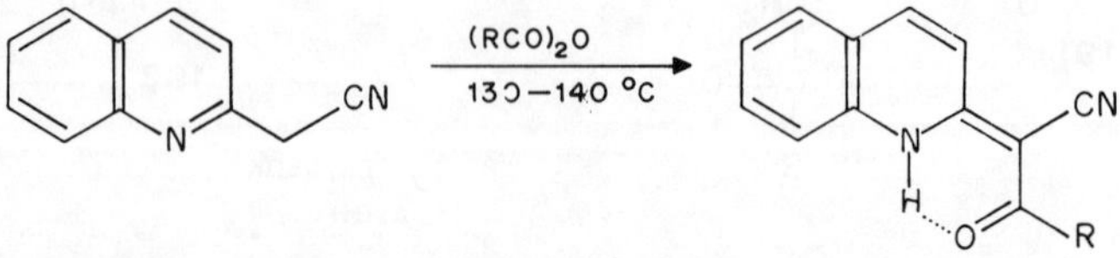

METHOD 10

Method 11. Some ketones in this class have been prepared by reactions of the oxazine **194**.[170,581]

METHOD 11

Method 12. Hydrolysis of the pyridoquinolines **195** and **196**. The product depends on the conditions used.[45]

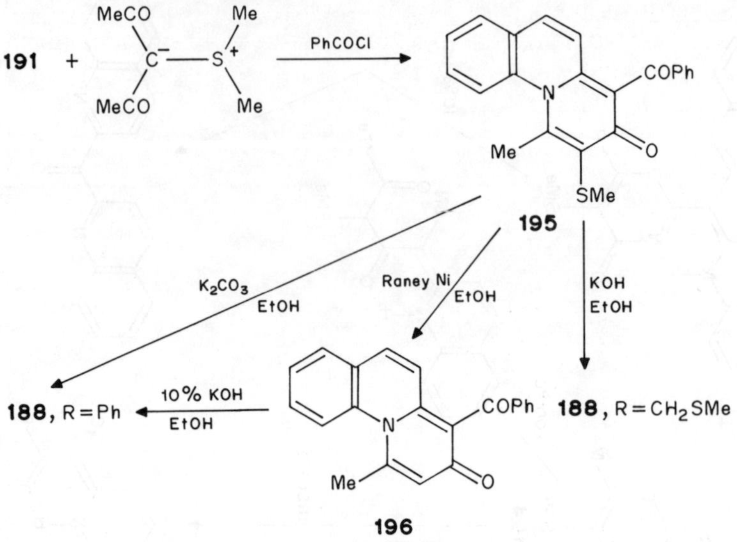

METHOD 12

Method 13. An *N*-debenzylation has been used[582] to produce ketone **188**, R = Ph.

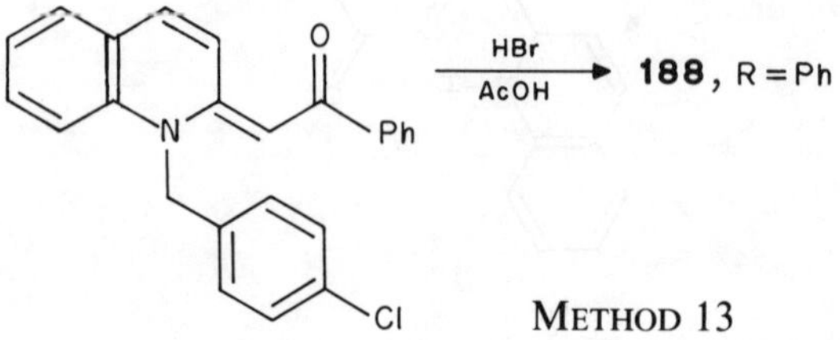

METHOD 13

Method 14. Enamine hydrolysis has been employed. This reaction also succeeded with 4-methylquinoline.[583]

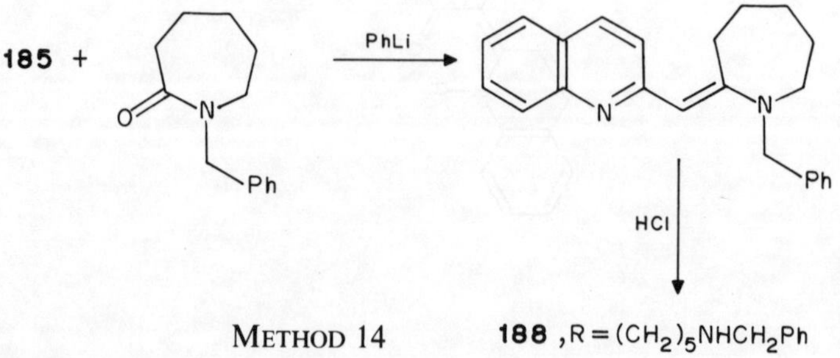

METHOD 14   **188**, R = (CH$_2$)$_5$NHCH$_2$Ph

Method 15. Enamines react like the enol ethers of Method 8 to give ketones.[557] The use of toluenesulphonyl chloride or enamines derived from amines other than morpholine gave lower yields.[584] Quinoline *N*-oxide **191** and *N*-cyclohexen-1-ylpiperidine gave 61.4% of the 2-substituted quinoline and 8% of the 4-substituted quinoline. The comparable morpholine enamine gave only the 2-quinoline (73.4%).[584] The same morpholine enamine and 2-methylquinoline *N*-oxide or 2-chloroquinoline *N*-oxide gave only the 4-substituted products **197**, R = Me, Cl. There was no evidence of chlorine displacement in the latter case.[561,585] One reaction was shown to go via the *O*-acyl *N*-oxide, and it was presumed that they all follow this mechanism.[571,586]

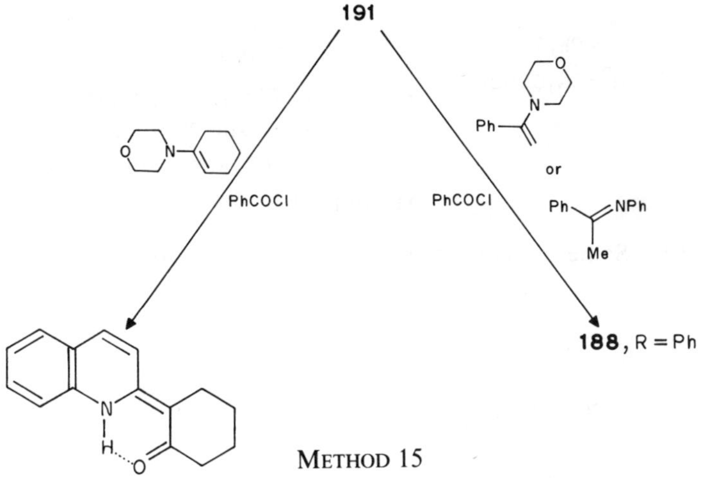

METHOD 15

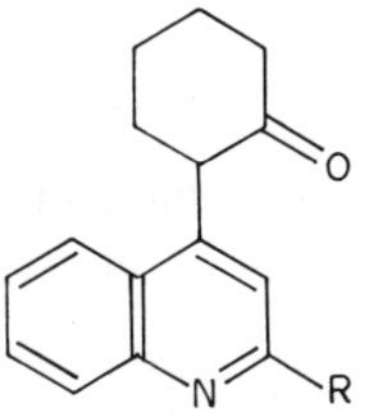

197

Method 16. Aromatic aldehydes have been induced to add on to the unsaturated ketone **198**.[587]

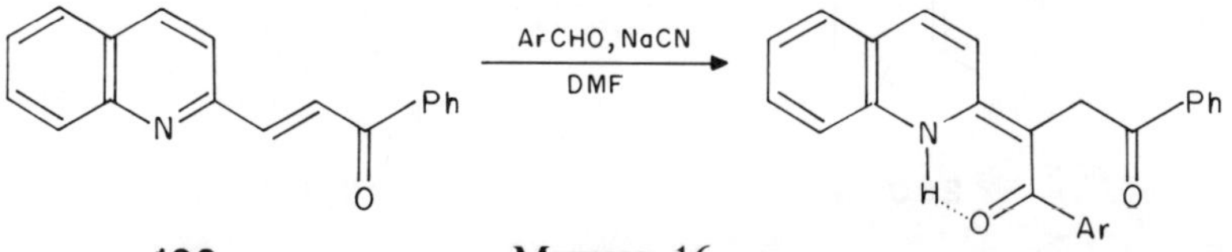

198   METHOD 16

Method 17. Activated acetylenes add to quinoline *N*-oxide **191** to give the 2-acylquinolines.[588-591] In two examples this reaction also gave small yields of the 3-ketomethylquinolines **199a**, and from 4-chloroquinoline *N*-oxide, **199b**. The betaine structures shown were assigned by comparison with an equivalent pyridine derivative.[588-590]

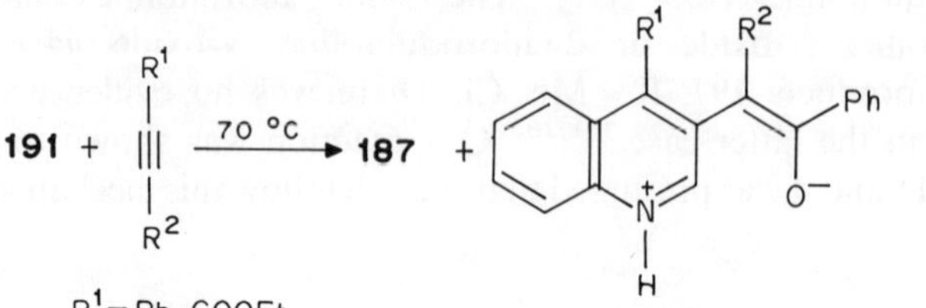

$$R^1 = Ph, COOEt$$
$$R^2 = CN, COOEt$$

199

a, $R^1 = H, R^2 = CN, 18\%$
b, $R^1 = Cl, R^2 = COOEt,$
        $4.5\%, m.p. 109-110\ °C$

METHOD 17

Method 18. Some quinophthalones have been hydrolysed to phenacyl-quinolines.[592,593]

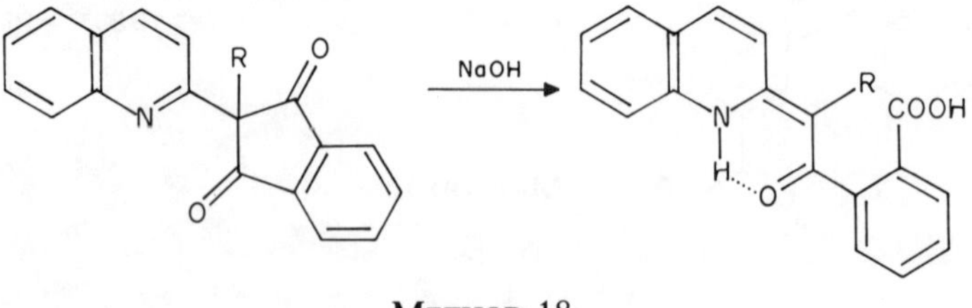

METHOD 18

Method 19. Quinazoline 3-oxide **200** and pentane-2,4-dione in refluxing benzene gave the rearranged ketones 3-acetyl-2-methylquinoline and **201** in low yields.[182]

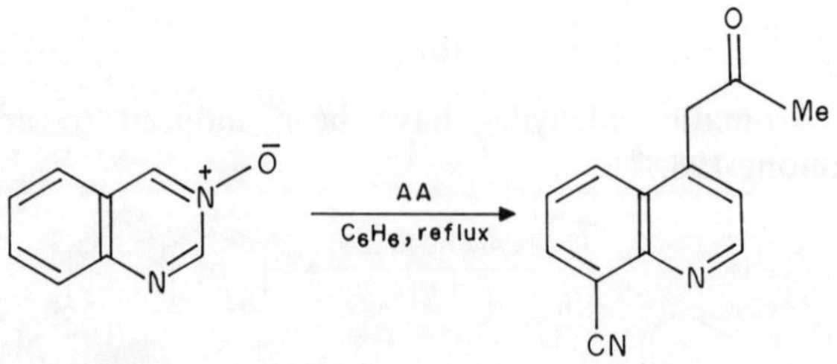

200   **201**

METHOD 19

4. 2-Ketomethylquinolines Monosubstituted at C-1

The ketonic ylids **202** were produced as shown in Scheme 27.[594]

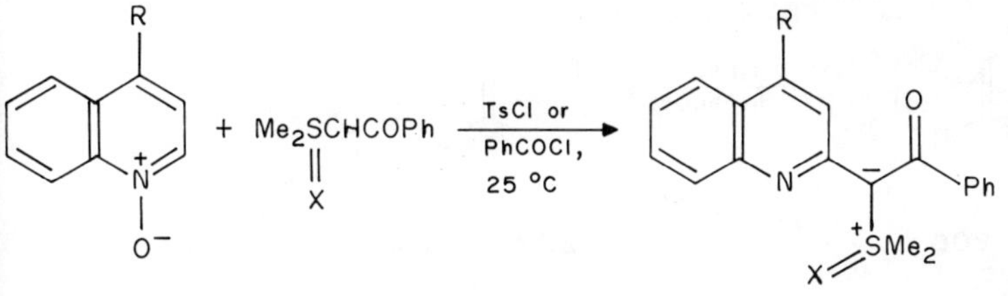

R = H, X = nothing, 53%, m.p. 199−200 °C
R = H, X = 0, 25%, m.p. 216−217 °C
R = Me, X = 0, 57%, m.p. 180−181 °C

SCHEME 27

When the nitrile ester **203** was warmed in trifluoroacetic acid, the ester group was assumed to hydrolyse to give the salt **204**. Partial decarboxylation was accompanied by reaction to give keto-acid **205**. The suggested mechanism is given in Scheme 28. Unfortunately, no experimental details or characterization data for structure **205** were given.[111]

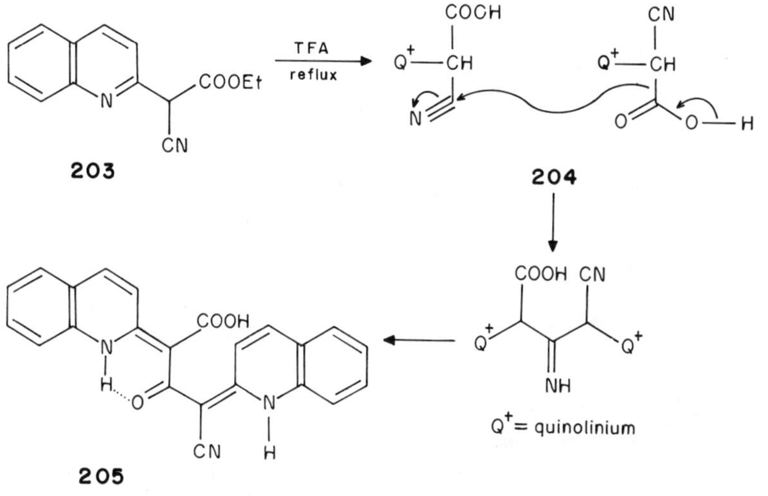

Q^+ = quinolinium

SCHEME 28

The ketones of Table 21 (p. 280) were prepared from the cyclopentanoquinoline **206** by Method 1, above.[595]

5. 2-Ketomethylquinolines Disubstituted at C-1

Quinoline **206** was monosubstituted by esters, but with benzoyl chloride the diketone **207** was produced.[595]

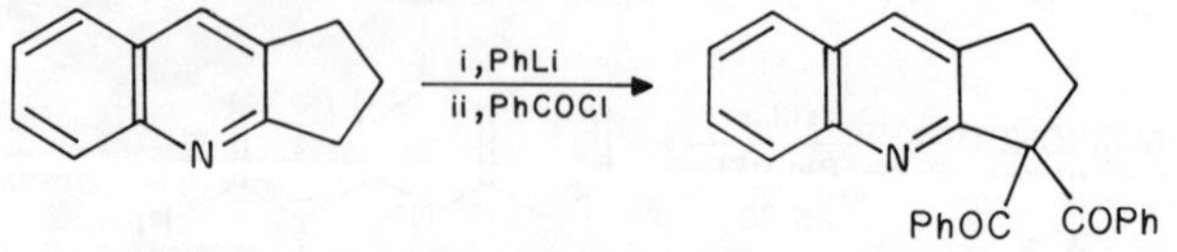
i, PhLi
ii, PhCOCl
N
N
PhOC
COPh

206 **207**, 55%, m.p. 164–165 °C

Method 2 also succeeded with isopropyl ketones to give compounds **208a, b, c**. Isopropyl methyl ketone gave a mixture of compounds **208b** (62%) and **208c** (13%) (g.l.c. yields). Compound **208c** was also prepared (49%) from ketone **208**, $R^1 = Me$, $R^2 = H$ with sodium hydride (2 equivs) and iodomethane (2 equivs).[565–567]

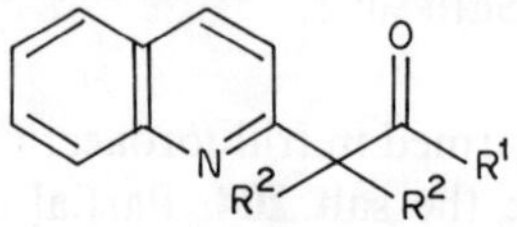
O
N
R^2
R^2
R^1

208

a, $R^1 = i\text{Pr}$, $R^2 = Me$, 78%, m.p. 93–94 °C
b.p. 131–132 °C/1.3 mm

b, $R^1 = i\text{Pr}$, $R^2 = H$

c, $R^1, R^2 = Me$, b.p. 116–118 °C/1.1 mm

Method 15 was applied to the substituted enamines **209**[557] and to ethyl 2-benzoylacetimidate[596]; Scheme 29.

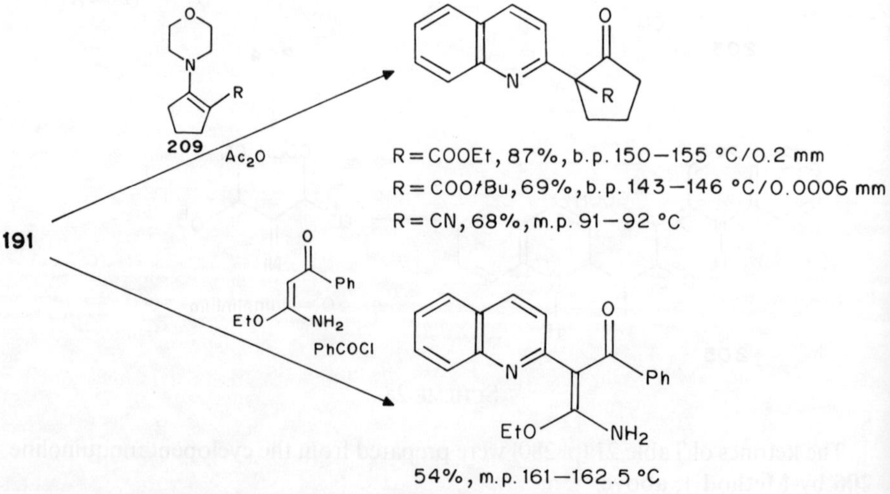
O
N
R
209
Ac_2O
191
O
Ph
EtO
NH_2
PhCOCl
O
N
R
R = COOEt, 87%, b.p. 150–155 °C/0.2 mm
R = COO*t*Bu, 69%, b.p. 143–146 °C/0.0006 mm
R = CN, 68%, m.p. 91–92 °C
O
N
Ph
EtO
NH_2
54%, m.p. 161–162.5 °C

SCHEME 29

Compounds **210** were prepared by chlorination or bromination in aqueous sodium hydroxide.[597,598]

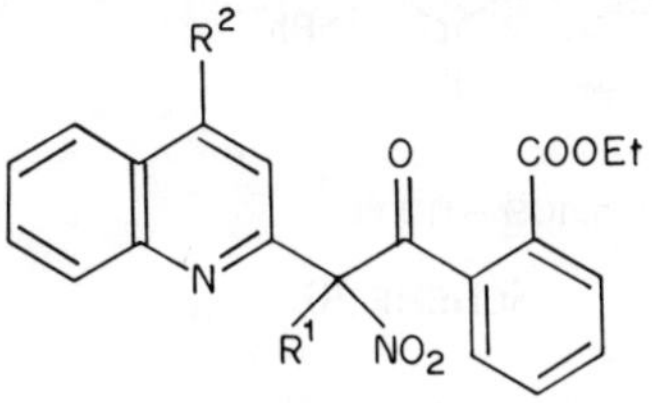

210

R^1 = Cl, R^2 = H, m.p. 133—134 °C
R^1 = Br, R^2 = H, 61.53%, m.p. 133—134 °C
R^1 = Br, R^2 = Me, used crude

Compound **211** was *C*-alkylated, but the product **212** decomposed on attempted purification.[557]

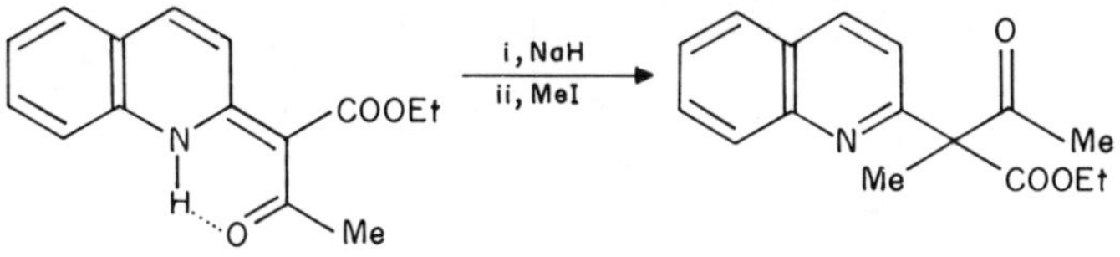

211   **212**

Ketone **213** was prepared from the appropriate nitrile and ethylmagnesium bromide.[599]

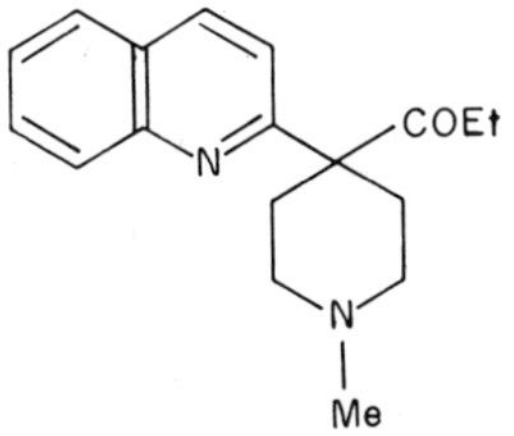

213, 2 Pic., m.p. 210—211 °C

A pinacol rearrangement produced 1,2-diphenyl-2,2-di(2-quinolyl)ethanone (see Section II.7, Scheme 25).

6. 2-Ketomethylquinolines With Double Bonds at C-1

Ketone **177** R = Ph was warmed with selenium dioxide in dioxane to give the dione **214**.[448] The α-bromoketone **215** oxidized to diketone **214** on standing in DMSO[131]; Scheme 30. Dione **214** gave a mixture of both possible monohydrazones on treatment with hydrazine hydrate; see Table 1.

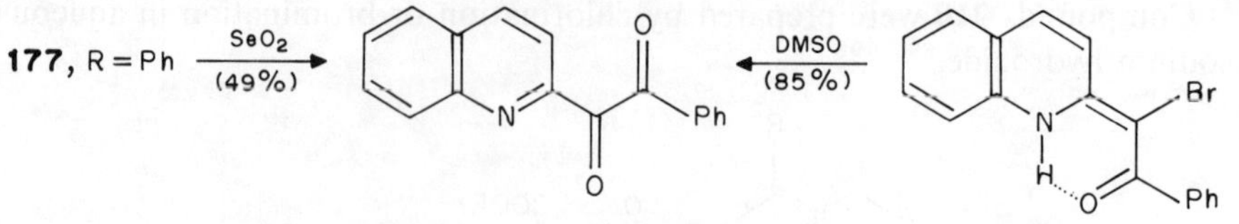

177, R = Ph $\xrightarrow[\text{(49%)}]{\text{SeO}_2}$

$\xleftarrow[\text{(85%)}]{\text{DMSO}}$

214, m.p. 109—110 °C

215

SCHEME 30

Diazonium salts substituted the α-carbon of quinolylpyruvates **216**.[600,601] Compound **216**, R^1 = H could be substituted by aromatic aldehydes in pyridine to give olefines **217**, Scheme 31.[602,603] The use of a stronger base (piperidine or EtONa) induced ring closure of compound **217** to give furanylquinolines[600].

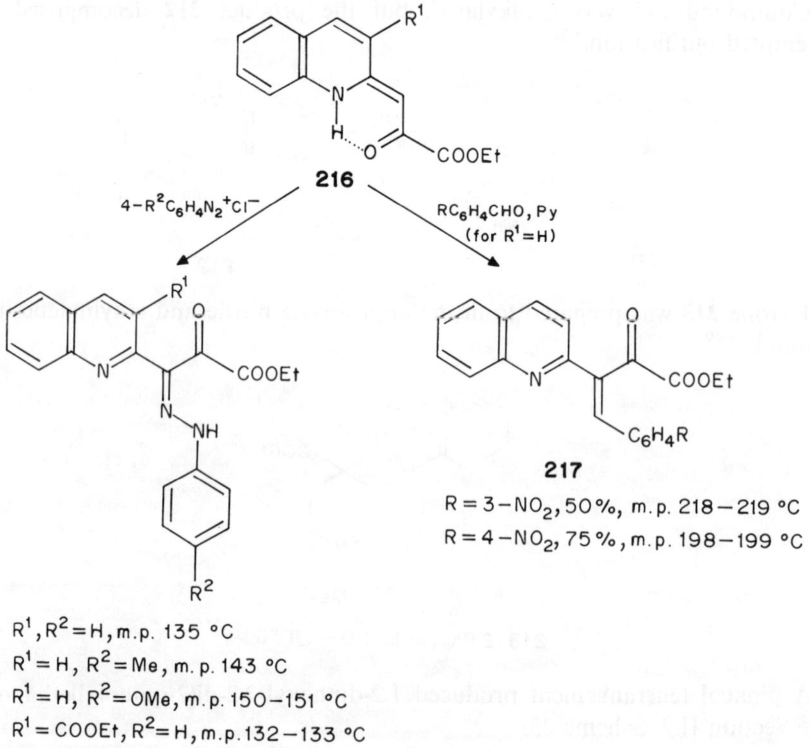

216

$4\text{-}R^2C_6H_4N_2^+Cl^-$

RC_6H_4CHO, Py
(for R^1=H)

217

R = 3–NO₂, 50%, m.p. 218–219 °C
R = 4–NO₂, 75%, m.p. 198–199 °C

R^1, R^2 = H, m.p. 135 °C
R^1 = H, R^2 = Me, m.p. 143 °C
R^1 = H, R^2 = OMe, m.p. 150–151 °C
R^1 = COOEt, R^2 = H, m.p. 132–133 °C

SCHEME 31

A different ring closure followed hydration of acetylenes **218** when $n = 2$ or 3 to give cycloalkenones **219**. If $n = 4$ or more, acyclic diketones were obtained; see Table 20.[604,655]

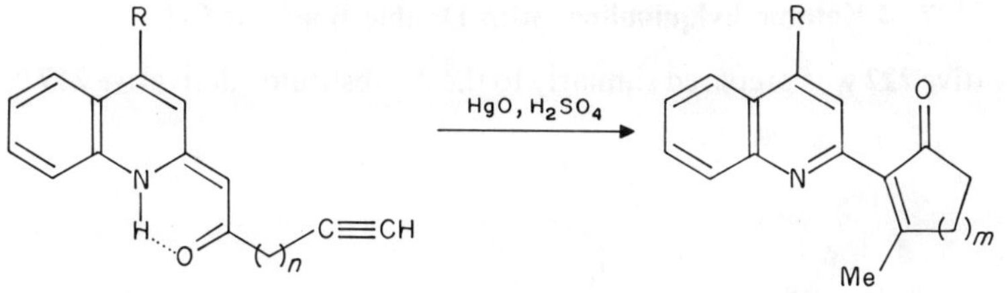

218 → **219**

HgO, H$_2$SO$_4$

R = H, m = 1, 83%, m.p. 103 °C
R = H, m = 2, 73%, m.p. 92 °C
b.p. 182 −186 °C / 0.4 mm
R = Me, m = 1, 68%, m.p. 104 °C
R = Me, m = 2, 83%, m.p. 104 °C

Several ketones **177**, R = alkyl or aryl reacted with p-toluenesulphonyl azide in the presence of base to give 3-keto-1,2,3-triazolo[1,5-a]quinolines.[605]

The appropriate acylquinolines have been α-oxaminated with nitrous acid to give the products **220**.[593,606]

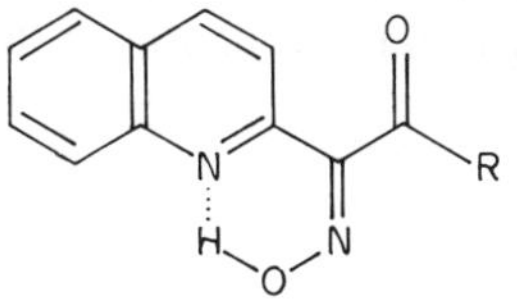

220, R = C$_6$H$_4$−2−COOH, m.p. 205 °C(d)
R = 2−Py, m.p. 175 °C

Ketones **177** with phenyl isothiocyanate and alkyl halides gave the derivatives **221**; Scheme 32.[607,608]

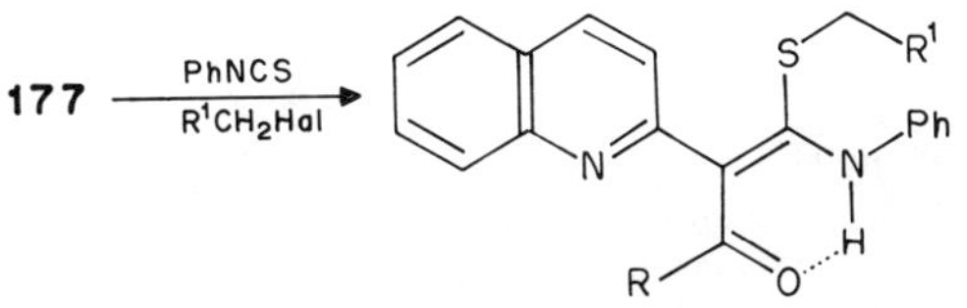

177 →
PhNCS
R^{1}CH$_2$Hal

221

R	R^1	Hal	Yield(%)	M.p. (°C)
Me	H	I	51	143−145
Me	COOEt	Br	40	100−103
Me	CN	Cl	38	175−178
iPr	COOEt	Br	42	102−103
Ph	COOEt	Br	58	118 −119

SCHEME 32

7. 4-Ketomethylquinolines with Double Bonds at C-1

The derivative **222** was prepared similarly to the 2-substituted derivative **217**.[602]

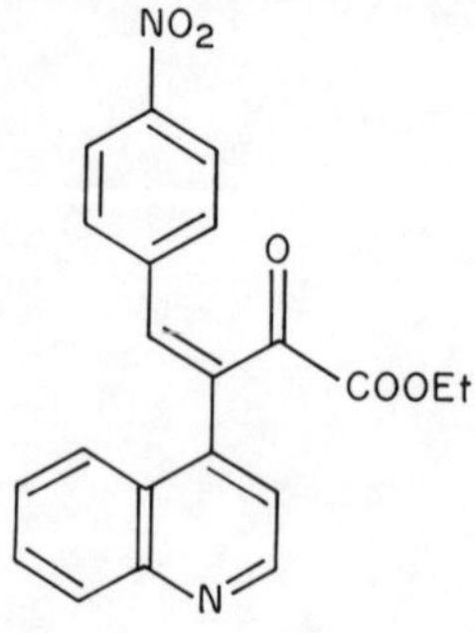

222, 50%, m.p. 210–212 °C

8. Quinophthalones and Related Ketones

The quinophthalones, Table 23 (p. 281), compounds of particular interest to the dye industry, are prepared by two main methods, illustrated in Scheme 33 for the parent of the series, **223**.

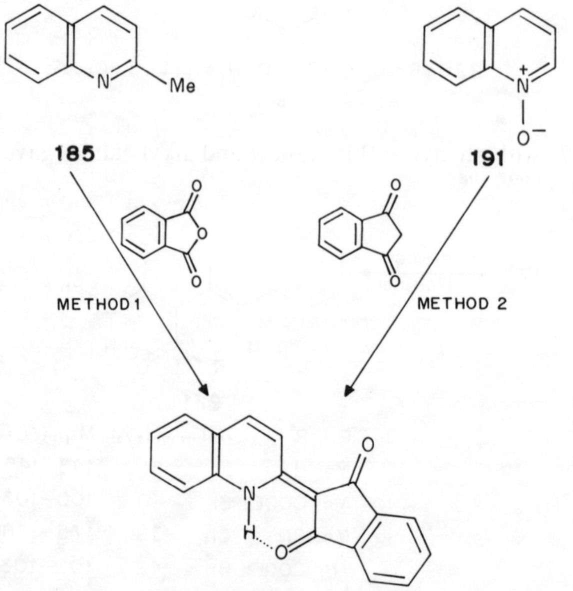

223

SCHEME 33

Early preparations of compound **223** used Method 1. The compound has been known as 'quinophthalone' since the nineteenth centry. Initial confusion about the structure[609] was resolved in the early years of the twentieth century.[592,593] The compound is sometimes called 'quinoline yellow'.[610,611] The

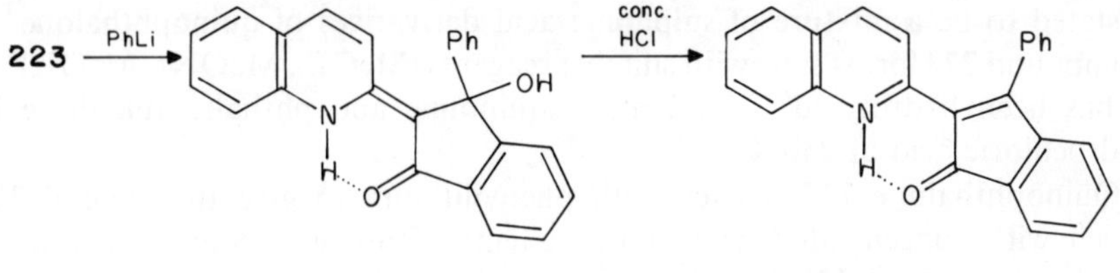

224, 100%, m.p. 164 °C   **225**, 96%, m.p. 220 °C

SCHEME 34

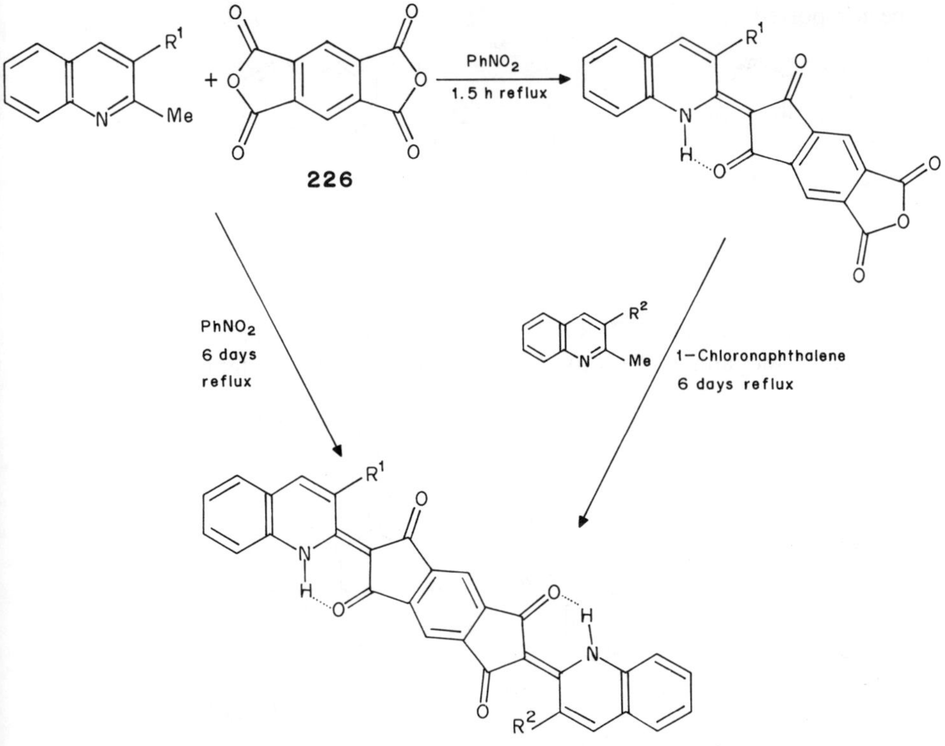

R^1, R^2 = H, 41%, m.p. >410 °C
R^1 = H, R^2 = OH, 52.4%, m.p. >410 °C
R^1, R^2 = OH, 71%, m.p. >410 °C

SCHEME 35

Colour Index gives C.I. 47000 'Solvent Yellow 33', which originally was quinophthalone, but is now a mixture of this and its 6-methyl derivative. However, Yamazaki and co-workers[612] have reported the preparation of sodium quinophthalone sulphonate (position of sulphonation not known). They called this product quinoline yellow, nomenclature in agreement with C.I. 47005, which is stated to be a mixture of sulphonic acid derivatives of quinophthalone.[613] Compound 223 forms salts with suitable reagents (MeOLi, MeONa, MeOK)[614]. It has been hydrolysed[615] to 2-methylquinoline and phthalic anhydride by hydrochloric acid at 240 °C.

Quinophthalone 223 reacted with phenyllithium to give the alcohol 224, which with concentrated acid at room temperature was converted into the unsaturated ketone 225; Scheme 34.[616]

Scheme 35 shows derivatives prepared from the dianhydride 226.[617] Similar compounds from compound 226 where one anhydride group forms an imide with an amine followed by reaction of the other with 2-methylquinolines have been reported.[618–623]

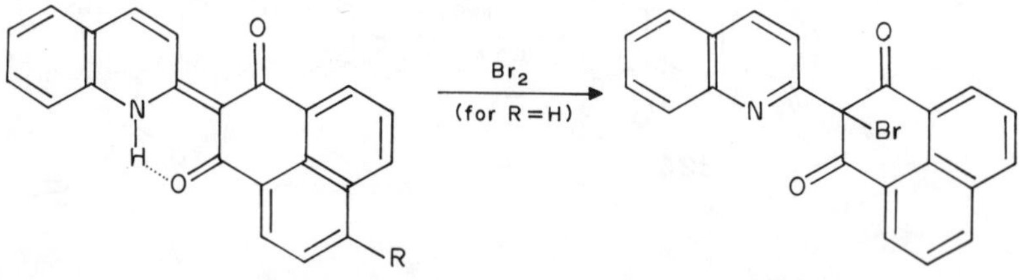

227, R=H, 62%, m.p. 254−255 °C   **228**, m.p. 239 °C

'Quinonaphthalone' 227, R = H was prepared from 2-methylquinoline and 1,8-naphthalic anhydride mixed with zinc chloride at temperatures over

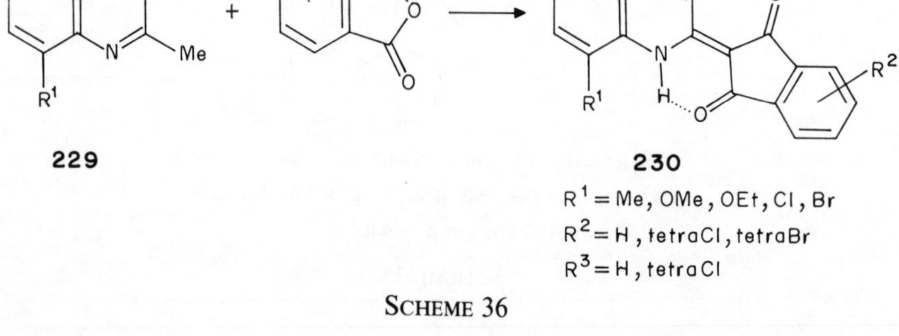

229   **230**

R¹ = Me, OMe, OEt, Cl, Br
R² = H, tetraCl, tetraBr
R³ = H, tetraCl

SCHEME 36

$200\,^{\circ}C.^{612,624}$ The derivatives **227**, R = H, Cl, Br were prepared in yields of over 60%.[625] The parent compound was brominated to give compound **228**.[626,627]

The 5-amino-2-methylquinolines **229** reacted with phthalic anhydrides in one or two steps to give the quinophthalones **230**; Scheme 36.[628] When aminosubstituted 3-hydroxy-2-methylquinoline-4-carboxylic acids were treated similarly, they decarboxylated at the reaction temperature $(210\,^{\circ}C)$.[629]

9. 3-Ketomethylquinolines

The betaines mentioned under Method 17, Section V.3 come into this class. The remaining members, which were synthesized by standard methods, appear in Table 24 (p. 284).

10. Quinolines with Ketomethyl Groups on the Benzene Ring

A. Position 5

The thione **231**, with hydroxylamine and sodium ethoxide, gave the oximino thione **232**.[375]

231   **232**, 45%, m.p. > 300 °C

B. Position 6

6-Quinolylacetic acid reacted with phthalic anhydride in aqueous buffer at $190–200\,^{\circ}C$ to give the keto-acid **233**.[704]

233, 64%, m.p. 225 °C

Table 20. 2-Ketomethylquinolines

Quinoline substituents(s) R²	R¹	Preparation			Yield(%)	M.p. (°C)	References
		Method	Base	Reagents			
None	H	1	PhLi	EtOAc	87.6	76–77	631–633
		1	PhLi	Ac₂O	19	b.p. 145–147/2.5 mm Pic. 182–183	634
		1	NaNH₂	AcOPh or	33	HCl 156–157	635
			KNH₂	AcOPh			
		2	KNH₂	MeCOMe, hv	90	71–74	565, 566
		3		HgSO₄, H₂SO₄	88	68–69	569
		4	BuLi	MeCN	69	68–70	636
		4	PhLi	MeCN	68	77–78	637*
		5		Ac₂CH₂ᵃ	43	78–80	555
		6			38	79.5–80.5	552, 573
		7	NaOH	MeCOMe		68–69	575
		8		CH₂=C(OEt)Me or CH₂=C(CN)Me	7	Pic. 183–184.5	578
		9		2-NO₂C₆H₄CH= CHCOCH₂COMe, SnCl₂, EtOH	88	Pic. 187	582 579, 580
Me	H			Q N-oxide, diketene, AcOH, 25°C	Trace	76–78	638

where $R^1 = Me$, $R^2 = H$ refer to the quinoline structure positions shown.

Me	H	4-Me	6			37	113–114	574
Me[b]	H	4-Ph			PhCOC$_6$H$_4$-2-NH$_2$, Ac$_2$CH$_2$, 150 °C		113–115	639
Me	H	8-OH	7	MeONa	Ac$_2$CH$_2$[a]	46.5	139–140	576
Me	H	4-OMe	6	PhLi	2-Et-Q, MeCN	59	68–69	574
Me	Me		4		Q-2-CH$_2$COMe, DMAD	40	112–114	637*
Me	COOMeC=CHCOOMe					3.4		636
Me	Et		5		MAA, 40 °C	69	119.5–120.5	640
Me	COOMe		5		EAA, Ac$_2$O	86	58.5–59	641
Me	COOEt						b.p. 155–164 /0.1 mm	557, 642
Me	1-NC$_5$H$_5$$^+ClO_4$$^-$	4-Me	6	EtONa	AcOEt	43	197–199	574
Me	1-NC$_5$H$_5$$^+ClO_4$$^-$	4-OMe	6		Ac$_2$O	66	230–232	574
Me	CN		1		Ac$_2$O, 150 °C	94	213–214	600
Me			10			36	215	600, 643
Me			11			85	212–213	170, 581
Me	1-NC$_5$H$_5$$^+ClO_4$$^-$		6		MeCOCH$_2$NC$_5$H$_5$$^+I^-$; NaClO$_4$		203–204	552, 573
CH$_2$Ph	Ph		5		(PhCH$_2$)$_2$CO	26.4	102–104	158
CH(COOH)-2-Q	CN				Q-2-CH(CN)COOEt, TFA			111
CF$_3$	H					24	219–221	644
CF$_3$	CN		11		(ClCH$_2$CO)$_2$O	99	196–197	170, 581
CH$_2$Cl	CN		10			77	130–132(d)	645
CHCl$_2$	H	4-CON-(CH$_2$CH$_2$)$_2$-N-CHCl$_2$			2-Me-Q-4-CON-(CH$_2$CH$_2$)$_2$NH, CHCl$_2$COCl, Et$_3$N			646
CH$_2$I	CN				Q-2-CH$_2$COCH$_2$Cl, NaI, acetone	83.3	202–203	645
CH(OMe)$_2$	H		2	iBuOK	MeCOCH(OMe)$_2$, hv	80	82	647

Table 20. (*Contd.*)

R¹	R²	Quinoline substituents(s)	Preparation			Yield(%)	M.p. (°C)	References
			Method	Base	Reagents			
CH_2SMe	H		12			40	68–69	45
CH_2NHtBu	H		4	BuLi	$tBuNHCH_2CN$		2HCl 135–140	648
CH_2NiPr / CH_2Ph	H		4	$tBuLi$	$iPrNCH_2CN$ / CH_2Ph	66	69–70	649
CH_2NiPr	H	3,4-diMe	4	BuLi	$iPrNCH_2CN$ / CH_2Ph	91	129–130	649
CH_2NtBu / CH_2Ph	H		4	$tBuLi$	$tBuNCH_2CN$ / CH_2Ph	82	100–102	649
CH_2NtBu / CH_2Ph	H	3,4-diMe	4	BuLi	$tBuNCH_2CN$ / CH_2Ph	87	129–130	648, 649
CH_2NtBu / CH_2Ph	H	3-Me-4-Ph-6-Cl	4	BuLi	$tBuNCH_2CN$ / CH_2Ph		Used crude	649
CH_2NtBu / CH_2Ph	H	4-Me	4	BuLi	$tBuNCH_2CN$ / CH_2Ph	92	126–127	648, 649
CH_2NtBu / CH_2Ph	H	6-Me	4	BuLi	$tBuNCH_2CN$ / CH_2Ph	94	111–112	648, 649
CH_2NtBu / CH_2Ph	H	7-Me	4	BuLi	$tBuNCH_2CN$ / CH_2Ph	95	108.5–110	648, 649

R	Ring subst.	R′	n	Base	Reagent	Yield (%)	b.p./m.p.	Ref.
CH₂NtBu	4-Ph	H	4	BuLi	tBuNCH₂CN	91	104–105	648, 649
CH₂Ph / CH₂NtBu	6-OMe	H	4	BuLi	CH₂Ph / tBuNCH₂CN	93	126–127	649
CH₂Ph / CH₂NtBu	6-NMe₂	H	4	BuLi	CH₂Ph / tBuNCH₂CN	87	108–109	649
CH₂Ph / CH₂NHV^c	3,4-diMe	H	4	BuLi	CH₂Ph / HVNCH₂CN		Used crude	649
CH₂Ph / Et		H	1	PhLi	CH₂Ph / EtCOOMe	94	b.p. 142–143 /1.4 mm	631, 632
			1	NaNH₂ or KNH₂	(EtCO)₂O or EtCOOPh	32	b.p. 172–175 /6.5 mm Pic. 181–182	635
			4	BuLi	EtCN	40	b.p. 150–154 /2 mm	636
			7	NaOH	MeCOEt	8	b.p. 75–85 /0.004 mm Pic. 179–183	575
Et		Me	2	NaNH₂	EtCOEt	70	Oil	567
			2	KNH₂	EtCOEt, hv	38	b.p. 131–132 /1.3 mm	565
Et		CN	10		(EtCO)₂O	92	168	643, 650, 651
Et		CN	11			2.6	164–165	581
CH₂CH₂Cl		CN						655
CH₂CH₂Br		CN						651
CH₂COPh		H			2-Ac-Q, PhCOOMe, NaH	56	138–139.5	633*
			2	NANH₂				633

Table 20. (*Contd.*)

R^1	R^2	Preparation Method	Preparation Base	Preparation Reagents	Yield(%)	M.p. (°C)	References
$CH_2COC_6H_4$-4-Cl	H			2-Ac-Q, 4-ClC_6H_4COOMe, NaH	62	158–168	633*
$CH_2COC_6H_3$-3,4-diCl	H			2-Ac-Q, 3,4-$diClC_6H_3COOMe$, NaH	64	186–188	633
$CH_2COC_6H_4$-4-OMe	H			2-Ac-Q, 4-$MeOC_6H_4COOMe$, NaH	45	145–147	633*
$CH_2COC_6H_2$-3,4,5-triOMe	H			2-Ac-Q, 3,4,5-$triMeOC_6H_2$-COOMe, NaH	30	158–160	633*
Pr	H	1	PhLi	PrCOOMe	60	32–33 b.p. 120–125/0.3 mm	631, 632
		1	$NaNH_2$ or KNH_2	PrCOOEt	14	b.p. 156/3 mm Pic. 181	635
		2	KNH_2	MeCOiBu, hv	62[d]		565
$(CH_2)_3Cl$	CN						651
$(CH_2)_3Br$	CN						651
CH(SMe)COMe	CN	11			13	135–136	170, 581
iPr	H	1	PhLi	iPrCOOMe	89.2	34–36 b.p. 152–154/2.0 mm Pic. 181–182	631, 632
		1	$NaNH_2$ KNH_2	iPrCOOEt or iPrCOOEt	36	b.p. 144/2.5 mm	635*
		2	KH	MeCOiPr	62		565

R	3	4	No.	Reagent	Reactant / Conditions	Yield (%)	m.p. or b.p. (°C)	Ref.
Bu	H		3		HgSO$_4$, H$_2$SO$_4$	70	b.p. 165–170/2 mm	569
CH$_2$CH$_2$C≡CH	H		4	PhLi	HC≡CCH$_2$CH$_2$CN	67–87	86–87; b.p. 170–175/0.5 mm; Pic. 177	652
CH$_2$CH$_2$C≡CH	H	4-Me	4	PhLi	HC≡CCH$_2$CH$_2$CN	75	104; b.p. 185/0.5 mm; Pic. 172	653
tBu	H		4	PhLi	tBuCN	93	65	551, 654
C$_5$H$_{11}$	H				Q-2-CH$_2$CO-(CH$_2$)$_3$C≡CH, H$_2$, Ni		b.p. 172/0.5 mm; Pic. 139	655
(CH$_2$)$_3$C≡CH	H		4	PhLi	HC≡C(CH$_2$)$_3$CN	79–90	36; b.p. 165/0.1 mm; HCl 134; Pic. 161	652
(CH$_2$)$_3$C≡CH	H	4-Me	4	PhLi	HC≡C(CH$_2$)$_3$CN	64	68; b.p. 180/1.2 mm; Pic. 157	653
CH$_2$CH$_2$C≡CCPh$_2$OH	H				Q-2-CH$_2$COCH$_2$-CH$_2$C≡CH, LiNH$_2$, Ph$_2$CO	50	148	656
(CH$_2$)$_5$NH$_2$	H		13	PhLi		73.3	129–130	583
			14			75.3	129–130	583
(CH$_2$)$_5$NHCOPh	H		14		Above, PhCOCl	11	81–82	583
(CH$_2$)$_5$NMe$_2^e$	H					65	b.p. 170/0.03 mm	583
C$_6$H$_{13}$	H				Q-2-CH$_2$CO-(CH$_2$)$_4$C≡CH, H$_2$, Ni		b.p. 180–183/0.5 mm; Pic. 141	655

Table 20. (*Contd.*)

R¹	R²	Quinoline substituents(s)	Preparation Method	Base	Reagents	Yield(%)	M.p. (°C)	References
$(CH_2)_4C\equiv CH$	H		4	PhLi	$HC\equiv C(CH_2)_4CN$	80–90	44 b.p. 165–173 /0.01 mm Pic. 160	652
$(CH_2)_4C\equiv CH$	H	4-Me	4	PhLi	$HC\equiv C(CH_2)_4CN$	58	58 b.p. 195 /1.0 mm Pic. 180	653
$(CH_2)_3C\equiv CCH_2NEt_2$	H		4		$Et_2NCH_2C\equiv C(CH_2)_3CN$	58	58 b.p. 200–205 /0.05 mm 2Pic. 150	656
$(CH_2)_4C\equiv CCH_2NMe_2$	H		4		$Me_2NCH_2C\equiv C(CH_2)_4CN$	55	b.p. 230/1 mm	656
$(CH_2)_4COMe$	H				$Q\text{-}2\text{-}CH_2COCH_2\text{-}CH_2C\equiv CH$, HgO, H_2SO_4	60	58 Pic. 143–144	655
$(CH_2)_3C\equiv CBu$	H		4		$BuC\equiv C(CH_2)_3CN$	75	b.p. 185–191 /0.02 mm Pic. 118	656
$(CH_2)_4C\equiv CBu$	H		4		$BuC\equiv C(CH_2)_4CN$	77	b.p. 222–225 /0.03 mm Pic. 143–144	656
$C_{11}H_{23}$	H				$Q\text{-}2\text{-}CH_2CO\text{-}(CH_2)_9C\equiv CH$, H_2, Ni		47–48 Pic. 127	655
$(CH_2)_9C\equiv CH$	H		4	PhLi	$HC\equiv C(CH_2)_9CN$	82–83	57 Pic. 118–119	652

$(CH_2)_9C\equiv CH$	4-Me	H	PhLi	4	$HC\equiv C(CH_2)_9CN$	50	52 b.p. 220 /1.2 mm Pic. 111	653
$(CH_2)_9COMe$		H			Q-2-CH_2CO-$(CH_2)_9C\equiv CH$, HgO, H_2SO_4	68	57 HCl 102 Pic. 130	655
$(CH_2)_9COMe$	4-Me	H			Q-2-CH_2CO-$(CH_2)_9C\equiv CH$, HgO, H_2SO_4	35	56 Pic. 137	604
$(CH_2)_9C\equiv CBu$		H		4	$BuC\equiv C(CH_2)_9CN$	87	b.p. 230–235 /0.05 mm Pic. 104–106	656
1,3-Dioxo-cyclopent-2-yl		H	KNH$_2$	5	2-Ac-cyclopentane-1,3-dione, Ac_2O	73	240–242	557
$-(CH_2)_3-$				2	cyclo$C_5H_{10}O$, hv	44	93–100 b.p. 163–167 /1.1 mm	565
				15	$O(CH_2CH_2)_2N$-cycloC_5H_7	82	99–100	557
				15	$O(CH_2CH_2)_2N$-cycloC_5H_7	32	HCl 168–170 109–110	657
$-(CH_2)_4-$			NaNH$_2$	2	$C_6H_{10}O$	60	120–121	567
				15	PhNHcycloC_6H_9 or cycloC_6H_{11}-NHcycloC_6H_9	40	120	586, 658
				8	EtOcycloC_6H_9	58	121–122	578
					4-Cl-2-(2'-oxocyclohexyl)-Q, H_2, Pd	73.4		584
$-(CH_2)_4-$	4-Me			15	$O(CH_2CH_2)_2N$-cycloC_6H_9	79.4	101–102.5	561*, 585, 586
$-(CH_2)_4-$	4-Cl			15	$O(CH_2CH_2)_2N$-cycloC_6H_9	78.1	97–99	585
				8	EtO-cycloC_6H_9	73	97–98	578

Table 20. (*Contd.*)

R¹	R²	Quinoline substituents(s)	Preparation			Yield(%)	M.p. (°C)	References
			Method	Base	Reagents			
	—(CH₂)₄—	4-OEt			4-Cl-2-(2'-oxocyclohexyl)-Q, EtONa	86	151–152	585
	—(CH₂)₄—	3-N(CH₂CH₂)₂O	15		3-Br- Q N-oxide, O(CH₂CH₂)₂N-cycloC₆H₉	16	153–154	659
	—(CH₂)₄—	3-N(CH₂CH₂)₂O-4-OMe	15		3-Br-4-MeO-Q N-oxide, O(CH₂CH₂)₂N-cycloC₆H₉	34	130–131	659
	—(CH₂)₅—		2	NaNH₂	C₇H₁₂O	60	f	567
	—(CH₂)₆—		2	NaNH₂	C₈H₁₄O	64	f	567
	(A)		15		(B)			660
	(A)	4-Me	15		(B)			660
	(A)	6-Me	15		(B)			660
	(A)	4-Cl	15		(B)			660
	(A)	4-NO₂	15		(B)			660
	(A)	4-CN	15		(B)			660
Ph	H		1	NaH	PhCOOMe	93	119–120 Pic. 177–178	661, 662
			1	PhLi	PhCOOMe or PhCOOEt	85.5	118	610, 611, 633
			1	EtOK	PhCOOEt	11.3	114–115	664
			1	KNH₂	PhCOOEt	60–65	116.4–117.1	635, 665
			2	NaNH₂	PhCOMe	33	120	568
			2	KNH₂	PhCOMe	14ᵈ		565
			3		PhCOMe, *hv*	81	116–117 120–122	105*, 569

(A) = —CH₂N(COOEt)CH₂CH₂—.
(B) = O(CH₂CH₂)₂NC=CHCH₂N(COOEt)CH₂CH₂.

			n			%	m.p.	Ref.
			4					637*
			6	NaOH	PhCOMe	74	116–117	552, 573
			7		PhCOMe	4	119–120 / Pic. 177	575
			15		O(CH$_2$CH$_2$)$_2$NC-(=CH$_2$)Ph or PhNH=C(Me)Ph		119–120	586
			12			64.1	118–119	45
			9			42	Pic. 174–175 / HBr 198	582
					Q-2-CHBrCHBrPh, KOH; H$_2$SO$_4$	17.4	116–117.5	666
Ph	H	6-Me	1	NaH	PhCOOMe	100	121–123 / Pic. 185–186	662
Ph	H	6-CF$_3$	1	NaH	PhCOOMe	100	151–152	662
Ph	H	3-Ph	1	NaNH$_2$	PhCOOMe	53	169–170	667
Ph	H	3-Ph-4-OPh	1	NaNH$_2$	PhCOOMe	51	185.5–187	667
Ph	H	6-F	1	NaH	PhCOOMe	94	132–133 / Pic. 162–163	662
Ph	H	6-Cl	1	NaH	PhCOOMe	39	147–148 / Pic. 184–185	662
Ph	H	6-Br	1	NaH	PhCOOMe	31	152–153 / Pic. 183–184	662
Ph	H	4-OMe	6			55	130–131	574
Ph	H	6-OMe	1	NaH	PhCOOMe	71	146–148 / Pic. 192–194	662
Ph	Me		2	KNH$_2$	PhCOOEt, $h\nu$	32	100–101.5	565
Ph	COPh		5		(PhCO)$_2$CH$_2$	53	188–189	668
Ph	CH$_2$COPh		16	NaCN	PhCHO	48	119	587
Ph	Br				Q-2-CH$_2$COPh, Br$_2$	90	74	131, 570
Ph	COOEt		5		PhCOCH$_2$COOEt		122–123	642
Ph	COOEt	4-Cl	16		PhC≡CCOOEt	8^g	148–149	588
Ph	N(CH$_2$)$_5$				Q-2-CH$_2$COPh, Br$_2$; C$_5$H$_{11}$N		HCl 223	669, 670
Ph	1-NC$_5$H$_5$$^+ClO_4$$^-$		6			76	229–230	552, 573
Ph	1-NC$_5$H$_5$$^+Br^-$		6			49	215–216	552, 573
Ph	1-NC$_5$H$_5$$^+ClO_4$$^-$	4-OMe	6			55		574

Table 20. (*Contd.*)

R^1	R^2	Quinoline substituent(s)	Method	Base	Reagents	Yield(%)	M.p. (°C)	References
Ph	CN		17; 5		$PhC{\equiv}CCN$ or $PhCOCH_2CN$, Ac_2O, 50 °C	11.3[g]	203–204	589–591
Ph	CN	4-Me	5		$PhCOCH_2CN$	29.4	204–206[h]	572
Ph	CN	4-OMe	5		$PhCOCH_2CN$	6.3	175–176[h]	572
$4\text{-}MeC_6H_4$	H		5		$PhCOCH_2CN$	19.1	278–280[h]	572
$4\text{-}MeC_6H_4$	H		1	KNH_2	$4\text{-}MeC_6H_4COOEt$	60	170–171	665
$4\text{-}MeC_6H_4$	H		4	PhLi	$4\text{-}MeC_6H_4CN$	68		637
$4\text{-}MeC_6H_4$	$1\text{-}NC_5H_5^+ClO_4^-$		6			62	172–173	552, 573
$4\text{-}MeC_6H_4$	$1\text{-}NC_5H_5^+ClO_4^-$	4-OMe	6			66	179–180	552, 573
$2\text{-}ClC_6H_4$	H		1	KNH_2	$2\text{-}ClC_6H_4COOEt$	35	115.9–117	574
$3,4\text{-}diClC_6H_3$	H		1	PhLi	$3,4\text{-}diClC_6H_3COOMe$	64	186–188	665
$4\text{-}ClC_6H_4$	H		1	NaH	$4\text{-}ClC_6H_4COOMe$	72	163–164[i]	633*
$4\text{-}ClC_6H_4$	H		3			81	145.2–146[i]	633*, 661
$4\text{-}ClC_6H_4$	H		6			65	162–163[i]	124
$4\text{-}ClC_6H_4$	$C_5H_{10}N$				Q-2-$CH_2COC_6H_4$-4-Cl, Br_2; $C_5H_{11}N$	29	135–136.5	522, 573
$4\text{-}ClC_6H_4$	$1\text{-}NC_5H_5^+ClO_4^-$		6	KNH_2		93	220–221	124
$4\text{-}BrC_6H_4$	H		1		$4\text{-}BrC_6H_4COOEt$	43	165.7–167.2	552, 573
$4\text{-}BrC_6H_4$	$1\text{-}NC_5H_5^+ClO_4^-$		6			62	167–168	665
$2,4\text{-}diHOC_6H_3$	H		6			81	207–208	552, 573
$2,4,6\text{-}triHOC_6H_2$	H							671, 672
$2\text{-}HO\text{-}4\text{-}AcOC_6H_3$	H							673
$2\text{-}HO\text{-}4\text{-}MeOC_6H_3$	H							671
								671

R	Substituent	4-Subst.	Base	No.	Reagent / Conditions	Yield (%)	m.p. (°C)	Refs.
$2\text{-HO-4-ProC}_6\text{H}_3$	H		KOH		$\text{Q-2-CH}_2\text{CO-(2,4-diHOC}_6\text{H}_3)$, PrI	84	148–149	671
$3,4,5\text{-triMeOC}_6\text{H}_2$	H		PhLi	1	$3,4,5\text{-triMeOC}_6\text{H}_2\text{-COOEt}$	30	158–168	633*
$4\text{-MeOC}_6\text{H}_4$	H		KNH_2	1	$4\text{-MeOC}_6\text{H}_4\text{COOEt}$	72	154.5–155	665
$4\text{-MeOC}_6\text{H}_4$	H		PhLi	4	$4\text{-MeOC}_6\text{H}_4\text{CN}$			637*
$4\text{-MeOC}_6\text{H}_4$	H			6		61	155–156	552, 573
$4\text{-MeOC}_6\text{H}_4$	H			6		57	148–150	552, 573
$4\text{-MeOC}_6\text{H}_4$	$\text{1-NC}_5\text{H}_5$			6			155	592*
$2\text{-C}_6\text{H}_4\text{COOH}$	H		NaOH	18		75	181–182	674
$2\text{-C}_6\text{H}_4\text{COOMe}$	NO_2			18	MeOH		193	674
$2\text{-C}_6\text{H}_4\text{COOMe}$	NO_2	4-Me		18	MeOH		176–177	674
$2\text{-C}_6\text{H}_4\text{COOMe}$	NO_2	4-Et		18	MeOH		250	593
$2\text{-C}_6\text{H}_4\text{COOEt}$	H			18	$\text{Q-2-CH}_2\text{COC}_6\text{H}_4\text{-2-COOH}$, EtOH, HCl	83	152–153	674
$2\text{-C}_6\text{H}_4\text{COOEt}$	NO_2	4-Me		18	EtOH		157–158	674
$2\text{-C}_6\text{H}_4\text{COOEt}$	NO_2	4-Et		18	EtOH		146–147	674
$4\text{-NO}_2\text{C}_6\text{H}_4$	$\text{1-NC}_5\text{H}_5^+\text{ClO}_4^-$			6		71	260–261(d)	552, 573
$4\text{-NH}_2\text{C}_6\text{H}_4$	H			6	Reduction of above or $\text{Q-2-CH}_2\text{COC}_6\text{H}_4\text{-4-NHAc}$, HCl	64	175–176	552, 573
$4\text{-AcNHC}_6\text{H}_4$	H							552
$4\text{-AcNHC}_6\text{H}_4$	H		PhLi	1	$4\text{-AcNHC}_6\text{H}_4\text{-COOEt}$	91	223–224	552
1-Naphthyl	H			6		36	149–151	552, 573
1-Naphthyl	$\text{1-NC}_5\text{H}_5^+\text{ClO}_4^-$			6		76	274–275(d)	552, 573
2-Naphthyl	H			6		50	196–197	552, 573
2-Naphthyl	$\text{1-NC}_5\text{H}_5^+\text{ClO}_4^-$			6		73	204–205	552, 573
2-Furyl	H		PhLi	1	$\text{C}_4\text{H}_3\text{O-2-COOMe}$	68.6	103–103.5; b.p. 185–190 /1.2 mm; Pic. 172–173	631
2-Furyl	H		KNH_2	1	$\text{C}_4\text{H}_3\text{O-2-COOEt}$	28	102.9–103.4	665
2-Furyl	$\text{1-NC}_5\text{H}_5^+\text{ClO}_4^-$			6		54	103–104	552, 573
2-Furyl	H			6		83	244–245(d)	552, 573
2-Furyl	CN			10	$(\text{C}_4\text{H}_3\text{OCO})_2\text{O}$	73	229	643

Table 20. (Contd.)

R¹	R²	Quinoline substituents(s)	Preparation			Yield(%)	M.p. (°C)	References
			Method	Base	Reagents			
2-Thienyl	H		1	PhLi	C_4H_3S-2-COOMe	75.5	125.5–126.5 b.p. 220–223/2.2 mm Pic. 159–160	631
2-Thienyl	$1\text{-NC}_5\text{H}_5^+\text{ClO}_4^-$		6			70	124–126	552, 573
2-Pyridyl			6			78	200–202	552, 573
2-Pyridyl	H		1	PhLi	Py-2-COOMe	58.4	152.5–154 Pic. 171–172	4, 631, 632
2-Pyridyl	CN		1	$NaNH_2$	Py-2-COOEt	39.9	158	606*
3-Pyridyl	H		10		$(C_5H_4NCO)_2O$	94	230	643
3-Pyridyl			1	PhLi	Py-3-COOMe	53.2	121–122 Pic. 215–216	4, 631
3-Pyridyl	CH_2COPh		16	NaCN	Py-3-CHO	63	142	587
4-Pyridyl	H		1	PhLi	Py-4-COOMe	77.1	147.3–147.8 Pic. 219–220	4, 631, 632
2-Quinolyl	H				See Table 1			
4-Quinolyl	H				See Table 4			
COOH	H		1	EtONa	$(COOEt)_2$		170(d)[i]	675
COOH					Q-2-CH_2COCOOEt, H_2O		198–199[i]	600
COOH					Q-2-CH_2COCOOEt, 6% H_2SO_4	90	167–168(d)[i]	676
COOH[j]	H	3-Ph, 7,8-diMe	1	EtOK	2-Me-7,8-diMeO-Q-3-COOEt, $(COOEt)_2$		241–242	677*
COOH	Ph							74
COOH			1	EtONa	$(COOEt)_2$		320–322	677*
COOH[k]	CN							600
COOMe	COOMe	4-OMe	17		DMAD	34	163–165	678

COOEt	H		1	EtOK	(COOEt)$_2$	80	130–132 HCl 225	600*, 676
			1	Na	(COOEt)$_2$ Q-2-CH$_2$COCN, EtOH, HCl	53	131–132 131–132	602 571, 644
COOEt	H	3-Ph	1	EtONa	(COOEt)$_2$	44	160 Pic. 145	667, 677*
COOEt	H	3-COOEt	1	K	(COOEt)$_2$		112	601*
COOEt	COOEt	4-Cl	17		DEAD	3.1	91–92	586, 678
						4.6	135–136[l]	
COOEt	Ph		1	EtONa	Q-2-CH$_2$Ph, (COOEt)$_2$		172	677*
COOEt[k]	CN		1	EtONa	Q-2-CH$_2$-(COOEt)$_2$		191	600
CONHNH$_2$	H				Q-2-CH$_2$-C(=NOH)-COOEt, N$_2$H$_4$			679*
CN	H		7		MeCOCN	36	176–177	571

* These references include carbonyl derivatives.

[a] The intermediate diketone always hydrolysed on work-up.

[b] This method and melting point correspond to the preparation of 3-acetyl-2-methyl-4-phenylquinoline[186], Table 2. Almost certainly the structure given here[639] is wrong.

[c] HV = Homoveratryl:

[d] G.l.c. yield.

[e] The patent claims that this compound was prepared by Method 14. It does not explain how this method could give a tertiary amine.

[f] Stated to be isomer mixtures, for which accurate melting points could not be determined.

[g] Obtained in admixture with the 3-acylquinoline (Table 2).

[h] Obtained as a by-product to the ylid; see text, Section V.3.

[i] Inconsistent literature melting points.

[j] The ester groups hydrolysed during work-up and the acid at position 3 decarboxylated.

[k] These two entries formed as a mixture. Presumably partial hydrolysis occurred on work-up.

[l] In view of the melting point difference, Ishiguro[678] suggests that Canonne[588] got something else. However, the experimental methods were remarkably similar.

Table 21. Cyclopentano[b]quinoline Derivatives

R¹	R²	Yield(%)	M.p. (°C)
Me	H	61.6	97–97.5(d)
Me	Me	39	101–103
Ph	H	32.4	135–136(d)
Ph	Me	81	193–194
4-MeC₆H₄	Me	65	178–179
3-ClC₆H₄	Me	74	199–200
4-MeOC₆H₄	Me	79	197–197.5
4-NO₂C₆H₄	Me	100	188–189
4-Py	H	67.7	177–178.5
4-Py	Me		182–184(d)

Table 22. Cyclic 2-Ketomethylquinolines

R	Quinoline substituent(s)	Method*	Reagents	Yield(%)	M.p. (°C)	References
CH₂	H	5		64	145	680
CH₂	4-Me	7		58	178–179	680
CH₂	4-Cl	7		65	173–174	680
CMe₂	H	7		79	186–187	555, 668, 680
CMe₂	4-Me	7		63	141–142	680
CMe₂	4-Cl	7		71	167–168	680
CMe₂	3-Br	7		83	249–250	680
		7	3-Br-Q N-oxide, dimedone, Ac₂O	78	145–146	680

*Method given in Section V.3.

Table 23. Quinophthalones

No.	R¹	R²	R³	Method*	Conditions	Yield (%)	M.p. (°C)	References
1	H	H	H	1	Heat	96	232	609
					Q-2-CH₂CO-2-C₆H₄-COOH, Conc. H₂SO₄			592
					Q-2-CH₂CO-2-C₆H₄-COOEt, EtONa	85	240	593
				1	triClC₆H₃, reflux	58.6	241–242	681
				1	PhNO₂, H₃PO₄	42	241–242	616
				1	ZnCl₂, 200°C		234–235	615
				1	1,2-diClC₆H₄, 183–184°C			611
				2	Ac₂O	68	241	169, 642, 668
					Q-2-Me, C₆H₄-1,2-diCOOEt, NaH	78	242–243	661
2	H	H	Me	2	Ac₂O	58	146–147	668
3	H	H	Ph	2	Ac₂O	83	167–168	668
4	H	H	4-BrC₆H₄	2	Ac₂O	56	138–139	668
5	H	H	4-MeOC₆H₄	2	Ac₂O	79	160–161	668
6	H	H	NO₂		Entry 1, HNO₃		150–151(d)	674
7	H	tetraCl	H	1	PhNO₂, reflux	50.3	>310	681
8	H	5-COOH	H	1	PhNO₂, reflux	48.2	>310	681
9	H	5-COOCO-6	H	1	PhNO₂, reflux	77.3	395–397(d)	617
10	H	5,6-diCOOMe	H		Entry 9, KOH; CH₂N₂	53	290	617
11	H	5-CON(Me)CO-6	H		Entry 9, MeNH₂	57.6	362–363	617
12	H	5-CON(Ph)CO-6	H		Entry 9, PhNH₂	88	387–388	617

Table 23. (*Contd.*)

No.	R^1	R^2	R^3	Method*	Conditions	Yield (%)	M.p. (°C)	References
13	4-Me		H	1, 2	Heat or Ac_2O	59	240–241	169, 681, 682
14	4-Me		NO_2	1	Entry 13, HNO_3		186–187	682
15	5,6,8-triMe		H	1	Heat		236	592
16	6-Me		H	1	$ZnCl_2$, 200 °C		203[a]	683
17	6,8-diMe		H	1	Heat		237[a]	592
18	8-Me		H	1	Heat		231	592
19	4-Et		H	1	Heat		235	592
20	4-Et		NO_2	1	Entry 19, HNO_3		197–198	682
21	4-Ph		H	1	$ZnCl_2$, 150–160 °C		141–142(d)	682
22	5,7-diCl-8-OH		H	1	Entry 23, aq. NaOH		270	442, 684
23	5,7-diCl-8-OAc		H	2	5,7-diCl-8-HO-Q N-oxide, Ac_2O	62.4	361–364	685
24	5,7-diCl-8-OAc		Cl	1	Entry 23, SO_2Cl_2		297–299	685
25	3-OH		H	1	$PhNO_2$, reflux	47.5	206–207	685
26	3-OH-4-Br		H	1	2-Me-3-HO-Q-4-COOH, 200 °C; Br_2, 120 °C	96	266–267	681
27	3-OH	tetraCl	H	1	$PhNO_2$, reflux			686
28	3-OH	5-OH	H		2-Me-3-HO-Q-4-COOH, 4-HO-C_6H_3-1,2-diCOOH, Na_2SO_4, 180 °C	76	>310	681
29	3-OH	5-OSO_2Me	H		Entry 28, $MeSO_2Cl$, Et_3N			687, 688[b]
30	3-OH	5-OSO_2Ph	H		2-Me-3-HO-Q-4-COOH, Na_2SO_4, 4-$PhSO_2O$-C_6H_3-1,2-diCOOH, 180 °C			687, 688[b]
31	3-OH-4-Br	5-OSO_2Ph	H	1	Entry 30, Br_2, AcOH			687, 688[b]
32	3-OH	5-COC_6H_4-2-COOH	H		Entry 30, Br_2, AcOH			689

No.	R^1	R^2	R^3		Conditions			Ref.
33	3-OH	5-COC$_6$Cl$_4$-2-COOH	H					689
34	3-OH	5-COC$_6$Cl$_4$-2-COOMe	H					689
35	3-OH-6,8-diCl	5-CO-(6-BrC$_6$H$_3$-2-COOMe)	H					689
36	3-OH-4-Br	5-CO-(3,6-diClC$_6$H$_2$-2-COOPr)	H					689
37	3-OH	3-COOH	H	1	C$_6$H$_3$Cl$_3$, 180–240 °C	93		690
38	3-OH	5-COOH	H	1	PhNO$_2$, reflux	60.5	>320	681
39	3-OH	5-COOCO-6	H	1	PhNO$_2$, reflux	66	>340	617
40	3-OH	5-COOCH$_2$CH$_2$OMe	H	1	2-Me-3-HO-Q-4-COOH, 1,2,4-benzenetricarboxylic acid anhydride; SOCl$_2$; HOCH$_2$CH$_2$OMe			686
41	3-OH	5-CON(Ph)CO-6	H	1				618–623
42	8-OH	H	H		Entry 43, aq. NaOH		340–343	685
43	8-OAc	H	H	2	8-HO-Q N-oxide, Ac$_2$O	61.5	257–258	685
44	8-OAc	H	Cl		Entry 43, SO$_2$Cl$_2$		176–177	685

[a] Inconsistent literature melting points.
[b] These patents list a further 89 derivatives with R^1 = 3-hydroxy, sometimes with additional methyl, bromo and/or chloro substituents; R^2 = various carboxylic acid and sulphonic acid esters; R^3 = H. None are characterized.
* Methods found in Section V.8.

Table 24. 3-Ketomethylquinolines

R	Quinoline substituent(s)	Preparation	Yield (%)	M.p. (°C)	References
Me		Q-3-CHO or Q-3-Ac, CH_2N_2		94	691*
Me	2-Me-4-COOH	Isatin, $(AcCH_2)_2$, K, 160–180 °C	32.9	280–281	58
Me	2-COOMe-4-Cl-8-OMe	3-CH_2C(Cl)=CH_2-4-Cl-8-MeO-Q-2-COOMe, conc. H_2SO_4, 0 °C, 10 min	73	173	692, 693
iPr	2,4-diOMe-7-OCH_2O-8 (Orixinone)	Orixine, 20% H_2SO_4, 100 °C		102–103	694
iPr	2,4,8-triOMe	2,4,8-triMeO-Q-3-CH_2CHOHiPr, CrO_3	43	66–68	695
Bu		Q-3-C≡CBu, $HgSO_4$, H_2SO_4	46	77–78	569
CH_2iPr	2,4,8-triOMe[a]	3-COCH$_2$iPr-4-HO-8-MeO-2-quinolone, CH_2N_2	1	62–63	630
Ph		Q-3-C≡CPh, $HgSO_4$, H_2SO_4	89	74–75	569
		3-Me-Q, LDA; PhCOOMe	52	b.p. 230–240/3 mm 70–71	696
		1-(4-ClC$_6$H$_4$CH$_2$)- or 1-(2,6-diClC$_6$H$_3$CH$_2$)-3-CH$_2$COPh-Q, HBr, AcOH, 180 °C	50	HClO$_4$ 194 Pic. 203–204	697
Ph	6-Me	3,6-diMe-Q, LDA; PhCOOMe	43	67–68	696
Ph	2-Ph-4-$CONH_2$	2-Ph-3-CH$_2$COPh-4-CONH$_2$-Q N-oxide, Fe, AcOH		214(d)	191

*This reference includes carbonyl derivatives.
[a] By-product to 4,8-diMeO-3-iPrCH$_2$COCH$_2$-2-quinolone; see Section XV.4.

Table 25. 4-Ketomethylquinolines

Structure: quinoline bearing at the 4-position a side chain R^2R^3C–C(=O)–R^1 (ring N as drawn).

R^1	R^2	R^3	Quinoline substituent(s)	Preparation			Yield (%)	M.p. (°C)	References
				Method	Base	Reagents			
Me	H	H	None	1	SDA[a]	EtOAc	64	71.8–72.6; b.p. 136–140 /0.7 mm	563
Me	H	H	2-CF_3	19		2-CF_3-Q-4-CHO, $EtNO_2$; NaH_2PO_2, Ni	85.6	Pic. 200.5–201.5	698,699
Me	H	H	8-CN				1.4	111–113	182
Et	H	H		1	SDA[a]	EtCOOEt	68	b.p. 134–135 /0.55 mm; Pic. 199–200	563
Et	Me	H		2	$NaNH_2$	EtCOEt	82	Oil	567
Et	Me	H	7-Cl	2	$NaNH_2$	EtCOEt	80	46–47	567
C_2F_5	H	H		1	SDA[a]	C_2F_5COOEt	65	255–256(d)	563
Pr	H	H		1	SDA[a]	PrCOOEt	65.9	b.p. 133–135 /0.5 mm; Pic. 177–178	563
iPr	H	H		1	SDA[a]	iPrCOOEt	58	45.4–46.6; b.p. 140–143 /0.8 mm; Pic. 191–192.5	563
iPr	Me	Me		2	$NaNH_2$	iPrCOiPr	50	47–49	567
iPr	Me	Me	7-Cl	2	$NaNH_2$	iPrCOiPr	44	115–116	567
Bu	H	H		3		$HgSO_4$, H_2SO_4	88	b.p. 147–151 /2 mm	569

Table 25. (Contd.)

R¹	R²	R³	Method	Base	Reagents	Yield (%)	M.p. (°C)	References
tBu		H	1	SDA[a]	tBuCOOEt	88	119–120	563
tBu		H	2	KNH₂	tBuCOMe, hv	70	Pic. 206–207.5	700
(CH₂)₅NH-CH₂Ph		H	16				108–110 / HCl 189–190	584
—(CH₂)₄—		H	2	NaNH₂	C₆H₁₀O	70	126–127	567
			2-Cl-Q-4-CHCO(CH₂)₄, H₂, Pd/C		C₆H₉NC₅H₁₀, PhCOCl	65.6	127–128	585
			15			8[b]	125	584
			5		(CH₂)₄COCH-COOEt, Ac₂O; H₂SO₄	0.7		584
—(CH₂)₄—		2-Me	15		C₆H₉N(CH₂CH₂)₂O, TsCl		106–106.5 TsOH 235–237	561
—(CH₂)₄—		2-Cl	15		C₆H₉N(CH₂CH₂)₂O, 54 PhCOCl		100–101	585*
—(CH₂)₄—		7-Cl	2	NaNH₂	C₆H₁₀O	65	105–106	567
—(CH₂)₅—		7-Cl	2	NaNH₂	C₇H₁₂O	92	127–128	567
—(CH₂)₅—		7-Cl	2	NaNH₂	C₇H₁₂O	90	89–90	567
—(CH₂)₆—		7-Cl	2	NaNH₂	C₈H₁₄O	93	110–111	567
—(CH₂)₆—		7-Cl	2	NaNH₂	C₈H₁₄O	87	96–97	567
Ph	H		1	NaH	PhCOOMe	98	116–117	661
			3		65% H₂SO₄	98	114	124, 669, 670
			3		HgSO₄, H₂SO₄	86	b.p. 180–185 /2 mm	569
Ph			1	SDA[a]	PhCOOEt	66.7	116.2–117.8	563*
	H	2-Me	2	NaNH₂	PhCOMe	45	117–118	567
Ph	H		1	NaNH₂	PhCOOEt	44	115.5–116.5	635
Ph	H	6-Cl	1	KNH₂	PhCOOEt	32	159–161	701
			1[c]			14	135–136	535

			Conditions	Yield (%)	mp (°C)	References
						307?
					142	702
Ph	CH$_2$COPh	H	Q-4-CHO, PhCOMe, aq. NaOH, 25 °C	79	142	702
Ph	Br	H	Q-4-CH$_2$COPh, Br$_2$, AcOH	75	115–120	124, 669, 670
Ph	C$_5$H$_{10}$N	H	Q-4-CHBrCOPh, C$_5$H$_{11}$N	11	HCl 219–219.5	124, 669, 670
4-MeOC$_6$H$_4$	H	H	1 · SDA^a · 4-MeOC$_6$H$_4$-COOEt	83.1	127.4–129 Pic. 182.5–183.5	563
2-C$_{10}$H$_7$	CH$_2$CO-2-C$_{10}$H$_7$	H	Q-4-CHO, 2-C$_{10}$H$_7$COMe, aq. NaOH, 25 °C	61	162	702
2-Furyl	H	H	1 · SDA^a · C$_4$H$_3$O-2-COOEt	81.8	100–101.4 Pic. 194–195.5(d)	563
2-Thienyl	H	H	1 · SDA^a · C$_4$H$_3$S-2-COOEt	68.5	84.2–84.4 Pic. 190–191(d)	563
2-Thienyl	CH$_2$CO-2-thienyl	H	Q-4-CHO, C$_4$H$_3$S-2-COMe, aq. NaOH, 25 °C	79	144	702
2-Pyridyl	H	H	1 · SDA^a · 2-Py-COOEt	79.5	88.4–89.6 Pic. 204–205(d)	563
COOH	H	H	Q-4-CH$_2$COCOOEt, 6% H$_2$SO$_4$		224–225(d)	676, 703*
COOH	=NNHC$_6$H$_5$		Q-4-CH$_2$COCOOEt, C$_6$H$_5$N$_2$Cl		174(d)	703
COOH	=NNHC$_6$H$_4$-4-Me		Q-4-CH$_2$COCOOEt, 4-MeC$_6$H$_4$N$_2$Cl		172(d)	703
COOEt	H	H	1 · EtOK(COOEt)$_2$ NaNH$_2$ or KNH$_2$ Na	75 60 48	197–198 193–193.5 Pic. 207–208	676, 703* 635 602*
COOEt	=NNC$_6$H$_4$-4-Me		Q-4-C(=NNHC$_6$H$_4$Me)COCOOH, EtOH		147	703
COOEt	=NNHC$_6$H$_4$-4-NO$_2$		Q-4-CH$_2$COCOOEt, 4-NO$_2$C$_6$H$_4$-CHO, C$_5$H$_{11}$N	50	210–212	602
CONHNH$_2$	H	H	Q-4-CH$_2$COCOOH, N$_2$H$_4$ Q-2-CH$_2$C(=NOH)COOEt, N$_2$H$_4$			679* 679*

*These references include carbonyl derivatives.

^{a}Sodium diisopropylamide.

^{b}Obtained with 61.4% of the 2-substituted quinoline; Table 20.

^{c}From methyl(1-PhCO-3-CH=CH$_2$-4-piperidyl)acetate. This product presumably arose from the N-benzoyl group. The expected 2-propanone (46%) was also obtained; Table 18.

C. Position 7

The oximino ketone **234** was prepared in a similar way to compound **232**.[375]

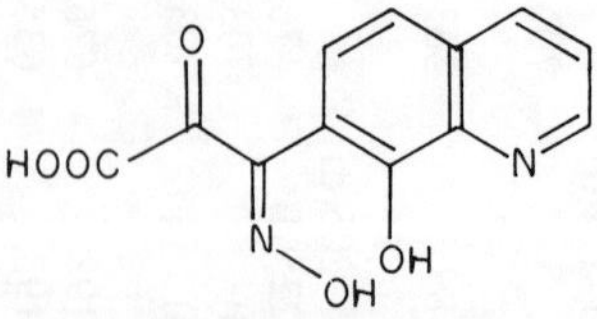

234, 48%, m.p. 170 °C (d)

Addition of resorcinol to 7-quinolylacetonitrile under the influence of boron trifluoride and hydrogen chloride gave the ketone **235a**, which with the appropriate acetobromoglucose in acetone containing sodium hydroxide gave **235b**.[705]

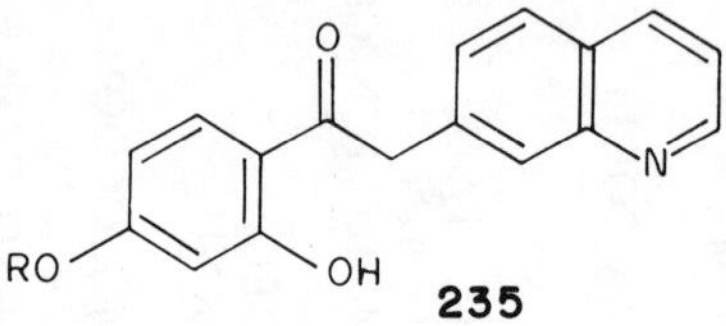

235

a, R = H, 90%

b, R =   13%

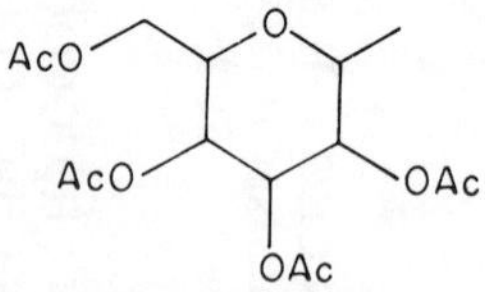

D. Position 8

Ketone **236a** was prepared from resorcinol and 8-quinolyl-acetonitrile in the presence of boron trifluoride. Acetic anhydride gave the mono- **236b** or di-acetyl, **236c** derivative according to the conditions.[706,707]

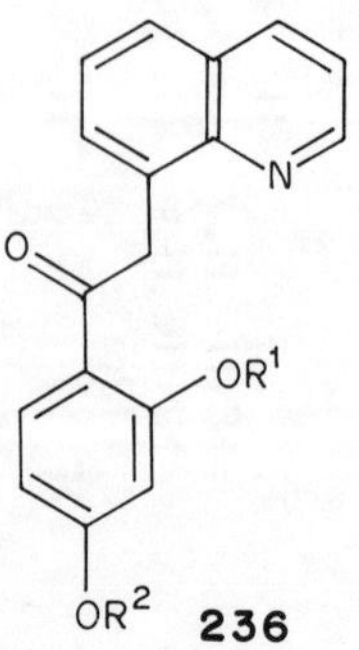

236

a, $R^1, R^2 = H$, 69%, m.p. 207 °C
b, $R^1 = H, R^2 = Ac$, 64%, m.p. 175 °C
c, $R^1, R^2 = Ac$, 62%, m.p 128 °C

VI. Ketoethyl Quinolines

1. 2-(Ketoethyl)quinolines

With ethyl acetoacetate in sodium ethoxide, 2-bromomethylquinoline reacted to give the ketone **237b**, which was hydrolysed in sodium ethoxide and decarboxylated to compound **237a**.[708,709] The same compound, **237a**, was claimed to have been made by hydrolysis of the appropriate terminal acetylene, but no characterization data were given.[710]

237

a, R = H, 70%, DNP, m.p. 196 °C

b, R = COOEt, 91%, DNP, m.p. 140 °C

Esters of 2-quinolylacetic acid were α-substituted with phenacyl bromide to give the keto-esters **238**. Hydrolysis and decarboxylation gave ketone **239**.[711] This corrected previous reports that phenacyl bromide alkylated the ring nitrogen.[712,713]

239, 88%, m.p. 78−80 °C

238, R = Me, n.m.r. reported

R = Et, 38%, m.p. 64−65 °C

The ketone **240** was prepared by electrolytic reduction of the appropriate chalcone in neutral solution. At high pH a one electron per molecule reaction gave a mixture which it was suggested may contain the three dimers shown in Scheme 37.[714]

Reaction between 2-methylquinoline and 1,2-diones or 1,2,3-triones at 140 °C gave the keto-alcohols **241**.[715]

Phenanthraquinone and 2-methyl- or 2,6-dimethylquinolines at 200 °C gave compounds of suggested structure **242**. These compounds failed to eliminate water on recrystallization from acetic anhydride, although benzil did give a chalcone with 2,3-dimethylquinoline at 150 °C, Section VI.5.[716]

Methyl and methylene ketones usually condense with quinoline aldehydes under basic conditions to give unsaturated ketones. Those from aryl methyl

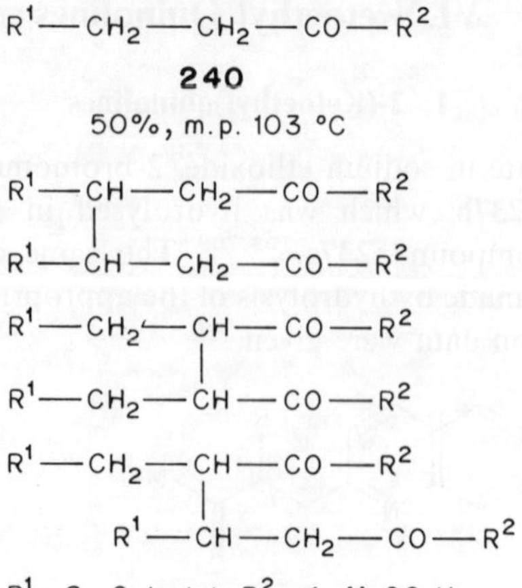

$$R^1—CH_2—CH_2—CO—R^2$$

240

50%, m.p. 103 °C

$$R^1—CH—CH_2—CO—R^2$$
$$R^1—CH—CH_2—CO—R^2$$

$$R^1—CH_2—CH—CO—R^2$$
$$R^1—CH_2—CH—CO—R^2$$

$$R^1—CH_2—CH—CO—R^2$$
$$R^1—CH—CH_2—CO—R^2$$

$R^1 = 2-$Quinolyl, $R^2 = 4-$MeOC$_6$H$_4$

SCHEME 37

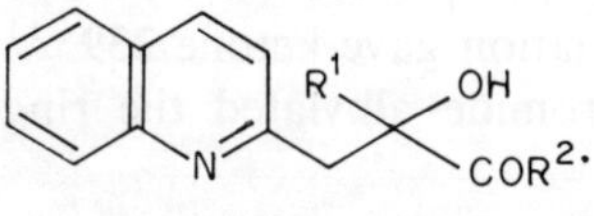

241

$R^1, R^2 = Ph, 70\%, m.p. 187-188$ °C
$R^1 = COPh, R^2 = Ph, 35\%, m.p. 258-260$ °C
$R^1 = COPh, R^2 = OEt, 8\%, m.p. 80-81$ °C

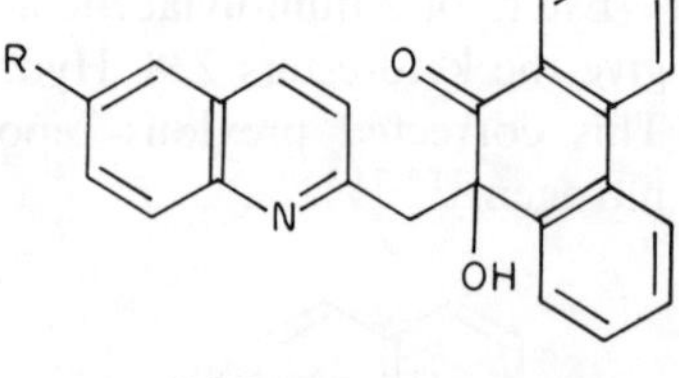

242, R = H, m.p. 169 °C
R = Me, m.p. 135 °C

ketones are described under chalcones, Section VI.5. Klosa[717] reacted quinoline-2-carboxaldehyde with acetone in potassium hydroxide to obtain the unsaturated ketone **243a**, while ethyl acetoacetate and diethylamine gave compound **243b**. The preparation of diketone **243c** from pentane-2,4-dione and of compounds **244** from cyclic ketones was catalysed by basic ion exchange resins.[718] Sometimes the intermediate keto-alcohols are stable enough to be

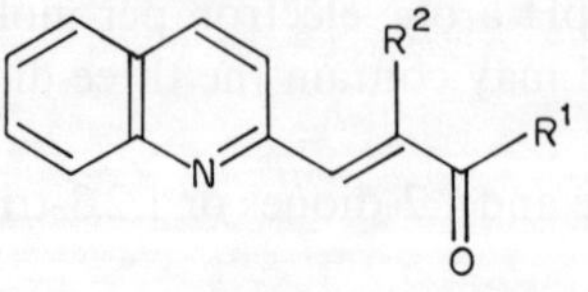

243

a, $R^1 = Me, R^2 = H, m.p. 118-120$ °C, Pic. m.p. 165-167 °C
b, $R^1 = Me, R^2 = COOEt, m.p. 206-208$ °C
c, $R^1 = Me, R^2 = COMe, 65\%, m.p. 134$ °C
d, $R^1 = Ph, R^2 = H$

244, $n = 1$, 65%
$n = 2$, 56%, m.p. 248 °C
$n = 3$, 53%, m.p. 146 °C

245

a, R = Me, 68%, m.p. 164–167 °C
b, R = Ph
c, R = 2–HOC$_6$H$_4$, 20%, m.p. 98 °C
d, R = 4–NO$_2$C$_6$H$_4$, 80%, m.p. 155.5–156.5 °C
e, R = 5–NO$_2$–2–thienyl, 35%, m.p. 163 °C
f, R = 3–Indolyl

isolated. Kwartler and Lindwall[719] reported that the quinoline aldehyde and acetone in sodium hydroxide or diethylamine at 0 °C gave the hydroxy-ketone **245a**, while acetophenone gave compound **245b** (78%, m.p. 114–116 °C). With acetophenone in aqueous sodium hydroxide Aryian and Mooney[720] got compound **245b** (m.p. 123 °C), which did not dehydrate to the chalcone at 60 °C. Klosa[717] working under similar conditions claimed to get the chalcone **243d**, m.p. 153–155 °C, Pic. m.p. 133–135 °C, which with bromine gave the dibromoketone **246**. Tsukerman's group[721] claimed to get compound **243d** (84%, m.p. 116 °C, Pic. m.p. 178–180 °C) plus a trace of the Michael addition product **247a** by adding 10% sodium hydroxide to a methanolic solution of the reactants. They suggested that Klosa's product was also **247a** in spite of his preparation of the dibromide, **246**. Tsukerman and co-workers[721] describe eleven further chalcones prepared by the same technique, but in two other cases, **245d, e**, obtained hydroxy-ketones. Both were dehydrated in acetic anhydride. Reaction of quinoline-2-carboxaldehyde with 2-hydroxyacetophenone in methanolic sodium methoxide at 0 °C gave, in the hands of Corvaisier[722] the hydroxy-ketone **245c** mixed with the corresponding chalcone. The diketone **247b** was obtained under catalysis by piperidine in ethanol.[723,724]

246, m.p. 172–174 °C

247 a, R = Ph, m.p. 154–155 °C
b, R = Benzothiazol–2–yl, 38.5%, m.p. 185 °C

At room temperature quinoline-2-carboxaldehyde and acetone/diethylamine gave compound **248**[719], but when a mixture of the aldehyde and acetonedicarboxylic acid was allowed to stand, followed by an acid work-up, the didehydrated derivative of compound **248** was obtained (60%, m.p. 198 °C).[725,726]

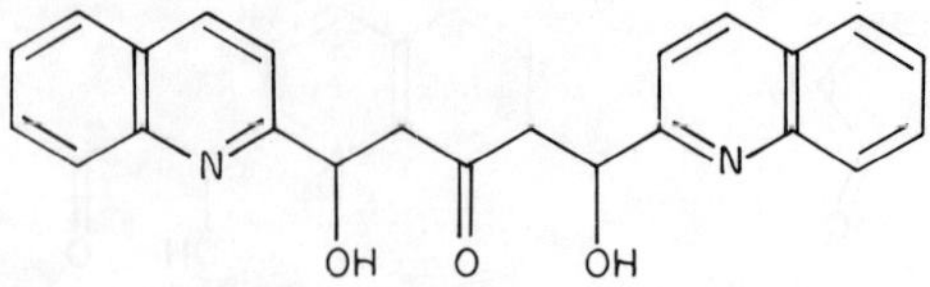

248, 32%, m.p. 208–210 °C

A reaction between quinoline-2-carboxaldehyde and 1-(4-dimethylamino-phenyl) but-1-en-3-one under chalcone formation conditions gave the dienone **249**.[727]

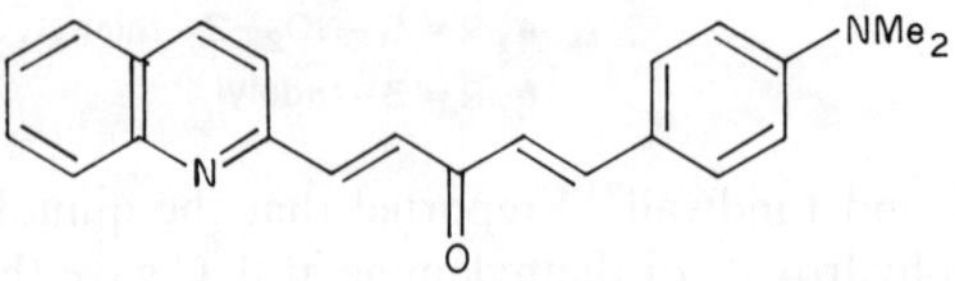

249, m.p. 174 °C

An attempt to make the chalcone from quinoline-2-carboxaldehyde and 3-acetylindole gave the hydroxy-ketone **245f**. This crude material was dehydrated by treatment with ethyl chloroformate followed by ammonia to give the required product, Table 31 (p. 306).[728] Michael addition of sulphinic acids to the chalcone **243**, $R^1 = 4\text{-MeOC}_6H_4$, $R^2 = H$ (EtOH, 25 °C) gave the saturated ketones **250**.[729]

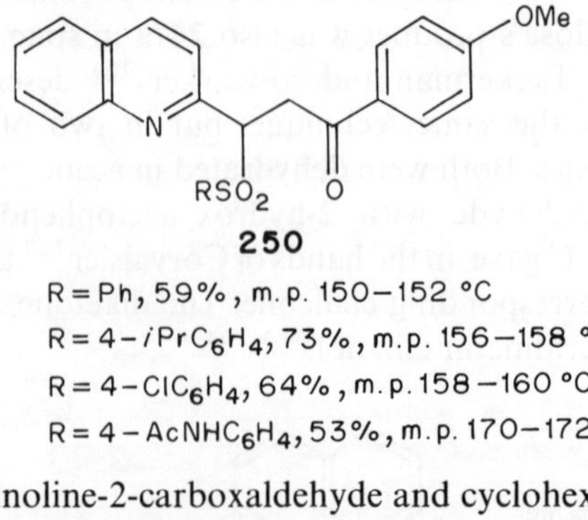

250

R = Ph, 59%, m.p. 150–152 °C

R = 4–iPrC$_6$H$_4$, 73%, m.p. 156–158 °C

R = 4–ClC$_6$H$_4$, 64%, m.p. 158–160 °C

R = 4–AcNHC$_6$H$_4$, 53%, m.p. 170–172 °C

A mixture of quinoline-2-carboxaldehyde and cyclohexanone with two drops of diethylamine reacted over 48 hours to give the hydroxy-ketone **251**.[730] Chalcone **243d** and hydrogen peroxide gave the epoxide **252**.[731]

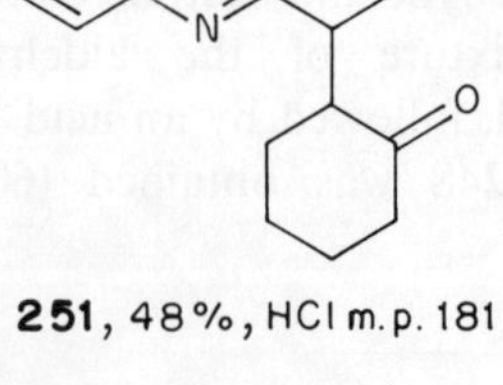

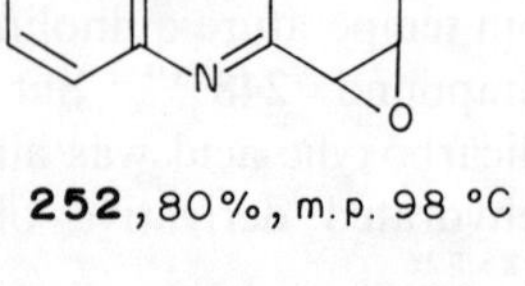

252, 80%, m.p. 98 °C

251, 48%, HCl m.p. 181 °C

Triphenylphosphoranylidene-2-propanone and 6-methylquinoline-2-carboxaldehyde in DMSO gave the unsaturated ketone **253**. Base catalysed addition of diethyl malonate followed by hydrolysis and decarboxylation gave the cyclohexane-1, 3-dione **254** which, with propionic anhydride, gave the trione **255**. The more reactive side chain ketone then gave the *O*-ethyloxime **256**, Scheme 38.[732]

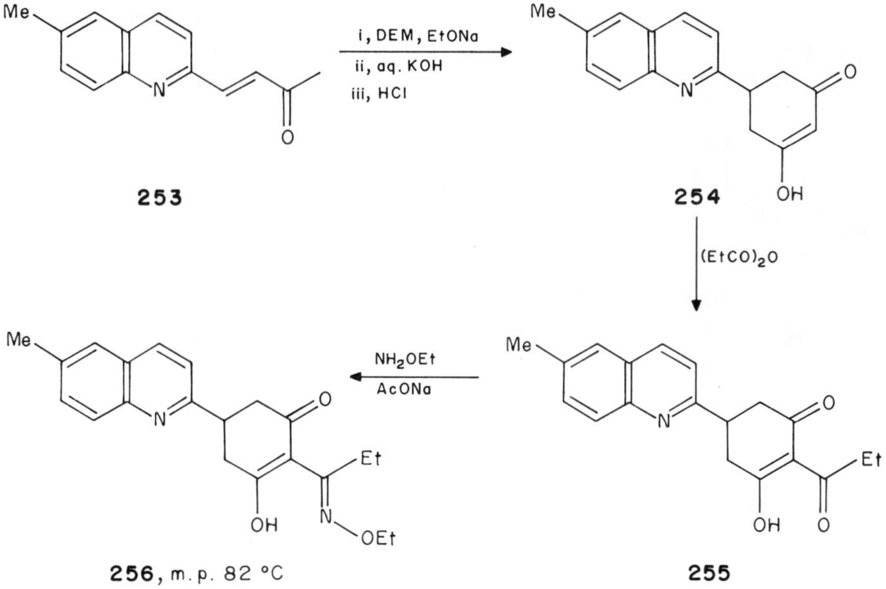
Me
N
O
253
i, DEM, EtONa
ii, aq. KOH
iii, HCl
254
OH
(EtCO)2O
NH2OEt
AcONa
Et
N
OEt
256, m.p. 82 °C
255

SCHEME 38

Finally, the reaction shown in Scheme 39 gave the ketones **257**.[26]

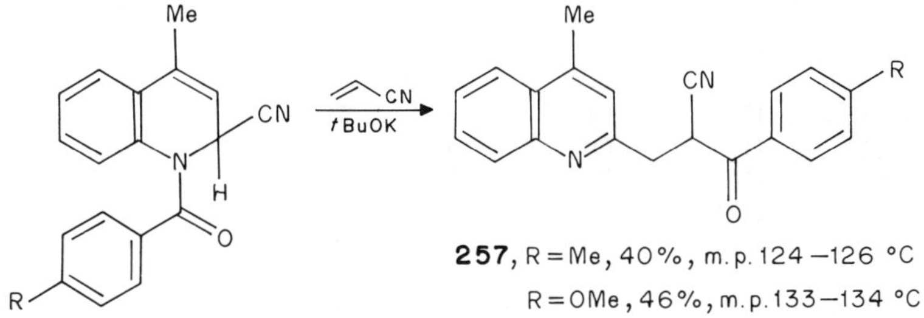
Me
CN
N
H
O
R
tBuOK
257, R = Me, 40%, m.p. 124–126 °C
R = OMe, 46%, m.p. 133–134 °C

SCHEME 39

2. 3-(Ketoethyl)quinolines

The 3-(3-oxobutyl)quinolines **259** have been prepared by hydrolysis of the vinyl chlorides **258** (Scheme 40 and Table 26).[733–738] Derivatives **260–262** of

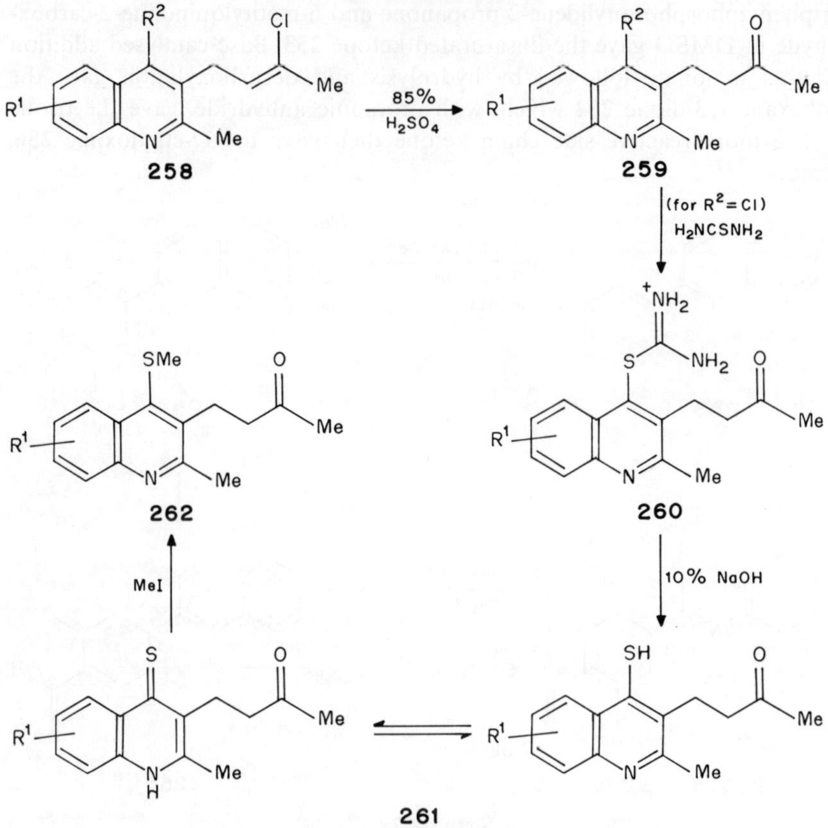

SCHEME 40

Table 26. 3-(3-Oxobutyl)quinolines (259)[733-738]

R[1]	R[2]	Yield (%)	M.p. (°C)
H	Cl	85	78
H	OMe	85.6	66
H	NH₂	91	—
6-Me	OMe	82.5	78
6-Cl	Cl	61.4	95
8-Cl	Cl	83	110
6-Br	Cl	80.4	98
8-Br	Cl	71	67
6-OMe	Cl	90	96
8-OMe	Cl	90	156
6-COOH	Cl	72.3	114
6-NH₂	OMe	—	92–93
6-NHAc	OMe	64.3	166–167

Table 27. 3-(3-Oxobutyl)quinolines (260–262)[738]

	R^1	Compound 260 Yield (%)	260 M.p. (°C)	Compound 261 Yield (%)	261 M.p. (°C)	Compound 262 Yield (%)	262 M.p. (°C)
a	H	88	178	80	188	89	84
b	6-Me			58	172		
c	8-Cl	90	175	82	129	79	98
d	6-Br			32	214		
e	6-OMe	80	160	65	143	90	82
f	8-OMe	94	179	75	150	85	132

compounds **259**, R^2 = Cl were also made.[738] Compounds **261** were shown as tautomeric mixtures, but the quinolinethione form appears to be preferred from the i.r. bands quoted for **261f**: 3220 (vN—H), 1730 (vC=O) and 1230 cm^{-1} (vC=S). The corresponding S-methyl derivative **262f** had no vC=S band.

Compound **263a** was prepared by a hydrolysis similar to that used for compounds **258 → 259**.[739] The quinoline **263**, R^1 = OH, R^2 = Me was chlorinated to give compound **263b**, and treated with thiourea to give compound **263c**.[740]

263

a, R^1, R^2 = Cl, 96%, m.p. 110–111 °C
b, R^1 = Cl, R^2 = Me, 98%
c, R^1 = SH, R^2 = Me, 97%

Depending on the nature of the substrates and conditions, ketones add to quinoline-3-carboxaldehyde to give hydroxy-ketones, Scheme 41, or chalcones, Section VI.5. Compounds **264–266** were reported by Jacoby and Zymalkowski.[456] Under carefully controlled conditions 2-hydroxyacetophenones gave hydroxy-ketones **267**; see Table 28, but with excess base, chalcones were formed.[741]

The diketo-diamide **268** rearranged and decarboxylated in phosphorus oxychloride/hydrogen chloride to give the ketone **269**.[742]

Quinoline-3-carboxaldehyde condensed with dehydroacetic acid in chloroform[743], with 3-oxoglutaric acid,[725,726] cyclopentanone,[456] or quinuclidin-3-one[456] to give unsaturated ketones **270**, **271**, **272** and **273**, respectively, in

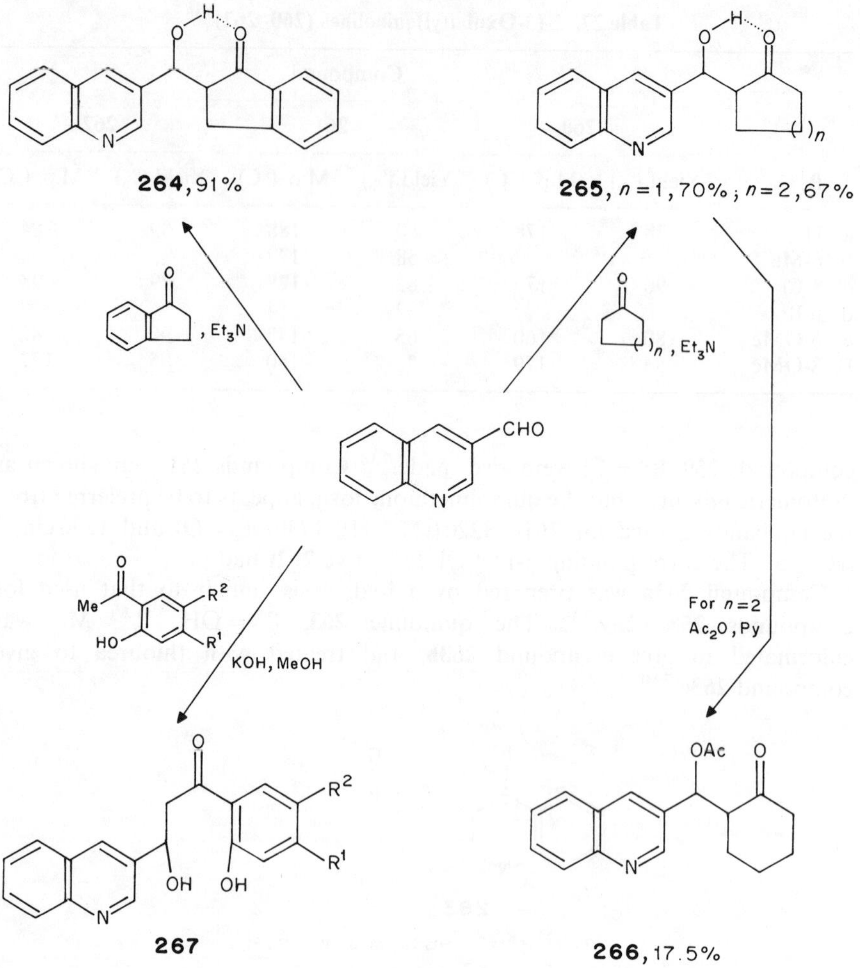

264, 91%

265, $n = 1, 70\%$; $n = 2, 67\%$

267

266, 17.5%

SCHEME 41

Table 28. Hydroxy-ketones (267)

R^1	R^2	M.p. (°C)
H	H	140–142
H	Cl	173–174
H	Br	158–160
H	Me	154–155
H	MeO	125–126
MeO	H	140–142

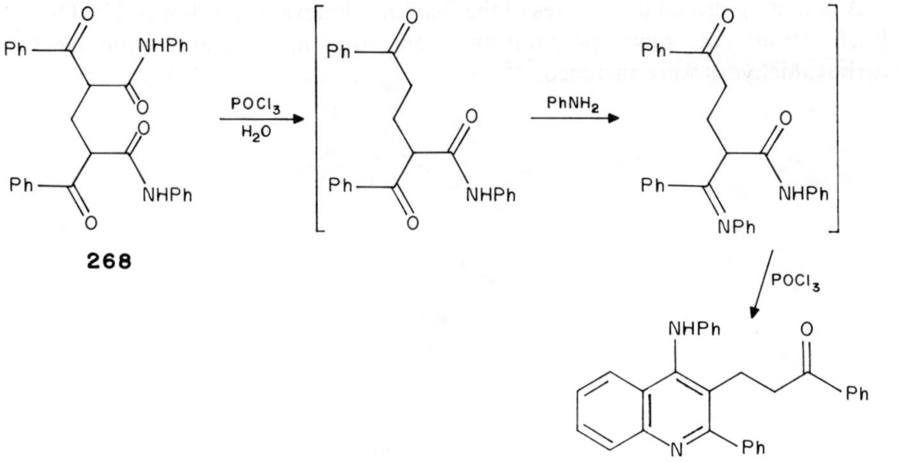

268

269, 22%, m.p. 152−154 °C
HCl, 166−169 °C

reactions reminiscent of chalcone formation. Cyclohexanone, however, gave the hydroxy-ketone **265**, $n = 2$. Attempted elimination of water gave either the acetyl ester **266** or the bis-quinolylidenecyclohexanone **274**.[456]

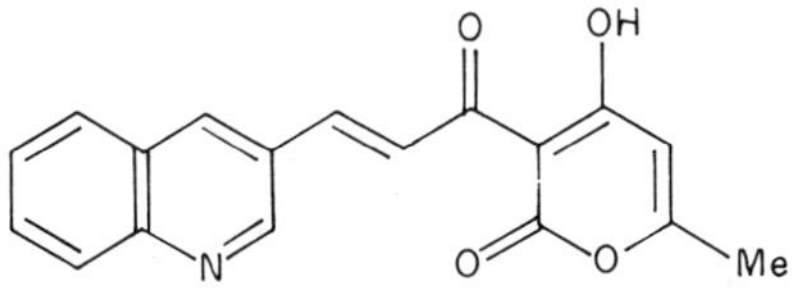

270, 55%, m.p. 205 °C

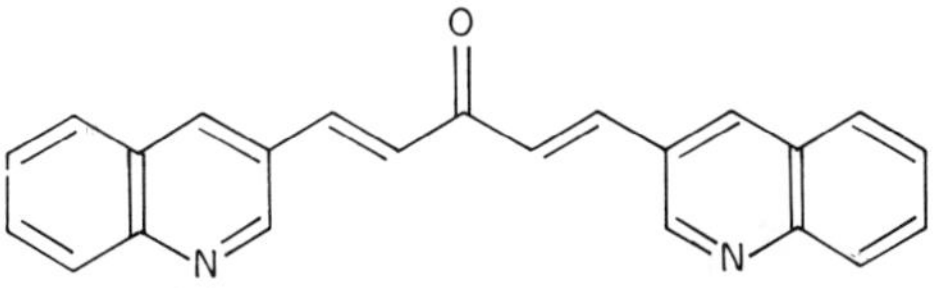

271, 93%, m.p. 242 °C

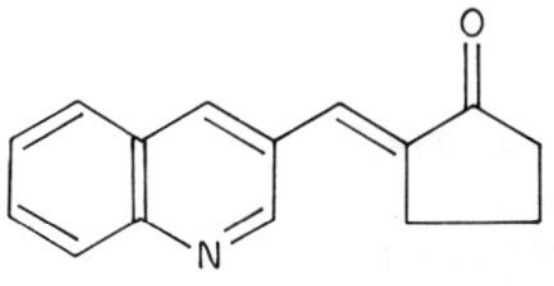

272, 70%

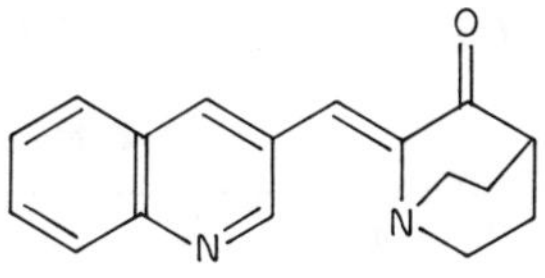

273, 74%

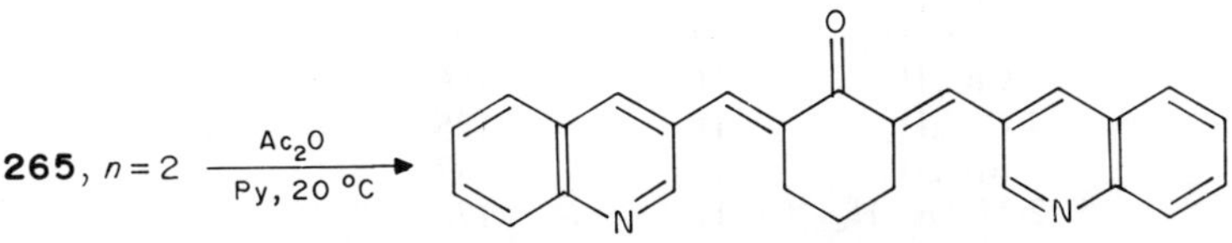

265, $n = 2$ $\xrightarrow[\text{Py, 20 °C}]{\text{Ac}_2\text{O}}$

274, 73%

A patent described derivatives of the 2-ketocyclohexane-1,3-diones **275**. Only brief details of their preparation from the appropriate quinoline-3-carboxaldehydes were included.[744]

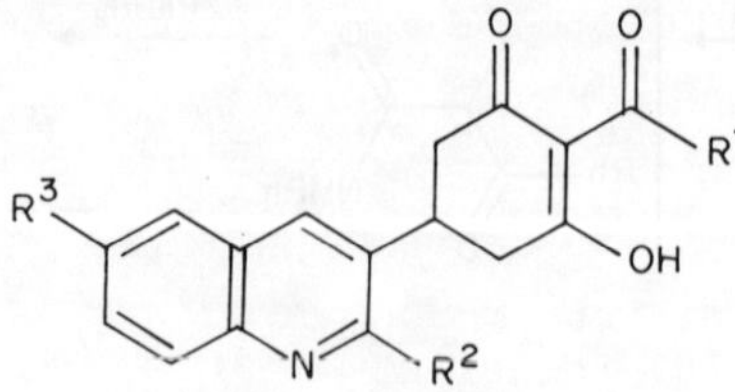

275

R^1 = Et or Pr for

R^2	R^3
H	H
Cl	H
OMe	H
OMe	OMe
SEt	H

3.　4-(Ketoethyl)quinolines

The 4-chloromethylquinolines **276** reacted with ketoamides **277** to give the derivatives **278** in 70–80% yields; Table 29.[745]

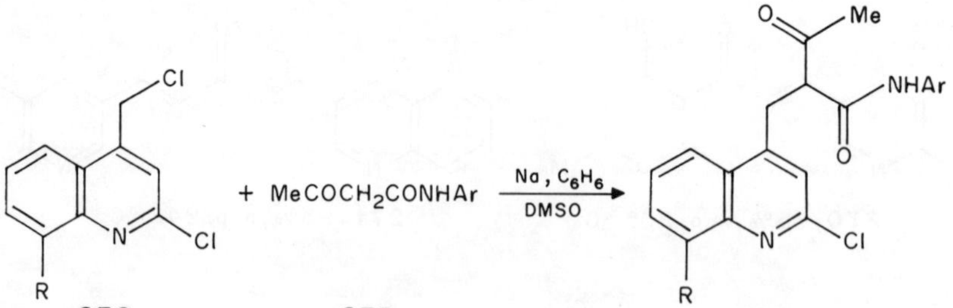

276　　　　　　　**277**　　　　　　　**278**

Table 29. Ketoamides (**278**)

Ar	R	M.p. (°C)
Ph	H	157
Ph	Me	185
2-MeC$_6$H$_4$	H	199
2-MeC$_6$H$_4$	Me	195
3-ClC$_6$H$_4$	H	166
4-ClC$_6$H$_4$	H	185
4-BrC$_6$H$_4$	H	187
2-MeOC$_6$H$_4$	H	178
4-MeOC$_6$H$_4$	H	189
1-C$_{10}$H$_7$	H	200

When quinoline-4-carboxaldehyde was treated with methanolic sodium hydroxide and an excess of the appropriate methyl ketone, Michael addition of a second molecule to the initially formed chalcone gave diones 279[702,719], but with esters of 3-ketoglutaric acid at pH 5–6 in alcohol this procedure gave the unsaturated ketones 280.[746] The same aldehyde and 2-hydroxyacetophenone in methanolic sodium methoxide at 0 °C gave a mixture of the chalcone (Section VI.5) and the hydroxy-ketone 281.[722]

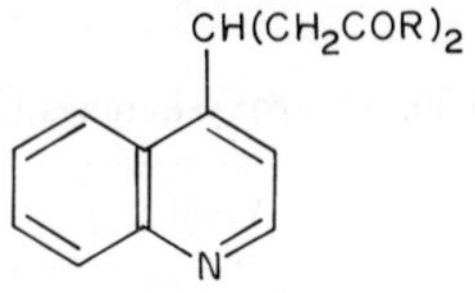

279

a, R = Ph, 87%, m.p. 144–146 °C

b, R = 2–Naphthyl, 61%, m.p. 162 °C

c, R = 2–Thienyl, 79%, m.p. 144 °C

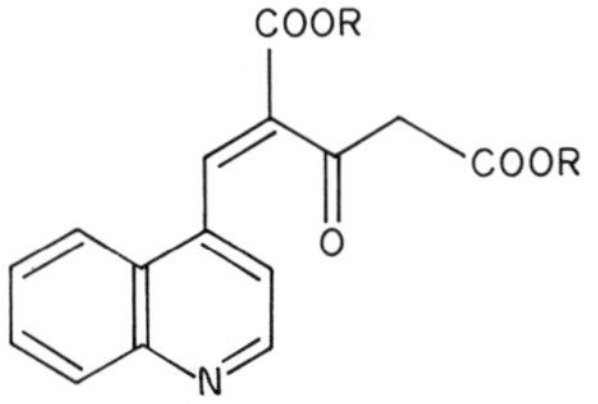

280, R = Me, 22%, m.p. 237 °C(d)

R = Et, 24%, m.p. 233 °C(d)

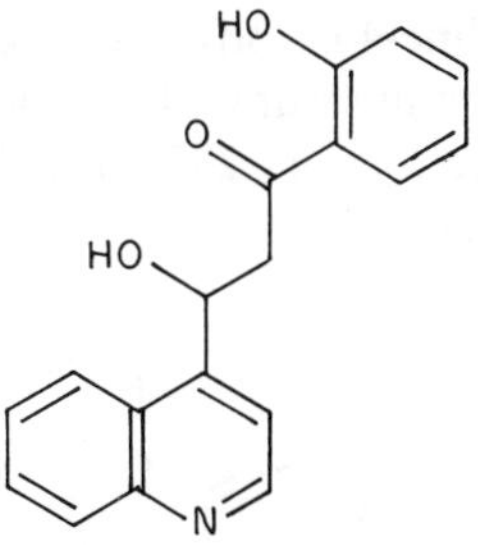

281, 10%, m.p. 140 °C

Reactions between the aldehydes **282** and cycloalkanones **283** gave hydroxy-ketones **284a, b, e**. Esterification (Ac₂O, Py) then gave the acetates **284c, d**, Table 30. Compound **284d** was deoxygenated to compound **285** as shown.[730]

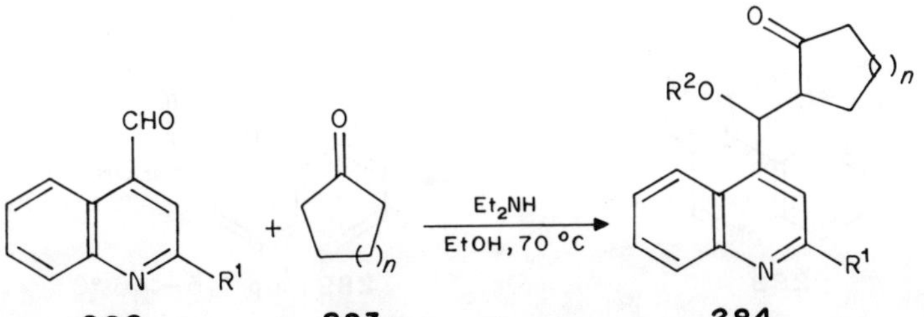

282   **283**   **284**

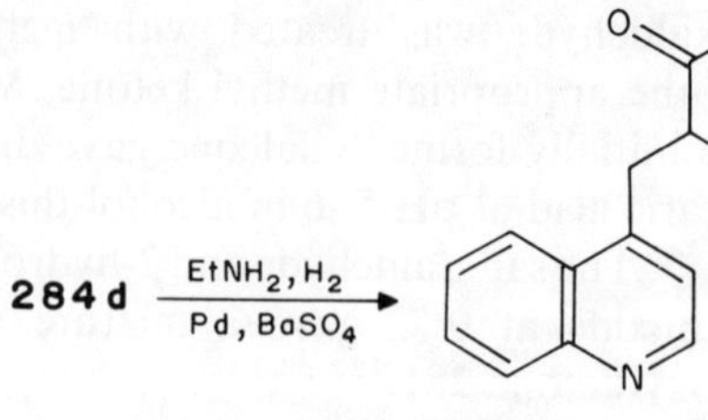

285, Pic. m.p. 210−211 °C

Table 30. Hydroxy-ketones (284)

	R^1	R^2	n	Yield (%)	M.p. (°C)
a	H	H	1		
b	H	H	2	89	90–110
c	H	Ac	1		150–151
d	H	Ac	2	35	173–174
e	Ph	H	2	70	HCl 161

Quinoline-4-carboxaldehyde condensed with dehydroacetic acid in toluene to give the unsaturated ketone **286**.[743]

The keto-ester **287** has been described in a patent as a synthetic intermediate.[747]

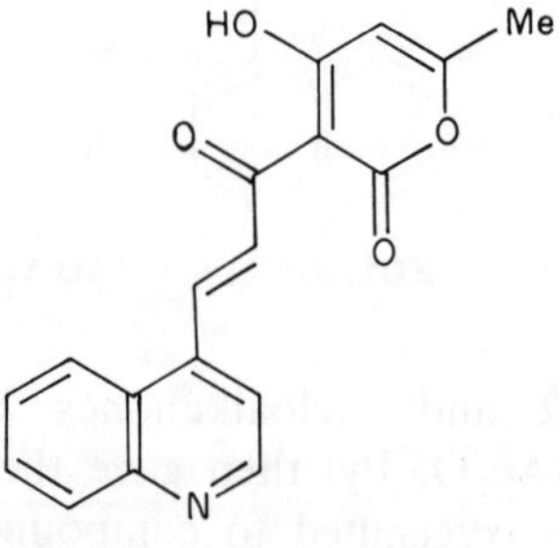

286, 53%, m.p. 202 °C

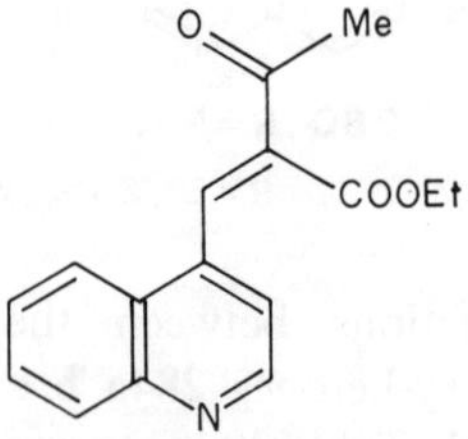

287

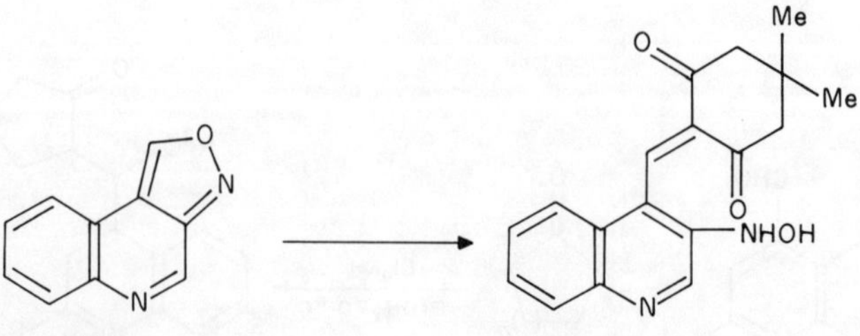

288

289, m.p. 145−146 °C

The isoxazoloquinoline **288** was formed from 3-azidoquinoline-4-carbox-aldehyde, and on treatment with dimedone was converted into the diketone **289**.[748]

Acid catalysed addition of acetophenones to the imine **290** gave the aminoketones **291**. Higher temperatures and longer heating caused ring closure to benzo[f]quinolines.[749]

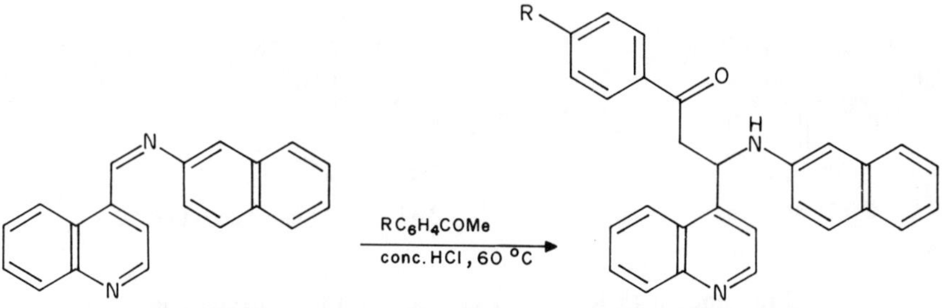

290

291, R = Me, 39%, m.p. 200 °C
R = Br, 21.1 %, m.p. 184 °C

Quinoline-4-carboxaldehyde and the salt **292** gave the ketone **293**.[743]

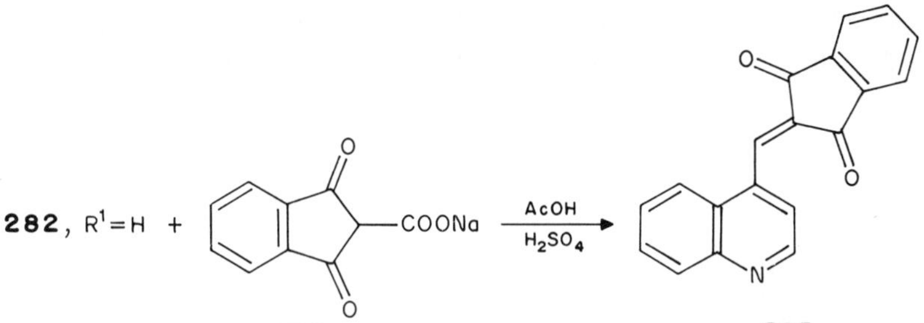

282, R¹=H

292

293

The dione **294** was prepared from quinoline-4-carboxaldehyde by the procedure shown in Scheme 38, Section VI.1. A modified process allowed the

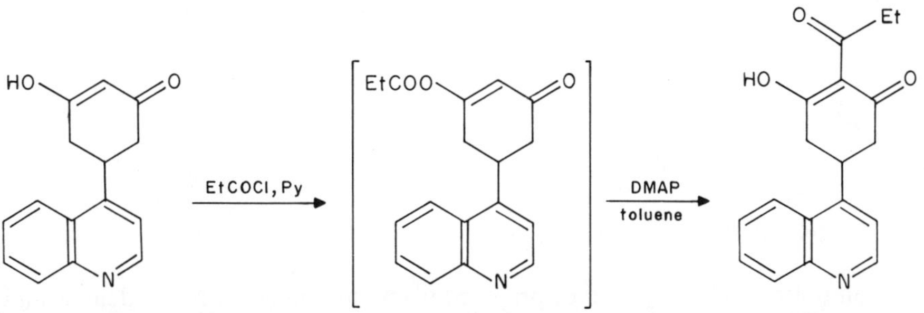

294, m.p. 190−195 °C

295, oil

preparation of the trione **295**, which gave an *O*-ethyloxime.[732] Compound **295** (Et changed to Pr) has been reported to be derivatized in a patent.[744]

Quinoline-4-carboxaldehyde and quinuclidin-3-one condensed to give ketone **296**, which was reduced to the saturated ketone **297**.[750]

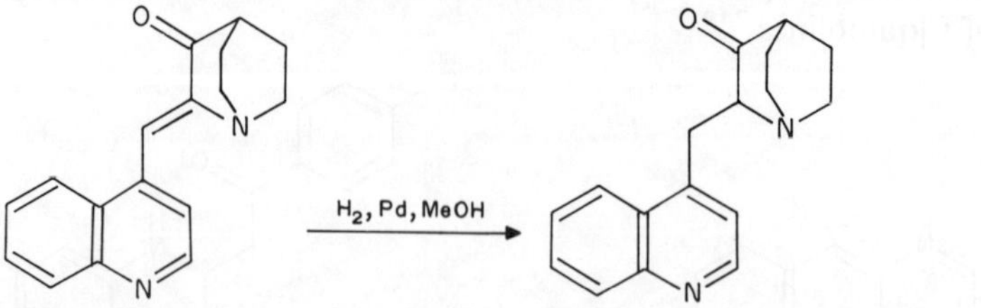

296, 50%, m.p. 153 °C   **297**, 60%, m.p. 125−126 °C

4. Quinolines with Ketoethyl Groups on the Benzene Ring

A. Position 6

Ethyl acetoacetate and 6-chloromethylquinoline gave the ketone **298**.[751] Compound **299** was prepared from the appropriate bromomethylquinoline and methyl acetoacetate.[752,753] Acetone condensed with quinoline-6-carboxaldehyde in the presence of base in the usual way to give the unsaturated ketone **300**, but ethyl acetoacetate with piperidine and ethanol at − 5 °C gave the diketo-diester **301**.[754]

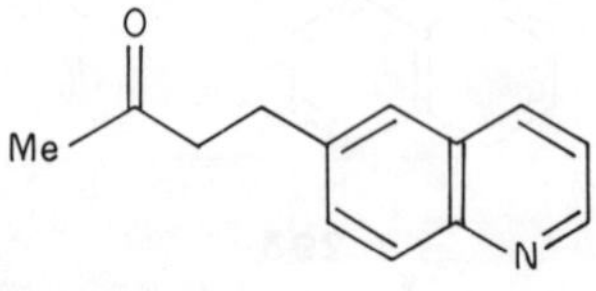

298, 28%, m.p. 56−57 °C

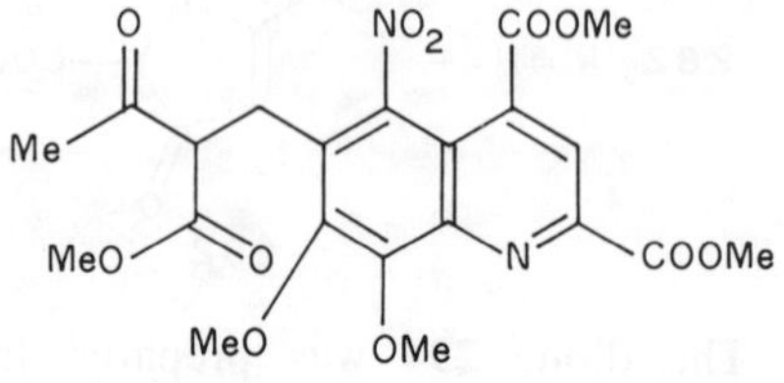

299, 91%, oil

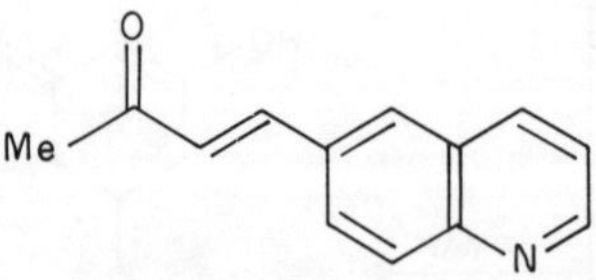

300, 33%, m.p. 112−112.2 °C

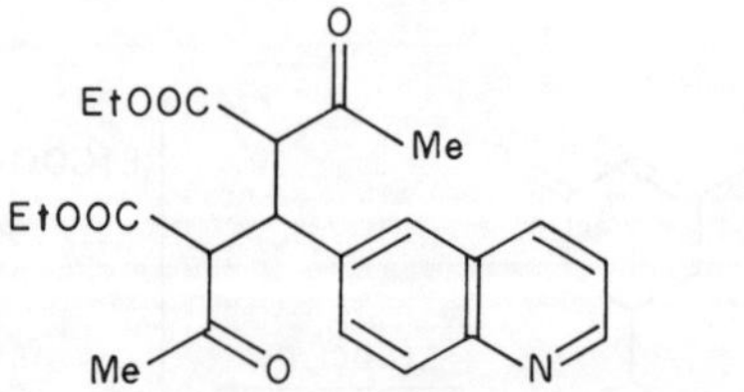

301, 33%, m.p. 151−152 °C

Compound **300** could also be prepared from quinoline-6-carboxaldehyde and triphenylphosphoranylidene-2-propanone and converted into the ketone **302a** by the method of Scheme 38, Section VI.1. The ketone **302a** with butyric anhydride gave compound **302b** (which gave an O-ethyloxime).[732]

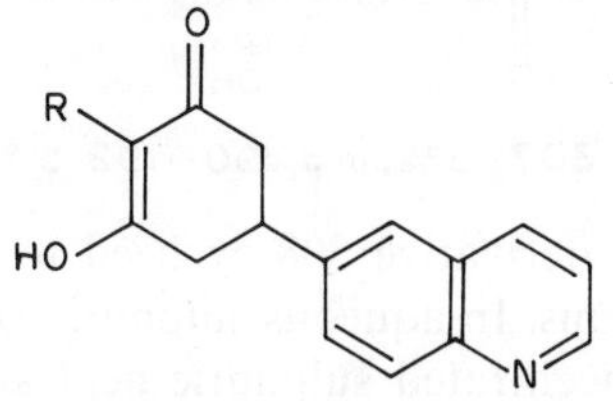

302 a, R = H
      b, R = COPr

Ketone **303** was prepared by the same method as ketone **291** (Section VI.3).[749]
A Wittig reaction on quinoline-6-carboxaldehyde gave the unsaturated keto-ester **304a**. Treatment with dimethylformamide dimethylacetal followed by aniline derivatives led to a series of compounds of general formula **304b**.[755]

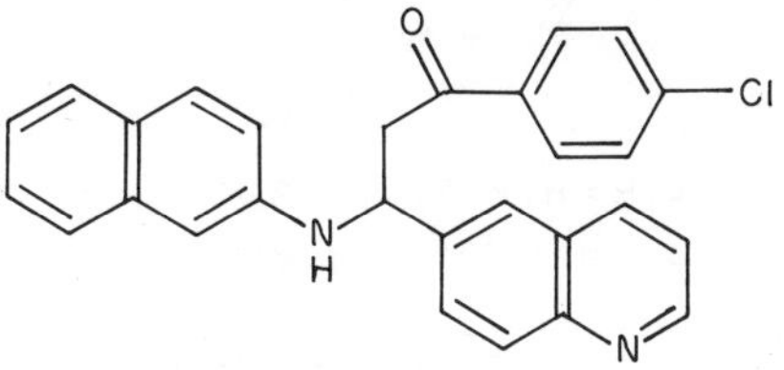

303, 18%, m.p. 180−182 °C

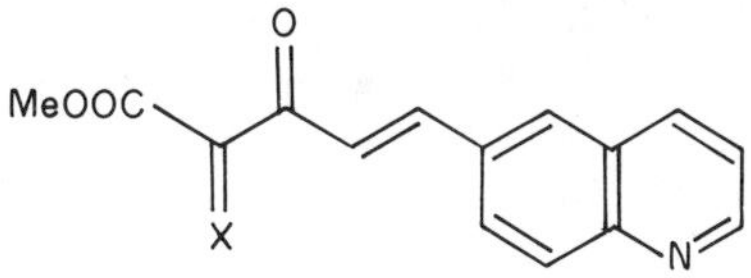

304 a, X = H_2
      b, X = $CHNHC_6H_4R$

B. Position 7

Hydrogenation of the nitro diketone **305** led to the expected ring closure to quinoline **306**. With excess hydrogen the 1,2,3,4-tetrahydro derivative of compound **306** was obtained; see Section X.16.[756]

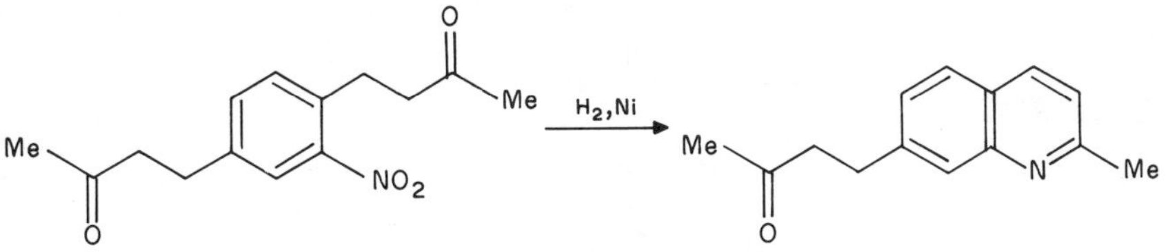

305

306, 77%, b.p. 140−143 °C/0.1 mm
Pic. m.p. 168−169 °C

A Mannich base (unspecified) was prepared from oxine (8-hydroxyquinoline) and reacted with the appropriate phosphorane in refluxing toluene to give the ketone **307** in low yield.[757]

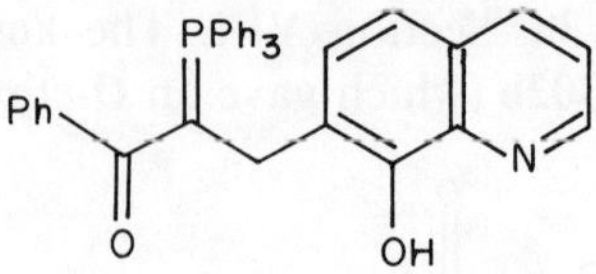

307, 5%, m.p. 190–192.5 °C

The phenylenediamine derivative **308** cyclized to a keto-acid or keto-ester depending on the conditions. In aqueous sulphuric acid at 95 °C the acid **309a** was obtained, but in concentrated sulphuric acid at 27–32 °C the ester **309b** was the product.[758]

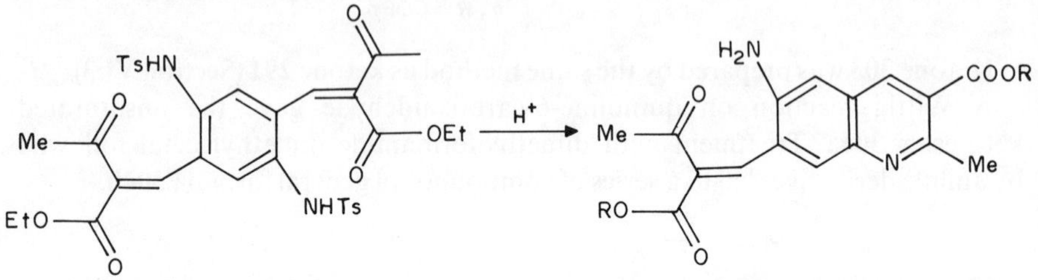

308

309

a, R = H, 65%, m.p. 200 °C(d)
b, R = Et, 63%, m.p. 219–220 °C(d)

Compounds similar to **304a** and **b** based on quinoline-7-carboxaldehyde have been reported.[755]

C. Position 8

Ketones **310a** and **b** were prepared from 5,6,7-trimethylquinoline-8-carboxaldehyde by procedures similar to those of Scheme 38, Section VI.1, and

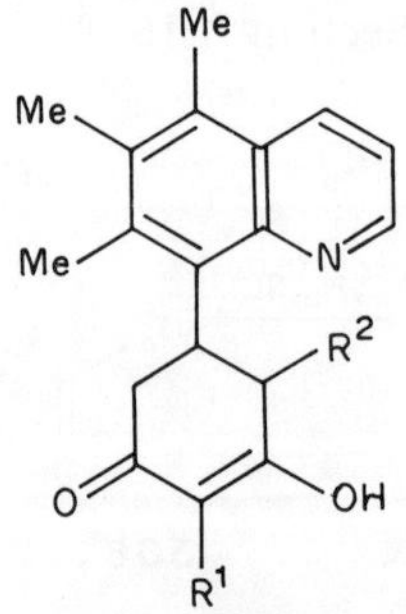

310

a, R¹, R² = H
b, R¹ = COPr, R² = H
c, R¹ = H, R² = COOMe
d, R¹ = COPr, R² = COOMe

ketone **310b** was converted into an *O*-ethyloxime. Additionally, compound **310c** was isolated and elaborated to compound **310d**.[732] A trione comparable to **310b** has been prepared from 3,7-dichloroquinoline-8-carboxaldehyde.[744]

The Mannich base derivative **311** gave the cyclohexenone **312a** by the reaction sequence shown. The acid **312a** was esterified with diazomethane to give ester **312b**.[438]

311

i, COCH₂COOEt
   |
   CH₂CH₂COOEt
ii, H₂O

312

a, R = H, m.p. 208–210 °C
b, R = Me, 34%, m.p. 88 °C

Compounds similar to **304a** and **b** based on quinoline-8-carboxaldehyde have been reported.[755]

5. Chalcones

Most known examples of quinoline chalcones have been prepared under standard conditions in which a small volume of aqueous sodium hydroxide (often a 50% solution) is added to the reactants in a large volume of methanol or similar solvent. Those from quinolinecarboxaldehydes and methyl ketones are in Tables 31–33; 2- and 5-acetylquinolines and aromatic aldehydes condense to give the chalcones of Tables 34 and 12 (p. 203) respectively.

Many of the known examples are derived from quinoline 2-, 3- and 4-carboxaldehydes and aryl methyl ketones, e.g. from **313** to form **314**. In some cases the use of excess methyl ketone led to a diketone such as compound **279**, Section VI.3. Another study showed that quinoline-2- and -4-carboxaldehydes gave chalcones in aqueous ethanolic sodium hydroxide, but in methanolic sodium methoxide mixtures of the chalcones with the hydroxy-ketones formed by addition were obtained; see Section VI.1.[722] In reactions of *o*-hydroxyacetophenones with quinoline-3-carboxaldehydes, a high concentration

313

CH₃COPh
NaOH

314

Table 31. Chalcones from Quinoline-2-carboxaldehydes

R	Quinoline substituent(s)	Catalyst	Yield (%)	M.p. (°C)	References
Ph		KOH	95	150–151	773
Ph	4-Me	NaOH	84	116	721
Ph		See text, Section VI.5			110
2-MeC$_6$H$_4$	8-Cl	KOH	80	121–122	774
3-MeC$_6$H$_4$	8-Cl	KOH	75	95–97	774
4-MeC$_6$H$_4$	8-Cl	KOH	95	118–120	774
4-MeC$_6$H$_4$		NaOH	70	149[a]	13
4-MeC$_6$H$_4$	8-Cl	KOH	60	178–179[a]	773
2-ClC$_6$H$_4$	8-Cl	KOH	64	158–160	774
2-Cl-5-HOC$_6$H$_3$		KOH	84	245–247	774
4-ClC$_6$H$_4$		NaOH		145–146	775
4-ClC$_6$H$_4$	8-Cl	KOH	95	124–125[a]	773
4-ClC$_6$H$_4$		EtONa	30	165[a]	729
4-BrC$_6$H$_4$	8-Cl	KOH	90	226–228	774
4-BrC$_6$H$_4$		KOH	85	138–139[a]	773
4-BrC$_6$H$_4$		NaOH	60	165.5[a]	13
4-BrC$_6$H$_4$	8-Cl	KOH	88	175–176	774
2-HOC$_6$H$_4$		KOH	30	125–126[a]	776
2-HOC$_6$H$_4$		NaOH	28	144[a]	772
2-HOC$_6$H$_4$[b]	8Cl	KOH	80	124–125	774
2-HO-5-MeC$_6$H$_3$		KOH	45	162–164	776
2-HOC$_6$H$_4$	6-OMe	KOH	84	151–153	777
2-HO-5-MeC$_6$H$_3$[b]	8-Cl	KOH	60	160–162	774
2-HO-5-MeC$_6$H$_3$[b]	6-OMe	KOH	62	171–173	777
2-HO-3-ClC$_6$H$_3$	8-Cl	KOH	75	98–100	774
2-HO-3-ClC$_6$H$_3$	6-OMe	KOH	80	189–191	777
2-HO-5-ClC$_6$H$_3$		KOH	35	156–157	776

Aryl group	Position	Base	Yield (%)	mp (°C)	Ref.
2-HO-5-ClC$_6$H$_3$[b]	8-Cl	KOH	80	158–160	774
2-HO-5-ClC$_6$H$_3$	6-OMe	KOH	69	176–177	777
2-HO-5-BrC$_6$H$_3$		KOH	36	146–147	776
2-HO-5-BrC$_6$H$_3$[b]	8-Cl	KOH	90	180–181	774
2-HO-5-BrC$_6$H$_3$	6-OMe	KOH	78	166–168	777
2,4-diHOC$_6$H$_3$		Ba(OH)$_2$	40	200–201	776
2,4-diHOC$_6$H$_3$[b]	8-Cl	KOH	80	177–178	774
2,4,6-triHOC$_6$H$_2$[b]	8-Cl	KOH	60	268–270	774
2,5-diHOC$_6$H$_3$		Ba(OH)$_2$	28	246–248	776
2,5-diHOC$_6$H$_3$[b]	8-Cl	KOH	80	246–248	774
2,6-diHOC$_6$H$_3$[b]	8-Cl	KOH	85	274–276	774
2-HO-4-MeOC$_6$H$_3$		Ba(OH)$_2$	45	138–140	776
2-HO-5-MeOC$_6$H$_3$	8-Cl	KOH	40	145–147	776
2-HO-5-MeOC$_6$H$_3$[b]	6-OMe	KOH	60	169–170	774
2-HO-5-MeOC$_6$H$_3$		Ba(OH)$_2$	67	146–148	777
2-HO-4-PhCH$_2$OC$_6$H$_3$	6-OMe	KOH	40	154–155	776
2-HO-4-PhCH$_2$OC$_6$H$_3$	8-Cl	KOH	56	191–193	777
2-HOC$_6$H$_3$-4-OCH$_2$COOH[b]		KOH	10	182–184	774
2-HOC$_6$H$_3$-4-OCH$_2$COOEt		KOH	30	292–293	776
2-HOC$_6$H$_3$-5-COOH		KOH	40	204–205	776
2-HOC$_6$H$_3$-5-COOH[b]	8-Cl	KOH	50	162–164	774
2-HO-3-NO$_2$C$_6$H$_2$-5-OCH$_2$COOH[b]	8-Cl	KOH	60	162–163	774
3-HOC$_6$H$_4$	8-Cl	KOH	95	208–210	774
4-HOC$_6$H$_4$		KOH	65	234–235	773
4-HOC$_6$H$_4$	8-Cl	KOH	43	174–175	774
2,4-diMeOC$_6$H$_3$		NaOH	70	96	721
2,4,6-triMeOC$_6$H$_2$		NaOH	83	152	721
4-MeOC$_6$H$_4$		NaOH or MeONa	86	133[a]	702, 721, 727, 729
4-MeOC$_6$H$_4$	8-Cl	KOH	60	195–196[a]	773
4-PhCH$_2$OC$_6$H$_4$		KOH	90	145–147	774
4-EtOOCCH$_2$OC$_6$H$_4$		KOH	70	131–132	773
C$_6$H$_4$-3-COOH		KOH	80	115–116	773
		KOH	85	189–191	773
2-NO$_2$C$_6$H$_4$	8-Cl	Ba(OH)$_2$	95	147–148	774
3-NO$_2$C$_6$H$_4$		KOH	70	163–164	773

Table 31. (*Contd.*)

R	Quinoline substituent(s)	Catalyst	Yield (%)	M.p. (°C)	References
3-$NO_2C_6H_4$	8-Cl	$Ba(OH)_2$	95	192–194	774
4-$NO_2C_6H_4$		NaOH	85	187	702, 721
4-$NO_2C_6H_4$	8-Cl	$Ba(OH)_2$	95	170–172	774
3-$NH_2C_6H_4$	8-Cl	KOH	95	134–136	774
3-NH_2-4-$MeOC_6H_3$	8-Cl	KOH	90	128–130	774
4-$NH_2C_6H_4$		NaOH	88	168.5	721
		NaOH	74	172	702
4-$NH_2C_6H_4$	8-Cl	KOH	80	216–218	774
4-$Me_2NHC_6H_4$		NaOH	78	187	721
3-$AcNHC_6H_4$		KOH	90	184–186	773
4-$AcNHC_6H_4$	8-Cl	KOH	80	238–240	774
3-CNC_6H_4		KOH	90	166–168	773
2-Furyl		NaOH	46	134	721
5-NO_2-2-furyl		H_2SO_4, AcOH	20	227–230	721
2-Thienyl		NaOH	88	142	721
5-NO_2-2-thienyl		H_2SO_4, AcOH	25	242–243	721
2-Selenophenyl					766
2-Pyrrolyl					766
1-Me-2-pyrrolyl					766
3-Indolyl		c	90	275	728
2-Pyridyl				114–116	778
3-Pyridyl				163.5–164.5	778
4-Pyridyl		NaOH	50	158	721, 778
2-Quinolyl		NaOH		138[a]	137
				160[a]	109
4-Quinolyl		ETONa	65	188	729
Ferrocenyl		EtONa	22	202	779

[a] Inconsistent literature melting points. The compound of Ref. 109 was prepared from quinoline-2-carboxaldehyde and diazomethane followed by lead oxide oxidation.

[b] These 2-hydroxyacetophenone derivatives were cyclized to chromones.

[c] From O-2-CHOHCH.CO-3-indolyl, ClCOOEt; NH_3.

Table 32. Chalcones from Quinoline-3-carboxaldehyde

R	Yield %	M.p. (°C)	References
Ph	76	149–150	780
2-HOC$_6$H$_4$	52	144–146	741
2-HO-5-MeC$_6$H$_3$	50	165–166	741
2-HO-5-ClC$_6$H$_3$	42	188–190	741
2-HO-5-BrC$_6$H$_3$	40	178–180	741
2-HO-4-MeOC$_6$H$_3$	44	172–174	741
2-HO-5-MeOC$_6$H$_3$	53	162–164	741

Table 33. Chalcones from Quinoline-4-carboxaldehyde

R	Catalyst	Yield (%)	M.p. (°C)	References
2-HOC$_6$H$_4$	KOH	62	141–142	741
	NaOH	40	142	722
2-HO-5-MeC$_6$H$_3$	KOH	52	150–152	741
2-HO-5-ClC$_6$H$_3$	KOH	54	178–180	741
2-HO-5-BrC$_6$H$_3$	KOH	52	164–165	741
2,5-diHOC$_6$H$_3$	Ba(OH)$_2$	28	144–146	741
2-HO-4-MeOC$_6$H$_3$	KOH	50	164–165	741
2-HO-5-MeOC$_6$H$_3$	KOH	48	146–147	741
4-NO$_2$C$_6$H$_4$	NaOH	81	206	702

of potassium hydroxide in methanol gave chalcones but a slightly lower concentration gave hydroxy-ketones.[741] A mixture of quinoline-2-carbox-aldehyde and 2-acetylquinoline treated under the standard conditions gave a mixture of the chalcone, Table 31 and the hydroxy-ketone, Table 1.[137] Some of the hydroxy-ketones formed in such reactions have been dehydrated to the chalcones by acetic anhydride.[721,728]

Quinoline-2-carboxaldehyde with D-2-acetyl-6-methyl-8-cyanomethylergoline gave a chalcone 315 in the normal way.[759]

Table 34. Chalcones from 2-Acetylquinoline

R	Catalyst	Yield (%)	M.p. (°C)	References
Ph	NaOH	97	132.5	137*
4-MeC$_6$H$_4$	NaOH	66	108	13
4-ClC$_6$H$_4$				13
2,4-diMeOC$_6$H$_3$	NaOH	95	141	137
2,3,4-triMeOC$_6$H$_2$				781
2,4,6-triMeOC$_6$H$_2$	NaOH	87	169–170	137
3,4-diMeOC$_6$H$_3$	NaOH	60	145–147	103
4-MeOC$_6$H$_4$	NaOH	95	138	137*
3,4-OCH$_2$OC$_6$H$_3$	NaOH	99	166–167	103
4-NO$_2$C$_6$H$_4$	NaOH	95	192–193	137*
4-NH$_2$C$_6$H$_4$	NaOH	50	174–176(d)	137
4-Me$_2$NHC$_6$H$_4$	NaOH	60	137–138	137
2-Furyl	NaOH	96	120	137*, 444
5-NO$_2$-2-furyl	conc. H$_2$SO$_4$	10	191–193(d)	137*
2-Thienyl	NaOH	94	132	137*
5-NO$_2$-2-thienyl	conc. H$_2$SO$_4$	38	203–204	137*
2-Seleninyl				766, 768
2-Pyrrolyl				769
1-Me-2-pyrrolyl				769
2-Quinolyl	See Table 31			

*These references include carbonyl derivatives.

315

Quinoline-2-carboxaldehyde gave bis-chalcones with 1,4-diacetylbenzene (80%, m.p. 246.5 °C)[760] and with 2,5-diacetylthiophene (61%, m.p. 236 °C).[761] The aldehyde **316** was reported to react with phenylethynylmagnesium bromide to give the chalcone **317** in low yield. It was suggested that the strongly basic Grignard reagent catalysed the rearrangement.[110]

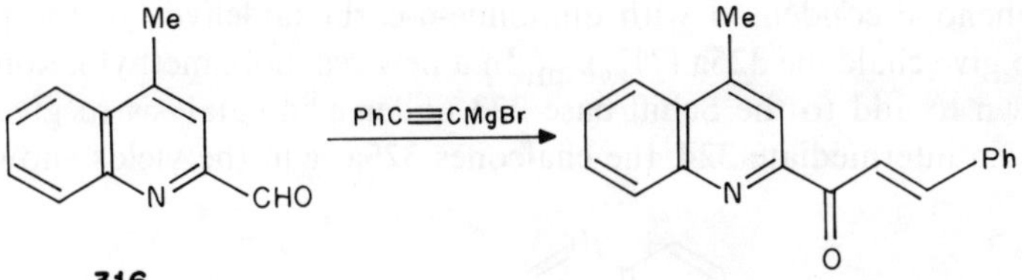

$$PhC\equiv CMgBr$$

316

317

Quinoline-2-carboxaldehyde with excess diazomethane in ether gave 2-acetylquinoline (73%), Table 1, while in methanol the main product was the propylene oxide. When the reaction was run in ether with the aldehyde in excess it gave 1,3-di-(2-quinolyl)-1-propanone, which could be oxidized (Pb$_2$O, HCl) to the chalcone.[109] Benzil condensed with 2,3-dimethylquinoline at 150 °C to give the chalcone **318**.[716] The quinoline-3-carboxaldehyde derivative **319** formed normally.[426]

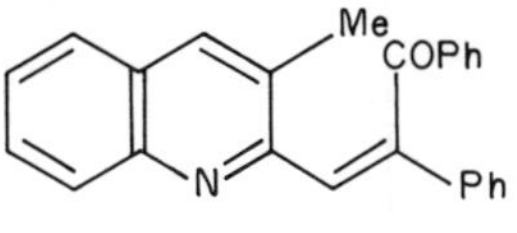

318, m.p. 248 °C

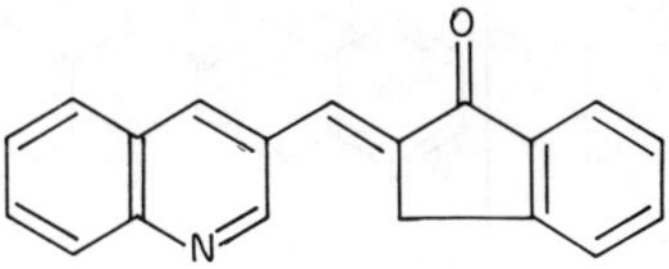

319, 80%

Other chalcones prepared in the standard way which do not fit into the various tables are **320a**[762], **320b**[392], **321**[444] and **322**.[763]

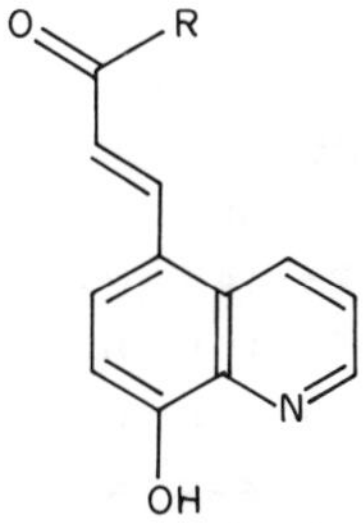

320

a, R = Ph, m.p. 172 °C
b, R = 4 − Py, m.p. 191 °C

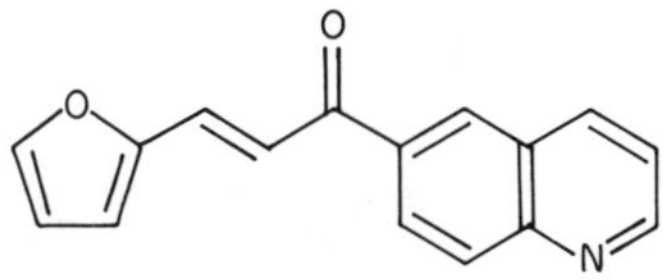

321, 90%, m.p. 100−102 °C

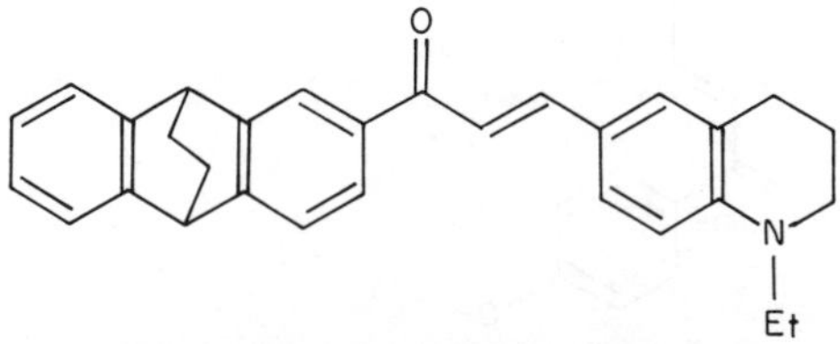

322, 64%, m.p. 70−72 °C

Acetophenone condensed with quinoline-6-carboxaldehyde in the presence of base to give chalcone **325a** (71%).[754] In a new reaction, methyl ketones have been shown to add to the Schiff base **323** under acid catalysis to give, it was assumed via intermediate **324**, the chalcones **325a–e** in the yields shown.[764]

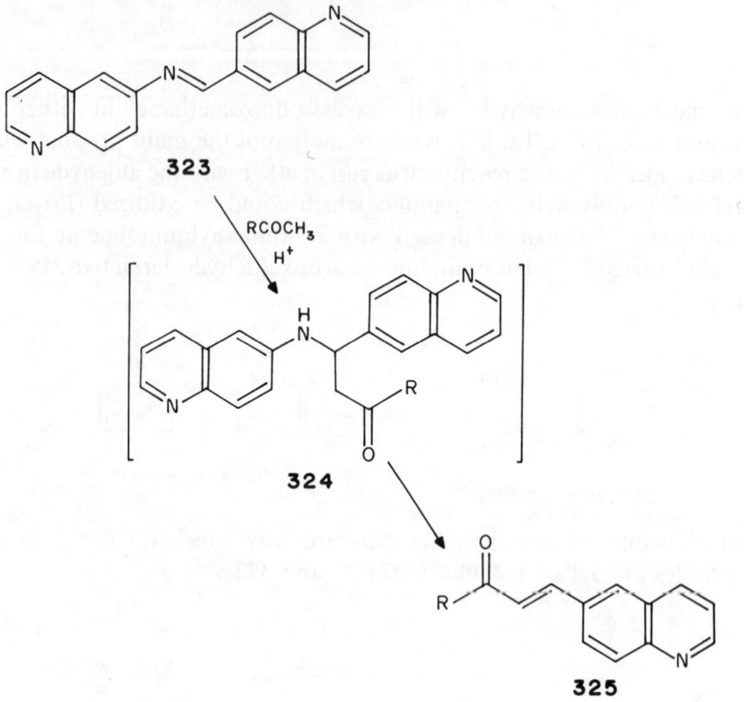

323

RCOCH₃
H⁺

324

325

R	%	M.p. (°C)
a Ph	23	145–146
b 4–MeC₆H₄	36	161–162
c 4–BrC₆H₄	18	199–200
d 4–NO₂C₆H₄	46	256–257
e 3–Py	41	184–185

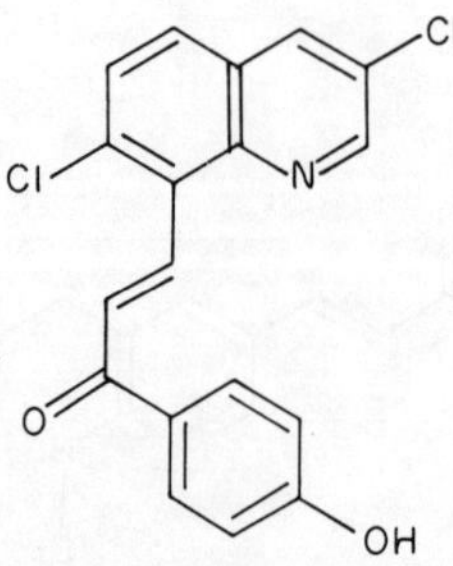

326, 70%, m.p. 230 °C

The chalcone **326** was prepared from the quinoline aldehyde and triphenylphosphonium-4-hydroxyacetophenone chloride in ethanol containing sodium ethoxide.[765]

Extensive studies have been made of the ultra-violet[766-771] and infra-red[767,769,771] absorption spectra and the dipole moments[772] of chalcones.

VII. Side Chain Ketone Groups Three or More Carbons Removed from the Quinoline Ring

Michael additions of acetophenone and propiophenone anions to 2-vinylquinoline gave ketones **327c** and **327d** respectively.[782,783] The reactions were catalysed by sodium in the absence of solvent. In a similar procedure the vinylquinoline was prepared *in situ* by addition of a quaternized Mannich base to a solution of an appropriate keto-ester in ethanolic sodium ethoxide to give keto-esters **327b** and **e**, hydrolysis of which gave ketones **327a** and **327c** respectively.[784]

327

	R^1	R^2	%	B.p.(°C/mm)	M.p.(°C)
a,	Me	H	33	180−200/0.01	
b,	Me	COOEt	44	185−196/0.4	
c,	Ph	H	64	170−190/0.1	66−68 Pic. 171−177
d,	Ph	Me		193−200/0.9	146−147
e,	Ph	COOEt	55	130−149/0.1	Pic.125−126

Michael addition of 2-methylquinoline to chalcone gave the ketone **328**.[785] Addition of 2-nitromethylquinolines to conjugated ketones in ethanol under the influence of triethylamine gave derivatives **329**.[786] Similarly, 2-phenacylquinoline added to benzalacetone; however, under the reaction conditions (EtONa) the product cyclized to give compound **330**. Elimination of water from compound **330** (85% H$_3$PO$_4$, 100 °C) gave the conjugated cyclohexenone (m.p. 149 °C). Reduction of the olefin bond then gave 3,5-diphenyl-4-(2-quinolyl)cyclohexanone (m.p. 200–203 °C).[663]

328, 60%, m.p. 115 °C

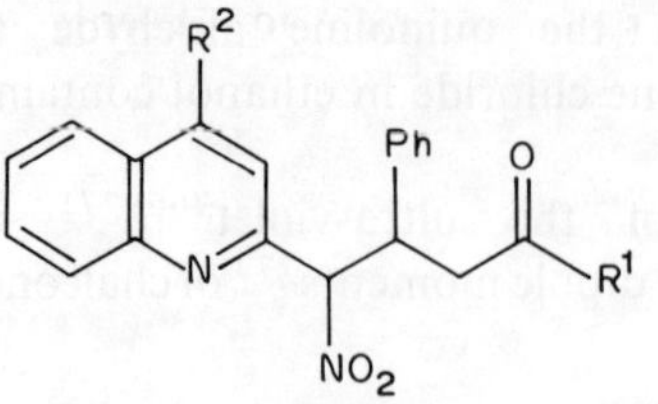

329

$R^1, R^2 = Me, m.p. 150-151 °C$
$R^1 = Ph, R^2 = H, 90\%, m.p. 163 °C$
$R^1 = Ph, R^2 = Me, 71\%, m.p. 151-152 °C$

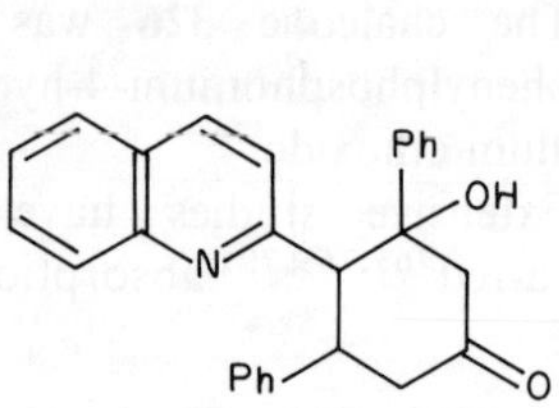

330, m.p. 203 °C

A mixture of 1-octene, acetone, 4-methylquinoline and silver nitrate was treated with sodium persulphate to give ketone **331** (45%), with 2-acetyl-4-methylquinoline (5%) as a by-product. Mechanisms to account for the formation of these products were suggested.[114]

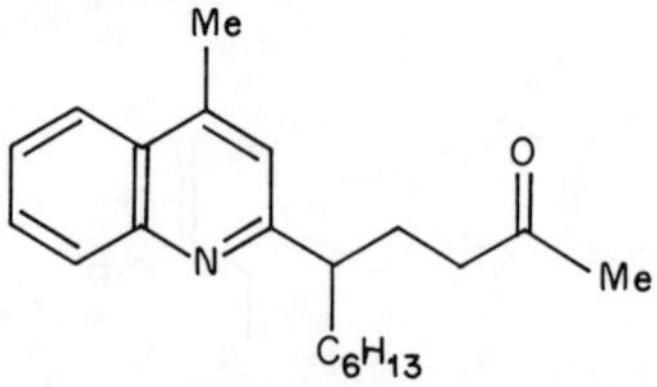

331

Heating 2,6-dimethylquinoline with p-benzoquinone led to the compound with suggested structure **332**.[716] The compound failed to eliminate water on recrystallization from acetic anhydride; cf. Section VI.1.

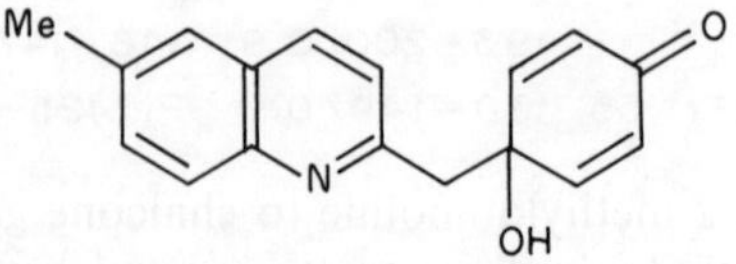

332, m.p. 137-139 °C

The reaction of quinoline N-oxide with 2-acetylcyclopentane-1, 3-dione gave the trione **333**, Table 20, which with ammonium acetate gave the enamine dione **334**.[557]

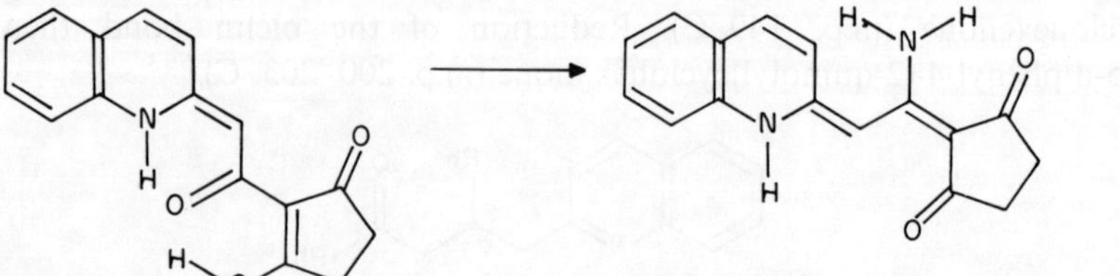

333

334, 66%, m.p. 185-186 °C

The reactions of Scheme 42 have been reported, but no experimental conditions or data are available.[787]

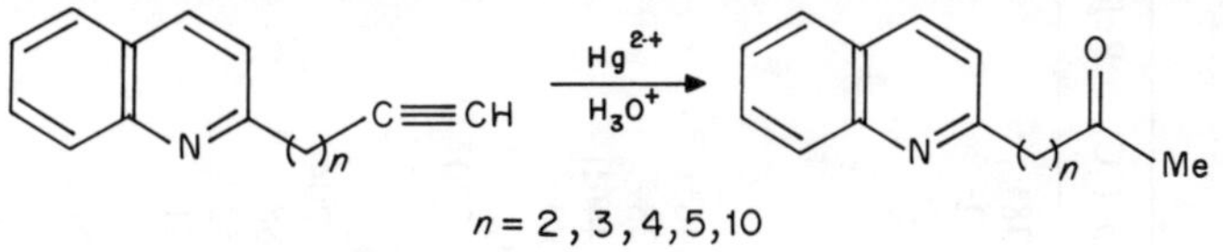

$n = 2, 3, 4, 5, 10$

SCHEME 42

Addition of 2,4-dimethylquinoline to chalcone in the presence of sodamide in liquid ammonia gave ketone **335**.[788]

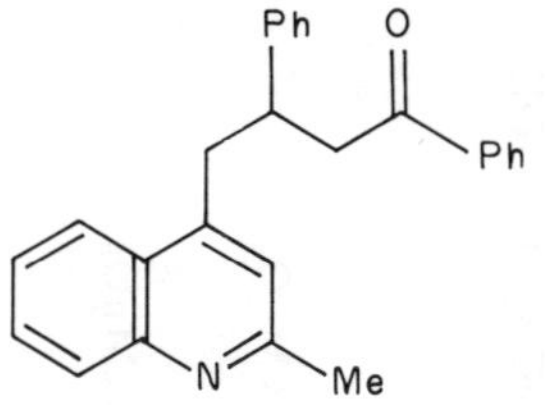

335, 80%, m.p. 107.5–109 °C

VIII. Ketoalkylenedihydroquinolines

1. 2-Ketomethylene-1,2-dihydroquinolines

Ketones of this type are collected in Table 35. General preparative routes are illustrated by the examples shown in Methods 1 to 4 (pp. 319–321). Methyl ketones in concentrated alkali attacked quinolinium ions at position 2 (0–20°C) to give, after permanganate oxidation, ketomethylene derivatives such as compound **336**. Warm dilute alkali was shown to cause attack at position 4 and mild oxidation then gave 4-ketomethylene derivatives, e.g. **337**. Pyrolysis of compounds **336** and **337** led to the known 2-acetonyl- and 4-acetonyl-quinolines, respectively.[582,789]

Acid chlorides and 1-alkyl-2-methylquinolinium ions react under basic conditions to give ketomethylene derivatives. The reaction is often carried out under the Schotten–Baumann procedure. The presumed intermediate **338** was isolated crude in benzene solution and subsequently treated with trifluoroacetic anhydride to give the trifluoroacetyl derivative **339**, Method 2.[790] The Schotten–Baumann procedure on 1-ethyl-2-methylquinolinium iodide was shown to give the enol ester **340**, which was assumed to be the result of O-acylation of the initially formed ketone **341**. Ester **340** was readily hydrolysed back to **341**.[791] This report corrected a much earlier one in which the structures were wrongly assigned.[792]

Table 35. 2-Ketomethylene-1,2-dihydroquinolines

R^1	R^2	R^3	Method	Reagents	Yield (%)	M.p. (°C)	References
Me	H	Me	3	1-Me-2-MeS-quinolinium Ms^-, $MeCOCH_2CO_2H$, Et_3N	76	137–138[a]	793
				1-Me-2-MeS-quinolinium Ms^-, $MeCOCH_2COCOOEt$, Et_3N	8	110–112[a]	812
							812
Me	H	Et	4		7[b]	145–146[a]	801
			3			137–138	793
						143–144(d)	813
Me	H	$CH_2C_6H_4$-2-Cl	1	$10M$-NaOH; $KMnO_4$	57	188–189 $HClO_4$ 219–220	582
Me	H	$CH_2C_6H_3$-2,4-diCl	1	$10M$-NaOH; $KMnO_4$	60	166–168 $HClO_4$ 203–204	582
Me	H	$CH_2C_6H_3$-3,4-diCl	1	$10M$-NaOH; $KMnO_4$	46	186–188	582
Me	H	$CH_2C_6H_4$-4-Cl	1	$10M$-NaOH; $KMnO_4$	60	150–152	582
Me	COMe	Me	3	1-Me-2-MeS-quinolinium I^-, Ac_2CH_2, NaH	88	139–141	794
Me	COOMe	Me	3	1-Me-2-MeS-quinolinium I^-, $MeCOCH_2COOMe$, NaH	98	160–161	794
Me	COPh	Me	3	1-Me-2-SO_3-quinolinium I^-, $PhCOCH_2COMe$, Et_3N	75.5	178–179	793
Me	COPh	Et	3	1-Et-2-SO_3-quinolinium I^-, $PhCOCH_2COMe$, Et_3N			793

R	R'	R''	n	Reagents	%	mp (°C)	Ref.
CF$_3$	H	Me	2	See text, Section VIII.1	57	187	790
CF$_3$	H	Et	2	TFAA, Py	63	205–206	814
CF$_3$	COCF$_3$	Me	2	TFAA, Py	47	193–194	815
CH$_2$Cl	H	Me	2	CCl$_3$COCl, Py	83	213	816
CCl$_3$	H	Me	2			170(d)	790
CH$_2$CN	H	Me		1-Me-2-ClCH$_2$COCH=1, 2-diH-Q, KCN		166	816
CH$_2$COPh	H	Me	2	PhCOCH$_2$COOEt	81	213–214	817
—(CH$_2$)$_3$CO—		Me	3	1-Me-2-MeS-quinolinium I$^-$, 1,3-cyclohexanedione, NaH		110–111	794
Ph	H	Me	2	PhCOCl, NaOH	38	112–114	818
Ph					38.5	107–108	554
Ph	H	Et	3		73	137–138	793
Ph	H	Et	3	PhCOCl, NaOH		137–139	793
Ph	H	CH$_2$C$_6$H$_4$-2-Cl	2	10m-NaOH; KMnO$_4$	33	186–187 / HBr 167–168	791
Ph	H	CH$_2$C$_6$H$_3$-2,4-diCl	1	10M-NaOH; KMnO$_4$	35	203–205 / HBr 190–191	582
Ph	H	CH$_2$C$_6$H$_3$-3,4-diCl	1	10M-NaOH; KMnO$_4$	35	199–201 / HBr 208–209	582
Ph	H	CH$_2$C$_6$H$_4$-4-Cl	1	10M-NaOH; KMnO$_4$	60	223–224.5	582
Ph	H	COPh	1	10m-NaOH; KMnO$_4$; 1,2-diH-1-PhCO-2-PhCOCH$_2$-Q, 2,2,6,6-tetramethylpiperidine N-oxide			819
Ph	Ph	Me	3	See text, Section VIII.1	91	218	796
Ph	COPh	Me	3	1-Me-2-SO$_3$-quinolinium I$^-$ PhCOCH$_2$COPh, Et$_3$N			793
Ph	N=NC$_6$H$_4$-4-NO$_2$	Me		1-Me-2-PhCOCH=1, 2-diH-Q, 4-NO$_2$C$_6$H$_4$N$_2$CL		224–225	820

Table 35. (*Contd.*)

R¹	R²	R³	Method	Reagents	Yield (%)	M.p. (°C)	References
C_6H_4-4-CCl_3	Me	H					821
C_6H_4-4-COOtBu	Et	H					822
C_6H_3-3,5-diNO_2	Me	H		1-Me-2-quinolylidene-acetophenone, $3,5$-di$NO_2.C_6H_3COCl$		320–321(d)	823
2-Benzothiazolyl	$COCH_2OEt$	Et	3	1-Et-2-EtS-quinolinium I^-, Benzothiazolyl-2-$COCH_2COCH_2OEt$		165	824

[a] Inconsistent literature melting points.
[b] Formed as a mixture with the 4-acetonylidene derivative, Table 36.

METHOD 1

METHOD 2

340   **341**

Anions from 1,3-diketones or 3-keto-esters have been shown to displace methylthio or sulphonic acid groups from position 2 of quinolinium salts as illustrated in Method 3.[793] In some examples diketones **342** were hydrolysed to simple ketomethylene derivatives **343**.[794] In one development of method 3, leaving groups in both reactants were employed to give the betaines **344**, Scheme 43;[795] in another, an intramolecular displacement occurred, **345** → **346**. The structure of compound **346** was confirmed by a conventional synthesis (Method 3) from 1-methyl-2-methylthioquinolinium tosylate and 1,2-diphenyl-ethanone.[796]

342

343

METHOD 3

In a recently discovered synthesis, N-methylquinolinium iodide **347** reacted with the Janovsky complex **348** at room temperature to give a mixture containing the ketomethylene derivative **349** (7%) and the isomer **350** (8%); Method 4. The Janovsky complex clearly releases an acetone anion, which attacks the salt **347** to give initially the dihydro derivatives of compounds **349** and **350**. Indeed, an earlier note from the same group[800] mentioned the isolation of the former of these. In the full paper it was suggested that the dinitrobenzene by-product was responsible for the subsequent oxidation.[801]

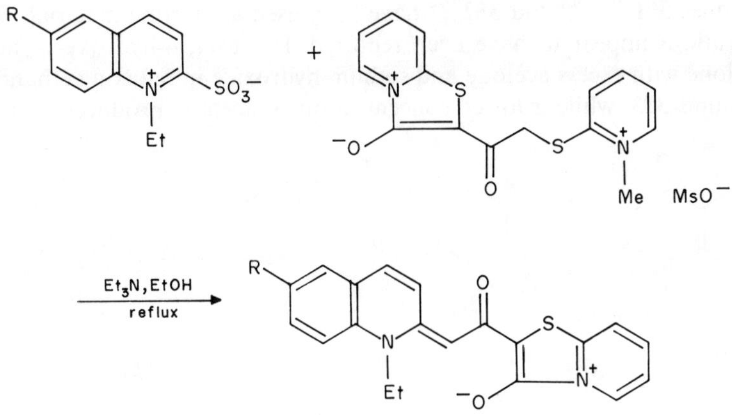

344, R=Me, m.p. 267 °C
R=OMe, m.p. 257 °C

SCHEME 43

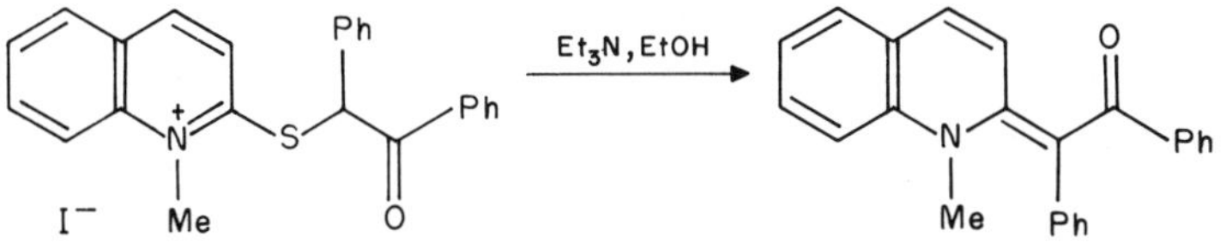

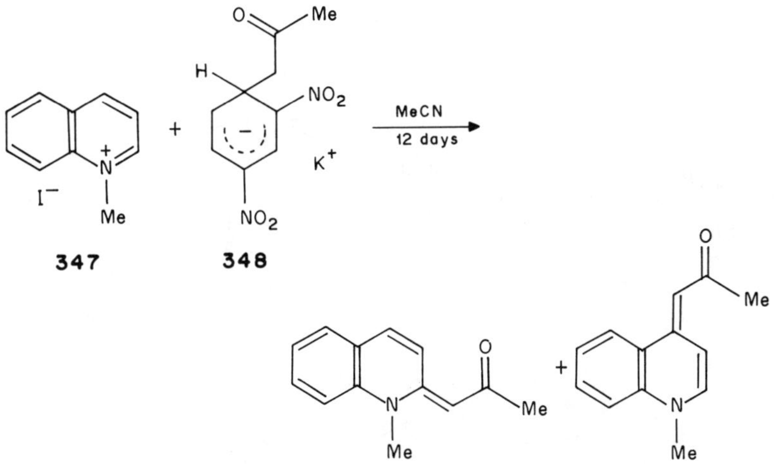

METHOD 4

Ketones **351**[797,798] and **352**[799] have been used as starting materials, but no preparations appear to have been reported. However, 4-hydroxy-1-phenyl-2-quinolone with excess acetone and sodium hydroxide in refluxing ethanol gave compound **353**, while a lower concentration of acetone produced compound **354**.[802]

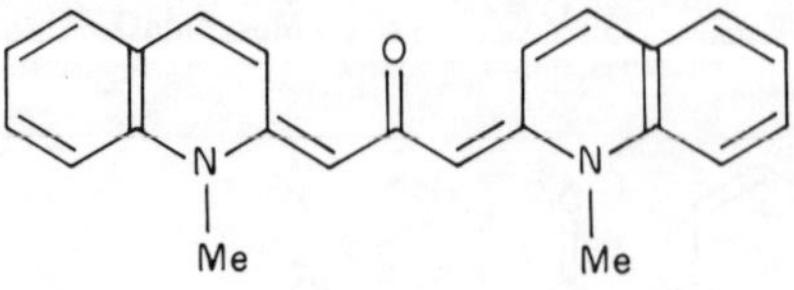

351

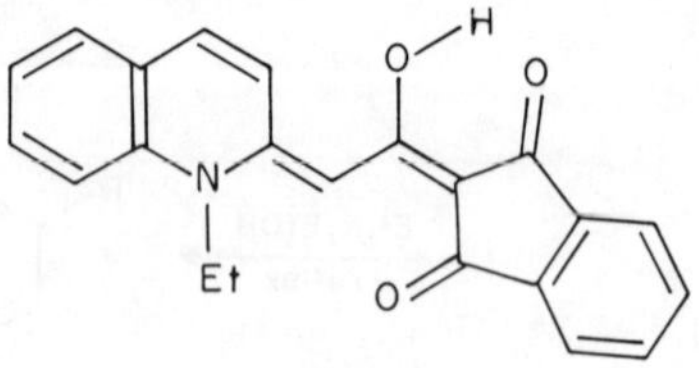

352

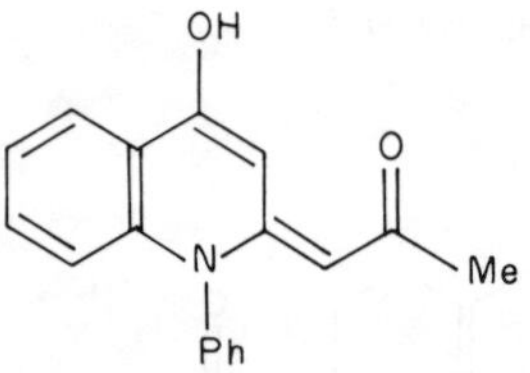

353, 68.5%, m.p. 264 °C(d) 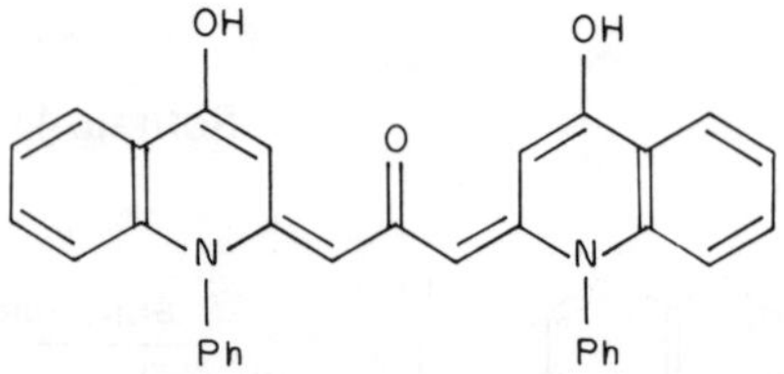   **354**, 77.1%, m.p. 242 °C(d)

2. Other Ketoalkylene-1,2-dihydroquinolines

Base catalysed addition of ketones to the vinylogous amidinium salt **355** gave dienones **356** and **357**.[803]

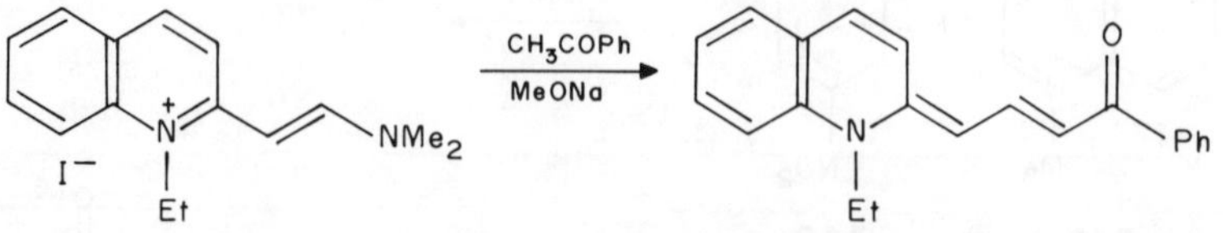

355   **356**, 57%, m.p. 156−157 °C

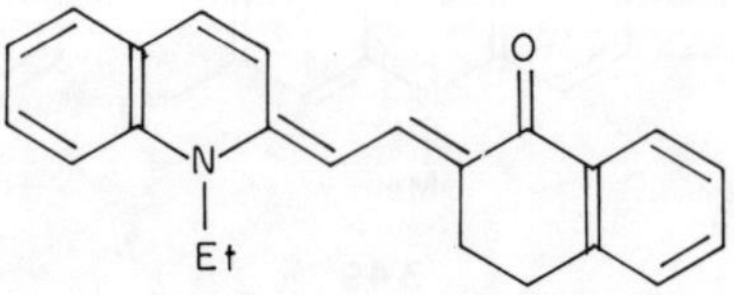

357, 49%, m.p. 150−151 °C

Active methylene compounds **359** and triethyl orthoformate reacted with the salt **358** to give ketones **360a** and **b** (Scheme 44) Compound **360a** was also obtained from the salt **358** and ethyl 2-formylacetoacetate. The sodium enolates **361** transformed the salt **358** into the ketones **360c** and **d**.[804]

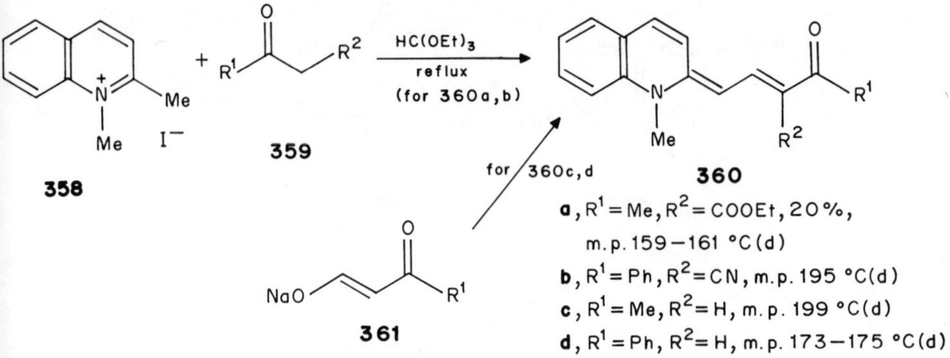
HC(OEt)3
reflux
(for 360a,b)
for 360c,d
358
359
360
361
a, R1 = Me, R2 = COOEt, 20%, m.p. 159–161 °C (d)
b, R1 = Ph, R2 = CN, m.p. 195 °C (d)
c, R1 = Me, R2 = H, m.p. 199 °C (d)
d, R1 = Ph, R2 = H, m.p. 173–175 °C (d)

SCHEME 44

A series of dyes was prepared by the method shown in Scheme 45.[805]

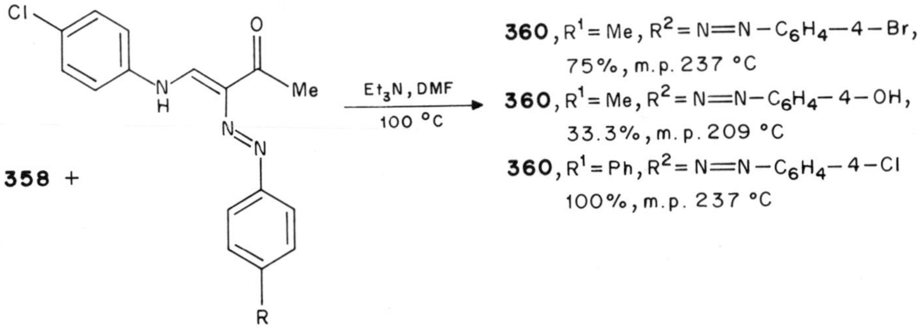
358 +
Et3N, DMF
100 °C
360, R1 = Me, R2 = N=N–C6H4–4–Br, 75%, m.p. 237 °C
360, R1 = Me, R2 = N=N–C6H4–4–OH, 33.3%, m.p. 209 °C
360, R1 = Ph, R2 = N=N–C6H4–4–Cl 100%, m.p. 237 °C

SCHEME 45

Nucleophilic addition to the salt **362** gave the ketones **363**.[806]

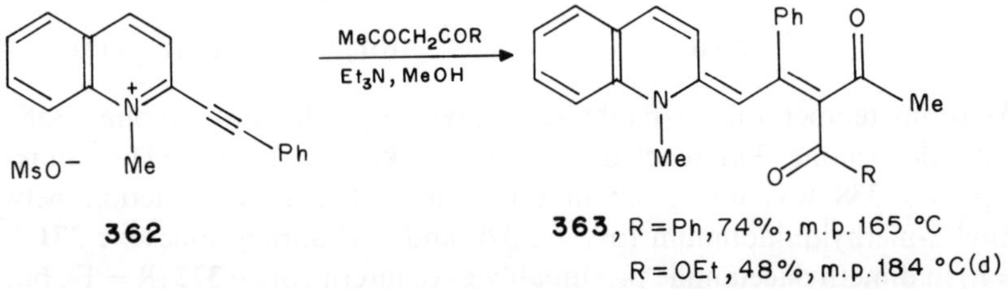
MeCOCH2COR
Et3N, MeOH
362
363, R = Ph, 74%, m.p. 165 °C
R = OEt, 48%, m.p. 184 °C (d)

In a development from Method 2, p. 319, the 2-methylquinolinium salts **364** reacted with ethoxymethylenediones **365** to give the conjugated ketones **366**.[807] Similarly, 1,2-dimethylquinolinium methanesulphonate reacted with chromone in refluxing ethanol containing sodium acetate to give compound **367**.[808]

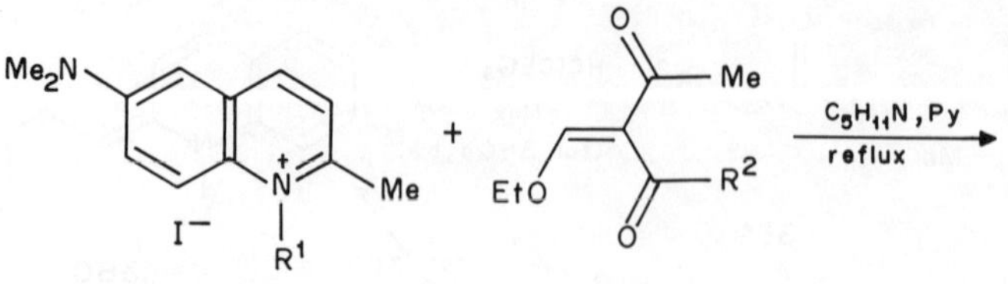

364   **365**

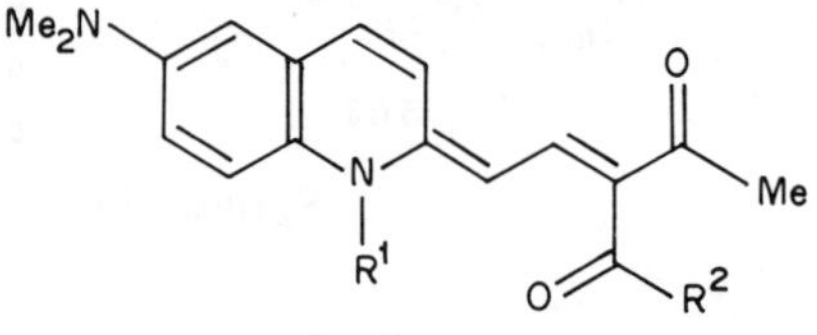

366, R^1, R^2 = Me, m.p. 267 °C
    R^1 = Me, R^2 = OEt, m.p. 261 °C
    R^1 = Et, R^2 = OEt, m.p. 172 °C

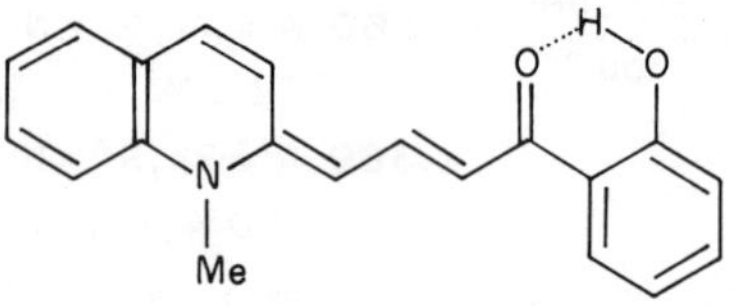

367, 46%, m.p. 249 °C

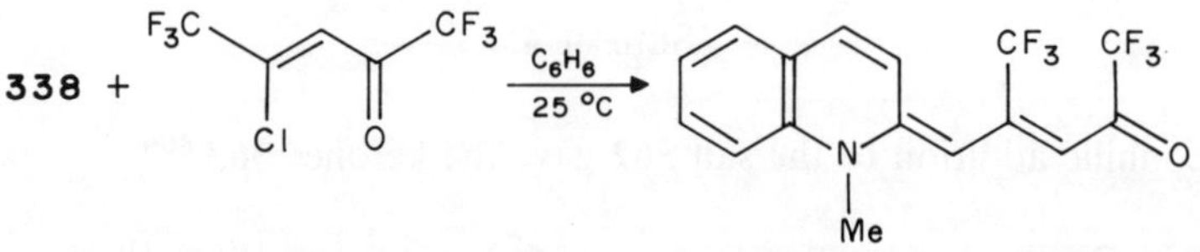

338 +

368   **369**, 55%, m.p. 170–171 °C

At room temperature, 1-methyl-2-methylene-1,2-dihydroquinoline **338** and the 3-chloroenone **368** reacted to give the ketone **369**.[809] (The source of compound **338** was not given in this paper.) The initial reaction between 1-ethyl-2-methylquinolinium tosylate **370** and perfluorocyclohexene **371** (with R = F) in dimethylacetamide presumably gave intermediate **372** (R = F), but the allylic fluorine atoms were hydrolysed during aqueous workup to give the ketone **373a**. Ketones **373b** and **373c** were prepared from 1,2-dichloro-octafluoro-cyclohexene and 1-chloro-2-cyano-octafluorocyclohexene respectively.[809]

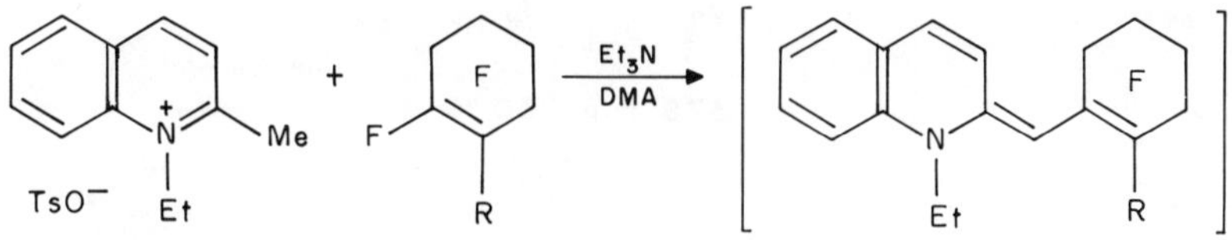

370   **371**   **372**

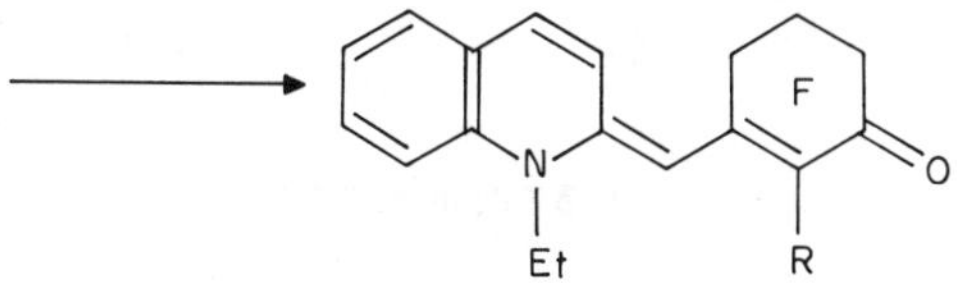

373

a, R = F, 2.5%, m.p. 156 −158 °C
b, R = Cl, 2.9%, m.p. 178 −179 °C
c, R = CN, 13.5%, m.p. 300 °C

Benzoylmalondialdehyde reacted with 1-ethyl-2-methylquinolinium tetra-fluoroborate in acetic anhydride with sodium acetate to give the dye **374**.[810]

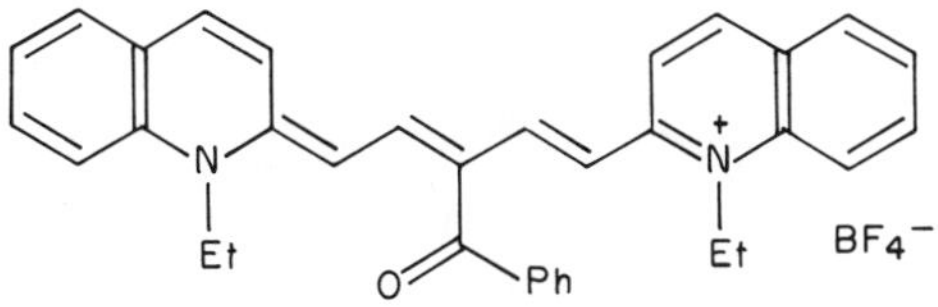

374, 5%, m.p. 245 −246 °C

The sidechain nitrogen of compound **375** was displaced by dimedone to give the dione **376**. Compound **377** was made similarly.[807]

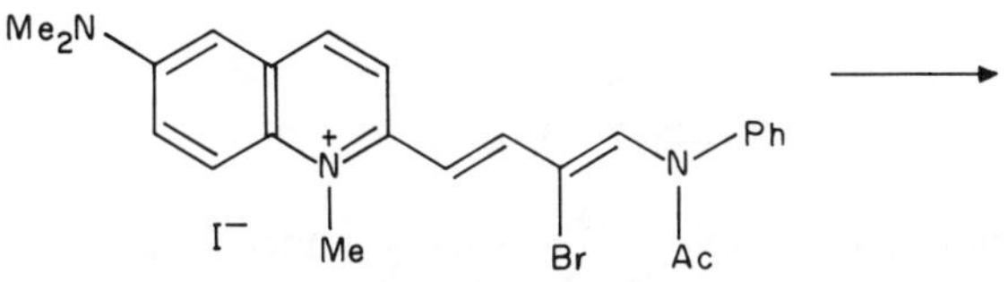

375

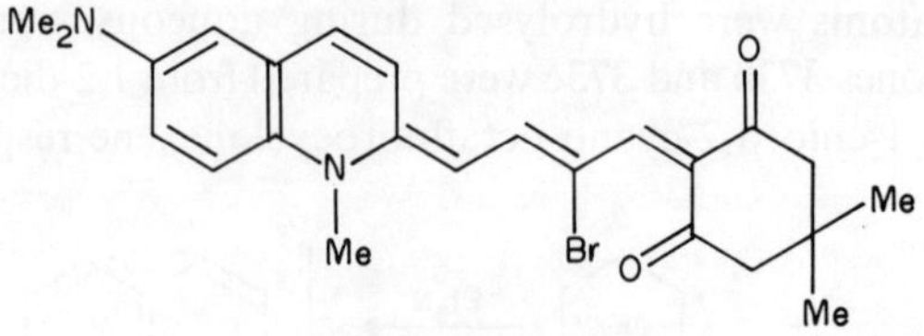

376, m.p. 230 °C

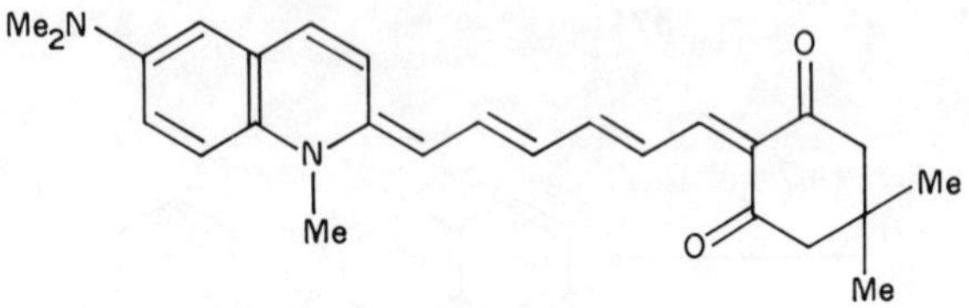

377, m.p. 237 °C

3. 4-Ketomethylene-1,4-dihydroquinolines

Most of the known examples of this class have been made by methods described above; see Table 36.

In the manner of Scheme 45 (p. 323), 1,4-dimethylquinolinium perchlorate reacted to give the dyes **378**.[805]

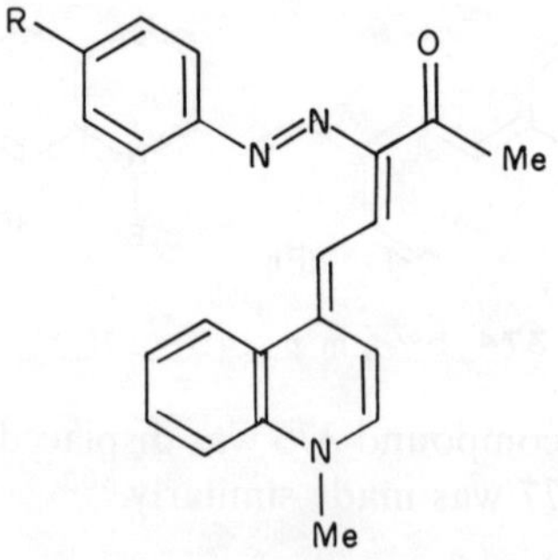

378

R = Cl, 60.5%, m.p. 212—213 °C
R = NO₂, 74.8%, m.p. 226 °C

Phenacylpyridinium bromide **379** reacted under base catalysis with the quinolinium derivative **380**. Displacement of the acetanilide group gave the dye **381**.[811]

Table 36. 4-Ketomethylene-1,4-dihydroquinolines

R^1	R^2	R^3	Method*	Reagents	Yield (%)	M.p. (°C)	References
Me	H	Me	4		8[a]	148–149 138–143(d)	801 813
Me	H	CH$_2$CH=CHPh	1	aq. NaOH, 80 °C	56	170–173	582
Me	H	CH$_2$C$_6$H$_3$-2,4-diCl	1	aq. NaOH, 80 °C	72	203–204 HClO$_4$ 189–190	582
Me	H	CH$_2$C$_6$H$_3$-3,4-diCl	1	aq. NaOH, 80 °C	66	188–189	582
Me	H	CH$_2$C$_6$H$_4$-4Cl	1	aq. NaOH, 80 °C	57	180–182 HClO$_4$ 130–132	582
Ph	H	Me	2	PhCOCl, NaOH	56	141–143(d)	554
Ph	H	CH$_2$Ph	1	aq. NaOH, 80 °C	67	176–177	582
Ph	H	CH$_2$CH$_2$Ph	1	aq. NaOH, 80 °C	61	164–165	582
Ph	H	CH$_2$C$_6$H$_4$-2-Cl	1	aq. NaOH, 80 °C		214–216 HBr 204–205	582
Ph	H	CH$_2$C$_6$H$_3$-2,4-diCl	1	aq. NaOH, 80 °C	76	176–178 HBr 184–186	582
Ph	H	CH$_2$C$_6$H$_4$-3-Cl	1	aq. NaOH, 80 °C	50	157–159 HBr 189–190.5	582
Ph	H	CH$_2$C$_6$H$_3$-3,4-diCl	1	aq. NaOH, 80 °C	50	159–161	582
Ph	H	CH$_2$C$_6$H$_4$-4-Cl	1	aq. NaOH, 80 °C	60	153–154 HBr 196–198	582
Ph	NO$_2$	Me	3	1-Me-4-MeS-quinolinium Ts$^-$, PhCOCH$_2$NO$_2$, Et$_3$N		170	825

* Methods given in Section VIII.1.

[a] Formed as a mixture with the 2-acetonylidene derivative; Table 35.

379   380   381, 40%, m.p. 195–196 °C(d)

IX. Partially Saturated Quinolines Carrying Ketone Groups at Position 1

A mixture of quinoline, dimethylaniline and 3-chloro-1-phenylpropenone reacted to give the ketone **382a**. Other electron rich aromatic systems could replace the dimethylaniline to give compounds **382b–d**.[826]

382

a, R = 4 – Dimethylaminophenyl, 58%, m.p. 231 – 232 °C
b, R = 3 – Indolyl, 50%, m.p. 195 – 196 °C
c, R = 1 – Methyl – 3 – indolyl, 40%, m.p. 186 – 187 °C
d, R = 1 – Methyl – 1, 2, 3, 4 – tetrahydro – 6 – quinolyl,
            70%, m.p. 222 – 223 °C

Ketone **383** appears in several patents as a photographic dye, but no preparative details are available.[828–830] A solution of quinoline and 3-butyn-2-one in nitromethane was assumed to form the intermediate **384**. Deprotonation of the solvent by the ylid led to attack at position 4 and isolation of ketone **385** in low yield.[827]

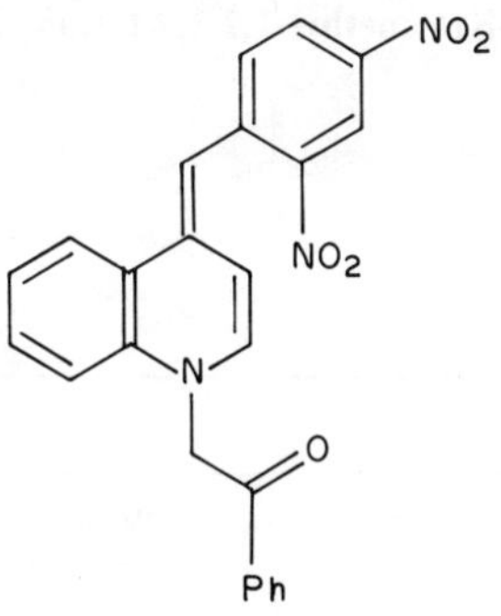

383

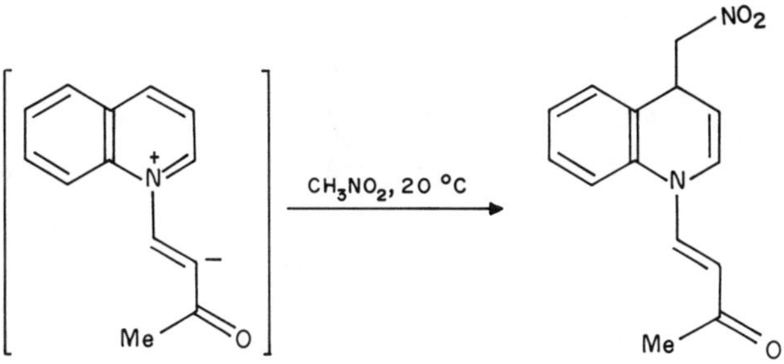

CH₃NO₂, 20 °C

384   **385**

2%, m.p. 112—114.5 °C

Appropriate chloro or bromo ketones with 1,2,3,4-tetrahydroquinoline gave the compounds of Table 37.

Benzoin and tetrahydroquinoline were heated in the presence of phosphorus pentoxide to give ketone **386**.[883] When the substituted butanone **387** was treated with 1,2,3,4-tetrahydroquinoline in ethanol for two hours at room temperature it gave the rearranged product **389**. It was assumed that the bromo amine gave the aziridinium ion **388** as it slowly dissolved, and subsequent attack by the secondary base occurred at the aziridinium ring carbon 3. Ketones prepared by similar procedures are in Table 38.

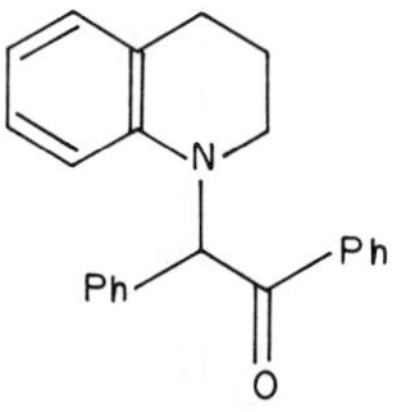

386

Table 37. 1-Ketomethyl-1,2,3,4-tetrahydroquinolines

R	M.p. (°C)	References
Me	39–41; b.p. 178/17 mm	831
2-Oxocyclohexyl	DNP 102–103	832
Ph	101–103	831
4-MeC$_6$H$_4$	90	833
4-FC$_6$H$_4$	94	833
4-ClC$_6$H$_4$	103	833
	106–108	839
4-BrC$_6$H$_4$	124	833
4-IC$_6$H$_4$	160	833
4-MeOC$_6$H$_4$	114	833
3-NO$_2$C$_6$H$_4$	136	833
2,3,4-triHOC$_6$H$_2$		834
3-Fluorenyl	167–169	835
1,2-Dihydro-5-acenaphthenyl	160–161	836
1,2,3,4-Tetrahydro-9-phenanthrenyl	149.5–150.5	837
3,4,5-TriMeOC$_6$H$_2$CONH-4-C$_6$H$_4$		838

Table 38. 1-(3-Oxopropyl)-1,2,3,4-tetrahydroquinolines

R^1	R^2	R^3	Yield (%)	M.p. (°C)	References
Me	Piperidino	H	48.5	126–127	840, 841
Me	Piperidino	OMe	39	124	842
Me	Morpholino	H	51	173	843
Me	Morpholino	OMe	40	126	842
Me	1,2,3,4-Tetrahydro-2-iso-quinolyl	H	43.7	107–109	840
Ph	Benzylmethylamino	H	45.5	150–153	841
Ph	Piperidino	OMe	85	160	842
Ph	Morpholino	OMe	68	143	842
Ph	1,2,3,4-Tetrahydro-2-iso-quinolyl	H	47	164	843

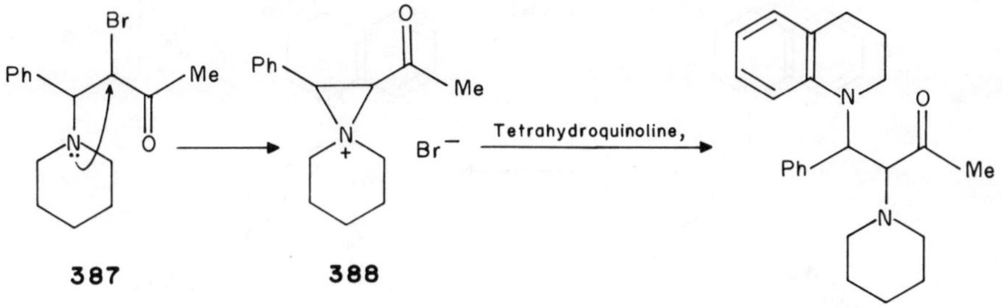

387   **388**

389

Acetonedicarboxylic acid reacted with dimethylformamide dimethylacetal followed by tetrahydroquinoline to give the ketone **390**.[845]

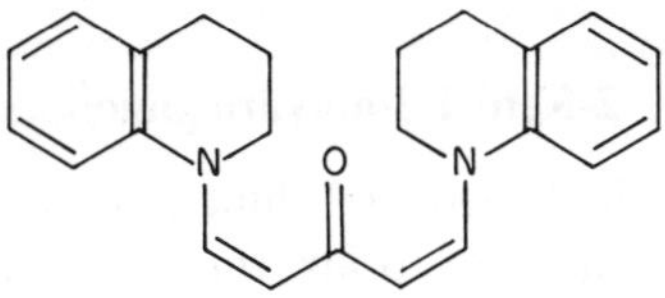

390

The enol ether **391** was heated in pyridine to give the unsaturated ketone **392**.[846]

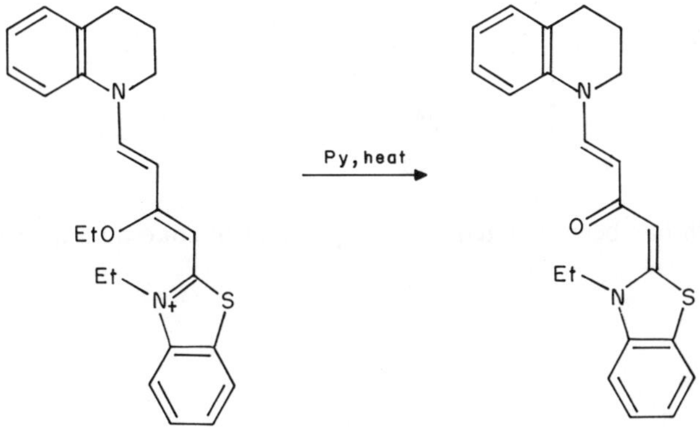

391   **392**, 75%, m.p. 195—196 °C

Tetrahydroquinoline and 4-chlorobutyronitrile gave compound **393**, which reacted with 1,3,5-trimethoxybenzene in the presence of hydrogen chloride to give the ketone **394**.[844]

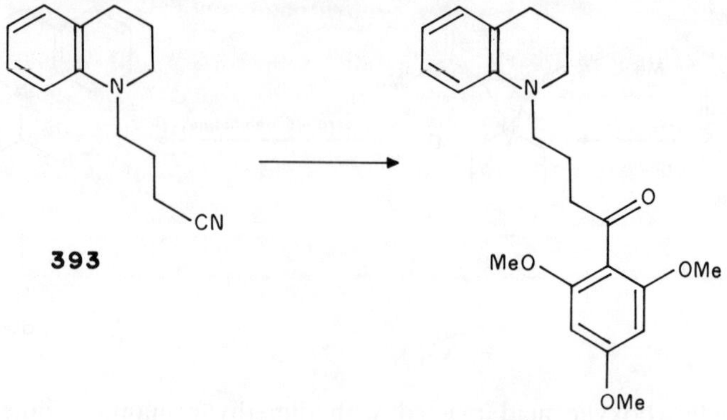

393

394

X. Quinoline Ketones with Partially Reduced Pyridine Rings

1. 2-Keto-1,2-dihydroquinolines

With aqueous potassium hydroxide 1,3-dimethyl-2-ethylquinolinium chloride gave the olefine **395**. On exposure to air, this was converted, via the epoxide **396**, into the unstable ketone **397** (10%). Although the ketone **397** was isolated crude, further exposure to the air caused hydrolysis to acetaldehyde and 1,3-dimethyl-2-quinolone.[847]

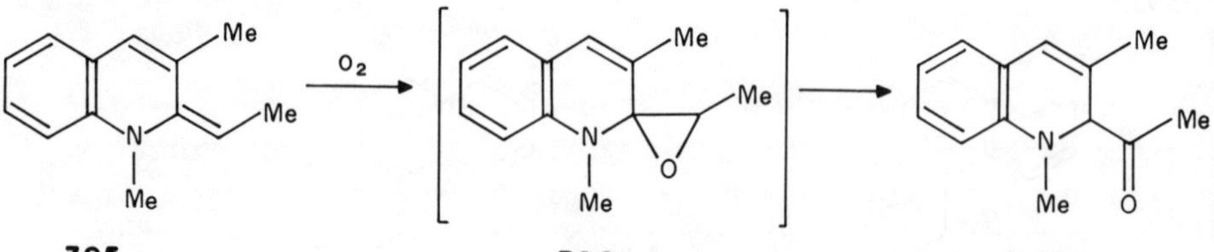

395   **396**   **397**

Compound **398** has been isolated as a by-product of coke manufacture.[905]

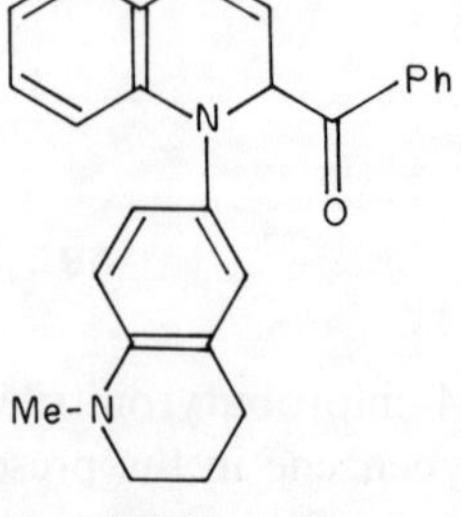

398

2. 3-Keto-1,2-dihydroquinolines

Attempted alkaline hydrolysis of the chloral derivative **399** gave a mixture of 3-(2-quinolyl)acrylic acid and 35–45% of the keto-acid **400a** (Scheme 46). The derivatives **400b–f** were prepared from the keto-acid **400a** by standard methods.[174] The *N*-ethyl derivative **400g** was prepared by treatment of compound **400c** with triethyloxonium tetrafluoroborate.[848]

	R^1	R^2	M.p. (°C)
a,	H	H	123–125 (d)
b,	H	PhCO	198–199 (d)
c,	Me	H	140–141
d,	Me	PhCO	140–141
e,	Et	H	110.5–111.5
f,	Et	Ac	131–132
g,	Me	Et	101–102 (88%)

SCHEME 46

The acid **400a** was resolved with (−)-brucine as part of the investigation of its structure. Oxidation of its sodium salt with potassium permanganate, chromic acid or by exposure to sunlight gave 3-acetylquinoline, sometimes accompanied by 3-acetylquinoline-2-carboxylic acid. Oxidation of the ester **400e** by potassium permanganate in pyridine gave ethyl 3-acetylquinoline-2-carboxylate; Table 2. The hydrolysis of compound **399** was first investigated at the end of the nineteenth century by Einhorn[849], but he thought that the product was 3-(2-quinolyl)lactic acid. Further, he believed the oxidation product to be 2-quinolylacetaldehyde. The true structures were only established after a painstaking investigation by Woodward and Kornfeld[174] completed in 1948.

Pyrolysis of the acid **400a** gave a mixture from which 3-acetylquinoline (36%) and its 1,4-dihydro-derivative **401** (8%) and 1,2,3,4-tetrahydro-derivative **402a** (13%) were isolated. The ketone **402a** decomposed rapidly.[70,174] Compound **401** could also be prepared by Raney nickel catalysed hydrogenation of 3-acetylquinoline, and was oxidized back to 3-acetylquinoline by potassium permanganate.[174,850] Reduction of 3-acetylquinoline with triethylammonium

$$400a \xrightarrow{200\ °C}$$

401

402 a, R = H
b, R = CHO

formate at 165–170 °C gave a mixture which included the ketones **401** (32%) and **402b** (1.7%). Formic acid at 140° converted compound **401** into the N-formyl derivative **402b** (66%).[850] Another preparation of compound **401** employed reduction of 3-acetylquinoline oxime and hydrolysis. It was noted that compound **401** failed to react with p-nitrophenylhydrazine.[851] The ester **400c** gave amide **403a** with methanolic ammonia at room temperature. The benzoyl derivative **400d** only reacted at 100 °C, when it debenzoylated and also gave the amide **403a**. Benzoyl chloride in triethylamine converted compound **403a** into the N-benzoyl derivative **403b**.[852]

403

a, R = H, 63.5%, m.p. 181.5 – 182.5 °C
b, R = PhCO, 15.5%, m.p. 230 – 231 °C

A careful study using ^{14}C labelling of C-2 and C-1′ of compound **399** showed that its base catalysed rearrangement involved a ring opening, ring closure procedure. C-1′ migrated into the nucleus and C-2 became the ketone carbonyl carbon of ketone **400a**.[853] The results disproved the mechanism proposed by Woodward and Kornfeld[174] but failed to distinguish between two other possibilities.

Michael addition of methyl vinyl ketone to the keto-amines **404** was followed by ring closure to give the hydroxy-ketones **405**; see Chapter 1, Section IV.1. Alcohol **405a** was dehydrated to compound **406a** in refluxing ethanolic sodium ethoxide and compound **405b** gave compound **406b** with potassium t-butoxide. Compound **406a** was oxidized by manganese dioxide at room temperature to 3-acetyl-6-chloro-4-phenylquinoline, Table 2.[200]

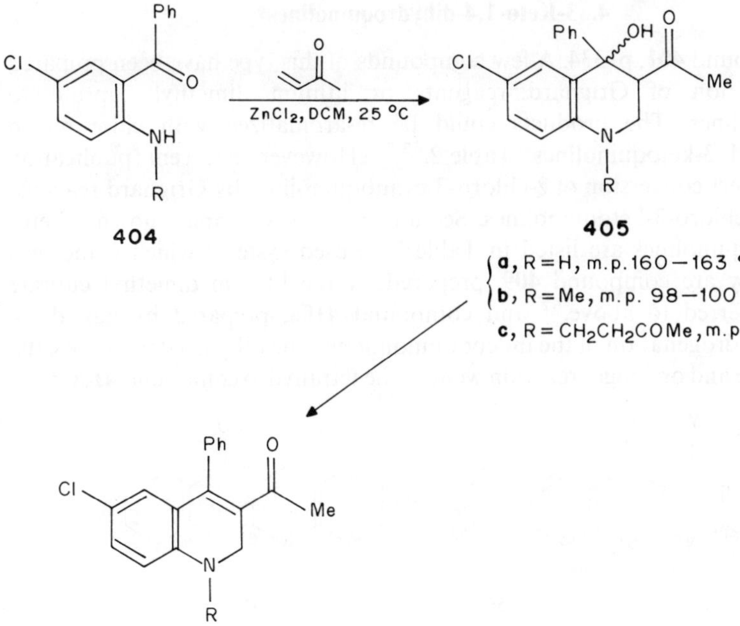

404

405

a, R = H, m.p. 160−163 °C
b, R = Me, m.p. 98−100 °C
c, R = CH₂CH₂COMe, m.p. 120−123 °C

406

a, R = H, m.p. 135−138 °C
b, R = Me, m.p. 82−85 °C

3. 6-Keto-1,2-dihydroquinolines

The reaction of formaldehyde with 2,2,4-trimethyl-1,2-dihydroquinoline gave a complex mixture from which small amounts of ketones **407** and **408** were isolated. Compound **408** was assumed to arise from an initially formed aldehyde (not detected) which condensed with the butanone used to elute the alumina column.[854]

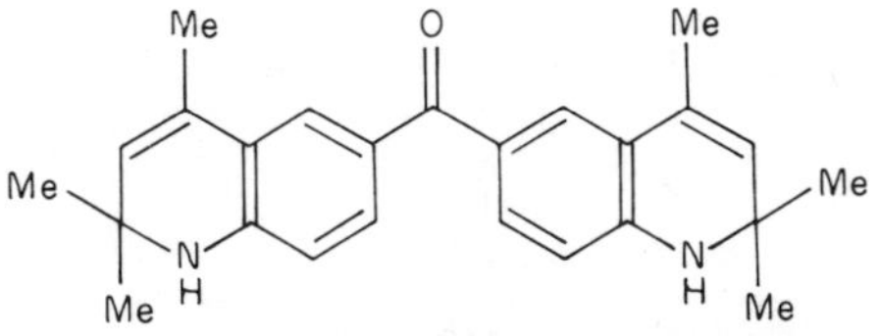

407, m.p. 229−231 °C

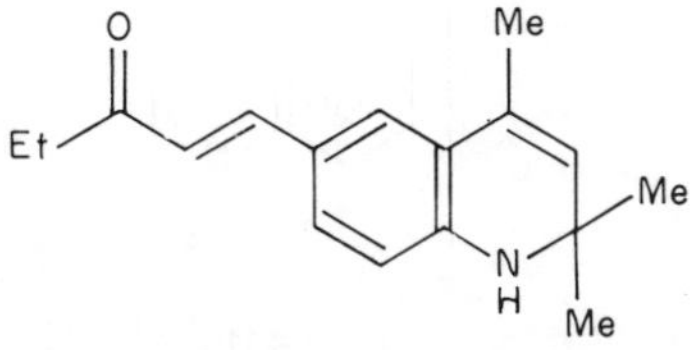

408, m.p. 142−144 °C

4. 3-Keto-1,4-dihydroquinolines

Note compound **401**, p. 334. A few compounds of this type have been prepared by 1,4-addition of Grignard reagents or lithium dimethyl cuprate to 3-ketoquinolines. The products could be re-aromatized with chloranil to 4-substituted 3-ketoquinolines, Table 2.[62,71] However, a recent publication reported direct conversion of 2-chloro-3-cyanoquinolines by Grignard reagents to 4-aryl-2-chloro-3-ketoquinolines; Section II.3.[72] The simple known 3-keto-1,4-dihydroquinolines are listed in Table 39. Fused systems which come into this category are compound **409**, prepared by the lithium dimethyl cuprate addition referred to above,[62] and compound **410a**, prepared by palladium catalysed hydrogenation of the indenoquinolinone. The dihydro derivative **410a** was unstable and on longer reaction went to the tetrahydro compound **410b**.[227]

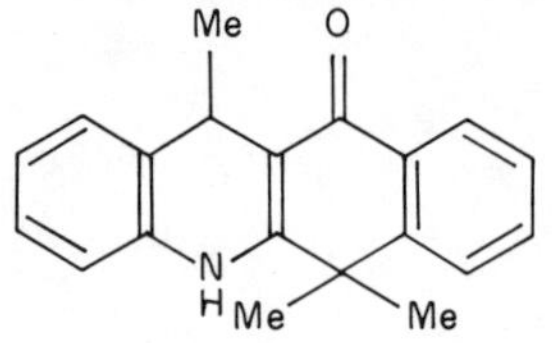

409, 89%, m.p. 255–256 °C

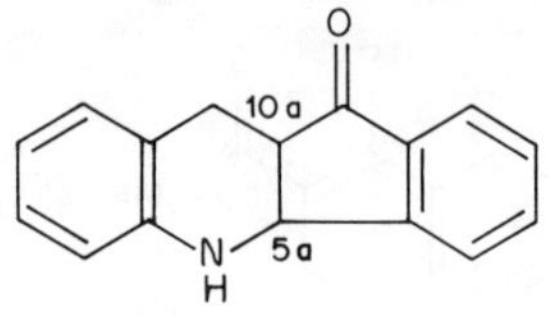

410

a, 5a–10a double bond,
18.1%, m.p. 215–221 °C

b, 19%, m.p. 140–143 °C

Quinolinium salts reacted with isonitriles and carboxylic acid salts in methanol at room temperature over several weeks. The initial products were assumed to have structures **411**, which underwent spontaneous O–C acyl migration to give the keto-amides **412**; Scheme 47.[848]

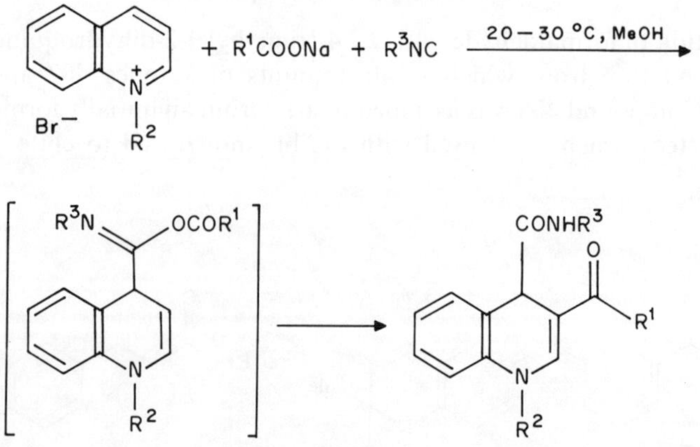

411

412

SCHEME 47

Table 39. 1,4-Dihydro-3-ketoquinolines

R^1	R^2	R^3	Other	Preparation	Yield (%)	M.p. (°C)	References
Me	H	H		3-Ac-Q, H_2, Ni	56	187	850
Me	H	H		3-Ac-Q, Et_3N, HCOOH	32		850
Me	H	H		1,2-diH-2-COOH-3-Ac-Q, 200 °C	8[a]	185–186	70
Me	H	COOH		1-MeCOCH_2CH_2-isatin, NaOH		238–240	889
Me	H	COOH	6-Me	1-MeCOCH_2CH_2-5-Me-isatin, NaOH		202–204	889
Me	H	COOH	8-Me	1-MeCOCH_2CH_2-7-Me-isatin, NaOH		230–232	889
Me	Me	CONHiPr		Me-quinolinium Br^-, iPrNC, AcONa	13	224–225	848, 890, 891
Me	Et	COOMe		1-Et-3-Ac-4-CN-1,4-diH-Q, HCl, MeOH	31	233–235	848
Me	Et	CONHiPr		Et-quinolinium Br^-, iPrNC, AcONa	24	208–209	848
Me	Et	CONHtBu		Et-quinolinium Br^-, tBuNC, AcONa	10	184–185	848, 891
Me	Et	CONHcyclo-C_6H_{11}		Et-quinolinium Br^-, cyclo-C_6H_{11}NC, AcONa	29	201–202	848
Me	Et	CONHcyclo-C_6H_{11}		1-Et-3-Ac-4-COOMe-1,4-diH-Q, MeONa; ClCOOMe; cyclo-$C_6H_{11}NH_2$	15	201–202	848
Me	Et	CONHC_6H_3-2,6-diMe		Et-quinolinium Br^-, 2,6-$Me_2C_6H_3$NC, AcONa	23	227–228	848, 890, 891
Me	Et	CN		1-Et-3-Ac-quinolinium Br^-, KCN, 20 °C	60	168–170	848
Me	$PhCH_2$	CONHiPr		1-$PhCH_2$-quinolinium Br^-, iPrNC, AcONa	40	206–208	848

Table 39. (*Contd.*)

R^1	R^2	R^3	Other	Preparation	Yield (%)	M.p. (°C)	References
Me	$PhCH_2$	$CONHtBu$		1-$PhCH_2$-quinolinium Br^-, tBuNC, AcONa	29	206–208	848
Me	$PhCH_2$	CONHcyclo-C_6H_{11}		1-$PhCH_2$-quinolinium Br^-, cyclo-C_6H_{11}NC, AcONa	57	229–230	848
CH_2Cl	$4-NO_2C_6H_4CH_2$	$CONHC_6H_3$-2,4-diOMe		1-(4-$NO_2C_6H_4CH_2$)-quinolinium Br^-, 2,4-diMeOC_6H_4NC, $ClCH_2$COONa		238–242	890, 891
COOMe	H	OMe	2-Me	1-HO-2-Me-3-COCOOMe-4-MeO-1,4-diH-Q, Zn, AcOH		231–232	892
				1-AcO-2-Me-3-COCOOMe-4-MeO-1,4-diH-Q, H_2, Pd			892
Et	$PhCH_2$	CONHEt		1-$PhCH_2$-quinolinium Br^-, EtNC, EtCOONa	37	203–205	848
Ph	H	Me		3-PhCO-Q, Me_2CuLi	79	187–188	71
Ph	H	$PhCH_2$		3-PhCO-Q, $PhCH_2$MgBr	10	195–200	62
Ph	H	Ph		3-PhCO-Q, PhMgBr	90	201–204	62
Ph	Et	CONHEt		1-Et-quinolinium Br^-, EtNC, PhCOONa	22	209–210.5	848
Ph	Et	CONHcyclo-C_6H_{11}		1-Et-quinolinium Br^-, cyclo-C_6H_{11}NC, PhCOONa	32	186–187	848, 890, 891
Ph	Et	CONHPh		1-Et-quinolinium Br^-, PhNC, PhCOONa	11	215–216	848
Ph	$PhCH_2$	CONHcyclo-C_6H_{11}		1-$PhCH_2$-quinolinium Br^-, cyclo-C_6H_{11}NC, PhCOONa	32	209–210	848
Ph	$PhCH_2$	$CONHCH_2Ph$		1-$PhCH_2$-quinolinium Br^-, $PhCH_2$NC, PhCOONa	28	182–183	848, 890, 891
2,4,6-tri-MeC_6H_2	H	Ph		3-(2,4,6-triMe-C_6H_2CO)-Q, PhMgBr	84.5	268–274	62
$4-NO_2C_6H_4$	Bu	$CONHC_6H_3$-3,5-diCl	6-OMe	1-Bu-6-MeO-quinolinium Br^-, 3,5-diClC_6H_3NC, AcONa		224–228 (d)	890, 891

[a]This product was a mixture with 3-acetylquinoline (36%) and 3-acetyl-1,2,3,4-tetrahydroquinoline (13%).

5. 6-Keto-1,4-dihydroquinolines

Compounds **413** were reported to be useful in diazo and photothermography imaging systems, but preparative details were not given.[855,856]

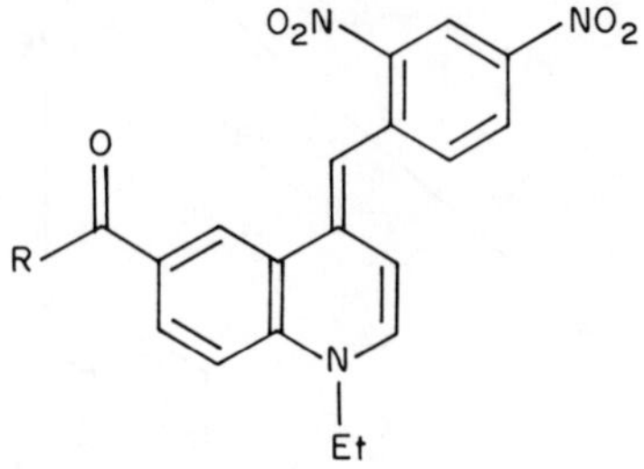

413, R = Me, Ph

6. 2-Keto-1,2,3,4-tetrahydroquinolines

A preliminary report announced the preparation of the ketone **414** by the cycloaddition shown in Scheme 48.[857]

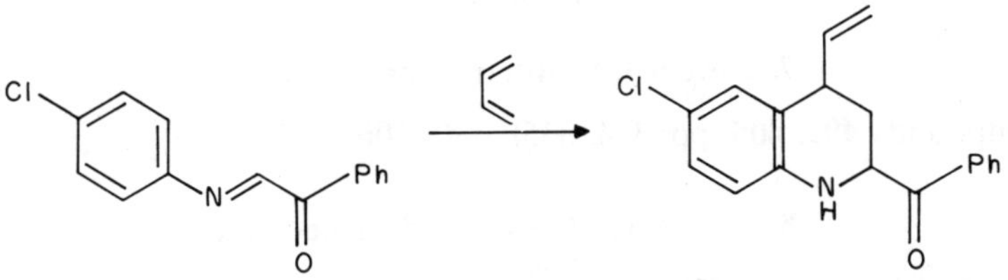

414, m. p. 115—116 °C

SCHEME 48

Treatment of the amido-acids **415** with acetic anhydride gave, via a common intermediate **416**, mixtures of oxazolo[3,4-*a*]quinolinones **417** and 2-acetyltetrahydroquinolines **418**, the latter formed by Dakin–West reactions; Scheme 49.[858]

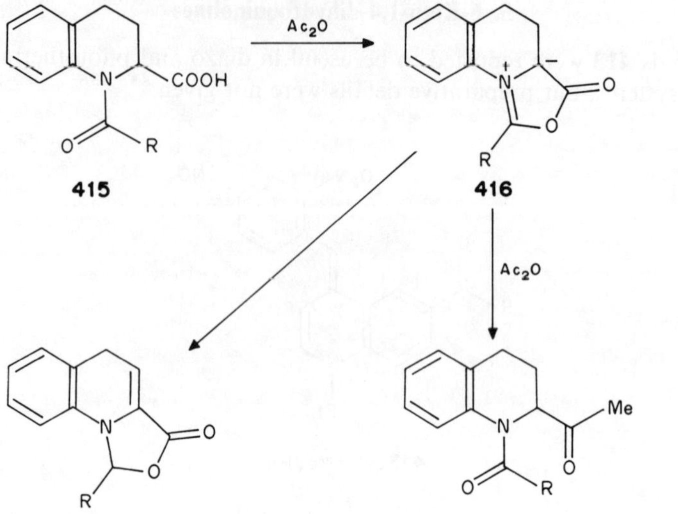

415 **416** **417** **418**

R	Yield (%)	Yield (%)	M.p. (°C)
Me	3.0	35.7	73 – 75
*t*Bu	73.0	7.0	oil
cycloC_6H_{11}	31.0	31.0	oil
Ph	41.2	3.3	109–110 (d)
4-MeOC_6H_4	25.5	10.1	oil
$PhCH_2OCONHCH_2CH_2$	16.2	13.5	103–104

SCHEME 49

7. 3-Keto-1,2,3,4-tetrahydroquinolines

See compounds **402**, **405** (pp. 334, 335) and **410b** (p. 336).

8. 4-Keto-1,2,3,4-tetrahydroquinolines

Oxidation of the olefine **419** by osmium tetroxide and periodic acid gave the hemiaminal **420**.[859]

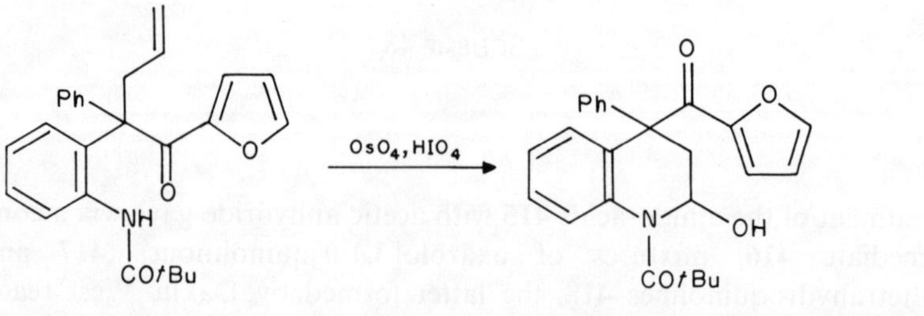

419 **420**, 91%, m.p. 147–149 °C

The tetrahydropyridazine **421** was heated in polyphosphoric acid to give a mixture of three ketones, Scheme 50. It was assumed that the products came from a common intermediate **422** which gave ketone **423a** via a [3, 3] sigmatropic rearrangement followed by hydrolysis. A retro Diels–Alder reaction of intermediate **422** produced an imine which hydrolysed to 1-phenylpropenone. This reacted with compound **423a** to give compound **423b**. The structure of compound **423a** was confirmed by its dehydrogenation to 4-benzoylquinoline and of compound **423b** by its formation from the unalkylated derivative **423a** and 1-phenylpropenone.[860]

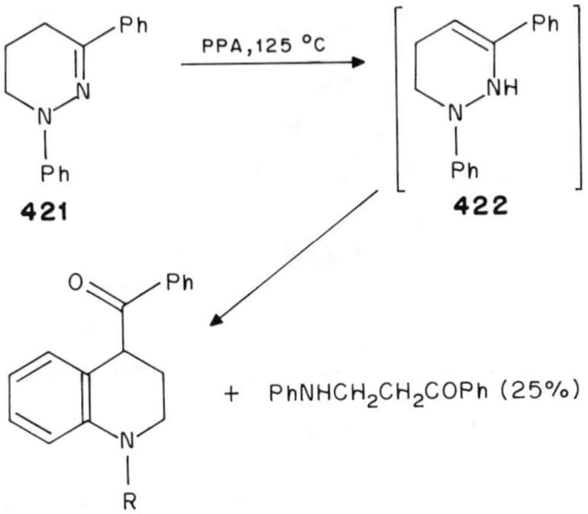

423

a, R = H, 35%, m.p. 125 °C

b, R = CH_2CH_2COPh, 5%, oil

SCHEME 50

Structures **424a**, longistrobine, and **424b**, isolongistrobine, were proposed for two alkaloids isolated from *Macrorungia longistrobus*.[861]

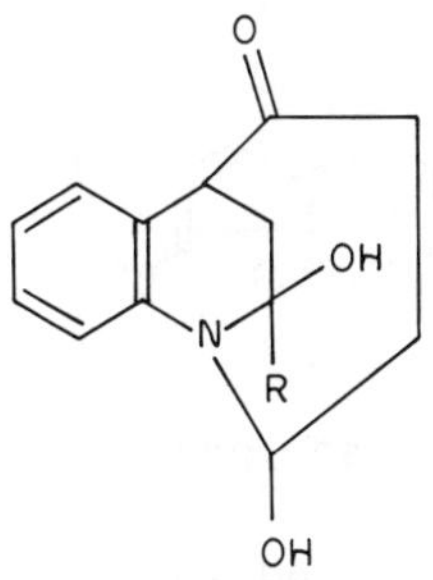

424

a, R = 1-Methyl-4-imidazolyl, m.p. 145–148 °C

b, R = 1-Methyl-5-imidazolyl, m.p. 132–136 °C

9. 5-Keto-1,2,3,4-tetrahydroquinoline

Ketone **425** was the substrate for a reaction described in patents,[862,863] but no source has been given.

425

10. 6-Keto- or 7-Keto-1,2,3,4-tetrahydroquinolines

The simple ketones at position 6, prepared by conventional reactions, are listed in Table 40.

In the absence of solvent, 1-acetyl-1,2,3,4-tetrahydroquinoline was treated with propionyl chloride and aluminium chloride to give, apparently, a mixture of the 6- and 7-ketones, **426**. The mixture was treated with isoamyl nitrite and the product was chromatographed to yield both the 6-, **427**, and the 7-, **428**, diketomonoximes.[864] However, 1-acetyl-4,4-dimethyl-1,2,3,4-tetrahydroquinoline is reported to react with acetyl chloride and aluminium chloride in carbon disulphide to give a mixture of the 6- and 8-ketones. The crude mixture was reduced ($NaBH_4$) and only the pure 6-alcohol was isolated.[865]

426   **427**, 20%   **428**, 25%

The phthalide **429** was hydrolysed and oxidized to the ketone **430**.[866]

Compounds **431a**, **432** and **433** were prepared by hydrogenation of the fully aromatic ketones over Raney nickel. Benzoylation of ketones **431a** and **432a** gave derivatives **431d** and **432b** respectively.[96] For the preparation of compound

Table 40. 6-Keto-1,2,3,4-tetrahydroquinolines

R	Quinoline substituents(s)	Preparation	M.p. (°C)	References
Me		1,2,3,4-tetraH-6-ClCH$_2$CO-Q, Fe, HCl (95% yield)	105–107; Pic. 125	281
Me	1,4,4-triMe	1,4,4-triMe-3,4-diH-6-Ac-2-quinolone, (CH$_2$OH)$_2$; LAH (45% yield)	Oil	893
Me	2-Me	2-Me-1,2,3,4-tetraH-6-ClCH$_2$CO-Q, Fe, HCl	69	281
Me	2-Me	2-MeCOCH$_2$CH$_2$-4-AcC$_6$H$_3$NHCOOCH$_2$Ph, Zn, 37% HBr, AcOH (88% yield)	HBr 208–210	894
Me[a]	1-Ac-4,4-diMe	Friedel–Crafts reaction		865
Me[b]	1-Ac-8-Me	Friedel–Crafts reaction	120	875, 876
CH$_2$Cl	1-Ac-8-Me	1-Ac-6-ClCH$_2$CO-1,2,3,4-tetraH-Q, 20% HCl (50% yield)	123–124	875–877
CH$_2$Cl	2-Me	Friedel–Crafts reaction	121; HCl 225–226	281
CH$_2$Cl		Friedel–Crafts reaction		281
CH$_2$Cl	1-Ac	Friedel–Crafts reaction (60% yield)	137	875–877
CH$_2$Cl[b]	1-Ac-8-Me	Friedel–Crafts Reaction	120	875, 876
CH$_2$Cl	1-N=O	6-ClCH$_2$CO-1,2,3,4-tetraH-Q, HNO$_2$	140	875, 876
CH$_2$Br	1-Ac	Friedel–Crafts reaction	134	875, 876
CH$_2$Br[b]	1-Ac-8-Me	Friedel–Crafts reaction	125–126	875, 876
CHBrCl	1-Ac	1-Ac-6-ClCH$_2$CO-1,2,3,4-tetraH-Q, Br$_2$ (25% yield)	179	875, 876
Et	1,4-diMe	Friedel–Crafts reaction		895
Et	1,4,4-triMe	Friedel–Crafts reaction		895
Et	4-Me	Friedel–Crafts reaction		895
Et	4,4-diMe	Friedel–Crafts reaction		895

Table 40. (*Contd.*)

R	Quinoline substituents(s)	Preparation	M.p. (°C)	References
Et^c	1-Ac	Friedel–Crafts reaction		864
CH_2CH_2COOH	1,4-diMe			895
CH_2CH_2COOH	1,4,4-triMe			896
CH_2CH_2COOH	4-Me			895
MeC$=$NOH				897
MeC$=$NOH	1-Ac	Mixture as in entry at top of this page, *i*AmONO, chromatography (20% yield)		864
MeC$=$NOH	1-Bu			897
$(CH_2)_3Cl$	1-Me	Friedel–Crafts reaction	158–160	898
$(CH_2)_3Cl$		Friedel–Crafts reaction	188–190	898
$(CH_2)_3$-S-(1-Me-1,2,3,4-tetrazol-5-yl)				898
$(CH_2)_3$-S-(1-Me-1,2,3,4-tetrazol-5-yl)	1-Me		110–113.5	898
$CHMeCH_2COOH$				896

CHMeCH$_2$COOH	1-Me	1-Me-1,2,3,4-tetraH-Q-6-COCHMeCN, HCl		899
CHMeCH$_2$COOH	1,4,4-triMe	Friedel–Crafts reaction		896
CHMeCH$_2$COOH	4-Me	Mannich reaction, MeI		895
CHMeCH$_2$-NMe$_3$$^+I^-$	1-Me			899
CHMeCH$_2$CN	1-Me	1-Me-1-2,3,4-tetraH-Q-6-COCHMeCH$_2$NMe$_3$$^+I^-$, KCN, MeOH		899
2-Oxotetrahydro-furan-4-yl	1-Et			896
Ph		6-PhCO-Q, H$_2$, Ni	113	281
Ph	8-Me	6-PhCO-8-Me-Q, H$_2$, Ni	118	281
2,4-Cl$_2$C$_6$H$_3$		6-(2,4-Cl$_2$C$_6$H$_3$)-CO-Q, H$_2$, Ni	137	281
2,5-Cl$_2$C$_6$H$_3$		6-(2,5-Cl$_2$C$_6$H$_3$)-CO-Q, H$_2$, Ni	d	281
3,4-Cl$_2$C$_6$H$_3$		6-(3,4-Cl$_2$C$_6$H$_3$)-CO-Q, H$_2$, Ni	153	281
4-ClC$_6$H$_4$		6-(4-ClC$_6$H$_4$)-CO-Q, H$_2$, Ni	156	281
2-ClC$_6$H$_4$-NMe$_2$C$_6$H$_3$	1-Bu	See text, Section X.10		866
2-COOH-3-Py / 3-COOH-2-Py		1,2,3,4-tetraH-Q, C$_6$H$_6$, AlCl$_3$, pyridine-2,3-dicarboxylic acid anhydride	e	900
2-COOH-3-Py / 3-COOH-2-Py	1-Me / 1-Me	1-Me-1,2,3,4-tetraH-Q, C$_6$H$_6$, AlCl$_3$, Pyridine-2,3-dicarboxylic acid anhydride	e	900, 901
1,2,3,4-tetraH-quinolin-6-yl		Di-2-quinolyl ketone, H$_2$, Raney Ni	197–197.5	96
1-Et-1,2,3,4-tetraH-quinolin-6-yl	1-Et			868

^{a} Obtained as a mixture with the 8-ketone, which was reduced before separation.
^{b} The paper recognizes that the ketone group could be at position 5, 6 and 7. The compounds are entered as 6-ketones as these seem most likely.
^{c} A mixture with the 7-propionyl ketone was obtained and used without purification.
^{d} Isolated as the N-nitrosamine; m.p. 135–136 °C.
^{e} Mixture obtained and used without separation.

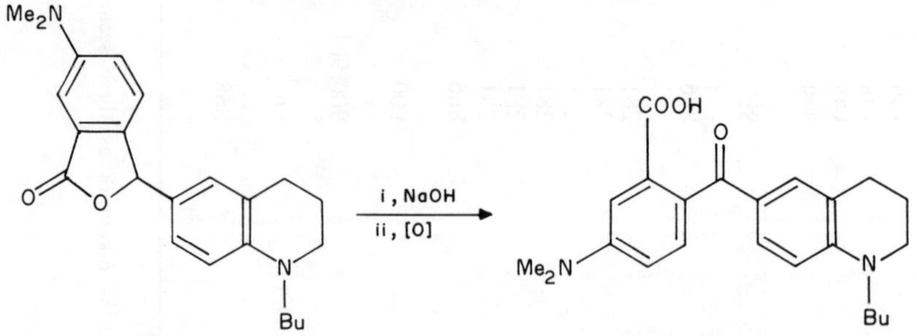

429   **430**

431b, the di(tetrahydroquinolyl)methane was oxidized with chloranil.[867]
Compound **431c** has been described as a starting material in a patent.[868]

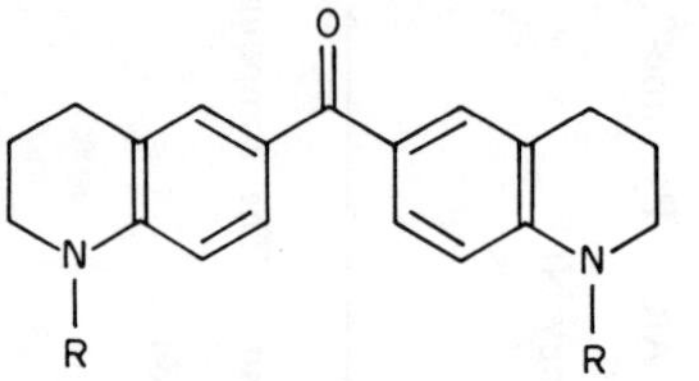

431

a, R=H, m.p. 197–197.5 °C
b, R=Me, 20%, m.p. 124–125 °C
c, R=Et
d, R=PhCO, m.p. 252–253 °C

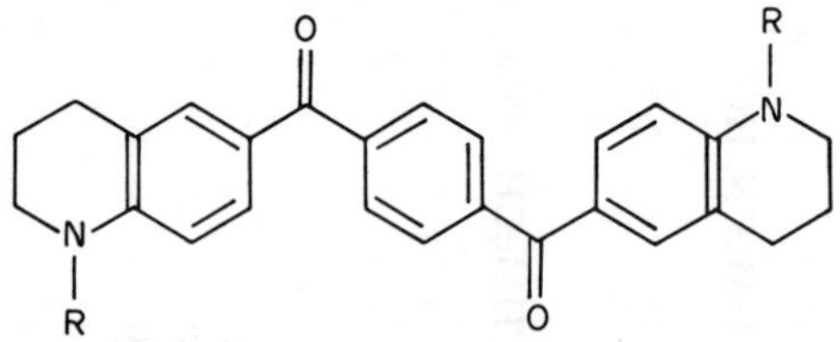

432

a, R=H, 100%, m.p. 274–276 °C
b, R=PhCO, m.p. 221–223 °C

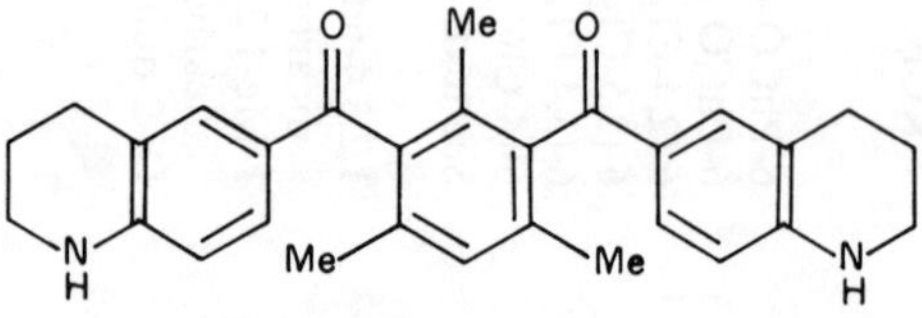

433, 100%, m.p. 312–313 °C

Pyridine-2,6-dicarboxylic acid chloride gave the diketone **434** via a
Friedel–Crafts reaction.[869]

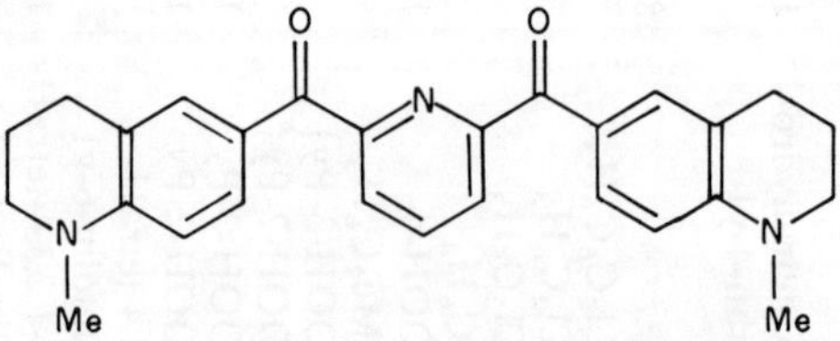

434, m.p. 242.5–244 °C

11. 8-Keto-1,2,3,4-tetrahydroquinolines

The ketal **435** was hydrolysed to the keto-acetal **436**.[870]

435   **436**

73%, b.p. 199 — 200 °C/12 mm

The pyrroloquinoline **437**, $R^1 = H$, was oxidized ($NaIO_4$) to the amido ketone **438a**, which was hydrolysed (HCl, EtOH) to ketone **438b**.[871,872] Bromoacetyl bromide gave compound **438c**. Ketones **438b** and **438d** were also prepared by addition of phenyllithium to the appropriate tetrahydroquinoline-8-carboxylic acid, and converted into compounds **438c** and **438e** respectively.[873] Other ketones **438**, $R^1 = Me$, Et, Br, OMe, OEt; $R^2 = H$, Ac, $COCH_2Br$ were prepared but not characterized. Compounds with $R^2 = COCH_2Br$ reacted with ammonia to give quinodiazepines having tranquillizer and anticonvulsant actions.[874]

437   **438**

	R^1	R^2	%	M.p. (°C)
a,	H	Ac	46	132 — 133.5
b,	H	H	89	68 — 69.5
c,	H	$COCH_2Br$	89	99 — 101
d,	Cl	H	39	98 — 100
e,	Cl	$COCH_2Br$		
f,	Cl	Ac		118 — 119

Hydrogenation of 6-chloro-8-benzoylquinoline over Raney nickel gave compound, **438d** which with acetic anhydride in pyridine gave the acetyl derivative **438f**.[443]

12. Ketones of Unknown Structure

In an early series of papers, 1-acetyl-6-methyl-1,2,3,4-tetrahydroquinoline was shown to undergo Friedel–Crafts reactions, but the position of the resulting

Table 41. 2-Ketomethyl-1,2-dihydroquinolines
See also Table 42

METHOD 1   METHOD 2

R¹	R²	R³	Preparation	Yield(%)	M.p. (°C)	References
Me	H	Me	See text, Section X.13			800
Me	H	CH₂C₆H₃-2,6-diCl	Method 1		122–124	789, 878
Me	Ac	COPh	Method 2	58	138	884
Me	COOEt	COPh	Method 2			902
Me	H	COPh	Method 2	17	150–151	903
CHCO(CH₂)₃	—(CH₂)₄—	CH₂C₆H₄-2-Cl	Method 1		115–120	789
	—(CH₂)₄—	CH₂C₆H₃-2,6-diCl	Method 1		146	789
Ph	H	Me	Method 1		120–122	789
Ph	H	CH₂C₆H₄-2-Cl	Method 1		144–146	789
Ph	H	CH₂C₆H₃-2,6-diCl	Method 1		90–93	789
Ph	H	CH₂C₆H₄-4-Cl	Method 1		108–110	789
Ph	H	COPh	Method 2	10.4	149–150.5	666, 903*
Ph	Ph	Me	Method 2		91–93	789
Ph	Ph	CH₂Ph	Method 1		130–131	789
Ph	Ph	CHPh₂	Method 1		150–152	789
Ph	Ph	CH₂C₆H₄-2-Cl	Method 1		102–104	789
Ph	Ph	CH₂C₆H₃-2,4-diCl	Method 1		130–133	789
Ph	Ph	CH₂C₆H₃-2,6-diCl	Method 1		173–175	789
Ph	Ph	CH₂C₆H₄-3-Cl	Method 1		126–127	789
Ph	Ph	CH₂C₆H₄-4-Cl	Method 1		96–97	789
Ph	Ph	CH₂C₆H₄-3-NO₂	Method 1		148–150	789
4-MeOC₆H₄	COOEt	CH₂C₆H₃-2,6-diCl	Method 1		154–155	789
Ph	4-MeOC₆H₄	COPh	Method 2	43.5	154–155	666, 902, 904

Ph	COOEt	COC_6H_4-4-Me	Method 2	25	147–148	904
Ph	COOEt	COC_6H_4-3-Cl	Method 2	64	141.5–142.8	904
Ph	COOEt	COC_6H_4-4-OMe	Method 2	49	137.1–138.5	904
4-MeC_6H_4	H	COPh	Method 2	31	142–143	903
4-FC_6H_4	COOEt	COPh	Method 2	38	169–170	904
4-$MeOC_6H_4$	COOEt	COPh	Method 2	40	157.7–158.1	904
3-$NO_2C_6H_4$	H	COPh	Method 2	12	171–172	903
4-$NO_2C_6H_4$	COOEt	COPh	Method 2	11	138.5–140	904
4-$NO_2C_6H_4$	COOEt	COC_6H_4-3-Cl	Method 2	40.3	132.5–133.5	904
2-Thienyl	H	COPh	Method 2	20	163–164	903*

*These references include carbonyl derivatives.

ketone was not determined.[875-877] These compounds are shown in general formula **439**.

439

R[1]	R[2]	%	M.p. (°C)
ClCH$_2$	Ac	100	132
BrCH$_2$	Ac		128
ClCH$_2$	H		122, HCl 218
BrClCH	Ac		143
Me	Ac	20	160

13. 2-Ketomethyl-1,2-dihydroquinolines

The quinolinium salt **440** gave the adduct **441** with acetophenone in the presence of a trace of sodium hydroxide at room temperature. This proved to be a general reaction; the same procedure was used for all the ketones prepared by Method 1 of Table 41.[789] Oxidation of these adducts gave ketomethylenequinolines; see Section VIII.1 and Table 35. Ketone **442** has been prepared from a Janovsky anion, which has been shown to deliver a hydride ion or acetone anion to a substrate, depending on the conditions; see Section VIII.1.[800]

A mixture of quinoline, acetophenone and benzoyl chloride, when left to stand for a month, gave the amido-ketone **443** (10.4%); 3-keto-esters reacted similarly, and this is Method 2 of Table 41. Compound **443** was reduced to compound **444** (82%). Hydrogenation of 1-phenyl-2-(2-quinolyl)ethanone over platinum in acidified ethanol followed by treatment with benzoyl chloride also gave compound **444**, but in only 9% yield.[666]

443   **444**, m.p. 151.5−152.4 °C

The quinolinium salts **445** reacted with silyl enol ethers, but the products were mixtures of the 2- (**446**) and 4- (**447**) ketomethyldihydroquinolines; see Table 42.[879]

445   **446**   **447**

Table 42. Ketones 446 and 447 Mixture Compositions

R^1	R^2	R^3	Yield	(%)
			446	**447**
Me	H	MeO	87	5
Et	Me	MeO	81	18
Et	Me	EtO	79	18
—(CH$_2$)$_4$—		MeO	85	12
Ph	H	Me	51	12
Ph	H	MeO	78	7
Ph	H	EtO	78	8
Ph	H	CCl$_3$CH$_2$O	69	8

14. 4-Ketomethyl-1,4-dihydroquinolines

The ketones **448**[880], **449**[878] and **450**[881] were made in a similar manner to compound **441**, p. 350, from the 1-alkylquinolinium salt, an active methylene compound and ethanolic sodium hydroxide at room temperature. The preparation of ketones **451a–c** employed sodium methoxide, and of ketone **451d**, sodium in dimethyl sulphoxide.[882]

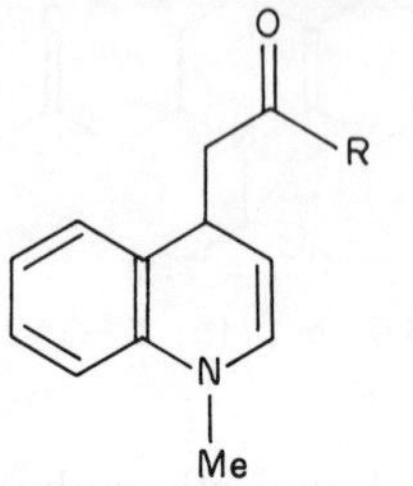

448

R	Yield (%)	M.p.(°C)
Me		oil
Cyclopentan-2-onyl	72	177
Cyclohexan-2-onyl		oil
Cyclohexane-2,5-dionyl	95	220 (d)
Inden-1-on-2-yl	98	137
Ph	100	133–134

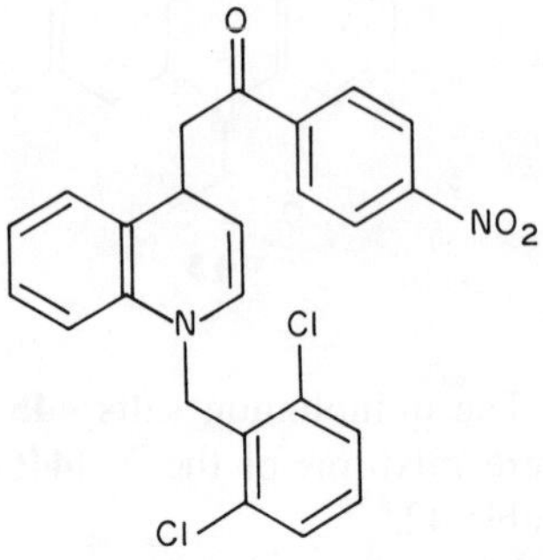

449, 66%, m.p. 141–142 °C

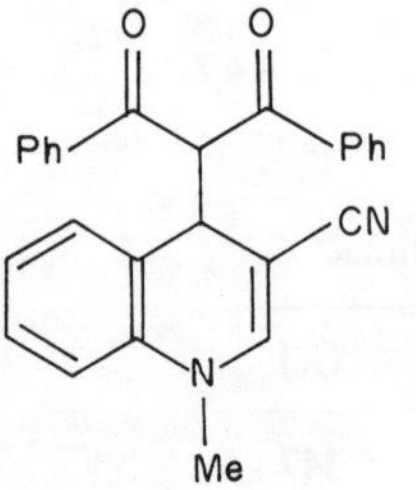

450, 81%, m.p. 174–175 °C

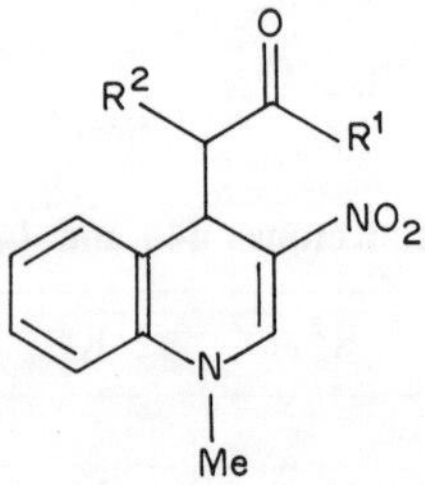

451

a, R^1 = Me, R^2 = COOEt, 60%, m.p. 106–110 °C (d)
b, R^1 = Me, R^2 = COPh, 45%, m.p. 188–189 °C
c, R^1 = Ph, R^2 = NO$_2$, 36%, m.p. 159–163 °C (d)

d, R^1, R^2 = 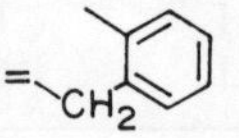 40%, m.p. 220 °C

15. 6-(Ketoethyl)-1,2-dihydroquinoline

See compound **408**, p. 335.

16. Ketoalkyl-1,2,3,4-tetrahydroquinolines

Hydrogenation of 2-acetonylquinoline (PtO$_2$, acidic ethanol) gave the 1,2,3,4-tetrahydro derivative (HCl m.p. 225 °C).[634] The diketone **452** was made by reduction of the equivalent 1,2-dihydroquinoline, Table 41, with hydrogen over platinum.[884] See also compound **444**, p. 351.

452, m.p. 148 °C

The allylic alcohol **453** was attacked by the anion from 2-methylcyclopentane-1,3-dione to give the ketone **454**.[885]

453   **454**, m.p. 115−117 °C

The ketones **455** have appeared in a patent.[886] The orange dye **456** appeared in two patents, but, again, no preparation was given.[887,888]

455, R = H, CO(CH$_2$)$_3$N(CH$_2$CH$_2$)$_2$O   **456**

Hydrogenation of the nitrodiketone **305**, Section VI.4.B, for a longer time and at higher pressure than was used previously gave the tetrahydroquinolyl ketone **457a**, from which derivatives **457b** and **c** were prepared. The benzoyl derivative, **457b**, was nitrated to give compound **457e**, which gave compound **457d** on hydrolysis.[756]

457

a, R^1, R^2 = H, 33%, m.p. 53—54 °C

b, R^1 = PhCO, R^2 = H, m.p. 73—75 °C

c, R^1 = 4—$NO_2C_6H_4CO$, R^2 = H, m.p. 121—122 °C

d, R^1 = H, R^2 = NO_2, m.p. 94—96 °C

e, R^1 = PhCO, R^2 = NO_2, m.p. 137.5—138.5 °C

XI. Quinoline Ketones with Partially or Fully Saturated Benzene Rings

1. 6-Keto-7,8-dihydroquinolines

The sulphone **458** underwent a Michael addition to chalcone **459** and cyclization to give the derivative **460**. In other similar examples the sulphur-containing groups were lost to give fully aromatic 6-ketoquinolines; see Section II.5.[98]

458   **459**   **460**, 59%, m.p. 170—172.5 °C

2. Keto-5,6,7,8-tetrahydroquinolines

Two ketones of this class are in Table 4 and one in Table 49. The 3-keto-2-methyl-5,6,7,8-tetrahydroquinolines are collected in Table 43. Most were prepared by one of Methods 1–3 as shown on p. 357. In Method 1 a

Table 43. 2-Methyl-, and 2-Styryl-3-keto-5,6,7,8-tetrahydroquinolines

$$R^3C_6H_4CHO \xrightarrow{\text{AcOH, reflux}}$$

(A: 2-methyl-3-keto-5,6,7,8-tetrahydroquinoline; B: 2-styryl-3-keto-5,6,7,8-tetrahydroquinoline)

R¹	R²	R³	Formula	Preparation	Yield (%)	M.p. (°C)	References
Me	H		A	Method 1		55; Pic. 155	913
				Method 2	52	b.p. 85/0.04 mm; Pic. 143–144	914
Me	5-Me-8-iPr		A	Method 3	64	46–47	915,916
				Method 2	69	b.p. 95–96/0.02 mm; Pic. 155–156	914
Me	H	NO₂	B			163–164	917
Me	6-Me	NO₂	A	Method 1		65	913
Me	6-Me	OMe	B			b.p. 165/15 mm; Pic. 168	917
Me	6-Me	NO₂	B			173; 213	917
Me	7-Me		A	Method 1		HCl 207; b.p. 169–170/16 mm; Pic. 169	913
CH₂Br	H		A	2-Me-3-COCH₂COOEt-5,6,7,8-tetraH-Q, Br₂, HBr, 40–50 °C	77.5	HBr 171–172.5	918

Table 43. (*Contd.*)

R^1	R^2	R^3	Formula	Preparation	Yield (%)	M.p. (°C)	References
CH_2OAc	H		A	2-Me-3-$COCH_2Br$-5,6,7,8-tetraH-Q, AcOK, EtOH	94	83.5–84.5	918
CH_2COOEt	H		A	2-Me-3-COOEt-5,6,7,8-tetraH-Q, EtOAc, NaH	31.5	53–54	918
Ph	H		A	Method 1		73	913
Ph	H	NO_2	B			181–182 HCl 210	917
Ph	6-Me		A	Method 1		76–77 Pic. 170	913
Ph	6-Me	NO_2	B			77	917
Ph	7-Me		A	Method 1		Pic. 177	919
Ph	7-Me	NO_2	B			186–187	917

hydroxymethylene-cyclohexanone **461** reacts with an acyclic enaminone **462** to give a product of general formula **463**. In Method 2, an enaminone **464**, derived from compound **461**, must undergo rearrangement before reaction with pentane-2,4-dione to give the ketone **463**, R^1 = Me. Reduction of the isoxazole **465** gave, presumably via the enaminone and nitrous acid oxidation, the ketone **463**, R^1 = Me, R^2 = H.

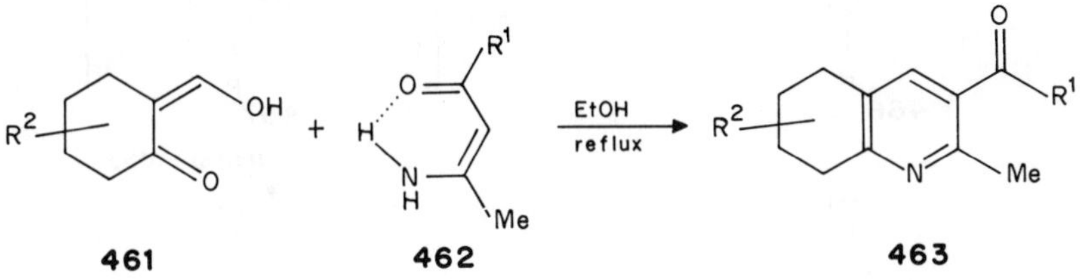

461   **462**   **463**

METHOD 1

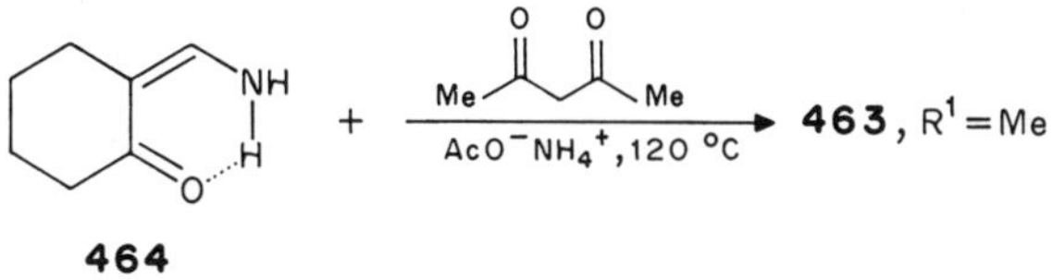

464   **463**, R^1 = Me

METHOD 2

465

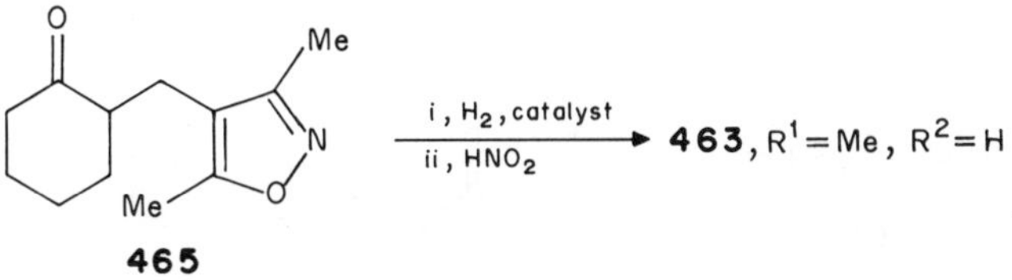

METHOD 3

The enamines **466** with benzoyl chloride gave the unstable quinolinium ketones **467**, which were deprotonated to ketones **468**. Further reactions of compounds **468**, Scheme 50, were used to prepare phenolic esters **469** with the 8-benzoylquinoline **470** as the common by-product.[907,908]

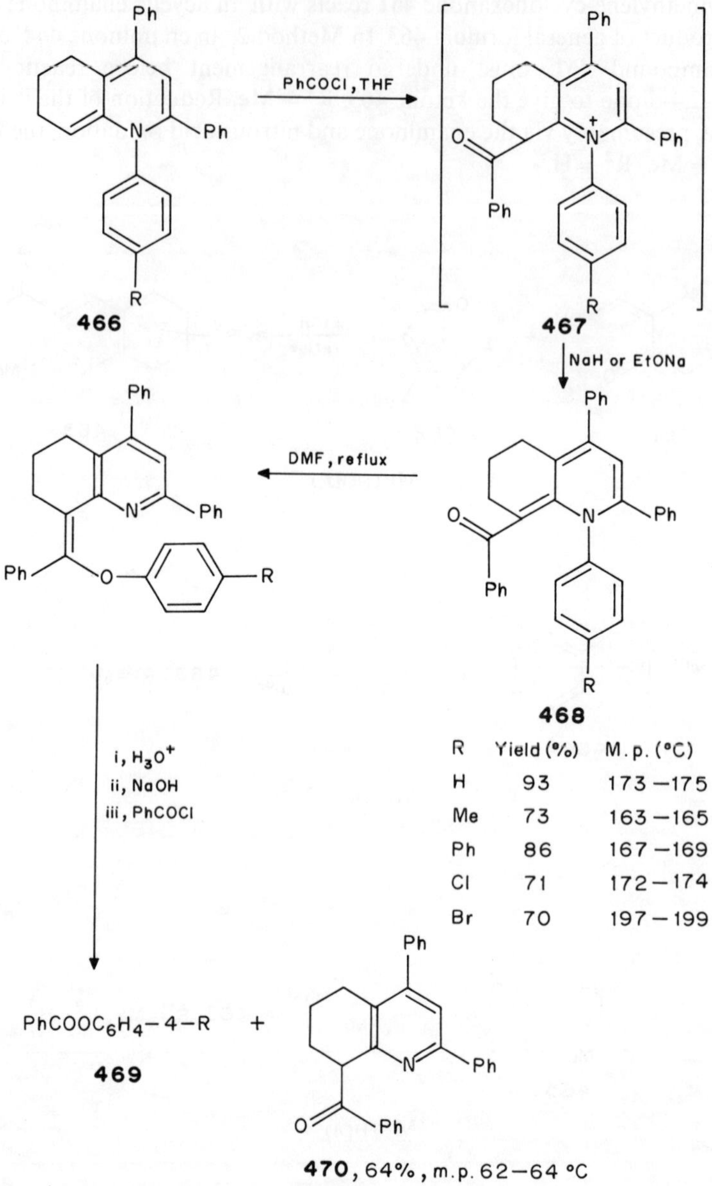

R	Yield (%)	M.p. (°C)
H	93	173−175
Me	73	163−165
Ph	86	167−169
Cl	71	172−174
Br	70	197−199

SCHEME 50

When the appropriate 5,6,7,8-tetrahydroquinolines were treated with a strong base followed by an ester, the ketones **471a**[909], **471b**[910] and **471c**[911] resulted.

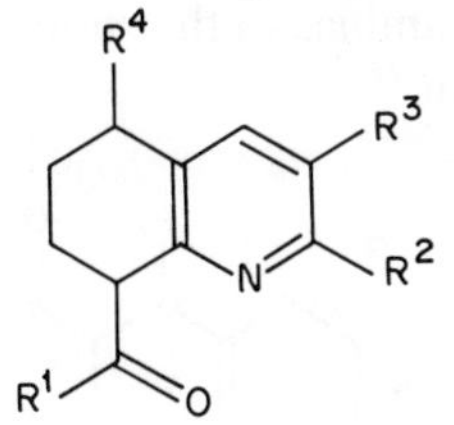

471

R¹ R² R³ R⁴

a, Me H   Me H   b.p. 130–132 °C / 0.6 mm
                        HCl m.p. 159 °C

b, Ph Ph H   H   2% m.p. 150–151 °C

c, Ph H   H   2-Pyridyl 52% m.p. 133–134 °C

The reaction of the appropriate quinolyl acetate with diethyl oxalate and potassium ethoxide gave the dihydroquinoline **472a**. If the quinolyl acetate was first hydrogenated (Pd) it gave the tetrahydroquinoline **472b**.[906]

The ketones **473** are included in a recent patent, with spectroscopic data, but the preparative method is conspicuous by its absence.[912]

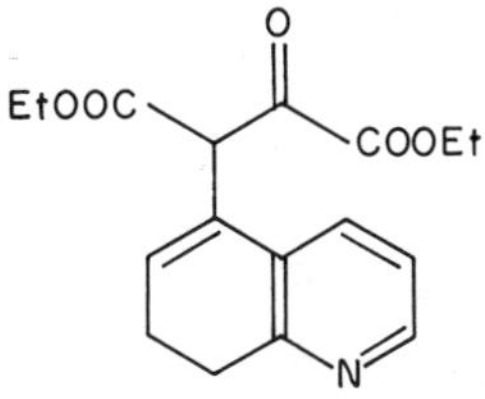

472

a, 75%, Oil

b, 5,6-bond saturated, 58%, m.p. 149 °C

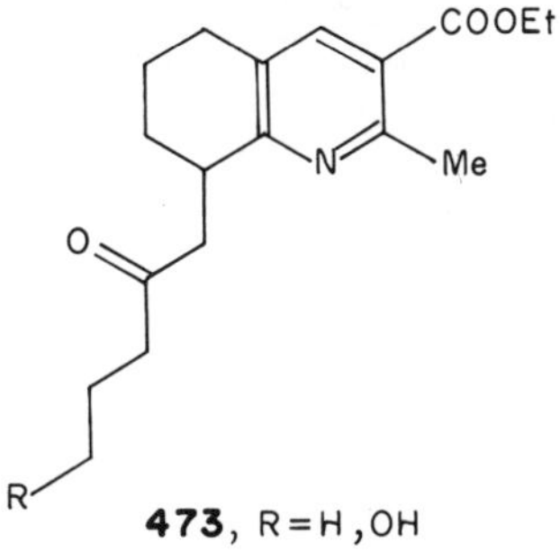

473, R = H, OH

XII. Quinolyl Ketones with Partial Saturation of Both Rings

Note the 1,5,6,7-tetrahydro-8-ketoquinolines (**468**) of Scheme 50, Section XI.2.

Deprotonation of 3-acetyl-4-methyl-1-phenyl-5,6,7,8-tetrahydroquinolinium perchlorate by sodium hydroxide gave compound **474** in high yield.[920]

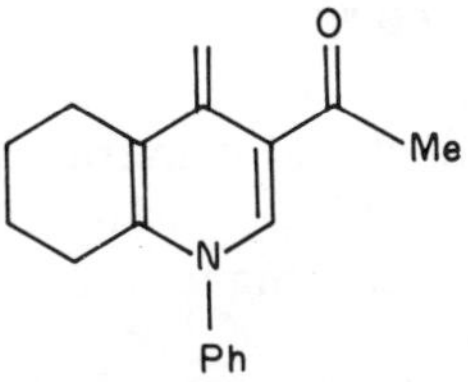

474, 89%, m.p. 80–81 °C

Conjugate addition of primary amines to the enone **475** was followed by ring closure to produce compound **476**.[921]

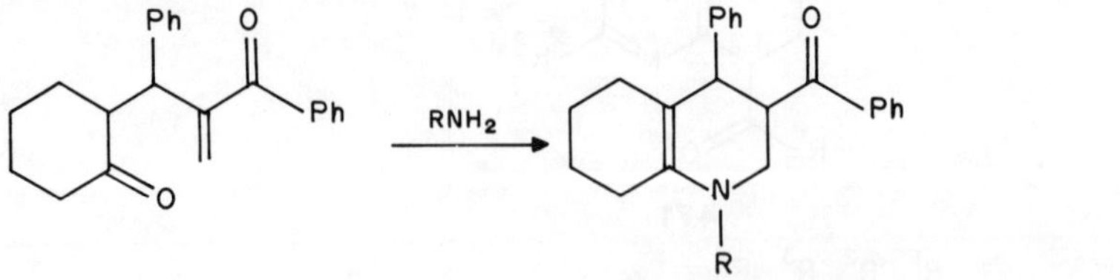

475

476, R=Me,49%, m.p.135–136 °C
R=CH₂Ph, 39%, m.p. 140–141 °C

Hydrogenation of the isoxazoles **477** over Raney nickel gave, presumably via intermediate enaminones, the quinolines **478**.[922,923] The carbinolamine **478c** gave the hexahydroquinoline **479** on shaking in ether with sodium hydroxide solution.[922]

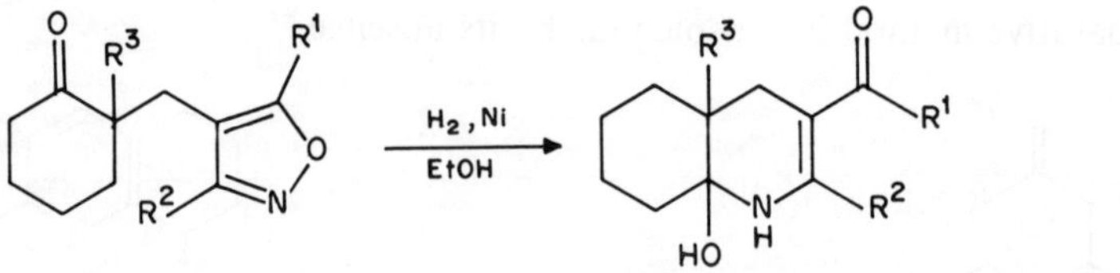

477

478 a, R¹=Me, R²=Et, R³=H
b, R¹=Bu, R²=Me, R³=H
c, R¹,R²=Me, R³=Ph

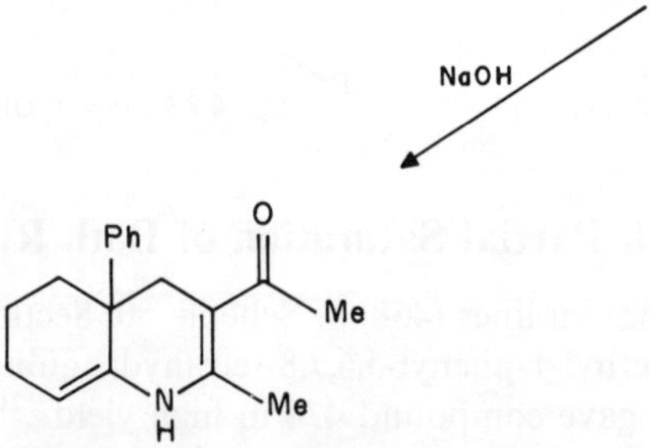

NaOH

479, 100%, m.p. 55–60 °C

The ketone **480** was prepared and converted into the enol ether salt **481**. With sodium ethoxide, the salt **481** gave an isolatable ketal which was reacted crude with aromatic amines in refluxing toluene to give the vinylogous amidine ketones **482**, Scheme 52.[924]

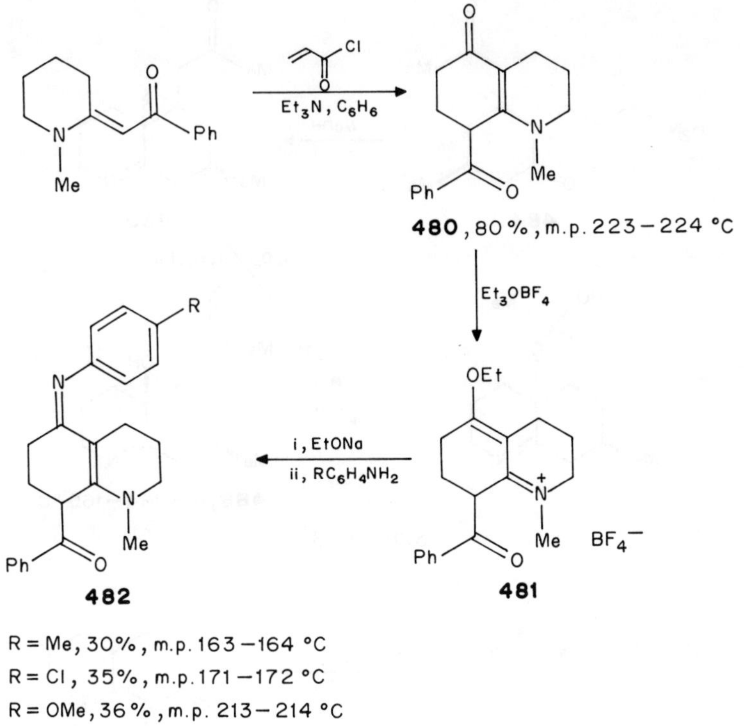
Cl
O
Et3N, C6H6
O
N
Me
Ph
480, 80%, m.p. 223–224 °C
Et3OBF4
OEt
i, EtONa
ii, RC6H4NH2
Me
BF4−
481
R
482
R = Me, 30%, m.p. 163–164 °C
R = Cl, 35%, m.p. 171–172 °C
R = OMe, 36%, m.p. 213–214 °C

SCHEME 52

Treatment of the appropriate octahydroquinoline with trifluoroacetic anhydride in pyridine gave the ketone **483**.[925] This was an intermediate in a synthesis of luciduline **508** (Section XIII.1).

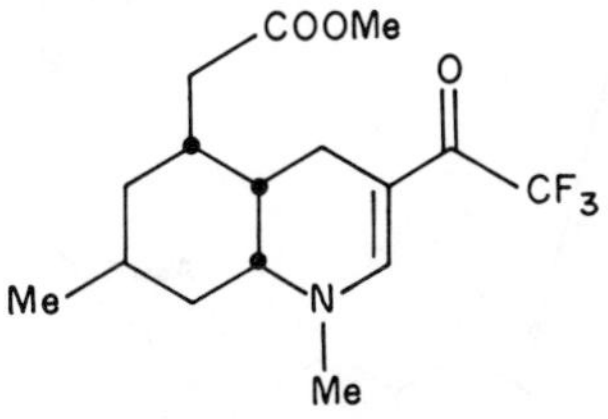
COOMe
O
CF3
Me
N
Me

483, 63%

The side chain ketones shown in Scheme 53 were intermediates in the synthesis of lycodoline.[926] The first product, **485**, was unstable and was not characterized. Later a one-pot procedure for the conversion of cyclohexanone **484** into compound **486** was developed.[927]

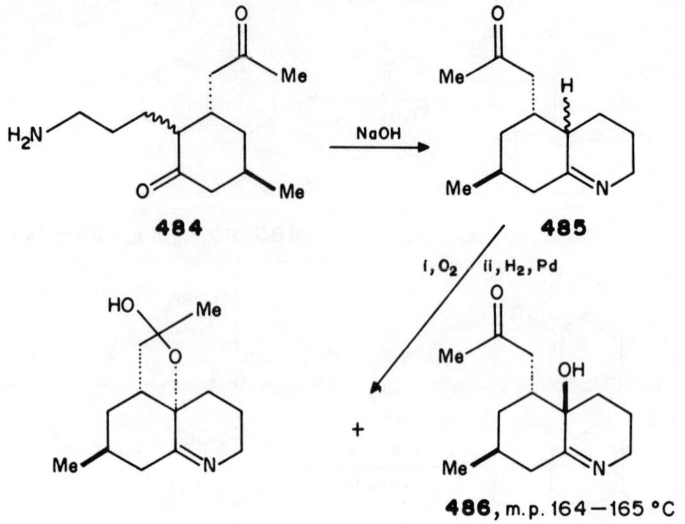

SCHEME 53

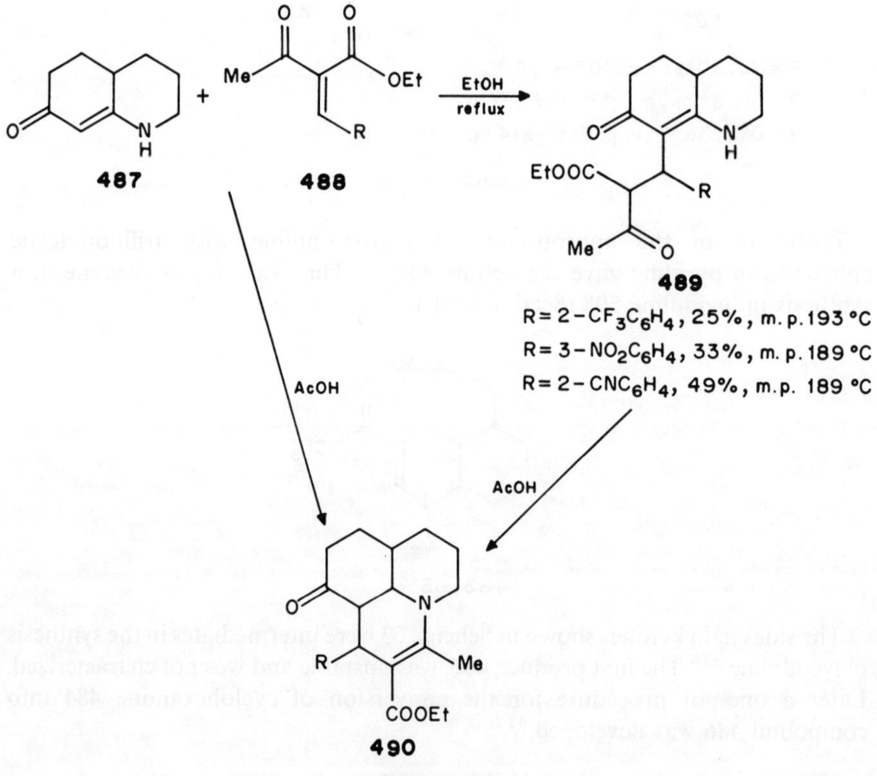

489

R = 2-CF$_3$C$_6$H$_4$, 25%, m.p. 193 °C
R = 3-NO$_2$C$_6$H$_4$, 33%, m.p. 189 °C
R = 2-CNC$_6$H$_4$, 49%, m.p. 189 °C

490

SCHEME 54

The enaminone **487** and the methyleneacetoacetic esters **488** gave the adducts **489** in hot alcohol. In acetic acid these were converted into the benzo[*i, j*]quinolizines **490**, Scheme 54.[928]

XIII. Decahydroquinolyl Ketones

1. Ketone at Position 4

A Grignard reaction on the appropriate 4-quinolone gave the acetylenic alcohol **491**, which was hydrated to the ketone **492**.[929,930]

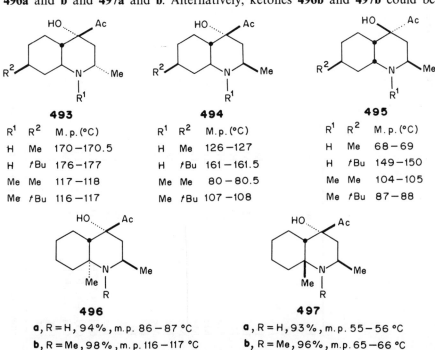

A series of papers describes the separation and characterization of several stereoisomers of compound **492** and similar ketones. These compounds are collected in Table 44.

Acetylenic hydrations also produced compounds **493**, **494** and **495**,[935,936] **496a** and **b** and **497a** and **b**. Alternatively, ketones **496b** and **497b** could be

493

R^1	R^2	M.p.(°C)
H	Me	170–170.5
H	*t*Bu	176–177
Me	Me	117–118
Me	*t*Bu	116–117

494

R^1	R^2	M.p.(°C)
H	Me	126–127
H	*t*Bu	161–161.5
Me	Me	80–80.5
Me	*t*Bu	107–108

495

R^1	R^2	M.p.(°C)
H	Me	68–69
H	*t*Bu	149–150
Me	Me	104–105
Me	*t*Bu	87–88

496

a, R = H, 94%, m.p. 86–87 °C

b, R = Me, 98%, m.p. 116–117 °C

497

a, R = H, 93%, m.p. 55–56 °C

b, R = Me, 96%, m.p. 65–66 °C

Table 44. 4-Ketodecahydroquinolines

R^1	R^2	R^3	Formula	Yield (%)	M.p. (°C)	HCl m.p. (°C)	References
H	H	H	A	89.9	117–118		931[a]
H	H	H	B	85.3	178–179	284–285	931[a]
H	H	Me	A	68.2	75–76	298(d)	932
H	H	Me	B		77–78	239–240	932
H	Me	H	A		152–153		932
Me	H	H	A	85.3	112–113		933
Me	H	H	B	75.2	138–139	222–223	933
Me	Me	H	A	70.6	63–64	211–212	932, 933
Me	H	Me	B		104–105	212–213	932, 933
Me	Me	Me	A		70–71		932
Et	H	H	A		131–132		933
Et	H	H	B				933
Pr	H	H	A				933
Pr	H	H	B	61.5	66–67	192–193	931[a]
$CH_2{=}CHCH_2$	Me	H	A	53.1	61–62	247–249	931[a]
$CH_2{=}CHCH_2$	H	Me	A	50.4		243–244	934
$CH_2{=}CHCH_2$	H	Me	B				934
iPr	H	H	A				934
iPr	H	H	B				931[a]

[a] Reference 931 is a pharmacological paper. Preparative details for these compounds have not yet appeared.

obtained by Eschweiler–Clark methylation of compounds **496a** and **497a** respectively.[937]

Details of n.m.r. studies on *cis*-**498**, R = PhCO and *trans*-**498**, R = MeCO, PhCO[938] and on the ketone **499**[939] have appeared. The bromomethyl ketone **500** was prepared by bromination of ketone **492** in 48% hydrobromic acid.[929,930] The preparation of a 4-acetyl-1-allyl-4-hydroxy-2-methyldecahydroquinoline from the secondary amine and allyl bromide has been recorded.[934]

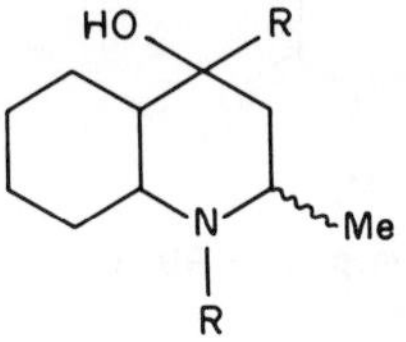

498

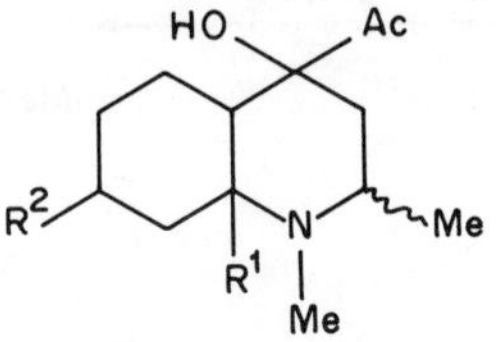

499, $R^1, R^2 = H$
$R^1 = H, R^2 = Me$
$R^1 = Me, R^2 = H$

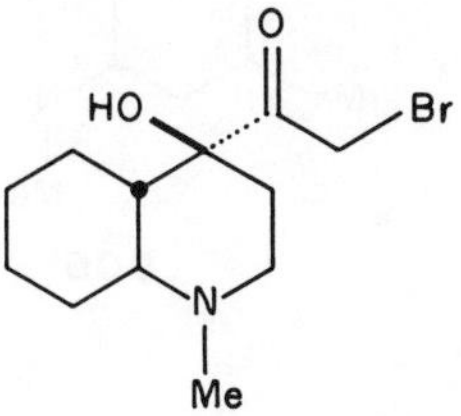

500, m.p. 214−216 °C

Hydration of the olefinic acetylene ester **501**, $R^1 = H$, $R^2 = Me$, $R^3 = Ac$ under the standard conditions gave a mixture from which the spiro compounds **502a** and **503a** were isolated. However, the alcohol **501**, R^1, $R^3 = H$, $R^2 = Me$ gave the diol **505**[940], but alcohol **504** gave the pyranones **502b** and **503b**.[941]

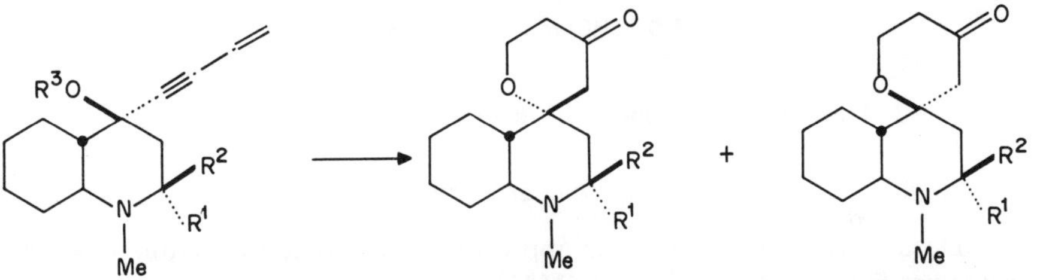

501

502 a
$R^1 = H, R^2 = Me, 34\%,$
m.p. 85−86 °C

503 a
$R^1 = H, R^2 = Me, 14\%,$
m.p. 103−104 °C

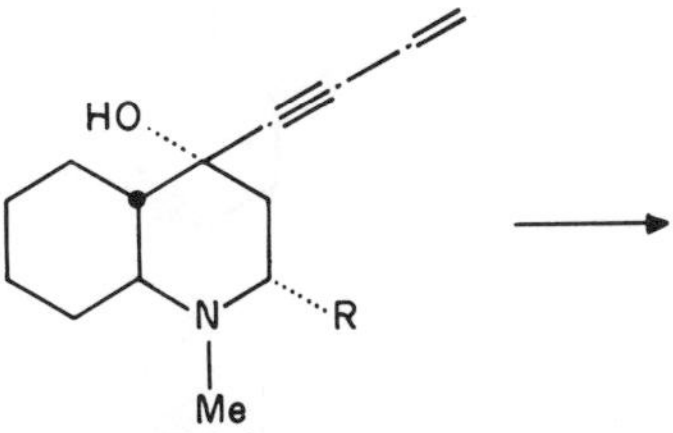

502 b
$R^1 = Me, R^2 = H, 3\%,$
m.p. 87−88 °C

503 b
$R^1 = Me, R^2 = H, 5\%,$
m.p. 129−130 °C

504

501, $R^1, R^3 = H, R^2 = Me$ ⟶

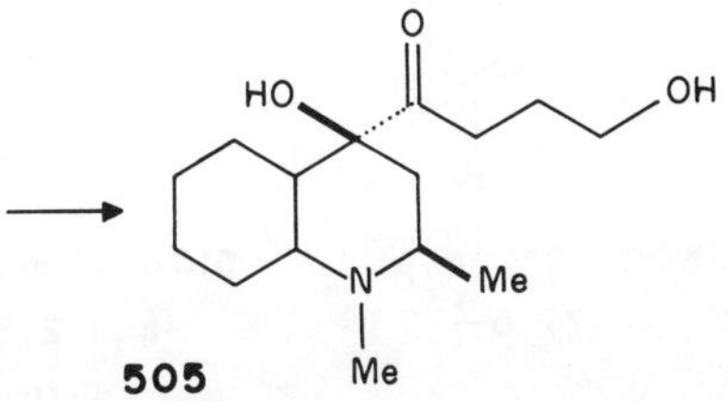

505

A synthesis of luciduline began with the quinolone **506**, which underwent a remarkable ring closure to the cyclic ketone **507**. Reduction to the alcohol–amine and reoxidation then gave the alkaloid **508**.[942] (See also compound **483**, Section XII).

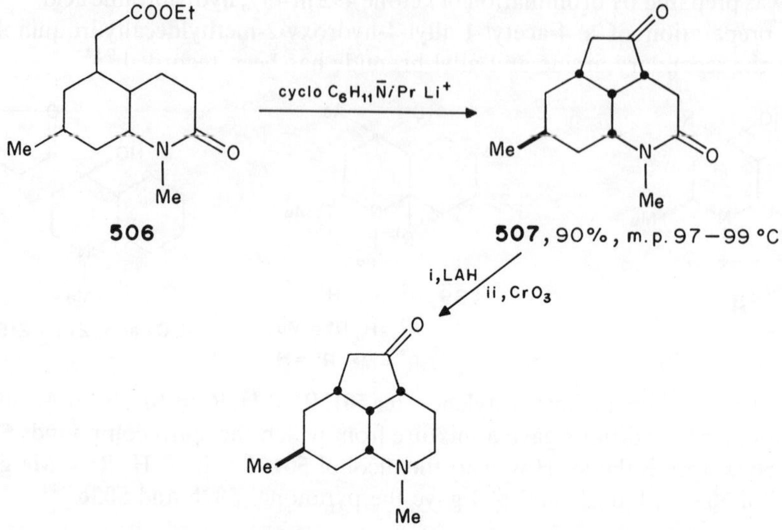

506

cyclo $C_6H_{11}\bar{N}/Pr\ Li^+$

507, 90%, m.p. 97–99 °C

i, LAH   ii, CrO₃

508, HCl m.p. 238–239 °C

2. Decahydroquinolines with Ketone Groups in Side Chains

A. Ketone at Position 1

trans-Decahydroquinoline and the appropriate bromomethyl ketone gave the derivatives **509**[839], **510**[837,943] and **511**[835].

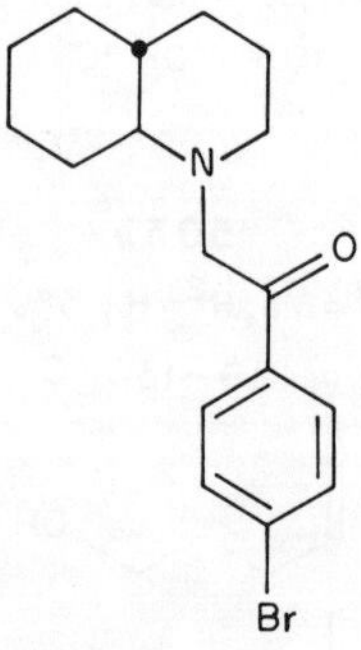

509, 77%, m.p.
77.5–78 °C

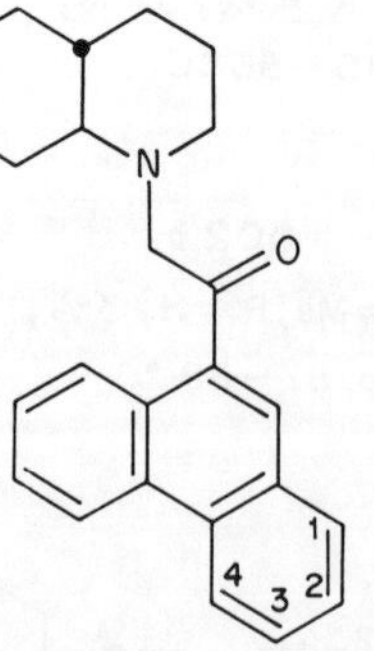

510 a, no m.p. given
b, 1,2,3,4-tetrahydro
derivative, m.p. 95–97 °C

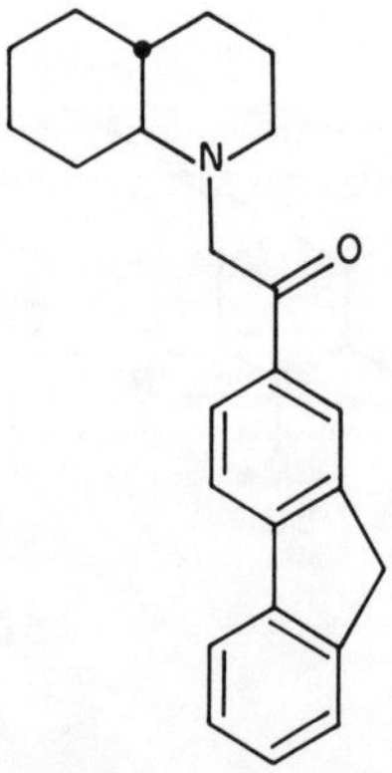

511, m.p. 104–106 °C

Benzoin and decahydroquinoline were heated in the presence of phosphorus pentoxide to give the ketone **512**.[883]

512

A series of 4-hydroxy-*trans*-decahydroquinolines carrying ketone-containing groups at position 1 has been prepared by standard alkylation procedures (e.g. secondary base, alkyl chloride, potassium carbonate and/or sodium bicarbonate with a trace of potassium iodide in an inert solvent at reflux). The alcohols **513** were esterified or converted into carbamates or thiocarbamates. In some examples the orientation of the 4-hydroxy group was deduced from i.r. and n.m.r. data. In some early reports the epimers were separated by chromatography and presented as form 1 and form 2.[944,945] The detailed stereochemistry was given in a recent paper.[946] In another report, only the epimer mixtures were described, and were submitted for pharmacological testing without separation.[947] The products are recorded in Table 45. Compound **514** was made by a Mannich reaction.[948]

513

514, HCl m.p. 260 °C

The anilino derivatives **515** were made from *trans*-decahydro-4-quinolone via metal hydride reduction of the Schiff bases. Alkylation, as above, then gave ketones **516**. In some examples the secondary nitrogen was converted into an

515

$Cl(CH_2)_nCOAr$

516

Table 45. 4-Hydroxy-1-ketoalkyl *trans*-Decahydroquinolines

R¹	R²	R³	n	orientation*	Base m.p. (°C)	Salt	M.p. (°C)	References
H	H	H	3	ax.	109	HCl	156	944
H	H	H	3	eq.	160	HCl	228	944
H	H	Me	3	ax.	96	HCl	176	944
H	H	Me	3	eq.	118	HCl	254	944
H	H	iPr	3	ax.	128	HCl	145	944
H	H	iPr	3	eq.		HCl	242	944
H	H	F	1	ax.	104	HCl	248	944
H	H	F	1	eq.		HCl	146	944
H	H	F	2	ax.	124	HCl	134	944
H	H	F	3	ax.	104	HCl	162	944
H	H	Cl	3	eq.	86	HCl	221	944
H	H	Cl	3	ax.	106	HCl	169	944
H	H	NMe₂	3	eq.	134	HCl	253	944
H	H	F	3	ax.	84	HCl	189	944
H	Ac	F	3	ax.		HCl	180	944
H	Ac	F	3	eq.		HCl	194	944
H	CONH₂	H	3	ax.		HCl	236–238	946
H	CONH₂	H	3	eq.		HCl	245–247	946

H	CONH₂	Me	3	ax.		HCl	238–240	946
H	CONH₂	Me	3	eq.		HCl	258–260	946
H	CONH₂	iPr	3	ax.		HCl	251–253	946
H	CONH₂	iPr	3	eq.		HCl	238–240	946
H	CONH₂	F	1	ax.	110	HCl	235–237	946
H	CONH₂	F	2	ax.	153–155	HCl	279–281	946
H	CONH₂	F	3	ax.		HCl	252–254	945, 946
H	CONH₂	Cl	3	eq.		HCl	243–244	945, 946
H	CONH₂	Cl	3	ax.		HCl	248–250	946
H	CONH₂	F	3	eq.		HCl	233–235	946
H	CONHMe	F	3	ax.		Oxalate	119–122	945, 946
H	CONHEt	F	3	ax.		Oxalate	146–150	945, 946
H	CONHPr	F	3	ax.		HCl	123–130	945, 946
H	CONHCH₂CH=CH₂	F	3	ax.		HCl	148–160	945, 946
H	CONHCH₂CH=CH₂	F	3	eq.	92	Fumarate	224–226	945, 946
H	CONHiPr	F	3	ax.		HCl	90	945, 946
H	CONHiPr	F	3	eq.		MeSO₃H	202–204	946
H	CONHcycloC₅H₉	F	3	ax.		HCl	209–210	945, 946
H	CONHcycloC₆H₁₁	F	3	ax.		HCl	120–125	945, 946
H	CONHCH₂Ph	F	3	ax.		HCl	112–115	945, 946
H	CONHPh	F	3	ax.		HCl	144–146	945, 946
H	CONHPh	F	3	eq.		HCl	232–234	945, 946
H	CONHC₆H₃-3,4-diMe	F	3	ax.		HCl	184–186	946
H	CONHC₆H₃-3,5-diMe	F	3	ax.		HCl	191–193	946
H	CONHC₆H₄-4-Me	F	3	ax.		HCl	175–179	946
H	CONHC₆H₄-4-Me	F	3	eq.		HCl	237–239	946
H	CONHC₆H₄-3-CF₃	F	3	ax.		HCl	208	946
H	CONHC₆H₄-2-CH₂CH=CH₂	F	3	ax.		HCl	189–191	946
H	CONHC₆H₄-4-F	F	3	ax.		HCl	120–123	945, 946
H	CONHC₆H₄-3-Cl	F	3	ax.		MeSO₃H	179–181	945, 946
						HCl	231–233	945

Table 45. (*Contd.*)

R¹	R²	R³	n	orientation*	Base m.p. (°C)	Salt	M.p. (°C)	References
H	CONHC₆H₄-3-Cl	F	3	eq.		HCl	231–233	946
H	CONHC₆H₃-3-Cl-4-Me	F	3	ax.		HCl	171–172	946
H	CONHC₆H₃-3-Cl-4-Me	F	3	eq.		HCl	233–235	946
H	CONHC₆H₃-3,4-diCl	F	3	ax.		HCl	130–132	946
H	CONHC₆H₄-4-Cl	F	3	ax.		HCl	191–193	945, 946
H	CONHC₆H₄-4-Cl	F	3	eq.		HCl	215–218	945, 946
H	CONHC₆H₄-4-Br	F	3	ax.		HCl	183–185	945
H	CONHC₆H₄-4-Br	F	3	eq.		HCl	222	946
H	CONHC₆H₄-4-OMe	F	3	ax.		HCl	149–151	945, 946
H	CONHC₆H₄-4-OMe	F	3	eq.		HCl	203–205	945, 946
H	CONHC₆H₄-2-OCH₂CH=CH₂	F	3	ax.		HCl	174–176	946
H	CONHC₆H₄-2-OCH₂C≡CH	F	3	ax.		HCl	131–133	946
H	CONHC₆H₄-4-NO₂	F	3	ax.		HCl	245–249	946
H	CSNH₂	F	3	ax.		HCl	223–225	946
H	CSNHMe	F	3	ax.		HCl	191–193	945, 946
H	CSNHPr	F	3	ax.		HCl	200–202	945, 946
H	CSNHCH₂CH=CH₂	F	3	ax.		HCl	180–182	945, 946
H	CSNHiPr	F	3	ax.		HCl	183–185	945, 946
H	CSNHPh	F	3	ax.		Fumarate	192–194	945, 946
H	CSNHC₆H₃-3-Me	F	3			Fumarate	184–185	945
H	CSNHC₆H₃-3-Cl-4-Me	F	3	ax.		Fumarate	187–189	946
H	CSNHC₆H₄-4-Cl	F	3	ax.		Fumarate	183–185	946
H	CSNHC₆H₄-4-Br	F	3	ax.		HCl	223–225	946
Pr	H	Me	3			HCl	186	947
Pr	H	F	3		70	HCl	164	947

Pr	Ac	F	3		Oxalate	157	947
Pr	COEt	F	3		HCl	169	947
Pr	CONH₂	F	3		HCl	224	947
CH₂CH=CH₂	H	F	3	120	HCl	135	947
CH₂CH=CH₂	Ac	F	3		Oxalate	147	947
CH₂CH=CH₂	COEt	F	3		HCl	152	947
CH₂CH=CH₂	CONH₂	F	3		HCl	229	947
CH₂C≡CH	H	H	1		HCl	223	947
CH₂C≡CH	H	H	2	100	HCl	98	947
CH₂C≡CH	H	H	3		HCl	95	947
CH₂C≡CH	H	Me	3		HCl	171	947
CH₂C≡CH	H	F	3		HCl	147	947
CH₂C≡CH	H	Cl	2	126	HCl	167	947
CH₂C≡CH	H	Cl	3		HCl	162	947
CH₂C≡CH	H	OMe	2	115	HCl	153	947
CH₂C≡CH	H	OMe	3		HCl	151	947
CH₂C≡CH	Ac	F	3		Fumarate	157	947
CH₂C≡CH	COEt	F	3		Fumarate	132	947

*ax. = axial; eq. = equatorial.

amide or a carbamate. The stereochemistry of the isomers of **516**, which were obtained by gas chromatography or fractional crystallization of the hydrochlorides, was determined from their i.r. and n.m.r. spectra.[949] In the corresponding patent, which carries more examples, the epimers were designated form a (having the lower R_f value) or form b according to their behaviour in a standard t.l.c procedure.[950] For these the stereochemical characteristics were not reported; see Table 46.

B. Ketone at Position 2

The enaminone **518** was produced from the thione **517** and reduced to the saturated ketone **519** as shown, Scheme 55.[951,952] Enaminone **520** and ketone **521** were made in a similar way to compounds **518** and **519** as part of a natural product synthesis. The C-2 of ketone **521** was epimerized by triethylamine.[953]

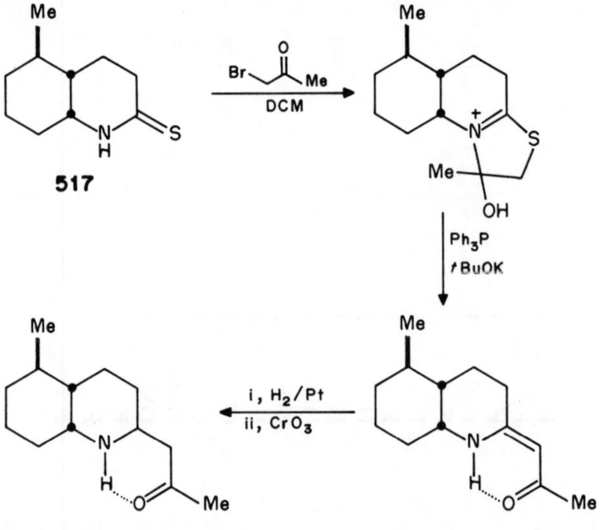

519, HCl m.p. 193 °C   **518**, 70%, oil

SCHEME 55

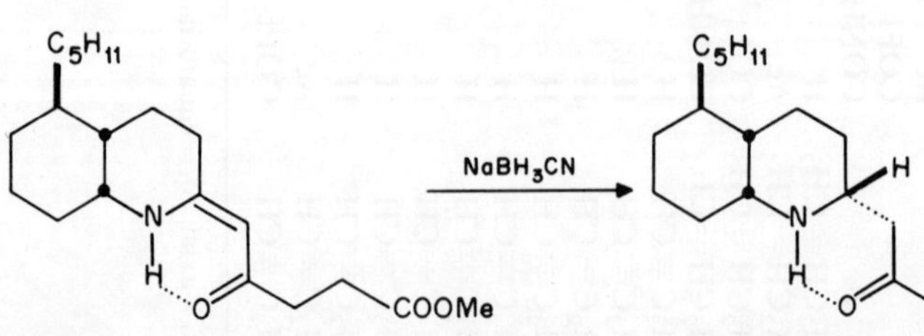

520, 81%, m.p. 52 °C   **521**, 99%

Table 46. 4-Amino-1-ketoalkyl *trans*-Decahydroquinolines

R¹	R²	R³	n	Form*	Salt	M.p. (°C)	References
Ph	H	Ph	1	ax.	HCl	230	949, 950
Ph	H	Ph	3	ax.	HCl	207	949, 950
Ph	Ac	Ph	1	ax.	HCl	233–235	949, 950
Ph	Ac	Ph	3	ax.	HCl	227–228	949, 950
Ph	EtCO	Ph	1	ax.	HCl	227–229	949, 950
Ph	EtCO	Ph	3	ax.	HCl	128–130	949, 950
4-MeC$_6$H$_4$	H	Ph	3	ax.	HCl	226	949, 950
4-MeC$_6$H$_4$	Ac	Ph	3	ax.	HCl	185–186	949, 950
4-MeC$_6$H$_4$	EtCO	Ph	3	ax.	HCl	214–215	949, 950
4-FC$_6$H$_4$	H	cycloC$_6$H$_{11}$	3	a	HCl	211–214	950
4-FC$_6$H$_4$	H	PhCH$_2$	3	a	HCl	255–256	950
4-FC$_6$H$_4$	H	Ph	1	ax.	HCl	256	949, 950
4-FC$_6$H$_4$	H	Ph	3	ax.	2HCl	205	949, 950
4-FC$_6$H$_4$	H	Ph	3	eq.	2HCl	229	949, 950
4-FC$_6$H$_4$	H	Ph	4	ax.	HCl	183	949, 950
4-FC$_6$H$_4$	H	4-MeC$_6$H$_4$	3	ax.	2HCl	191	949, 950
4-FC$_6$H$_4$	H	4-MeC$_6$H$_4$	3	eq.	2HCl	236	949, 950
4-FC$_6$H$_4$	H	4-MeOC$_6$H$_4$	3	ax.	2HCl	180	949, 950
4-FC$_6$H$_4$	HCO	Ph	3	ax.	HCl	209	949
4-FC$_6$H$_4$	PhCH$_2$CO	Ph	3	ax.	HCl	195	949
4-FC$_6$H$_4$	4-ClC$_6$H$_4$CH$_2$CH$_2$CO	Ph	3	ax.	HCl	118	949

Table 46. (*Contd.*)

R^1	R^2	R^3	n	Form*	Salt	M.p. (°C)	References
4-FC$_6$H$_4$	Ac	Ph	1	ax.	HCl	234–236	949, 950
4-FC$_6$H$_4$	Ac	Ph	3	eq.	Oxalate	198–200	949
4-FC$_6$H$_4$	Ac	Ph	3	a	HCl	168–170	949, 950
4-FC$_6$H$_4$	Ac	Ph	3	b	HCl	188–190	950
4-FC$_6$H$_4$	Ac	Ph	4	ax.	HCl	212–214	949, 950
4-FC$_6$H$_4$	EtCO	cycloC$_6$H$_{11}$	3	ax.	HCl	154–156	949
4-FC$_6$H$_4$	EtCO	cycloC$_6$H$_{11}$	3	eq.	Oxalate	178–180	949, 950
4-FC$_6$H$_4$	EtCO	PhCH$_2$	3	a	HCl	147–149	950
4-FC$_6$H$_4$	EtCO	Ph	1	ax.	HCl	225–227	949, 950
4-FC$_6$H$_4$	EtCO	Ph	3	ax.	HCl	158–160	949, 950
4-FC$_6$H$_4$	EtCO	Ph	3	b	Oxalate	161–162	949, 950
4-FC$_6$H$_4$	EtCO	Ph	4	ax.	HCl	162–164	949, 950
4-FC$_6$H$_4$	EtCO	4-MeC$_6$H$_4$	3	ax.	HCl	201–203	949, 950
4-FC$_6$H$_4$	EtCO	4-MeOC$_6$H$_4$	3	ax.	HCl	193–195	949, 950
4-FC$_6$H$_4$	EtCO	1-C$_{10}$H$_7$	3	ax.	HCl	122–124	949, 950
4-FC$_6$H$_4$	C$_3$H$_7$CO	Ph	3	ax.	HCl	174–175	949, 950
4-FC$_6$H$_4$	C$_4$H$_9$CO	Ph	3	ax.	HCl	158–160	949, 950
4-FC$_6$H$_4$	PhCO	Ph	3	ax.	HCl	139–141	949, 950
4-FC$_6$H$_4$	2-Furoyl	Ph	3	ax.	HCl	230–231	949, 950
4-FC$_6$H$_4$	Nicotinoyl	Ph	3	ax.	2HCl	211–213	949, 950
4-FC$_6$H$_4$	Isonicotinoyl	Ph	3	ax.	2HCl	213–215	949, 950
4-FC$_6$H$_4$	COOMe	Ph	3	a		194–196	950
4-FC$_6$H$_4$	COOEt	Ph	3	a		134–136	950
4-FC$_6$H$_4$	COOCH$_2$CH$_2$OMe	Ph	3	a		159–161	950
4-FC$_6$H$_4$	COOBu	Ph	3	a		181–183	950
4-FC$_6$H$_4$	COOsBu	Ph	3	a		178–180	950
4-FC$_6$H$_4$	COOPh	Ph	3	a		185–187	950
4-ClC$_6$H$_4$	H	Ph	3	ax.	HCl	248	949, 950

4-ClC$_6$H$_4$	H	4-MeOC$_6$H$_4$	3	ax.	2HCl	225	949, 950
4-ClC$_6$H$_4$	Ac	Ph	3	ax.	HCl	193–194	949, 950
4-ClC$_6$H$_4$	Ac	4-MeOC$_6$H$_4$	3	ax.	HCl	193–195	949, 950
4-ClC$_6$H$_4$	EtCO	Ph	3	ax.	HCl	130–131	949, 950
4-ClC$_6$H$_4$	EtCO	4-MeOC$_6$H$_4$	3	ax.	HCl	213–215	949, 950
4-BrC$_6$H$_4$	H	Ph	3	ax.	HCl	242	949, 950
4-BrC$_6$H$_4$	Ac	Ph	3	ax.	HCl	219–221	949, 950
4-BrC$_6$H$_4$	EtCO	Ph	3	ax.	HCl	143–145	949, 950
4-MeOC$_6$H$_4$	H	Ph	3	ax.	HCl	231	949, 950
4-MeOC$_6$H$_4$	Ac	Ph	3	ax.	HCl	184–186	949, 950
4-MeOC$_6$H$_4$	EtCO	Ph	3	ax.	HCl	214–216	949, 950
2-Thienyl	H	Ph	3	ax.	2HCl	168	949, 950
2-Thienyl	Ac	Ph	3	ax.	HCl	205–207	949, 950
2-Thienyl	EtCO	Ph	3	ax.	HCl	111–113	949

*ax. = axial, eq. = equatorial; notes a and b relate to t.l.c. R_f values where the orientation has not been determined.

The ketone **522** was prepared as shown in Scheme 56 during a synthetic procedure, but no details were given in the preliminary report.[954]

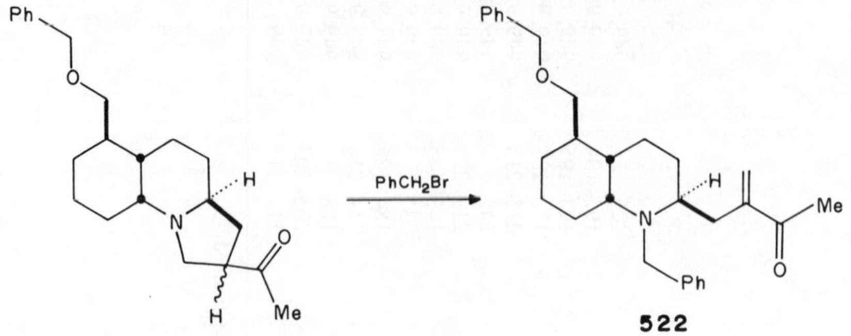

SCHEME 56

C. Ketone at Position 4

See compounds **502** and **503**, p. 365.

XIV. Ketoquinolones

1. 3-Keto-2-quinolones

Ketones of this class. are listed in Table 47. Friedländer synthesis between 2-aminobenzaldehyde **523** and ethyl acetoacetate at 160 °C without solvent gave very poor yields of the ketoquinolone **524** and the ester **525**.[182,955] However,

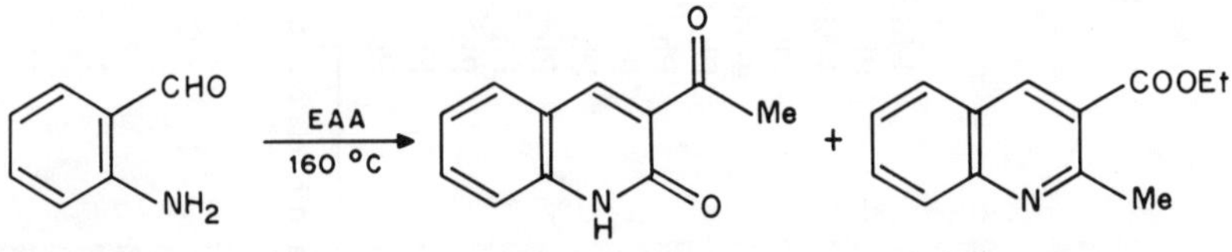

523   **524**, 0.9%   **525**, 2.3%

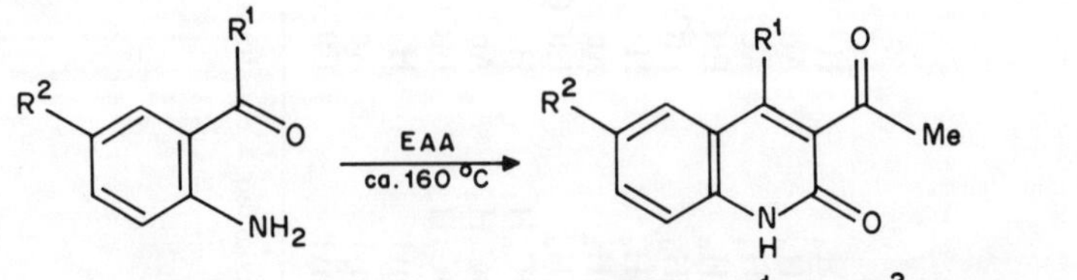

526

527 a, R^1=Me, R^2=NO$_2$, 61%

   b, R^1=Ph, R^2=Cl, 81%

4-substituted quinolones are often available in good yields from aminoketones **526**, e.g. compounds **527a**[956] and **527b**[205]. Improved yields from 2-aminobenzaldehyde have been obtained by reaction with pyrazolones rather than keto-esters. For example, pyrazolone **528** gave 8% of ketoquinolone **524**, but the main product was the pyrazolone **529**, Scheme 57. Other pyrazolones gave phenylhydrazones of 3-keto-2-quinolones in high yields, but these were not hydrolysed.[957]

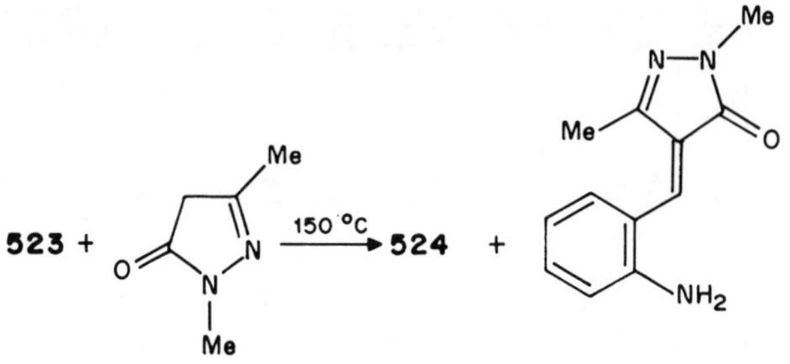

528 8% **529**, 70%

SCHEME 57

Quinolone-4-carboxylic acids have been prepared from isatins, e.g. **530** → **531**.[958]

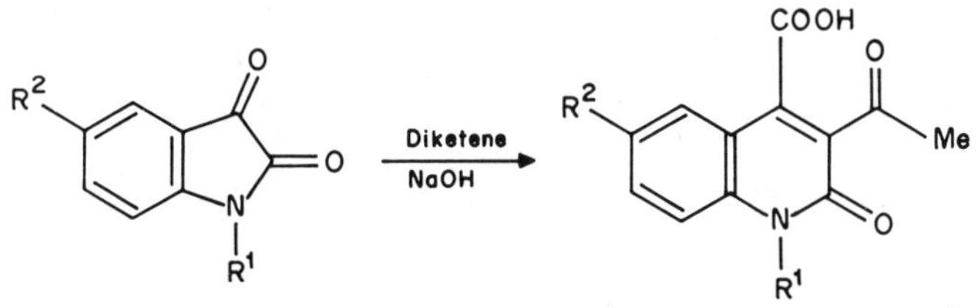

530 **531**, R¹, R² = H, Me

When the diketone **532** was heated in polyphosphoric acid the cyclopentenone ring formed and the methoxy group hydrolysed to give quinolone **533**.[212]

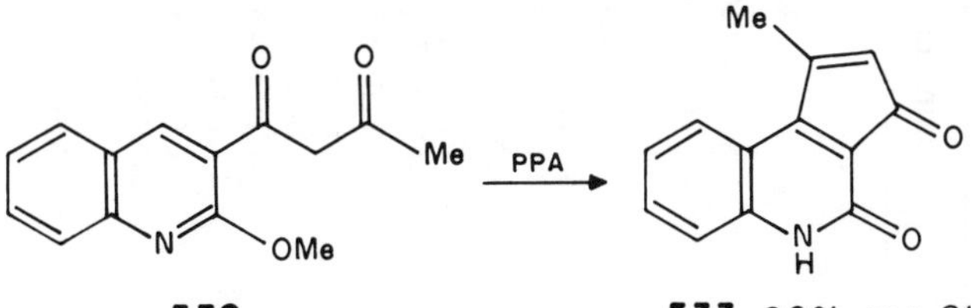

532 **533**, 90%, m.p. 214–216 °C

The pyranone ring of naphthopyranoquinolone **534** was opened with refluxing 30% potassium hydroxide to give a mixture of ketones **535a** and **b**.

Table 47. 3-Keto-2-quinolones

R	Quinoline substituent(s)	Preparation	Yield (%)	M.p. (°C)	References
Me	None	Friedländer synthesis	0.9ᵃ	237	182, 955
		2-NH₂C₆H₄CHO, 1,3-diMe-5-pyrazolone, 150°C	8ᵇ	246	957*, 1007*
Me	1-Me-4-COOH	1-Me-isatin, diketene, NaOH	8	245–247	958
Me	1-CH₂CH₂NMe₂-4-Ph-6-Cl	3-Ac-4-Ph-6-Cl-2-quinolone, Me₂NCH₂CH₂Cl, NaH, DMF	38ᶜ	156–157	205
Me	4-Me-6-NO₂	Friedländer synthesis	61	340–341(d)	956
Me	4-CH(OMe)₂	Friedländer synthesis	94	193	1008
Me	4-Et	Friedländer synthesis		198–199	1009
Me	4-Ph	Friedländer synthesis	60	251–252	639*
Me	4-Ph-6-Cl	Friedländer synthesis	81	274–275	205
Me	4-(2-FC₆H₄)-6-Cl	2-NH₂-5-ClC₆H₃COC₆H₄-2-F, diketene	94	254–256	188*
Me	5,7-diBr-8-OH				1010
Me	6-OMe	Friedländer synthesis		297	1011*
Me	8-OMe	Friedländer synthesis		183	1012*
Me	6-CHO-7-NH₂	Friedländer synthesis	95		1013
Me	4-COOH	Isatin, diketene, NaOH	43	288–289	958
Me	4-COOH-6-Me	5-Me-isatin, diketene, NaOH	47	280–281	958
Me	4-NMe₂	PhN=C=O, Me₂NC≡CCOMe	70	228	1014
CH₂Ph		Q-2-SO₂CHPhCO-3, aq. NaOH	93	255–256.5	206
CF₃	4-Me	3-CF₃CHOH-4-Me-2-quinolone, CrO₃, AcOH		220	1015
Et	1-Me	Friedel–Crafts reaction	25.8ᵈ	130–132	1016, 1017
CH=CHPh		2-NH₂C₆H₄CHO, 1-Ph-3-Me-5-pyrazolone, 140°C; PhCHO		269	1007

			Yield (%)	mp (°C)	Ref.
Pr	1-Me	Friedel–Crafts reaction	9.5[d]	91–92	1016
Bu	1-Me	Friedel–Crafts reaction	2.6	83–84	1016
tBu	1-Ph-5-Me-7-NEt$_2$			>270	1018
Ph	1-Me	Friedländer synthesis			955, 957*
Ph	1-Me	Friedel–Crafts synthesis: (PhCO)$_2$O, H$_2$SO$_4$	3.6	140–142	1017
		Friedel–Crafts synthesis: PhCOCl, AlCl$_3$	2.3		1019
Ph	1-Me-4-CH$_2$Ph	3-PhCO-4-PhCH$_2$-2-quinolone, MeI, KOH	78	178–179	1020
Ph	4-CH$_2$Ph	Friedländer synthesis	77	263.5–265.5	1020
Ph	4-Et	Friedländer synthesis		213	1009
Ph	4-Ph	Friedländer synthesis	70	259–260	639
Ph	6-OMe	Friedländer synthesis		293	1011*
Ph	8-OMe	Friedländer synthesis		206	1021*
Ph	6-CHO-7-NH$_2$	Friedländer synthesis	95	278–279(d)	1022
Ph	6-CHO-7-NHAc	Above, Ac$_2$O		320(d)	1013, 1022
C$_6$H$_4$-2-NO$_2$	4-Me	Friedländer synthesis		238–240	1023
C$_6$H$_4$-2-NH$_2$	4-Me	3-(COC$_6$H$_4$-2-NO$_2$)-4-Me-2-quinolone, H$_2$, Pd		277–279	1023
C$_6$H$_4$-2-NHAc[e] C$_6$H$_4$-2-NAc$_2$ }	Me Me }	1-HO-3-(COC$_6$H$_4$-2-NO$_2$)-4-Me-2-quinolone, H$_2$, Ni; Ac$_2$O		{274–276 218–220}	1023

*These references include carbonyl derivatives.

[a] Formed as a mixture with ethyl 2-methylquinoline-3-carboxylate (2.3%).

[b] By-product to the main reaction; see Section XIV.1.

[c] The reaction product is reported to be either N-alkylated, as shown, or O-alkylated; Table 2.

[d] Formed as a mixture with the 6-ketone; Table 51.

[e] Mixture obtained.

Compound **535a** could be converted into the O-ethyl derivative **535b** with sodium ethoxide in ethanol.[959]

534

KOH, EtOH
reflux

535 a, R = H, 22.6%, m.p. 238—242 °C
   b, R = Et, 41.6%, m.p. 225—226 °C

Some 7-amino-6-formyl-3-keto-2-quinolones are described in Chapter 1, Section V.4 and included in Table 47.

2. 4-Hydroxy-3-keto-2-quinolones

This large class of compounds is listed in Table 48. The ^{13}C chemical shifts shown on the formula of 3-acetyl-4-hydroxy-1-methyl-2-quinolone **536** were determined in deuterochloroform. It was shown that they confirmed the tautomeric structure of the compound. No signals for the alternative 2-hydroxy-4-one form were seen.[960] The mass spectral fragmentations of this and several similar compounds have been studied.[961]

536

Figures Indicate ^{13}C Chemical Shifts

Photochemical conversion of the oxindole **537** to the quinolone **538** was observed and assumed to go through the stages shown in Scheme 58.[962]

The 3-keto group is often introduced into these compounds by a Friedel–Crafts reaction or a Fries rearrangement of a 4-acyloxy-2-quinolone; See Table 48. Many examples have been prepared by ring closures, e.g. **539** → **540**.[963,964] In many cases the keto-amide is prepared *in situ* from the anthranilic ester and diketene[965,966] or a keto-ester.[967–969]

Isatotic anhydrides **541**, R^1 = Me or Et reacted with keto-esters, R^2 = alkyl or aryl to give ketones **542**, although yields were low. The enol protons showed chemical shifts of 12–15 ppm.[970]

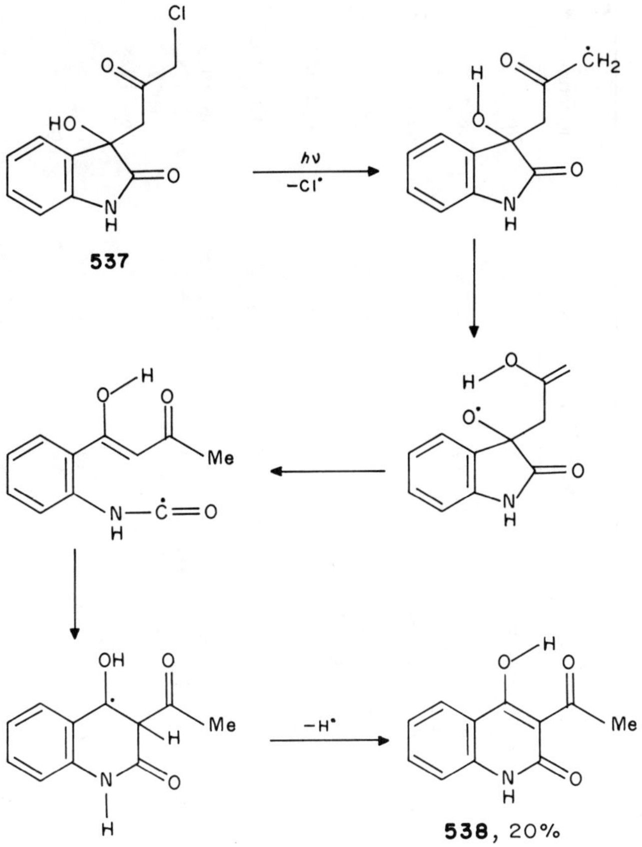

SCHEME 58

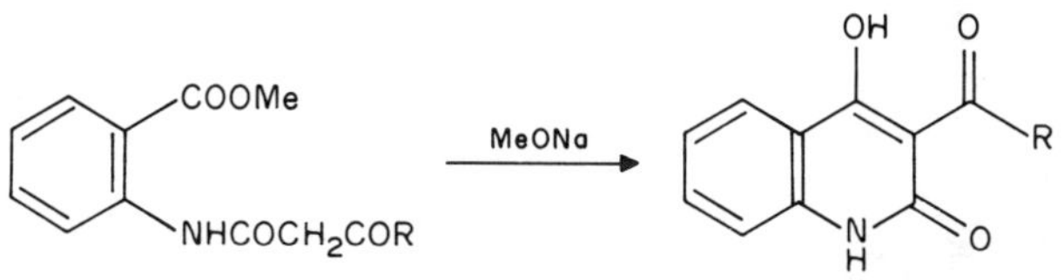

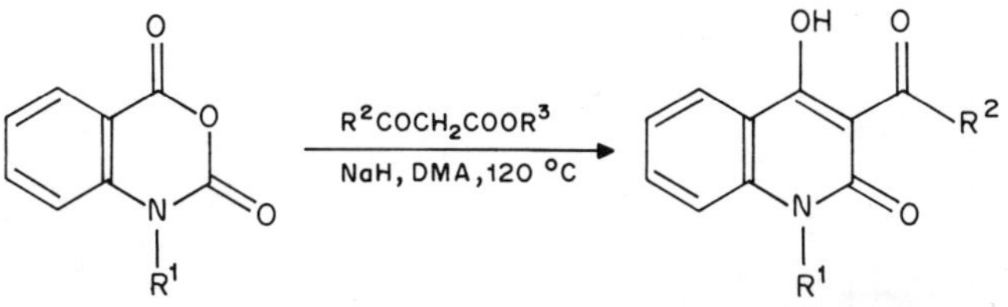

Table 48. 4-Hydroxy-3-keto-2-quinolones

Entry	R^1	R^2	Other substituent(s)	Preparation	Yield (%)	M.p. (°C)	References
1	Me	H	None	2-MeOOCC$_6$H$_4$NHCOCH$_2$COMe, MeONa	95		963
				2-MeOOCC$_6$H$_4$NH$_2$, diketone, 120 °C	96	259	966*
				2-MeOOCC$_6$H$_4$NH$_2$, diketone, NaOH			978
				2-H$_2$NC$_6$H$_4$COOH, diketene, Ac$_2$O; MeONa	83	249–251	965
				PhNHCOCH(COMe)COOEt, 210 °C	56	256	1024, 1025
				4-HO-2-quinolone, AcOH, P$_2$O$_5$, H$_3$PO$_4$	80	255–257	1026*, 1027
				4-AcO-2-quinolone, AlCl$_3$	65	255–256	946, 1028*
				2-EtOOCC$_6$H$_4$NHCOCH$_2$COMe, 50% KOH, 90 °C	30	245–250	977, 978, 1029
				PhNH$_2$, AcCH(COOEt)$_2$, Ph$_2$O or PhNO$_2$	22		966, 1028
				Friedel–Crafts reaction	39	255–256	964
				3-N$_2$-4-Ac-4-HO-2-quinolone, C$_6$H$_6$, reflux	28	254–257	976
2	Me	H	1-Me	Friedel–Crafts reaction	56.5	143–145	964
				See text, Section XIV.2, Scheme 59	52	144–145	974*
				1-Me-isatoic anhydride, EAA	9	143–146	970
				1-Me-4-AcO-2-quinolone, AlCl$_3$			964

No.						mp	Ref.
3	Me	H	1-Me-6-Br	1-Me-3-Ac-4-HO-2-quinolone, Br$_2$, AcOH	75	197	1030
4	Me	H	5-Me	2-MeOOC-3-MeC$_6$H$_3$-NHCOCH$_2$COMe, MeONa			963
5	Me	H	6-Me	4-MeC$_6$H$_4$NH$_2$, AcCH(COOEt)$_2$, 220–235 °C	25	290–296	967
6	Me	H	6,8-diMe	2,4-diMeC$_6$H$_3$NH$_2$, AcCH(COOEt)$_2$	22	285–290.5	967
7	Me	H	7,8-diMe	2,3-diMeC$_6$H$_3$NH$_2$, AcCH(COOEt)$_2$	32	276–281.5	967
				2-MeOOC-5,6-dimeC$_6$H$_2$-NHCOCH$_2$COMe, MeONa			963
8	Me	H	8-Me	2-MeC$_6$H$_4$NH$_2$, AcCH(COOEt)$_2$	31	249–251.5	967
9	Me	H	1-Et	Friedel–Crafts reaction	26	115–117	964
10	Me	H	6-Et	4-EtC$_6$H$_4$NH$_2$, AcCH(COOEt)$_2$		259–262	967
11	Me	H	6-Bu	4-BuC$_6$H$_4$NH$_2$, AcCH(COOEt)$_2$	27	218–220	967
12	Me	H	1-Ph	Ph$_2$NH, CH$_2$(COOEt)$_2$, heat; 2м-NaOH	86	234	972
				1-Ph-3-COCH$_2$COOEt-4-HO-2-quinolone, NaOH	85.6	234	1031
				1-Ph-4-HO-2-quinolone, Ac$_2$O, BF$_3$	31		972
13	Me	H	6-F	4-FC$_6$H$_4$NH$_2$, AcCH(COOEt)$_2$	17	282–287	967
14	Me	H	5-Cl	3-ClC$_6$H$_4$NH$_2$, AcCH(COOEt)$_2$, PhNO$_2$	9.7[a]	295	971
15	Me	H	6-Cl	4-ClC$_6$H$_4$NH$_2$, AcCH(COOEt)$_2$	18	268–275	967

Table 48. (*Contd.*)

Entry	R¹	R²	Other substituent(s)	Preparation	Yield (%)	M.p. (°C)	References
16	Me	H	6,8-diCl	2-NH$_2$-3,5-diClC$_6$H$_2$COOH, diketene, Ac$_2$O; MeONa	84	303–306	965
17	Me	H	7-Cl	3-ClC$_6$H$_4$NH$_2$, AcCH(COOEt)$_2$, PhNO$_2$	42.1[a]	283–286	971
18	Me	H	7-Cl-8-Me	2-Me-3-ClC$_6$H$_3$NH$_2$, AcCH(COOEt)$_2$	12	266–268.5	967
19	Me	H	6-Br	4-BrC$_6$H$_4$NH$_2$, AcCH(COOEt)$_2$	11	281–286.5	967
20	Me	H	6-OMe	4-MeOC$_6$H$_4$NH$_2$, AcCH(COOEt)$_2$, PhNO$_2$	63	287–289	971
				4-MeOC$_6$H$_4$NH$_2$, AcCH(COOEt)$_2$	34	281–284	967
21	Me	H	6,8-diOMe	2,4-diMeOC$_6$H$_3$NH$_2$, CH$_2$(COOEt)$_2$, Ph$_2$O; NaOH		280–282	973
				2,4-diMeOC$_6$H$_3$NH$_2$, AcCH(COOEt)$_2$, PhNO$_2$		280–282	973
22	Me	H	7-OMe	3-MeOC$_6$H$_4$NH$_2$, AcCH(COOEt)$_2$	42	287–291.5	967
23	Me	H	8-OMe	2-MeOC$_6$H$_4$NH$_2$, AcCH(COOEt)$_2$	9	267–273	967
24	Me	H	6-OCH$_2$Ph	4-PhCH$_2$OC$_6$H$_4$NH$_2$, AcCH(COOEt)$_2$	32	240–246	967
25	Me	H	6-COOEt	4-EtOOCC$_6$H$_4$NH$_2$, AcCH(COOEt)$_2$	31	225–231	967
26	Me	H	7-NO$_2$	2-NH$_2$-4-NO$_2$C$_6$H$_3$COOH, diketene, Ac$_2$O; MeONa	73.6	288–290	965
27	CH$_2$(4-HO-2-quinolon-1-yl)	H	1-CH$_2$COOH	See text, Section XV.1		213(d)	1032, 1033, 986

No.	R		N-1	Preparation	Yield (%)	m.p. (°C)	Ref.
28	CH$_2$(4-MeO-2-quinolon-1-yl)	Me	1-CH$_2$COOH	See text, Section XV.1	91	230–232	1032, 1033, 986
29	CF$_3$	H	1-Me	1-Me-isatoic anhydride, CF$_3$COCH$_2$COOEt	13	167–169	970
30	CF$_3$	H	1-Et	1-Et-isatoic anhydride, CF$_3$COCH$_2$COOEt	5.5	147–150	970
31	CHCl$_2$	H	1-Me	See text, Section XIV.2, Sceheme 59	51	188–199	975
32	Et	H		4-HO-2-quinolone, EtCOOH, P$_2$O$_5$, H$_3$PO$_4$	72	224–225	1026*, 1027
33	Et	H	1-Me	Friedel–Crafts reaction	38	225–226	964
34	CH$_2$CHPh$_2$	H	1-Me	Friedel–Crafts reaction	45.4	150–151	964
				1-Me-3-COCH=CHPh-4-HO-2-quinolone, PhMgBr	45	136–137	1034
35	CH$_2$CH(Ph)SPh	H	1-Me	1-Me-3-COCH=CHPh-4-HO-2-quinolone, PhSH	74	143–144	1034
36	CH$_2$CH(Ph)SPh	H	1-Et	1-Et-3-COCH=CHPh-4-HO-2-quinolone, PhSH	72	135–136	1034
37	CH$_2$COOEt	H	1-Ph	Ph$_2$NH, Me$_2$NCOCH$_2$-COOEt, POCl$_3$, 95–100 °C	10	140–141	1031
38	CH$_2$CH$_2$NEt$_2$	H	1-Ph	Mannich reaction	100	310	1035
39	CH$_2$CH$_2$N(CH$_2$)$_5$	H	1-Ph	Mannich reaction	90	256	1035
40	CH$_2$CH$_2$N(CH$_2$-CH$_2$)$_2$O	H	1-Ph	Mannich reaction	100	227	1035
41	CH$_2$CH$_2$NHPh	H	1-Ph	Entry 39, PhNH$_2$, H$_2$O, EtOH, HCl	62	224	1035
42	CH$_2$CH(Ph)SO$_2$Ph	H	1-Ph	Entry 41, PhSO$_2$Cl, Py	60	211	1035
43	CH=CH$_2$	H	1-Me	1-Me-3Ac-4-HO-2-quinolone, HCHO	82	295–300	1034
44	CH=CH$_2$	H	1-Et	1-Et-3-Ac-4-HO-2-quinolone, HCHO	79	268–270	1034
45	CH=CH$_2$	H	1-Ph	1-Ph-3-Ac-4-HO-2-quinolone, HCHO	80	327–330	1034
46	CH=CHPh	H	1-Me	1-Me-3-Ac-4-HO-2-quinolone, PhCHO	85	170–171	1034

Table 48. (*Contd.*)

Entry	R^1	R^2	Other substituent(s)	Preparation	Yield (%)	M.p. (°C)	References
47	CH=CHPh	H	1-Et	1-Et-3-Ac-4-HO-2-quinolone, PhCHO	85	135–136	1034
48	CH=CHPh	H	1-Ph	1-Ph-3-Ac-4-HO-2-quinolone, PhCHO	83	306–310	1034
49	CH=CHC$_6$H$_4$-4-OMe	H	1-Me	1-Me-3-Ac-4-HO-2-quinolone, 4-MeOC$_6$H$_4$CHO	80	172–173	1034
50	CH=CHC$_6$H$_4$-4-OMe	H	1-Et	1-Et-3-Ac-4-HO-2-quinolone, 4-MeOC$_6$H$_4$CHO	82	138	1034
51	CH=CHC$_6$H$_4$-4-OMe	H	1-Ph	1-Ph-3-Ac-4-HO-2-quinolone, 4-MeOC$_6$H$_4$CHO	81	260	1034
52	Pr	H		4-HO-2-quinolone, PrCOOH, P$_2$O$_5$, H$_3$PO$_4$	70	217–219	1026*, 1027
				EtOOCC$_6$H$_4$NHCOCH$_2$-COPr, Na	59	218–219	964
53	Pr	H	1-Me	Friedel–Crafts reaction	25	97–98	964
				Friedel–Crafts reaction	21	95–96	964
				1-Me-isatoic anhydride, PrCOCH$_2$COOEt	17		970
54	CH$_2$COMe	H	1-Me	1-Me-3-Ac-4-HO-2-quinolone, EtOAc, Na	76	142–143	1030
55	CH$_2$COMe	H	1-Et	1-Et-3-Ac-4-HO-2-quinolone, EtOAc, Na	74	143–144	1030
56	CH$_2$COMe	H	1-Ph	1-Ph-3-Ac-4-HO-2-quinolone, EtOAc, Na	65	176–177	1030
57	CH$_2$COCOOEt	H	6-Me	Entry 5, (COOEt)$_2$, EtONa			967
58	CH$_2$COCOOEt	H	8-Me	Entry 8, (COOEt)$_2$, EtONa			967
59	CH$_2$COCOOEt	H	6-Et	Entry 10, (COOEt)$_2$, EtONa			967
60	CH$_2$COCOOEt	H	6-Bu	Entry 11, (COOEt)$_2$, EtONa			967
61	CH$_2$COCOOEt	H	6-OMe	Entry 20, (COOEt)$_2$, EtONa			967

Entry					Yield (%)	mp	Ref.
62	CH$_2$COCOOEt	H	7-OMe	Entry 22, (COOEt)$_2$, EtONa			967
63	CH$_2$COCOOEt	H	8-OMe	Entry 23, (COOEt)$_2$, EtONa			967
64	CH$_2$COCOOEt	H	6-F	Entry 12, (COOEt)$_2$, EtONa			967
65	CH$_2$COCOOEt	H	6-Cl	Entry 15, (COOEt)$_2$, EtONa			967
66	CH$_2$COCOOEt	H	7-Cl-8-Me	Entry 18, (COOEt)$_2$, EtONa			967
67	CH$_2$COCOOEt	H	6-Br	Entry 19, (COOEt)$_2$, EtONa			967
68	CH$_2$COCOOEt	H	6-OCH$_2$Ph	Entry 24, (COOEt)$_2$, EtONa			967
69	CH$_2$COCOOEt	H	6-COOEt	Entry 25, (COOEt)$_2$, EtONa			967
70	CH$_2$COCOOEt	H	7,8-diMe	Entry 7, (COOEt)$_2$, EtONa			967
71	CH$_2$COCOOEt	H	6,8-diMe	Entry 6, (COOEt)$_2$, EtONa	59	221–224	630
72	iPr	H		Friedel–Crafts reaction 4-HO-2-quinolone, iPrCOOH, P$_2$O$_5$, H$_3$PO$_4$	18	222–224	1026
73	iPr	H	1-Me	1-Me-isatoic anhydride, iPrCOCH$_2$COOEt	22	94–96	970
74	iPr	H	8-OMe	Friedel–Crafts reaction	75.2	169–170	630
75	iPr	Me		3-iPrCO-4-HO-2-quinolone, CH$_2$N$_2$	13.8[b]	158–160	630
76	iPr	Me	1-Me-8-OMe	3-iPrCO-4-HO-8-MeO-2-quinolone, CH$_2$N$_2$	10[c]	68–69	630
77	iPr	Me	8-OMe	3-iPrCO-4-HO-8-MeO-2-quinolone, CH$_2$N$_2$	17.5[c]	189–191	630
78	CH=C(Me)NHEt	H	1-Et	1-Et-3-COCH$_2$COMe-4-HO-2-quinolone, EtNH$_2$	62	167	1030
79	CH=C(Me)NHEt	H	1-Ph	1-Ph-3-COCH$_2$COMe-4-HO-2-quinolone, EtNH$_2$	88	219–220	1030
80	CH=C(Me)NHBu	H	1-Et	1-Et-3-COCH$_2$COMe-4-HO-2-quinolone, BuNH$_2$	71	153	1030
81	CH=C(Me)NHBu	H	1-Ph	1-Ph-3-COCH$_2$COMe-4-HO-2-quinolone, BuNH$_2$	84	218–219	1030
82	CH=C(Me)-NHcycloC$_6$H$_{11}$	H	1-Me	1-Me-3-COCH$_2$COMe-4-HO-2-quinolone, cycloC$_6$H$_{11}$NH$_2$	75	176–177	1030

Table 48. (*Contd.*)

Entry	R^1	R^2	Other substituent(s)	Preparation	Yield (%)	M.p. (°C)	References
83	CH=C(Me)-NHCH$_2$Ph	H	1-Me	1-Me-3-COCH$_2$COMe-4-HO-2-quinolone, PhCH$_2$NH$_2$	72	184–185	1035
84	CH=C(Me)NHPh	H	1-Me	1-Me-3-COCH$_2$COMe-4-HO-2-quinolone, PhNH$_2$	73	163–164	1030
85	CH=C(Me)NHPh	H	1-Et	1-Et-3-COCH$_2$COMe-4-HO-2-quinolone, PhNH$_2$	80	180	1030
86	CH=C(Me)NHPh	H	1-Ph	1-Ph-3-COCH$_2$COMe-4-HO-2-quinolone, PhNH$_2$	83	240	1030
87	CH=C(Me)-NHC$_6$H$_4$-4-Me	H	1-Me	1-Me-3-COCH$_2$COMe-4-HO-2-quinolone, 4-MeC$_6$H$_4$NH$_2$	70	186–187	1030
88	CH=C(Me)N(CH$_2$)$_5$	H	1-Me	1-Me-3-COCH$_2$COMe-4-HO-2-quinolone, C$_5$H$_{11}$N	59	274	1030
89	Bu	H		4-HO-2-quinolone, BuCOOH, P$_2$O$_5$, H$_3$PO$_4$	55	211–212	1026
90	CH=CHCH=CHPh	H	1-Me	1-Me-3-Ac-4-HO-2-quinolone, PhCH=CHCHO	76	182–183	1034
91	CH=CHCH=CHPh	H	1-Et	1-Et-3-Ac-4-HO-2-quinolone, PhCH=CHCHO	80	157	1034
92	CH=CHCH=CHPh	H	1-Ph	1-Ph-3-Ac-4-HO-2-quinolone, PhCH=CHCHO	75	220–262	1034
93	CH$_2$iPr	H		4-HO-2-quinolone, iPrCH$_2$COOH, P$_2$O$_5$, H$_3$PO$_4$	98	197–198	1026*, 1027
				Friedel–Crafts reaction EtOOCC$_6$H$_4$NHCOCH$_2$-COCH$_2$iPr, Na	14	196–197	964, 1036 964

No.	R			Method	Yield (%)	m.p./b.p. (°C)	Ref.
94	CH$_2$iPr	H		Friedel–Crafts reaction	16	b.p. 150–158/0.04 mm	964
95	CH$_2$iPr	H	1-Me	Friedel–Crafts reaction	54	142–143	1037
			8-OMe	2-MeOC$_6$H$_4$NH$_2$, iPrCH$_2$COCH(COOEt)$_2$, Ph$_2$O	23.4		1037
96	CH$_2$iPr	Me		3-iPrCHOH-4-MeO-2-quinolone, CrO$_3$	66	186–187	1036
97	CHBriPr	H		Entry 93, CH$_2$N$_2$	43	Crude	1036
98	CH=CMe$_2$^{c}	H		Entry 93, PyBr$_3$			1036
				9-Me$_2$C=CHCOO-2-quinolone, AlCl$_3$			1028
99	C$_5$H$_{11}$	H		4-HO-2-quinolone, C$_5$H$_{11}$COOH, P$_2$O$_5$, H$_3$PO$_4$	48	184–185	1026*, 1027
				EtOOCC$_6$H$_4$NHCOCH$_2$-COC$_5$H$_{11}$, Na	14	183–184	964
100	C$_6$H$_{13}$	H		Friedel–Crafts reaction	12	166	964
				4-HO-2-quinolone, C$_6$H$_{13}$COOH, P$_2$O$_5$, H$_3$PO$_4$	44		1026, 1027
101	C$_7$H$_{15}$	H		4-HO-2-quinolone, C$_7$H$_{15}$COOH, P$_2$O$_5$, H$_3$PO$_4$	40	174–174.5	1026*, 1027
				Friedel–Crafts reaction	17	168–169	964
				EtOOCC$_6$H$_4$NHCOCH$_2$-COC$_7$H$_{15}$, Na	Trace		964
102	C$_9$H$_{19}$	H		Friedel–Crafts reaction	10	127–130	964
				2-RNH-3-PhCO-4-quinolone, aq. HCl, reflux (for R=4-MeC$_6$H$_4$, 4-HOC$_6$H$_4$, cycloC$_6$H$_{11}$, 1-naphthyl)	80–90	262	1038
103	Ph	H		2-ArNH-3-PhCO-4-quinolone, aq. HCl (for Ar=Ph, substituted Ph)	70	253	1039

Table 48. (*Contd.*)

Entry	R^1	R^2	Other substituent(s)	Preparation	Yield (%)	M.p. (°C)	References
				$EtOOCC_6H_4NHCOCH_2\text{-}COPh$, Na	26	258	964
				Friedel–Crafts reaction 4-PhCOO-2-quinolone, $AlCl_3$	12		964
104	Ph	H	1-Me	Friedel–Crafts reaction	36.5	182–184	964
				1-Me-4-PhCOO-2-quinolone, $AlCl_3$			964
105		H	1-Me-6-Cl	1-Me-isatoic anhydride, $PhCOCH_2COOEt$	3.5[d]	178–180	970
106		H	1-Me-6-OCH_2O-7	1-Me-6-Cl-isatoic anhydride, $PhCOCH_2COOEt$	5.2[d]	200–202	970
				1-Me-6-OCH_2O-7-isatoic anhydride, $PhCOCH_2$-COOEt	5[d]	239–243	970
107	Ph	H	6-Me			295–296	968
108	Ph	H	6,7-diMe			292	968
109	Ph	H	6,8-diMe			>300	968
110	Ph	H	7-Me			>300	968
111	Ph	H	8-Me			273–274	968
112	Ph	H	1-Et	Friedel–Crafts reaction		135–136	964
				1-Et-isatoic anhydride, $PhCOCH_2COOEt$	2.4[d]	101–104	970
113	Ph	H	6-iPr	Friedel–Crafts reaction 4-HO-5-iPr-2-quinolone, NaH, DMF, PhCOCl		270	968
114	Ph	H	6-Bu			230–231	968
115	Ph	H	6-C_5H_{11}			218–219.5	968
116	Ph	H	6-C_6H_{13}			206–207	968

No.	R (2-)	N	Ring subst.	Reagents / Method	m.p. (°C)	Ref.
117	Ph	H	6-C_8H_{17}		207–208.5	968
118	Ph	H	6-Cl	Friedel–Crafts reaction	>300	968
119	Ph	H	6,8-diCl		258–259	968
120	Ph	H	7-Cl		>300	968
121	Ph	H	6,7-diOMe		>300	968
122	Ph	H	7-OMe		259–261	969*
123	Ph	H	5-COOH	3-MeOC_6H_4NH_2, PhCOCH(COOEt)$_2$, Ph$_2$O; 3-PhCO-4-HO-5-COOMe-2-quinolone, NaOH, H$_2$O, DMSO	>300	968
124	Ph	H	6-COOH		>300	968
125	Ph	H	7-COOH		>300	968
126	Ph	H	8-COOH		286–288	968
127	Ph	H	5-COOMe	2,3-diCOOMeC_6H_3NH_2, PhCOCH_2COOH, DCCl; MeONa	263–265	968
128	Ph	H	6-COOMe		279–281	968
129	Ph	H	7-COOMe		285–287	968
130	Ph	H	8-COOMe		184–186	968
131	Ph	H	7-NO_2		>300	968
132	2-MeC_6H_4	H			292–295	968
133	2,4-diMeC_6H_3	H			253–255	968
134	3-$CF_3C_6H_4$	H			258–260	968
135	4-MeC_6H_4	H			290–292	968
136	4-EtC_6H_4	H	6-C_6H_{13}		231–233	968
137	4-iPrC_6H_4	H			286–288	968
138	4-$C_5H_{11}C_6H_4$	H			199–202	968
139	4-$C_5H_{11}C_6H_4$	H	6-C_6H_{13}		180–182	968
140	2-ClC_6H_4	H			>300	968
141	2-ClC_6H_4	H	6,8-diMe		>300	968
142	2-ClC_6H_4	H	6-iPr		224–225	968
143	2-ClC_6H_4	H	6-C_6H_{13}		184–185	968
144	2,4-diClC_6H_3	H	6-C_6H_{13}		294–297	968
145	3-ClC_6H_4	H	6-C_6H_{13}		268	968

Table 48. (*Contd.*)

Entry	R^1	R^2	Other substituent(s)	Preparation	Yield (%)	M.p. (°C)	References
146	3,4-diClC$_6$H$_3$	H				285–290	968
147	4-ClC$_6$H$_4$	H				>300	968
148	2-MeOC$_6$H$_4$	H				275–276	968
149	2,4-diMeOC$_6$H$_3$	H				260–262	968
150	2,3,4-triMeOC$_6$H$_2$	H				247–249	968
151	2,3,4-triMeOC$_6$H$_2$	H	6-Me			231–233	968
152	2,3,4-triMeOC$_6$H$_2$	H	8-Me			244–246	968
153	2,3,4-triMeOC$_6$H$_2$	H	7-NO$_2$			>300	968
154	2,4,5-triMeOC$_6$H$_2$	H				228–230	968
155	3,4-diMeOC$_6$H$_3$	H		2-COOEtC$_6$H$_4$NH$_2$, 3,4-diMeOC$_6$H$_3$COCH$_2$-COOEt, EtONa		235–236	968
156	3,4-diMeOC$_6$H$_3$	H	6-Me			282–284	968
157	3,4-diMeOC$_6$H$_3$	H	8-Me			259–261	968
158	3,4-diMeOC$_6$H$_3$	H	7-NO$_2$			285–290	968
159	3,4,5-triMeOC$_6$H$_2$	H				199–200	968
160	4-MeOC$_6$H$_4$	H				275–280	968
161	4-MeOC$_6$H$_4$	H	6-C$_6$H$_{13}$			203–206	968
162	2-EtOC$_6$H$_4$	H				265–267	968
163	2-iPrOC$_6$H$_4$	H				260–262	968
164	4-COOHC$_6$H$_4$	H				>300	968
165	4-COOMeC$_6$H$_4$	H				>300	968
166	4-NO$_2$C$_6$H$_4$	H				>300	968
167	4-NO$_2$C$_6$H$_4$	H	6,8-diMe			>300	968
168	4-NO$_2$C$_6$H$_4$	H	6-iPr			>300	968
169	4-NO$_2$C$_6$H$_4$	H	6-C$_6$H$_{13}$			264–265	968

170	2-HO-1-naphthyl	H	1-Me	8-Methyl-13,14-dioxo-8H,13H,14H-naphtho-[1^1,2^1:5,6]pyrano[2,3-b]-quinoline, KOH, EtOH] HCl	87.2	320–322	959
171	5-Me-1,3,4-oxadizaol-2-yl	Ac	1-Me	Entry 172, Ac_2O	74	228	975
172	5-Tetrazolyl	H	1-Me	Entry 31, NaN_3, DMF	60	232	975

*These references include carbonyl derivatives.
[a]These two isomers were obtained as a mixture.
[b]Obtained as a mixture with 1-Me-3-iPrCOCH$_2$-4-MeO-2-quinolone; see text, Section XV.4.
[c]These two compounds were obtained as a mixture.
[d]The major product was the ethyl 1-alkyl-2-phenyl-4-quinolone-3-carboxylate.
N.B. Only a few examples of the preparations of the compounds of Ref. 968 are given in the patent, but the rest are said to be made by one or other of these methods.

Anilines **543** condensed with diethyl 2-acetylmalonate, usually without solvent at temperatures over 220 °C or in boiling nitrobenzene, to give 3-acetyl-4-hydroxy-2-quinolones **544**.[967,971] In one case diethyl 2-benzoylmalonate was used similarly.[969] Anilines heated with excess diethyl malonate reacted with two molecules to give the pyranoquinolines **545**. Hydrolysis then gave ketones **544**.[972–974] The N-methyl derivative **545**, $R^1 = Me$, $R^2 = H$ was prepared similarly and treated with sulphuryl chloride to give the dichloromethyl ketone **546**, which was elaborated to the heterocyclic ketones **547** and **548**, Scheme 59.[975]

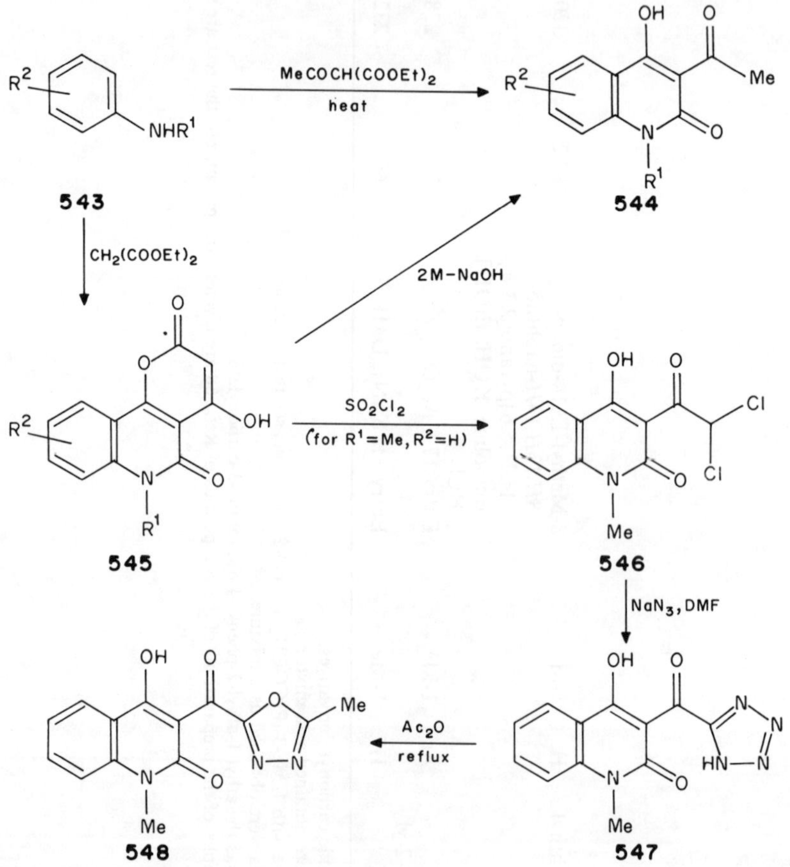

SCHEME 59

In nitrobenzene at 230 °C 3-chloroaniline and diethyl 2-acetylmalonate gave a mixture of ketoquinolones **549a** (10%) and **549b** (42%). It was stated that both compounds were more acidic than phenol. Compound **549a** dissolved in sodium carbonate solution, but compound **549b** gave an insoluble salt, making separation easy.[971]

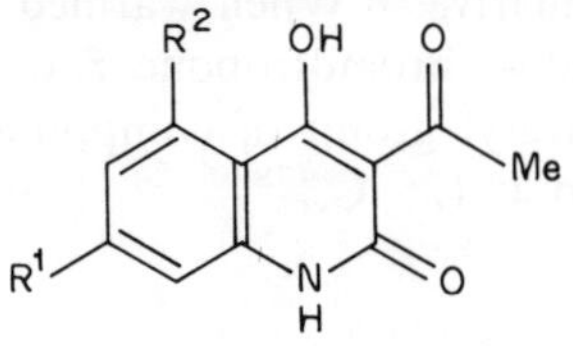

549 a, R[1]=H, R[2]=Cl
      b, R[1]=Cl, R[2]=H

Diazotization of the aminobenzazepine **550** in aqueous conditions gave the 4-acetyl-2-quinolone **554** in a reaction presumed to go via the intermediates **551**, R = H and **553**. Compound **554** rearranged to 3-acetyl-4-hydroxy-2-quinolone **552** in boiling benzene. When the original diazotization was run in methanol, the intermediate **551**, R = Me could be isolated and thermolysed directly to **552**, Scheme 60.[976]

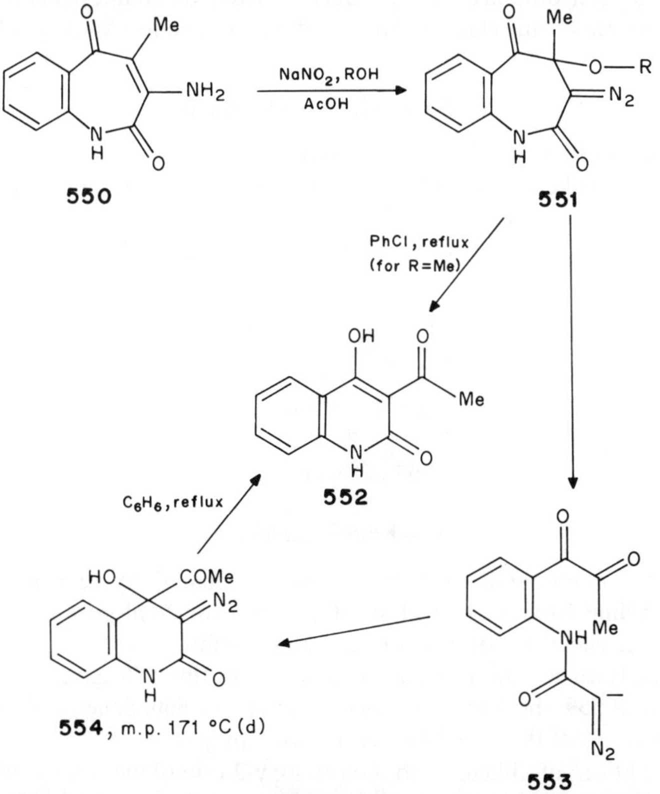

SCHEME 60

Diketene and 2-aminobenzoic acid reacted in acetic anhydride to give the benzoxazine **555**. In strong base this rearranged to the quinolone **552**. Some derivatives were made similarly.[965] When warmed with alcoholic potassium hydroxide, 2-acetoacetamido-7-bromotropone **556** rearranged to compound **552**, Scheme 61.[977] The acetyl group of compound **552** was hydrolysed in concentrated sulphuric acid at 125 °C.[978]

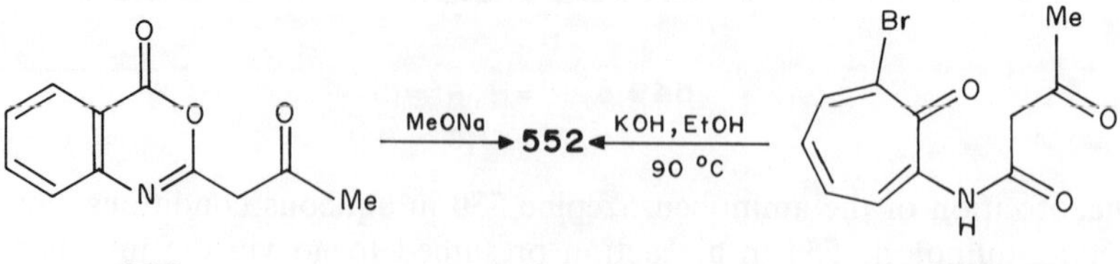

555   **556**

SCHEME 61

As described in Section XV.4, 3-isobutyryl-4-methoxy-2-quinolone **678** (p. 445) has been obtained as a product of a diazomethane reaction.[630] One further example of this class is compound **647**, in Section XV.1 (p. 439).

3. 3-Keto-2,4-quinolinedione

Benzoyl chloride reacted with 3-benzyl-4-hydroxy-2-quinolone in aqueous buffer to give the ketone **557**. On strong heating (185 °C) or in refluxing pyridine this rearranged to 4-benzoyloxy-3-benzyl-2-quinolone.[979] This ester structure had previously been falsely assigned to the ketone **557**.[980]

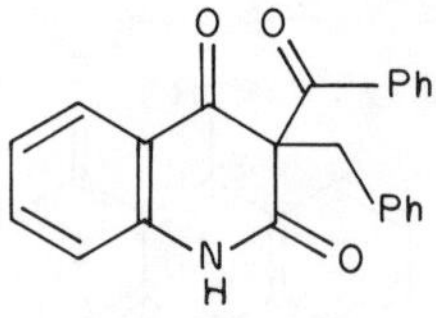

557, 91%, m.p. 195 °C

4. 4-Keto-2-quinolones

Most of the compounds in Table 49 were prepared by rearrangements of 4-ketoquinoline 1-oxides, hydrolysis of 2-ethoxy- or 2-chloro-4-ketoquinolines or by further chemistry on preformed 4-keto-2-quinolones.

Base catalysed addition of diazoketones to isatins **558** gave, via isolatable intermediates **559**, the 3-hydroxy-4-ketoquinolones **560**, Scheme 62. Ether and ester derivatives of the 3-hydroxy group were prepared.[981]

Cycloaddition of diketene to 4-acetyloxy-2-quinolone under ultra-violet irradiation gave, after acid treatment, a mixture of compounds **561** and **562**. Both gave the ketone **564** with sodium methoxide. This reaction was assumed

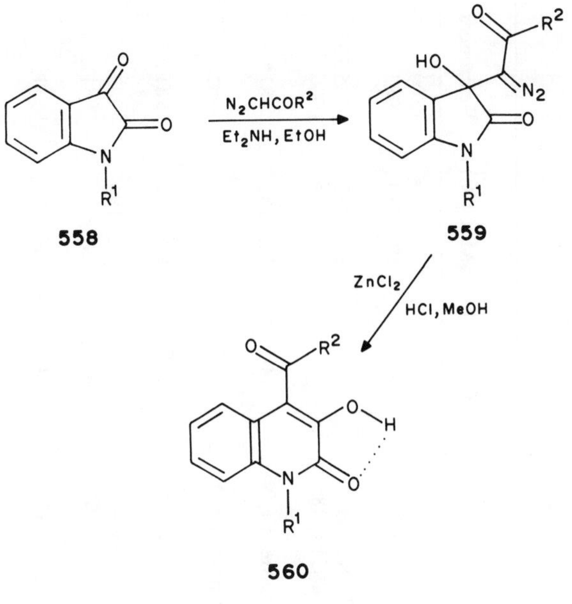

$R^1 = H, Me; R^2 = Me, Ph$

SCHEME 62

to occur via the keto-ester **563**, which apparently decarbomethoxylated without the need for acidification, Scheme 63.[982]

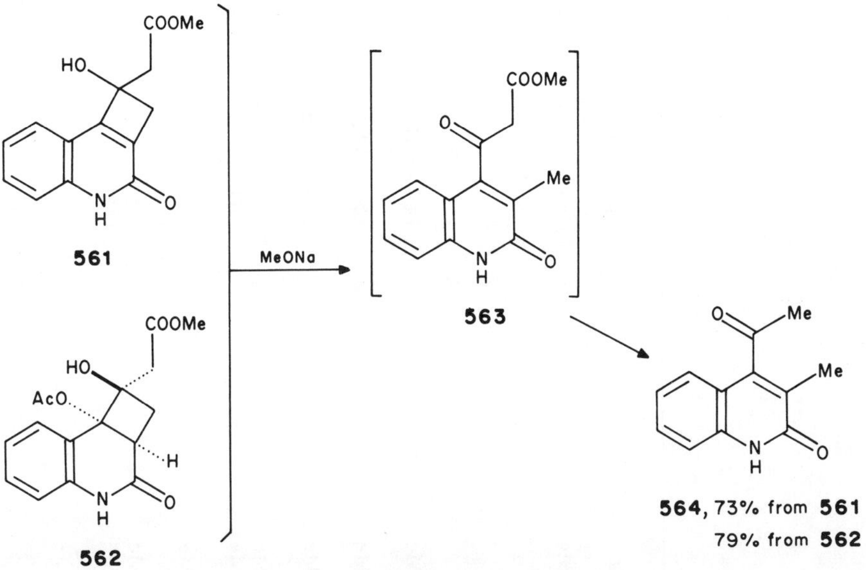

SCHEME 63

Table 49. 4-Keto-2-quinolones

R	Quinoline substituent(s)	Preparation	Yield (%)	M.p. (°C)	References
Me	None	2-EtO-4-Ac-Q, 48% HBr	92		300
		4-Me-2,5-dioxo-2,5-dihydro-1H-1-benzazepine, HCl	35	191–193	983
		4-Ac-Q N-oxide, TsCl, NaOH		194–195	235,300
		2-EtO-Q-4-COCH$_2$COOEt, 48% HBr			300
Me	1-Me-3-OH	1-Me-isatin, N$_2$CHCOMe, Et$_2$NH; ZnCl$_2$, HCl	60	185–187	981*
Me	1-Me-3-OAc	1-Me-3-HO-4-Ac-2-quinolone, Ac$_2$O, AcONa	100	160–162	981
Me	1-Me-3-OMe	1-Me-3-HO-4-Ac-2-quinolone, CH$_2$N$_2$	100	129–131	981
Me	3-Me	See Scheme 62, Section XIV.4	79	193–194	982
Me	3-OH	Isatin, N$_2$CHCOMe, Et$_2$NH; ZnCl$_2$, HCl	48	248	981
Me	3-OAc	3-HO-4-Ac-2-quinolone, Ac$_2$O, AcONa	90	213–215	981
Me	3-OMe	3-HO-4-Ac-2-quinolone, CH$_2$N$_2$	90	190–192	981
Me	6-OMe	2-EtO-6-MeO-4-Q-COCH$_2$COOEt, conc. HCl	92		285
Me		2-EtO-4-Ac-6-MeO-Q, conc. HCl	90	201–202	285

			Yield (%)	mp	Ref.
Me		3,4-diH-4-Ac-6-MeO-2-quinolone, CrO$_3$, AcOH	72	196–199	1040
		2-EtO-4-Ac-6-MeO-Q, 47% HBr	50	193–194	301
		2-Cl-6-MeO-Q-4-COCH(COOEt)$_2$, 15% HCl			275
Me	1-Ph[a]	2-Ph-4-Ac-Q N-Oxide, $h\nu$	11.6	160–162	984
Me	3-Ph[a]			>300	984
Me	1-(4-ClC$_6$H$_4$CO)-6-OMe	4-MeOC$_6$H$_4$N(COC$_6$H$_4$-4-Cl)-COCH$_2$COCOOMe, HF, 60–65 °C			1041
Me	6-NO$_2$	4-Ac-2-quinolone, (CH$_2$OH)$_2$; KNO$_3$, H$_2$SO$_4$	60	300(d)	284
CH$_2$Br		2-EtO-Q-4-COCHBrCOOH, 48% HBr	80	190–192	258
COCH$_2$CH$_2$CH(CH$_2$CH$_2$)$_2$O		2-EtO-Q-4-COOEt, EtOOCCH$_2$-CH$_2$CH(CH$_2$CH$_2$)$_2$O, EtONa; HCl	47.5	179–180	271
COCCH$_2$CH(CH$_2$CH$_2$)$_2$O =NOH		Above, C$_5$H$_{11}$ONO, EtONa	86	213	271
Ph		4-PhCO-Q N-oxide, TsCl, NaOH	84	263	235, 300
Ph	1-Me-3-Ph-6-OMe	1-Me-4-COOEt-6-MeO-2-quinolone, PhLi	44.2	191–194	283
Ph	1-Me-3-OH	1-Me-isatin, N$_2$CHCOPh, Et$_2$NH; ZnCl$_2$, HCl	100	207–209	981
Ph	1-Me-3-OAc	1-Me-3-HO-4-PhCO-2-quinolone, Ac$_2$O, AcONa	90	182–184	981
Ph	1-Me-3-OMe	1-Me-3-HO-4-PhCO-2-quinolone, CH$_2$N$_2$	100	113–114	981
Ph	6-Me	2-Cl-4-PhCO-6-Me-Q, HCl	96	274	284
Ph	3-OH	Isatin, N$_2$CHCOPh, Et$_2$NH; ZnCl$_2$, HCl	56	240–241	981

Table 49. (*Contd.*)

R	Quinoline substituent(s)	Preparation	Yield (%)	M.p. (°C)	References
Ph	3-OAc	3-HO-4-PhCO-2-quinolone, Ac_2O, H_2SO_4	90	225–227	981
Ph	3-OMe	3-HO-4-PhCO-2-quinolone, CH_2N_2	90	240–242	981
Ph	6-OMe	2-Cl-4-PhCO-6-MeO-Q, conc. HCl	90	208	285
4-ClC_6H_4	3-OH	Isatin, 4-$ClC_6H_4COCHN_2$	50	225	1042
2-Pyridyl		2-Cl-4-(2-Py)-Q, 15% H_2SO_4	85	259–260(d)	327
2-Pyridyl	6-Me	2-Cl-4-(2-Py)-6-Me-Q, conc. HCl	95	280(d)	328
2-Pyridyl	6-OMe	2-Cl-4-(2-Py)-6-MeO-Q, 15% H_2SO_4	95	217–218(d)	329
2-Pyridyl	7-OMe	2-Cl-4-(2-Py)-7-MeO-Q, 15% H_2SO_4	95	236–238(d)	329
2-Quinolyl		See Table 1			
2-(5,6,7,8-tetraH-Q)		2-Cl-Q-4-CO[2-(5,6,7,8-tetraH-Q)], conc. HCl	91	259–261(d)	38*

* These references include carbonyl derivatives.
a These two ketones were obtained as a mixture with several non-ketonic products.

The rearrangement of benzazepinedione **565** to quinolone **566** was achieved, but no mechanistic explanation was offered.[983]

565   **566**

Photolysis of the *N*-oxide **567** gave small quantities of ketones **568** and **569** along with several other products.[984]

567   **568**   **569**

Oppenauer oxidation of the appropriate secondary alcohols gave ketones **570a** and **570c**. Reduction (H$_2$, Pd, AcOH) then gave dihydro derivatives **570b** and **570d** respectively.[545] Likewise, compound **570d** has been prepared by oxidation of the appropriate 3,4-dihydro-2-quinolone alcohol. The ketone **570d** was then brominated and heated to form the 3,4-double bond of quinolone **570c**.[985,986]

570

a, R=H, 3,4-unsaturated, m.p. 192—193.5 °C
b, R=H, 3,4-saturated, m.p. 172—174 °C
c, R=OMe, 3,4-unsaturated, 54%, m.p. 185—186 °C
d, R=OMe, 3,4-saturated, 50%, m.p. 163—164 °C

5. 5-Keto- and 6-Keto-2-quinolones

These compounds, all prepared by standard methods, are listed in Tables 50 and 51, respectively.

Table 50. 5-Keto-2-quinolones

R	Quinoline substituent(s)	Preparation	Yield (%)	M.p. (°C)	References
Me	None	5-Ac-8-MeSO₃-2-quinolone, aq. KOH, Pd, H₂	33	252	1043
Me	1-Me-8-OH	Friedel–Crafts reaction	88	94	1043
Me	1-Me-8-OMe				1044
Me	8-OH	5-Ac-8-MeO-2-quinolone, Py.HCl, 200 °C	93	>260	359, 1043
Me	8-OSO₂Me	5-Ac-8-HO-2-quinolone, MeSO₂Cl, KOH	66	202	1043
Me	8-OMe	Friedel–Crafts reaction	75	216	1043, 1045
Me	8-OCH₂Ph	5-Ac-8-HO-2-quinolone, PhCH₂I, Na₂CO₃	66	170	359, 1043
		5-Ac-8-PhCH₂O-Q N-oxide, Ac₂O	56.8	174–176	1046, 1047
CH₂Cl	1-Me-8-OH	Friedel–Crafts reaction	57	287–289(d)	1046, 1048
CH₂Cl	1-Me-8-OMe	Friedel–Crafts reaction	35	204–205.5	1049
CH₂Cl	8-OH	8-OCOCH₂Cl-2-quinolone, AlCl₃		285–287(d)	1044, 1050–1052
		5-COCH₂Cl-8-OCOCH₂Cl-2-quinolone, 10% HCl or 5% KOH			1047

CH$_2$Cl	8-OCOCH$_2$Cl	Friedel–Crafts reaction		239–241(d)	1049, 1051
CH$_2$Cl	8-OMe	Friedel–Crafts reaction		243–244	1047
CH$_2$Cl	8-OCH$_2$Ph	Friedel–Crafts reaction			1048
CH$_2$Br	8-OCH$_2$Ph	5-Ac-8-PhCH$_2$O-2-quinolone, Br$_2$, BF$_3$, reflux	75.7	203–205	359
CH(OH)$_2$	8-OCH$_2$Ph	5-Ac-8-PhCH$_2$O-2-quinolone, NBS	45.8		359
CH(OH)$_2$	8-OCH$_2$Ph	5-Ac-8-PhCH$_2$O-2-quinolone, SeO$_2$, dioxane	88.5	132–147	359
CHO	1-Me-8-OH	1-Me-5-Ac-8-HO-2-quinolone, SeO$_2$ dioxane			1053
CHO	8-OH	5-Ac-8-HO-2-quinolone, SeO$_2$, dioxane			1053
CHO	8-OMe	5-Ac-8-MeO-2-quinolone, SeO$_2$, dioxane			1053
CH$_2$NH$_2$	1-Me-8-OH				1054
CH$_2$NH$_2$	8-OH				1050
CH$_2$NH$_2$	8-OMe				1054
CH$_2$NHCH$_2$Ph	8-OH	Method A	46	HCl 278–279(d)	1047, 1049, 1050
CH$_2$NHCH$_2$Ph	8-OMe				1055
CH$_2$NHEt	8-OCH$_2$Ph				1056
CH$_2$NHCHMePh	8-OH	Method A		HCl 246–249	1049, 1057
CH$_2$NHCH$_2$CH$_2$Ph	8-OH	Method A			1050
CH$_2$NHCH$_2$CH=CH$_2$	8-OH	Method A	32	HCl 277–279	1058
CH$_2$NHCH$_2$CH=CH$_2$	8-OMe	Method A	35	HCl 218–220	1058
CH$_2$NHiPr	1-Me-8-OH	Method A or 1-Me-5-iPrNHCH$_2$-CO-8-MeO-2-quinolone, 47% HBr	33	136–138(d)	1046, 1047, 1050, 1059
CH$_2$NHiPr	1-Me-8-OMe	Method A	49	HCl 286–288(d)	1046, 1056
CH$_2$NHiPr	8-OH	Method A			1047, 1049, 1050, 1052
CH$_2$NHiPr	8-OMe	Method A	49	HCl 230–231	1047, 1050, 1056
CH$_2$NHCHMeCH$_2$Ph	8-OH	Method A			1060
CH$_2$NHsBu	8-OH	Method A	26	HCl 289–291(d)	1044, 1047, 1050

Table 50. (*Contd.*)

R	Quinoline substituent(s)	Preparation	Yield (%)	M.p. (°C)	References
CH₂NHCH₂CMe₂Ph	8-OH	Method A or 5-PhCMe₂CH₂NH-CH₂CO-2-quinolone, 47% HBr			1059
CH₂NHCMe₂CH₂Ph	8-OH	Method A	28	HCl 246–247	1047, 1049, 1050
CH₂NHtBu	8-OH	Method A or 5-tBuNHCH₂CO-2-quinolone, 47% HBr	45	HCl 291–293(d)	1047, 1049, 1059
CH₂NHtBu	8-OMe	Method A			1056
CH₂NHcycloC₆H₁₁	8-OH	Method A	25	HCl 294–296	1047, 1057
CH₂N(CH₂)₅	8-OH	Method A		HCl 239–241(d)	1049
CH₂N(CH₂)₅	8-OMe				1055
CH₂N(CH₂CH₂)₂CHPh	8-OMe	Method A			1061
CH₂N(CH₂CH₂)₂O	8-OH	Method A		239–239.5	1049, 1057
CH₂N(CH₂CH₂)₂NH	8-OMe				1055
CH₂N(CH₂CH₂)₂NMe	8-OMe	Method A			1061
CH₂NHC₆H₄-4-Me	8-OH	Method A			1060
CHBrMe	8-OH	Friedel–Crafts reaction		229–231	1047
CHMeNHCH₂CH₂OH	8-OH	Method A			1062
CHMeNHCH₂CH₂CN	8-OH	Method A			1062
CHMeNHCH₂-CH=CH₂	8-OH	Method A	22	HCl 253–254	1058
CHMeNHiPr	8-OH		41	HCl 227–228	1047, 1050
CHMeNHtBu	8-OH	Method A	25	HCl 264–265	1047
CHMeNHCHMe₂CH₂Ph	8-OH				1063
Pr	1-Et-3-COOEt-4-OH	1-Et-3-COOEt-4-HO-2-quinolone-6-COCl, PrLi		163–165	1064
Pr	1-CH₂CH=CH₂-3-COOEt-4-OH-7-Me			112–114	1064
Pr	8-OH	Friedel–Crafts reaction			1045
Pr	8-OMe				1044
CHClEt	8-OH				1065

CHClEt	8-OMe	Friedel–Crafts reaction	70		1061
CHBrEt	8-OH	Friedel–Crafts reaction		218–219(d)	1044, 1066, 1067
CHBrEt	8-OMe	5-COPr-8-HO-2-quinolone, Br₂			1047
CHBrEt	8-SH	Friedel–Crafts reaction			1048
CHEtNH₂	8-OH				371
CHEtNH₂	8-OMe				1054
CHEtNHCH₂Ph	8-OH	Method A			1056
CHEtNHCH₂Ph	8-OMe	Method A	27	HCl 241–243	1050, 1068
CHEtNHEt	8-OH	Method A			1056
CHEtNHCH₂CH₂Ph	8-OH	Method A	37	HCl 232–234	1047, 1069
CHEtNHCH₂CH=CH₂	8-OH	Method A		HCl 200–201(d)	1066
CHEtNHiPr	8-OCOiPr	Method A	28	HCl 250–253	1058
CHEtNHiPr	8-OH	Method A or 5-iPrNHCHEtCO-2-quinolone, 47% HBr	55	136–137	1059, 1066, 1067
CHEtNHsBu	8-OH			HCl 245–247	1047, 1069
CHEtNHtBu	8-OH	Method A			1055
CHEtNHtBu	8-OMe	Method A	35	HCl 212–214(d)	1047, 1050, 1057, 1066
CHEtN(CH₂CH₂)₂-CHCH₂Ph	8-OMe	Method A			1050
CHEtN(CH₂CH₂)₂O	8-OH	Method A			1056
CHEtN(CH₂CH₂)₂O	8-OMe	Method A			1070
CHEtN(CH₂CH₂)₂-NCH₂Ph	8-OH	Method A		HCl 179–182(d)	1066
CHEtN(CH₂CH₂)₂-NCH₂Ph	8-OMe	Method A			1061
COEt	8-OH	5-COPr-8-HO-2-quinolone, SeO₂, dioxane			1053
CH₂CH₂COOH	4-Me-8-OMe				1071
CH₂CH₂COOH	8-OMe				1071
CHPrNHMe	8-OMe				1055
CHPrNHCH₂CH=CH₂	8-OH	Method A			1062

Table 51. 6-Keto-2-quinolones

R	Quinoline substituent(s)	Preparation	M.p. (°C)	References
Me	1-Me	Friedel–Crafts reaction (37.5% yield)	164	1016, 1017
Me	1,4-diMe	Friedel–Crafts reaction		895
Me	1-Me-3-CH$_2$CH=CMe$_2$-4-OH	4-AcC$_6$H$_4$NHMe, Me$_2$C=CHCH$_2$CH(COOEt)$_2$		1072
CH$_2$Cl		Friedel–Crafts reaction (50% yield)	275–277	989, 1073
CH$_2$NC$_5$H$_5$$^+Cl^-$		6-ClCH$_2$CO-2-quinolone, Py	>300	989, 990, 1074
CH$_2$N(CH$_2$CH$_2$)$_2$NH				1075, 1076
CH$_2$N(CH$_2$CH$_2$)$_2$NC$_6$H$_3$-2,3-diMe				1077
CH$_2$N(CH$_2$CH$_2$)$_2$NC$_6$H$_3$-2,3-diOMe				1077
CH$_2$N(CH$_2$CH$_2$)$_2$NCOPh		6-[HN(CH$_2$CH$_2$)$_2$NCH$_2$CO]-2-quinolone, PhCOCl	HCl 212–214(d)	1078[a], 1079[a]
CH$_2$N(CH$_2$CH$_2$)$_2$NCOC$_6$H$_4$-3-Cl		6-[HN(CH$_2$CH$_2$)$_2$NHCH$_2$CO]-2-quinolone, 3-ClC$_6$H$_4$COCl	HCl 212–215(d)	1075, 1078[a], 1079[a]
CH$_2$N(CH$_2$CH$_2$)$_2$NCOC$_6$H$_2$-3,4,5-triOMe		6-[NH(CH$_2$CH$_2$)$_2$NCH$_2$CO]-2-quinolone, 3,4,5-triMeOC$_6$H$_2$COCl	HCl 201–204(d)	1078[a], 1079[a]
Et		Friedel–Crafts reaction		1080
Et	1-Me	Friedel–Crafts reaction (16.1% yield); 1-Me-3,4-diH-6-EtCO-2-quinolone, DDQ (39% yield)	172–174	1016, 1017; 864
Et	1,4-diMe	Friedel–Crafts reaction		895
C-Me=	4-Me	Friedel–Crafts reaction		895
N-OH	1-Me	1-Me-6-EtCO-2-quinolone, iPrCH$_2$CH$_2$ONO		864

R	Quinolone	Method	mp (°C)	Refs
Pr	1-Me	Friedel–Crafts reaction (2.4% yield)[b]	145–146	1016
Pr	1,7-diMe-3-COOEt-4-OH	1,7-diMe-6-PrCO-isatotic anhydride, CH₂(COOEt)₂, NaH, DMA	130–131	415, 1064
Pr	1-Et-3-COOH-4-OH-7-Me	1-Et-3-COOtBu-4-HO-6-PrCO-7-Me-4-quinolone, HClO₄	157–158	415, 1064
Pr	1-Et-3-COOEt-4-OH	1-Et-3-COOEt-4-HO-6-COCl-2-quinolone, PrLi	163–165	415, 1064
Pr	1-Et-3-COOEt-4-OH-7-Me	1-Et-3-CN-4-HO-6-PrCO-7-Me-2-quinolone, EtOH, NH₄Cl	125–126	415, 1064
Pr		1-Et-3-COOEt-4-HO-6-CHOHPr-7-Me-2-quinolone, CrO₃, Py		415, 1064
Pr		2-COOEt-4-PrCO-5-Me-C₆H₂N(Et)COCH₂COOEt, EtONa		415, 1064
Pr	1-Et-3-COOtBu-4-OH-7-Me	1-Et-6-PrCO-7-Me-isatotic anhydride, CH₂(COOtBu)₂, NaH, DMA	124–126	415, 1064
Pr	1-Et-3-CN-4-OH-7-Me	1-Et-6-PrCO-7-Me-isatotic anhydride, CNCH₂COOET, NaH, DMA	244–245	415
Pr	1-Et-3-CONH₂-4-OH-7-Me	1-Et-3-CN-4-HO-6-PrCO-7-Me-2-quinolone, 90% H₂SO₄, 75°C	233–235	415
Pr	1-Pr-3-COOEt-4-OH-7-Me	1-Pr-6-PrCO-7-Me-isatoic anhydride, CH₂(COOEt)₂, NaH, DMA	116–117	415, 1064
Pr	1-CH₂CH=CH₂-3-COOEt-4-OH-7-Me	1-CH₂=CHCH₂-6-PrCO-7-Me-isatoic anhydride, CH₂(COOEt)₂, NaH, DMA	112–114	415, 1064
Pr	1-Bu-3-COOEt-4-OH-7-Me	1-Bu-6-PrCO-7-Me-isatotic anhydride, CH₂(COOEt)₂, NaH, DMA	104–105	415, 1064
CH₂CH₂COOH	1,4-diMe	Friedel–Crafts reaction		895
Ph	1-Me	Friedel–Crafts reaction (28.3% yield)	128–129	1017, 1019

[a] Refs. 1078 and 1079 include several similar substituted piperazinylmethyl ketones.
[b] Separated from mixture with the 3-ketone (Table 47).

6. 6-Keto-4-hydroxy-2-quinolones

Ketone substituted isatoic anhydrides react with malonic esters, cyanoacetates etc. under similar conditions to the keto-ester reactions shown in **541 → 542** to give 3-substituted-4-hydroxy-6-keto-2-quinolones; see Table 51.[415] In a variation of the standard preparation of 4-hydroxy-2-quinolones from anilines and diethyl malonates, the phenolic diester **571** condensed with 4-aminoaceto-phenone to give the 6-acetylquinolone **572**.[987]

571   **572**, m.p. 316−317 °C

7. 8-Keto-2-quinolones

When 2-quinolone was treated with chloroacetyl chloride and aluminium chloride in dichloromethane, a mixture of the 6-keto- (Table 51) and 8-keto-, **573**, quinolones was obtained. Both compounds were converted into pyridinium ketones as shown for the preparation of compound **574**.[988−990]

573   **574**

11.6%, m.p. 177.5−179 °C   100%, m.p. 261.5−265 °C

The hydrazone **575** and ethyl acetoacetate gave an isolatable quinolone acid hydrazone which was hydrolysed to the keto-acid **576**.[432]

575   **576**, m.p. >280 °C

                                                 methyl ester m.p. 234 °C

8. 2-Keto-4-quinolones

The diazoketones **577** rearranged on refluxing in toluene for nine days to the ketones **579**. When the reactions were run in benzene the intermediate enamines **578** were isolated.[120]

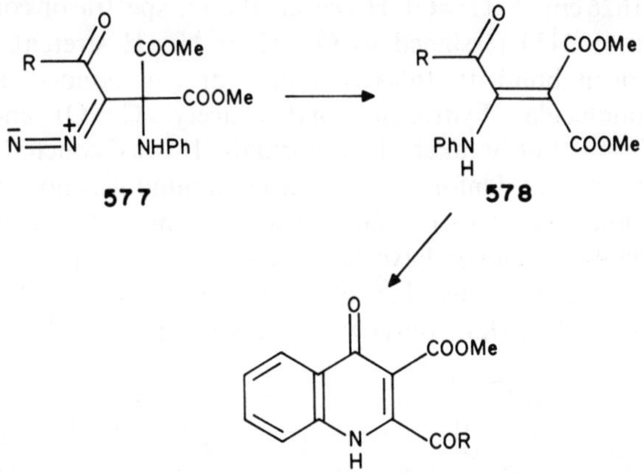

577   **578**

579

R = Me, 54%, m.p. 202 °C
R = Ph, 47%, m.p. 240−241 °C

Phenylpropiolic acid reacted with 2-phenylisatogen (**580**) in refluxing xylene to give the 2-benzoyl-4-quinolone **581**.[991]

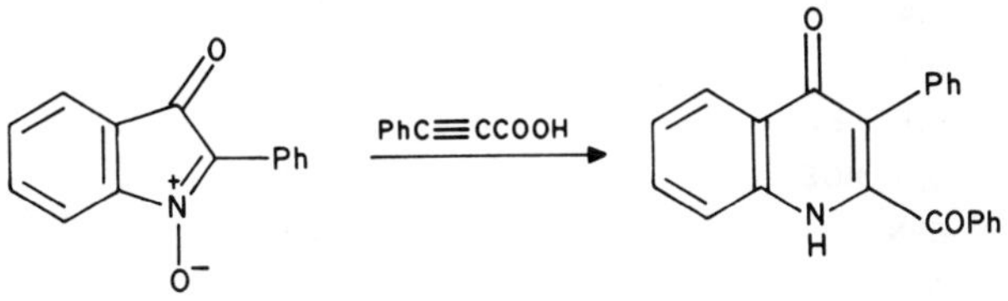

580   **581**, 45%, m.p. 326−327 °C

The heterocycle **582** was hydrolysed by hot aqueous sodium hydroxide to the 2-acetyl-4-quinolone **583**.[992]

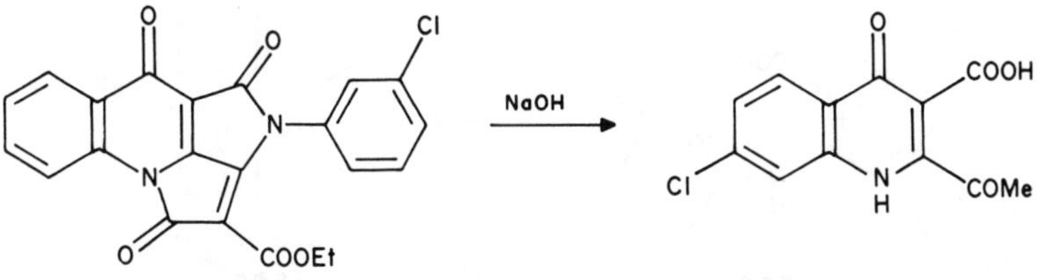

582   **583**, m.p. 306−309 °C

9. 3-Keto-4-quinolones

Table 52 is devoted to this group of compounds. Several 3-keto-4-quinolones were shown to exist essentially in the quinolone form by u.v. and i.r. spectroscopy. Their i.r. spectra (KBr) were characterized by bands at ca. 3250 (N—H) and 1626 cm^{-1} (C=O). However, the i.r. spectra of compounds **584** showed bands at 3434 (assigned to O—H or N—H stretch), 2680 (O—H strongly hydrogen bonded), 1660(w) (non hydrogen bonded acetyl C=O), 1646(s) (intramolecular hydrogen bonded acetyl C=O) and 1632 cm^{-1} (quinolone C=O, but weaker than normal). It was concluded that both tautomers were present. Unfortunately, the compound was not soluble enough to allow solution spectra to be run.[20] The pK_a values for deprotonation of several 3-acetyl-4-quinolones have been determined in aqueous dioxane; see Scheme 64. The authors treated them all as hydrogen bonded hydroxyquinolines, **584a**, and gave i.r. data (KBr only) which may support this.[993]

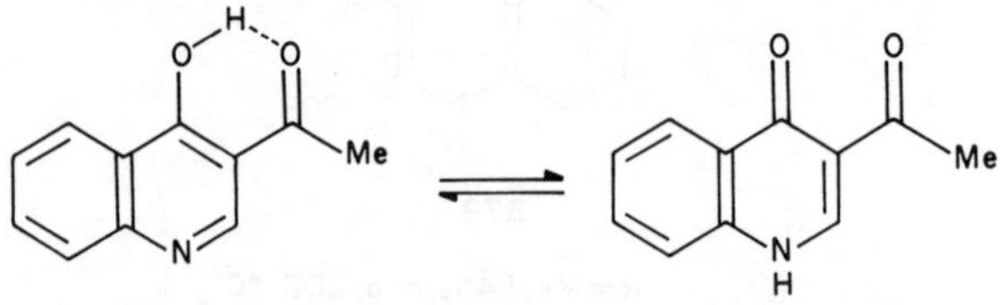

Substituents	pK_a (−H$^+$)
—	10.14
5,8-diMe	11.41
6-Me	10.35
6,8-diEt	10.64
5,8-diOEt	10.92
8-NO$_2$	8.84
6-NHAc	9.84

SCHEME 64

Enaminones **585**, from methyl anthranilate and ketoacetylenes, were cyclized in base to the 4-quinolones **586** (Scheme 65 and Table 52).[20]

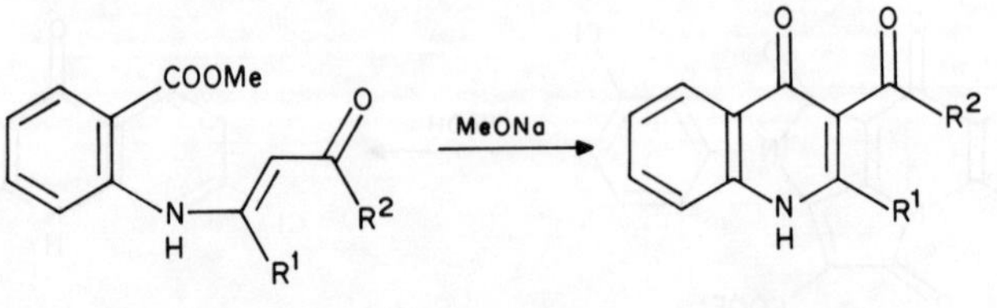

R^1 = H, Ph, PhCO; R^2 = Me, Ph

SCHEME 65

Table 52. 3-Keto-4-quinolones

R	Quinoline substituent(s)	Preparation	Yield (%)	M.p. (°C)	References
Me	None	PhNH₂, EtOCH=C(Ac)COOEt, Ph₂, Ph₂O, 250°C	88	241–243	1081
		2-MeOOCC₆H₄NHCH=CHAc, MeONa	71	240–244	20
		PhNHCH=C(Ac)COOEt, Ac₂O, Conc.H₂SO₄	62	244	1082*
		PhNHCH=C(Ac)COOEt, paraffin oil, 250–260°C	40	243–244	993, 997, 1083
Me	1-Me	3-Ac-4-quinolone, Me₂SO₄, KOH		231–232	1084
Me	1-Me-2-Ph	1-Me-isatoic anhydride, PhCOCH₂COMe	17	195–197	970
Me	1-Me-7-(4-pyridyl)[a]	3-Ac-7-(4-pyridyl)-4-quinolone, Me₂SO₄		255–257	1085
Me	1-Me-6-NO₂	1-Me-3-Ac-4-quinolone, HNO₃, H₂SO₄		256–257	1084
Me	2-Me	Friedel–Crafts reaction		255	353*
		1-AcO-2-Me-3-Ac-4-quinolone, H₂, Pd		255	1086
		1-AcO-2-Me-3-Ac-6-Cl-4-quinolone, H₂, Pd		255	1086
		See text, Section XIV.9	25	259–260	999
Me	2-Me-6-Cl	1-HO-2-Me-3-Ac-6-Cl-4-quinolone, Zn, AcOH		310(d)	1087
Me	5-Me[b]	3-MeC₆H₄NHCH=C(Ac)COOEt, Ac₂O, Conc. H₂SO₄		222	1082
Me	5,6,8-triMe	3,5-diMeC₆H₃NH₂, EtOCH=C(Ac)COOEt			1081
Me	5,7-diMe	2,5-diMeC₆H₃NHCH=C(Ac)COOEt, heat			997
Me	5,8-diMe				993
Me	6-Me	4-MeC₆H₄NHCH=C(Ac)COOEt, Ac₂O, Conc. H₂SO₄	36	258	1082
		4-MeC₆H₄NH₂, EtOCH=C(Ac)COOEt			993, 997
Me	6,7-diMe	3,4-diMeC₆H₃NH₂, EtOCH=C(Ac)COOEt			997

Table 52. (*Contd.*)

R	Quinoline substituent(s)	Preparation	Yield (%)	M.p. (°C)	References
Me	6,8-diMe	2,4-diMeC$_6$H$_3$NHCH=C(Ac)COOEt, Ph$_2$O	65	251	1082*
Me	6-Me-7-Cl	3-Cl-4-MeC$_6$H$_3$NH$_2$, EtOCH=C(Ac)COOEt			997
Me	7-Me[b]	3-MeC$_6$H$_4$NHCH=C(Ac)COOEt, Ph$_2$O		279	1082
Me		3-MeC$_6$H$_4$NH$_2$, EtOCH=C(Ac)COOEt			997
Me	7,8-diMe	2-Cl-3-Ac-7,8-diMe-4-quinolone, Zn, AcOH	62	256–259	967
Me		2,3-diMeC$_6$H$_3$NH$_2$, EtOCH=C(Ac)COOEt			967
Me	8-Me	2-MeC$_6$H$_4$NHCH=C(Ac)COOEt, Ac$_2$O, Conc. H$_2$SO$_4$		273–274	1082
Me		2-MeC$_6$H$_4$NH$_2$, EtOCH=C(Ac)COOEt			997
Me	1-Et	3-Ac-4-quinolone, Et$_3$PO$_4$, K$_2$CO$_3$	75	150–152	1081
Me	1-Et-5,6,8-triMe				1081
Me	1-Et-7-Me				1081
Me	1-Et-7-(3-pyridyl)[a]	3-Ac-7-(3-pyridyl)-4-quinolone, TsOEt, K$_2$CO$_3$	80	231–233	1085
Me	1-Et-7-(4-pyridyl)[a]	3-Ac-7-(4-pyridyl)-4-quinolone, TsOEt, K$_2$CO$_3$	90	197–200	1085
Me	1-Et-6-Cl	3-Ac-6-Cl-4-quinolone, Et$_3$PO$_4$, K$_2$CO$_3$			1081
Me	1-Et-6-OMe				1081
Me	1-Et-6-OCH$_2$O-7				1081
Me	6-Et	4-EtC$_6$H$_4$NH$_2$, EtOCH=C(Ac)COOEt			997
Me	6,8-diEt	2,4-diEtC$_6$H$_3$NHCH=C(Ac)COOEt, heat			993
Me	7-Et	3-EtC$_6$H$_4$NH$_2$, EtOCH=C(Ac)COOEt			997
Me	7-cycloC$_6$H$_{11}$	3-cycloC$_6$H$_{11}$C$_6$H$_4$NHCH=C(Ac)COOEt, Ph$_2$O	100	274–278(d)	1088
Me	2-Ph	PhNHC(Ph)=C(Ac)COOEt, Ph$_2$O, reflux	80	246–248	195*, 995*
		2-MeOOCC$_6$H$_4$NHC(Ph)=CHAc, MeONa	80	248–249	20
		PhN=C(Ph)C(COOMe)=C(NH$_2$)Me, Conc. H$_2$SO$_4$	80		194
		PhC(Cl)=NPh, EAA, Na	12	282–284	197
		PhN=C(Ph)C(COOMe)=C(NH$_2$)Me, PPA, 165–175 °C		250–251	1089*
		PhN=C(Ph)C(COOEt)=C(NH$_2$)Me, PPA, 165–175 °C			1089

Me	2-Ph-5&7-Me	3-MeC$_6$H$_4$C(Cl)=NPh, EAA, Na		220–240	197
		One pure isomer from the mixture		277–280	197
Me	2-Ph-6-Me	4-MeC$_6$H$_4$N=C(Ph)CH(Ac)COOEt, Ac$_2$O, H$_2$SO$_4$	9.5	263	196
		4-MeC$_6$H$_4$C(Cl)=NPh, EAA, Na		255	197
		4-MeC$_6$H$_4$N=C(Ph)C(COMe)=C(NH$_2$)Me, PPA, 165–175 °C		263–264	1089
Me	2-Ph-6,8-diMe	2,4-diMeC$_6$H$_3$N=C(Ph)C(COMe)=C(NH$_2$)Me, PPA, 165°–175 °C		216–217	1089
Me	2-Ph-8-Me	2-MeC$_6$H$_4$C(Cl)=NPh, EAA, Na	9.5	215	197
		2-MeC$_6$H$_4$N=C(Ph)C(COMe or Et)=C(NH$_2$)Me, PPA, 165–175 °C		211–212	1089
Me	2,8-diPh	2-PhC$_6$H$_4$N=C(Ph)C(COMe)=C(NH$_2$)Me, PPA	80	206–207	1090
Me	2-Ph-5(or 7)-Cl	3-ClC$_6$H$_4$N=C(Ph)CH(Ac)COOEt, Ac$_2$O, H$_2$SO$_4$	44	312–313	196
Me	2-Ph-5&7-Cl	3-ClC$_6$H$_4$C(Cl)=NPh, EAA, Na	7.5	292–300	197
Me	2-Ph-6-Cl	4-ClC$_6$H$_4$C(Cl)=NPh, EAA, Na	12.5	298–300	197
Me	2-Ph-8-Cl	2-ClC$_6$H$_4$C(Cl)=NPh, EAA, Na	12	284–286	197
Me	2-Ph-6-OMe	4-MeOC$_6$H$_4$N=C(Ph)NHPh, EAA, Na	12	270	197
		4-MeOC$_6$H$_4$N=C(Ph)C(COMe)=C(NH$_2$)Me, PPA		290–292	1089
Me	2-Ph-6-OEt	4-EtOC$_6$H$_4$N=C(Ph)NHPh, EAA, Na		295	197
Me	2-(2-MeC$_6$H$_4$)	PhN=C(2-MeC$_6$H$_4$)C(COMe)=C(NH$_2$)Me, PPA		256–257	194, 1089, 1091*
Me	2-(3-MeC$_6$H$_4$)	PhNHC(3-MeC$_6$H$_4$)=C(Ac)COOEt, Ph$_2$O	10	279–280	1091*
Me	2-(4-MeC$_6$H$_4$)	PhN=C(4-MeC$_6$H$_4$)C(COMe)=C(NHMe)Me, conc. H$_2$SO$_4$	80	239–240	194
		PhN=C(4-MeC$_6$H$_4$)C(COMe)=C(NH$_2$)Me, PPA	70	239–240	194, 1089*
		PhNHC(4-MeC$_6$H$_4$)=C(Ac)COOEt, Ph$_2$O	12	>297^c	1091*
Me	2-(4-MeC$_6$H$_4$)-6-Me	4-MeC$_6$H$_4$N=C(4-MeC$_6$H$_4$)C(COMe)=C(NH$_2$)Me, PPA	70	236–237	194
Me	2-(4-ClC$_6$H$_4$)	PhNHC(4-ClC$_6$H$_4$)=C(Ac)COOEt, Ph$_2$O	15	284	1091
Me	2-(3-BrC$_6$H$_4$)	PhNHC(3-BrC$_6$H$_4$)=C(Ac)COOEt, Ph$_2$O	13	257–259	1091*
Me	2-(4-BrC$_6$H$_4$)	PhNHC(4-BrC$_6$H$_4$)=C(Ac)COOEt, Ph$_2$O	5	280–282	1091

Table 52. (*Contd.*)

R	Quinoline substituent(s)	Preparation	Yield (%)	M.p. (°C)	References
Me	2-(4-MeOC₆H₄)	PhN=(4-MeOC₆H₄)C(COOMe)=C(NH₂)Me, PPA	66	202–203	194
Me	2-(4-MeOC₆H₄)-8-Me	2-MeC₆H₄N=C(4-MeOC₆H₄)C(COOMe)=C(NH₂)Me, PPA	25 70	236–237^d 192–194	198* 194
Me	2-(4-MeOC₆H₄)-6-OMe	2-MeOC₆H₄N=C(4-MeOC₆H₄)C(COOMe)=C(NH₂)Me, PPA	6 70	246–247^d 229–231	198* 194
Me	2-(4-MeOC₆H₄)-6-OEt		12 5	295–296^d 271–171^d	198* 198*
Me	6-Ph	4-PhC₆H₄NH₂; EtOCH=C(Ac)COOEt			997
Me	7-Ph	3-PhC₆H₄NH₂; EtOCH=C(Ac)COOEt			997
Me	8-Ph	2-PhC₆H₄NH₂; EtOCH=C(Ac)COOEt			997
Me	7-(3-Pyridyl)^a	(3-Pyridyl)-C₆H₄-3-NHCH=C(Ac)COOEt^e, Dowtherm			1085
Me	7-(4-Pyridyl)^a	(4-Pyridyl)-C₆H₄-3-NHCH=C(Ac)COOEt^e, Dowtherm	54	>300	1085
Me	6-F	4-FC₆H₄NH₂, EtOCH=C(Ac)COOEt			997
Me	2-Cl-7,8-diMe	See text, Section XIV.9	56	124–126	967
Me	5-Cl^b	3-ClC₆H₄NHCH=C(Ac)COOEt, Ac₂O, conc. H₂SO₄		248	1082
Me	5-Cl-8-OMe	2-MeO-5-ClC₆H₃NH₂, EtOCH=C(Ac)COOEt, 100 °C	80	250	1092

R	Substituent	Conditions	Yield (%)	mp (°C)	Refs.
Me	6-Cl	4-ClC$_6$H$_4$NH$_2$, EtOCH=C(Ac)COOEt, Ph$_2$, Ph$_2$O, 250 °C	68	>300	997, 1081
		4-ClC$_6$H$_4$NHCH=C(Ac)COOEt, Ac$_2$O, conc. H$_2$SO$_4$	52	>300	1082, 1093
Me	7-Cl[b]	3-ClC$_6$H$_4$NHCH=C(Ac)COOEt, Ph$_2$O	90.5	315	1082, 1094
		3-ClC$_6$H$_4$NH$_2$, EtOCH=C(Ac)COOEt			997
Me	8-Cl	2-ClC$_6$H$_4$NHCH=C(Ac)COOEt, Ph$_2$O	60	294	1092, 1093
Me	5-Br-8-OH	2-OH-5-BrC$_6$H$_3$NH$_2$, EtOCH=C(Ac)COOEt	20	300	1092
Me	6-Br	4-BrC$_6$H$_4$NHCH=C(Ac)COOEt, Ph$_2$O	50	>300	204
Me	7-OH	3-HOC$_6$H$_4$NH$_2$, EtOCH=C(Ac)COOEt	30	275	997, 1081
Me	8-OH	2-HOC$_6$H$_4$NH$_2$, EtOCH=C(Ac)COOEt, 100 °C; Ph$_2$O, reflux	28	269	1092
Me	5,7-diOMe	3-ClC$_6$H$_4$NH$_2$, EtOCH=C(Ac)COOEt	40	232.5–235	997
Me	6-OMe	4-MeOC$_6$H$_4$NHCH=C(Ac)COOEt, Ph$_2$O	90	290	1082
		4-MeOC$_6$H$_4$NH$_2$, EtOCH=C(Ac)COOEt			997
Me	6-OMe-8-NO$_2$	2-NO$_2$-4-MeOC$_6$H$_3$NH$_2$, EtOCH=C(Ac)COOEt, Dowtherm	39	285	203
Me	7-OMe	3-MeOC$_6$H$_4$NH$_2$, EtOCH=C(Ac)COOEt	65	289–290	997
Me	8-OMe[f]	2-MeOC$_6$H$_4$NH$_2$, EtOCH=C(Ac)COOEt	82	267	997, 1092
		2-MeOC$_6$H$_4$NHCH=C(Ac)COOEt, Ph$_2$O			1082*
		2-MeOC$_6$H$_4$NH$_2$, EAA, HC(OEt)$_3$, Ph$_2$O			1095
Me	5,8-diOEt	2,5-diEtOC$_6$H$_3$NHCH=C(Ac)COOEt, heat	52	278	993
Me	6-OEt	4-EtOC$_6$H$_4$NHCH=C(Ac)COOEt, Ph$_2$O	90.5	>300	1082
Me	8-OEt	2-EtOC$_6$H$_4$NHCH=C(Ac)COOEt, Ph$_2$O		315	1082
Me	7-Cl[b]	4-ClC$_6$H$_4$NHCH=C(Ac)COOEt, Ac$_2$O, conc. H$_2$SO$_4$			1082, 1093
		3-ClC$_6$H$_4$NHCH=C(Ac)COOEt, Ph$_2$O			1082, 1094
		3-ClC$_6$H$_4$NH$_2$, EtOCH=C(Ac)COOEt			997

Table 52. (*Contd.*)

R	Quinoline substituent(s)	Preparation	Yield (%)	M.p. (°C)	References
Me	6-OCH$_2$O-7	3-OCH$_2$O-4-C$_6$H$_3$NH$_2$, EtOCH=C(Ac)COOEt			997, 1081
Me	7-SMe	3-MeSC$_6$H$_4$NH$_2$, EtOCH=C(Ac)COOEt			997
Me	6-COOH	4-COOHC$_6$H$_4$NH$_2$, EtOCH=C(Ac)COOEt			997
Me	6-NO$_2$				204, 1081
Me	6-NO$_2$-8-OMe	2-MeO-4-NO$_2$C$_6$H$_3$NH$_2$, EtOCH=C(Ac)COOEt	70	255	1092
Me	7-NO$_2$				204
Me	8-NO$_2$				204, 993
Me	2-NHPh	2-NO$_2$C$_6$H$_4$NHCH=C(Ac)COOEt, heat			1039
Me	2-(NHC$_6$H$_3$-2,5-diMe)	2-NH$_2$C$_6$H$_4$COOH, AcCH$_2$CONHPh, xylene, reflux	20–30	256–257	1039
Me	2-(4-MeC$_6$H$_4$)NH	2-NH$_2$C$_6$H$_4$COOH, AcCH$_2$CONHC$_6$H$_3$-2,5-diMe	20–30	232–233	1039
Me	2-(NHC$_6$H$_4$-2-Cl)	2-NH$_2$C$_6$H$_4$COOH, AcCH$_2$CONHC$_6$H$_4$-4-Me	20–30	241–243	1039
Me	2-(NHC$_6$H$_3$-2,5-diCl)	2-NH$_2$C$_6$H$_4$COOH, AcCH$_2$CONHC$_6$H$_4$-2-Cl	20–30	273–274	1039
Me	2-(NHC$_6$H$_4$-4-Cl)	2-NH$_2$C$_6$H$_4$COOH, AcCH$_2$CONHC$_6$H$_3$-2,5-diCl	20–30	252–254	1039
Me	2-(NHC$_6$H$_4$-4-OMe)	2-NH$_2$C$_6$H$_4$COOH, AcCH$_2$CONHC$_6$H$_4$-4-Cl	20–30	246–248	1039
Me	2-(NHC$_6$H$_4$-4-OEt)	2-NH$_2$C$_6$H$_4$COOH, AcCH$_2$CONHC$_6$H$_4$-4-OMe	20–30	259–260	1039
Me	2-(NHC$_6$H$_4$-4-NO$_2$)	2-NH$_2$C$_6$H$_4$COOH, AcCH$_2$CONHC$_6$H$_4$-4-OEt	20–30	227–228	1039
Me	2-NH-(2-naphthyl)	2-NH$_2$C$_6$H$_4$COOH, AcCH$_2$CONHC$_6$H$_4$-4-NO$_2$	20–30	245–247	1039
Me	6-NHAc	2-NH$_2$C$_6$H$_4$COOH, AcCH$_2$CONHC$_{10}$H$_7$	20–30	267–268	1039
Me	6-N=N-Ph	4-AcNHC$_6$H$_4$NHCH=C(Ac)COOEt, heat			993
Me		PhN=NC$_6$H$_4$-4-NHCH=C(Ac)COOEt, Ph$_2$O, reflux	85	240	1096
Me	6-N=N-(C$_6$H$_4$-4-Me)	4-MeC$_6$H$_4$N=NC$_6$H$_4$-4-NHCH=C(Ac)COOEt, Ph$_2$O	82	295	1096
Me	6-N=N-(C$_6$H$_4$-4-Cl)	4-ClC$_6$H$_4$N=NC$_6$H$_4$-4-NHCH=C(Ac)COOEt, Ph$_2$O	80	341–342	1096
Me	6-N=N-(C$_6$H$_4$-4-Br)	4-BrC$_6$H$_4$N=NC$_6$H$_4$-4-NHCH=C(Ac)COOEt, Ph$_2$O	81	325–326	1096

Me	6-N=N-(C₆H₄-4-NO₂)	4-NO₂C₆H₄N=NC₆H₄-4-NHCH=C(Ac)COOEt, Ph₂O	80	>360	1096
Me	6-N=N-(C₆H₄-4-NMe₂)	4-Me₂NC₆H₄N=NC₆H₄-4-NHCH=C(Ac)COOEt, Ph₂O	84	>360	1096
Me	6-N=N-(C₆H₄-4-NEt₂)	4-Et₂NC₆H₄N=NC₆H₄-4-NHCH=C(Ac)COOEt, Ph₂O	82	274–275	1096
CH₂Cl		PhNHCH=C(COCH₂Cl)CONHPh, AlCl₃ or H₂SO₄	63	255(d)	1097
CH₂Cl	6-Cl	4-ClC₆H₄NHCH=C(COCH₂Cl)CONHPh, AlCl₃ or H₂SO₄	31	260(d)	1097
CH₂OMe	1-Me	See text, Section XIV.9		208–210	1001
CHO	2-Ph	2-Ph-3-Ac-4-quinolone, SeO₂, EtOH		206	195, 995
CH₂SOMe	7-F	3-COOEt-7-F-4-quinolone, NaCH₂SOMe, DMSO	36	202–204	1098–1100
CH₂SOMe	6,7-diOMe	3-COOEt-6,7-diMeO-4-quinolone, NaCH₂SOMe	60	223–225	1098–1101
CH₂SOMe	6-OCH₂O-7	3-COOEt-(6-OCH₂O-7)-4-quinolone, NaCH₂SOMe	71	201.5–203	1098–1101
COOH	1-Me-2-COOH	See text, Section XIV.9		236–138(d)	1001
COOMe	2-Me	1-MeO-2-Me-3-COCOOMe-4-quinolone, aq. KOH	35	213–215	215
COOEt	2-Me	2-Me-3-CHOHCOOEt-4-quinolone, MnO₂	64	215–216	215
COOEt	1-Et-2-Me	2-Me-3-COCOOEt-4-quinolone, EtOTl; EtI	12^g	163–164	215
CH₂NH₂	1-Me	1-Me-3-Ac-4-quinolone, Br₂; (CH₂)₆N₄		HBr 284	411
CH₂NC₅H₅⁺I⁻	5,6,8-triMe	3-Ac-5,6,8-triMe-4-quinolone, I₂, Py	60	243–244	1081
CH₂NC₅H₅⁺I⁻	8-Me	3-Ac-8-Me-4-quinolone, I₂, Py	71	250–251	1081
CH₂NC₅H₅⁺I⁻	1-Et	1-Et-3-Ac-4-quinolone, I₂, Py	79	251–252	1081
CH₂NC₅H₅⁺I⁻	1-Et-5,6,8-triMe	1-Et-3-Ac-5,6,8-triMe-4-quinolone, I₂, Py	39	250–251	1081
CH₂NC₅H₅⁺I⁻	1-Et-7-Me	1-Et-3-Ac-7-Me-4-quinolone, I₂, Py	91.5	225–227	1081
CH₂NC₅H₅⁺I⁻	1-Et-6-Cl	1-Et-3-Ac-6-Cl-4-quinolone, I₂, Py	90	272–274(d)	1081
CH₂NC₅H₅⁺I⁻	1-Et-6-OMe	1-Et-3-Ac-6-MeO-4-quinolone, I₂, Py	83	257–258	1081
CH₂NC₅H₅⁺I⁻	1-Et-6-OCH₂O-7	1-Et-3-Ac-(6-OCH₂O-7)-4-quinolone, I₂, Py	97.5	274–275	1081, 1102
CH₂NC₅H₅⁺I⁻	7-OH	3-Ac-7-HO-4-quinolone, I₂, Py	40	245–247	1081
CH₂NC₅H₅⁺I⁻	6-OCH₂O-7	3-Ac-(6-OCH₂O-7)-4-quinolone, I₂, Py	66	265–266	1081
CH₂NC₅H₅⁺I⁻	6-NO₂	3-Ac-6-NO₂-4-quinolone, I₂, Py	71	273–274	1081

Table 52. (*Contd.*)

R	Quinoline substituent(s)	Preparation	Yield (%)	M.p. (°C)	References
CH=CHPh	6-Br	3-Ac-6-Br-4-quinolone, PhCHO, NaOH	[h]	270–271	1103
CH₂COCOOEt	H	3-Ac-4-quinolone, (COOEt)₂, EtONa			997
CH₂COCOOEt	5,7-diMe	3-Ac-5,7-diMe-4-quinolone, (COOEt)₂, EtONa			997
CH₂COCOOEt	6-Me	3-Ac-6-Me-4-quinolone, (COOEt)₂, EtONa			997
CH₂COCOOEt	6,7-diMe	3-Ac-6,7-diMe-4-quinolone, (COOEt)₂, EtONa			997
CH₂COCOOEt	6-Me-7-Cl	3-Ac-6-Me-7-Cl-4-quinolone, (COOEt)₂, EtONa			997
CH₂COCOOEt	7-Me	3-Ac-7-Me-4-quinolone, (COOEt)₂, EtONa			997
CH₂COCOOEt	8-Me	3-Ac-8-Me-4-quinolone, (COOEt)₂, EtONa			997
CH₂COCOOEt	6-Et	3-Ac-6-Et-4-quinolone, (COOEt)₂, EtONa			997
CH₂COCOOEt	7-Et	3-Ac-7-Et-4-quinolone, (COOEt)₂, EtONa			997
CH₂COCOOEt	6-Ph	3-Ac-6-Ph-4-quinolone, (COOEt)₂, EtONa			997
CH₂COCOOEt	7-Ph	3-Ac-7-Ph-4-quinolone, (COOEt)₂, EtONa			997
CH₂COCOOEt	8-Ph	3-Ac-8-Ph-4-quinolone, (COOEt)₂, EtONa			997
CH₂COCOOEt	6-F	3-Ac-6-F-4-quinolone, (COOEt)₂, EtONa			997
CH₂COCOOEt	6-Cl	3-Ac-6-Cl-4-quinolone, (COOEt)₂, EtONa			997
CH₂COCOOEt	7-Cl	3-Ac-7-Cl-4-quinolone, (COOEt)₂, EtONa			997
CH₂COCOOEt	7-OH	3-Ac-7-HO-4-quinolone, (COOEt)₂, EtONa			997
CH₂COCOOEt	5,7-diOMe	3-Ac-5,7-diMeO-4-quinolone, (COOEt)₂, EtONa			997
CH₂COCOOEt	6-OMe	3-Ac-6-MeO-4-quinolone, (COOEt)₂, EtONa			997
CH₂COCOOEt	7-OMe	3-Ac-7-MeO-4-quinolone, (COOEt)₂, EtONa			997
CH₂COCOOEt	8-OMe	3-Ac-8-MeO-4-quinolone, (COOEt)₂, EtONa	60.6	293–294	997
CH₂COCOOEt	6-OCH₂O-7	3-Ac-(6-OCH₂O-7)-4-quinolone, (COOEt)₂, EtONa			997
CH₂COCOOEt	7-SMe	3-Ac-7-MeS-4-quinolone, (COOEt)₂, EtONa			997
CH₂COCOOEt	6-COOH	3-Ac-6-COOH-4-quinolone, (COOEt)₂, EtONa			997
CH₂COCH=CHPh		3-Ac-4-quinolone, (PhCH=CHCO)₂O, Et₃N, reflux		210–214(d)	1103
Ph	1-Me-2-CH₂Ph[i]	2-PhCH₂-3-PhCO-4-quinolone, MeONa, MeI	75	178–179	218, 1104
Ph	1-Me-2-CHMePh[i]	2-PhCH₂-3-PhCO-4-quinolone, 2NaH, MeI, DME	46	171–172	218, 1104

Ph	1-Me-2-CHEtPh[i]	1-Me-2-PhCH$_2$-3-PhCO-4-quinolone, EtI, MeONa	19	216–218	1104
Ph	1-Me-2-Ph	1-Me-isatoic anhydride, PhCOCH$_2$COPh	100	287–289	970
Ph	2-Me	PhN=C(Ph)-O-C(Me)=CHOOEt, Ph$_2$O, reflux	48	284	195*, 995
		PhN=C(Me)CH(COPh)COOEt, Ac$_2$O, H$_2$SO$_4$			196*
		Friedel–Crafts reaction			195, 995
		PhN=C(Me)CH(COPh)COOEt, Ph$_2$O		289	1105
		2-Me-4-PhCOO-Q, AlCl$_3$, 200 °C			994
Ph	2,5-diMe[j]	PhNH$_2$, PhCOCH(Ac)COOEt, HCl; Ph$_2$O		284	994, 1106*
		3-MeC$_6$H$_4$N=C(Me)CH(COPh)COOEt, Ac$_2$O, H$_2$SO$_4$			196
Ph	2,7-diMe[j]	4-MeC$_6$H$_4$NH$_2$, PhCOCH(Ac)COOEt, HCl; Ph$_2$O		>290	196
Ph	2,6-diMe	2-MeC$_6$H$_4$N=C(Me)CH(COPh)COOEt, Ac$_2$O, H$_2$SO$_4$	30	297–298	994, 1106*
Ph	2,8-diMe	2-MeC$_6$H$_4$NH$_2$, PhCOCH(Ac)COOEt, HCl; Ph$_2$O			196
Ph	2-CH$_2$Ph[i]	PhNHC(CH$_2$Ph)=C(COPh)COOEt, Ph$_2$O, reflux	27	289	994, 1106
Ph	2-COPh	2-MeOOCC$_6$H$_4$NHC(COPh)=CHCOPh, MeONa	86	260–262	218, 1104
Ph	2-Ph	PhNHC(Ph)=C(COPh)COOMe, 200 °C	79	250–251.5	20
		2-MeOOCC$_6$H$_4$NHC(Ph)=CHCOPh, MeONa	75	280	1108
		1,2-diPh-3-PhCO-4,5-pyrrolidinedione, 250 °C	74	287–289	20
		2-Ph-3-C(Ph)=NPh-furan-4,5-dione, 140 °C	65	280	219
		1,2-diPh-3-PhCO-4,5-pyrrolidinedione, PhNC	60	280	1109
		2-NH$_2$C$_6$H$_4$COOH, PhCOCH$_2$CONHPh, xylene	30	280–282	1000
		1,2-diPh-3-PhCO-4,5-pyrrolidinedione, gas phase, >600 °C		310	1039, 1107
Ph	2-Ph-6-Me	1-(4-MeC$_6$H$_4$)-2-Ph-3-COPh-4,5-pyrrolidinedione, 250 °C	79	295	219, 1109
Ph	2-COOEt	1-Ph-2-COOEt-3-COPh-4,5-pyrrolidinedione, 160 °C	40	233–236	221
Ph	2-NHPh	(PhNH)$_2$C=C(COPh)COOEt, Ph$_2$O, reflux		303–304	1110
Ph	2-NH-cycloC$_6$H$_{11}$	2-NH$_2$C$_6$H$_4$COOH, PhCOCH$_2$CONHcycloC$_6$H$_{11}$, xylene	14	312–314	1038
Ph	2-NH-(C$_6$H$_4$-4-Me)	2-NH$_2$C$_6$H$_4$COOH, PhCOCH$_2$CONHC$_6$H$_4$-4-Me, xylene	5	303–304	1038

Quinoline Ketones

Table 52. (Contd.)

R	Quinoline substituent(s)	Preparation	Yield (%)	M.p. (°C)	References
Ph	2-NH-(C$_6$H$_4$-4-OH)	2-NH$_2$C$_6$H$_4$COOH, PhCOCH$_2$CONHC$_6$H$_4$-4-OH, xylene	11	343–345	1038
Ph	2-NH-(1-naphthyl)	2-NH$_2$C$_6$H$_4$COOH, PhCOCH$_2$CONHC$_{10}$H$_7$, xylene	13	312–316	1038
4-MeOC$_6$H$_4$	2-Me	PhN=C(Cl)C$_6$H$_4$-4-OMe, EAA, Na		235–237	194
4-Me$_2$NC$_6$H$_4$	1-Me-2-CHMePh				1104

*These references include carbonyl derivatives.

[a] Several similar derivatives with alkyl substituted pyridines are reported in the patent, but not characterized.[1085]

[b] The structural evidence for these isomers is not clear.

[c] The high m.p. for this compound may indicate that rearrangement has occurred; see Ref. 195.

[d] The melting points of these compounds do not correspond with others reported. It is probable that the rearranged compounds were obtained; see text, Section XIV.9.

[e] Other esters were also used.

[f] With NBS this gave[1095] a monobromo derivative, m.p. 247°C.

[g] Formed as a mixture with 2-Me-4-EtO-Q-3-COCOOEt; Table 2.

[h] By-product from the pyrano[3,2-c]quinoline formation.

[i] A further 25 similar 2-substituted-3-benzoyl-4-quinolones are listed in the patent, but not characterized.[1104]

[j] Isolated as a mixture.

The products of reaction between the sodium salt of ethyl acetoacetate and phenylbenzimido chloride **587** have been the cause of controversy. Early reports[197,994] were corrected by Singh and Nair;[195,995] see Scheme 66. The reaction gave a mixture of intermediates, separated by chromatography and formulated as compounds **588** and **589**. The evidence provided for structure **588** is not conclusive; it is possible that the true intermediate was a rearrangement product from compound **589** with the Ph and Me groups reversed. However,

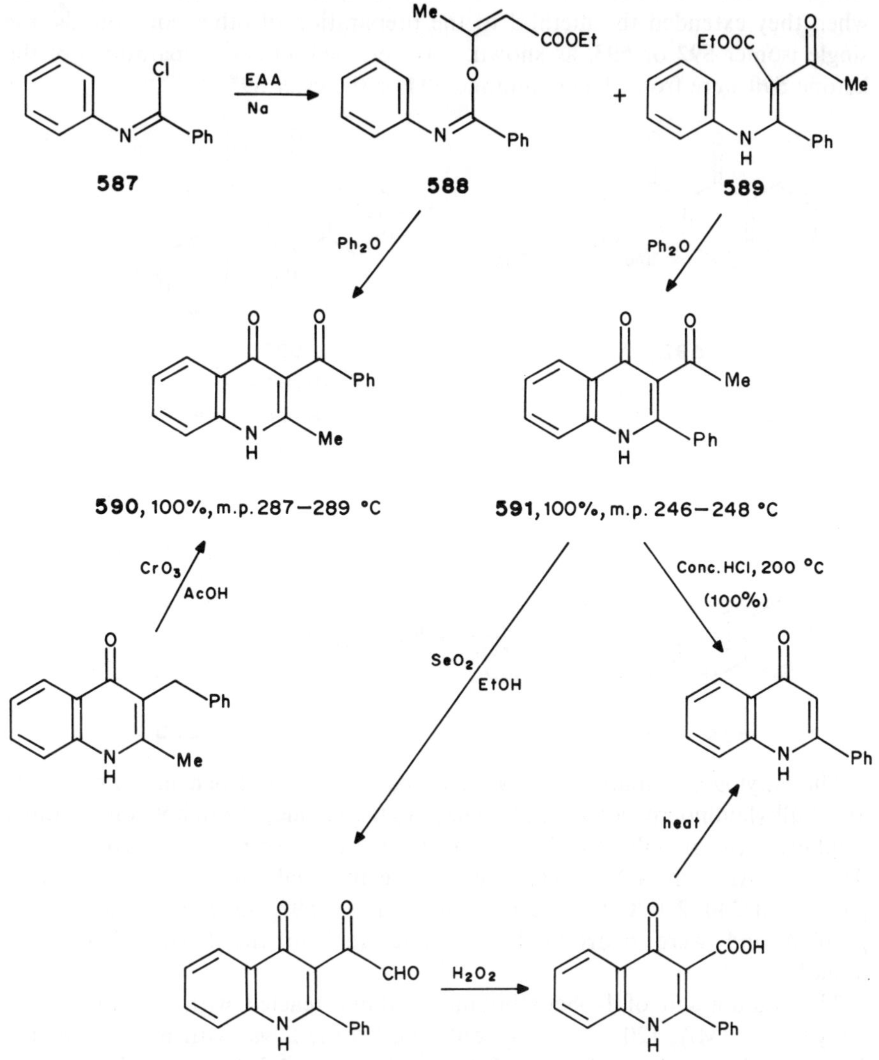

SCHEME 66

both intermediates cyclized in high yield in refluxing diphenyl ether to give the ketones **590** and **591**. Alternatively, ring closure was performed on the mixture and the quinolones were separated from each other by their solubility in potassium hydroxide solution. The structures were then unequivocally established by the other reactions shown in Scheme 66. The hydrolysis of a 3-acetyl group as shown for compound **591**, with acid in a sealed tube above 200 °C, is a general method.[194] Singh and Nair's findings were confirmed by Anderson and Staskun,[194] who also made compounds **590** and **591**. However, when they extended the method to the preparation of other compounds, the single isomer **592** or **593**, as shown, was obtained. Other preparations of the ketone **590** were from the enaminone **594** or the ester **595**.[195,994,996]

592

593

R^1 = H, R^2 = 2 – Me
R^1 = 8 – Me, R^2 = 4 – OMe
R^1 = 6 – OMe, R^2 = 4 – OMe

594

i, Na
ii, PhCOCl
iii, Ph$_2$O

590

AlCl$_3$
200 °C

595

The vinylogous amidines **596** were prepared from aryl benzimido chlorides and 3-alkylaminocrotonates, and cyclized to the ketones **598** in hot concentrated sulphuric acid or polyphosphoric acid. In two cases compounds **596**, R^1 = H, R^2 = Me; Ar = Ph, 4-MeC$_6$H$_4$, the starting materials, were heated in liquid paraffin at 240–250 °C to give the intermediates **597**, but these could not be purified and were hydrolysed by dilute acid to the known 3-acetyl-4-quinolones.[194]

The sodium salt of 1-phenylbutane-1,3-dione reacted with 1-methylisatoic anhydride (DMA, 120 °C) to give only the ketone **599a**, with no trace of the 3-benzoyl-2-methylquinolone; 1,3-diphenylpropane-1,3-dione similarly gave compound **599b**.[970]

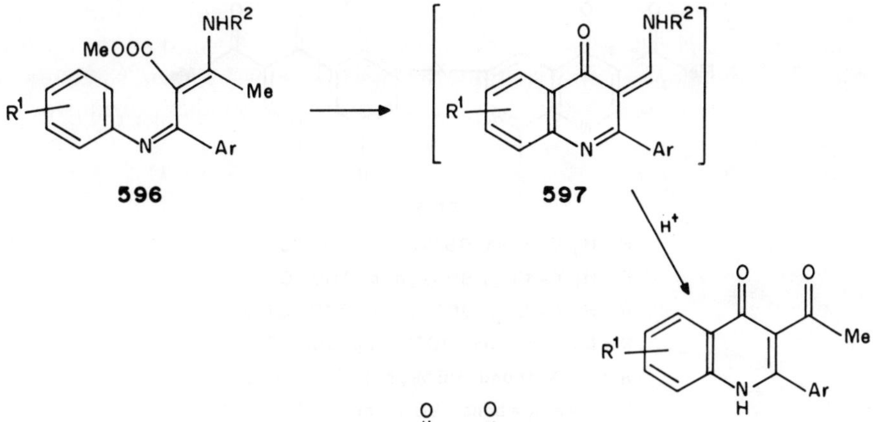

596

597

598

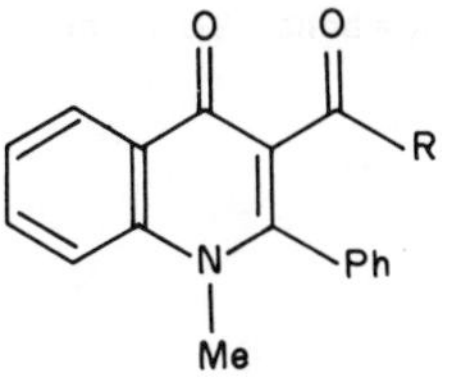

599 a, R = Me

b, R = Ph

Compounds unsubstituted at C-2 are prepared from anilines and ethyl ethoxymethyleneacetoacetate, i.e. **601** from **600**.[993,997] The diketones **602** were prepared as intermediates to pyrano[3,2-c]quinolines.[997] From the appropriate benzidines or other diamines, by a method similar to that used for **600**→**601**, were obtained intermediates which cyclized in refluxing diphenyl ether to give the diketones **603**.[998] The pyran ring of pyranoquinoline **604** was opened with thionyl chloride, after which 10% ethanolic sodium hydroxide gave the 2-chloro-4-quinolone **605**.[967]

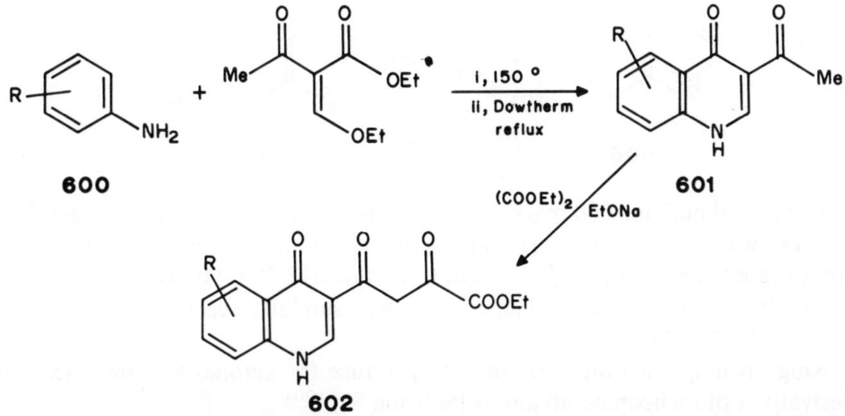

600

601

602

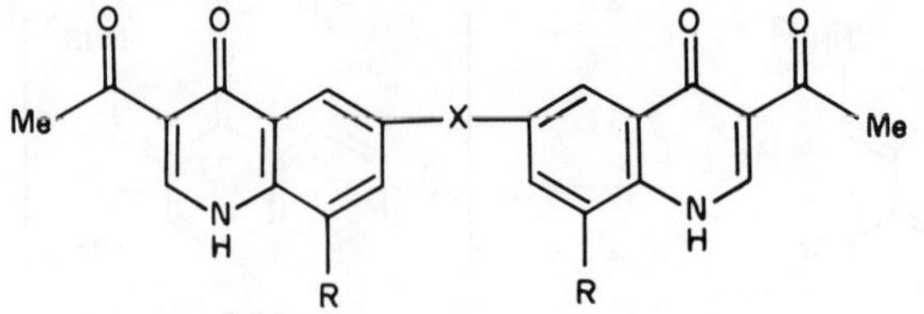

603

R = H, X = bond, 98%, m.p. 360 °C
R = H, X = CH₂, 89%, m.p. 360 °C
R = H, X = SO₂, 96%, m.p. 270 °C (d)
R = Me, X = bond, 90%, m.p. 360 °C
R = Cl, X = bond, 96%, m.p. 275 °C (d)
R = OMe, X = bond, 90%, m.p. 275 °C (d)

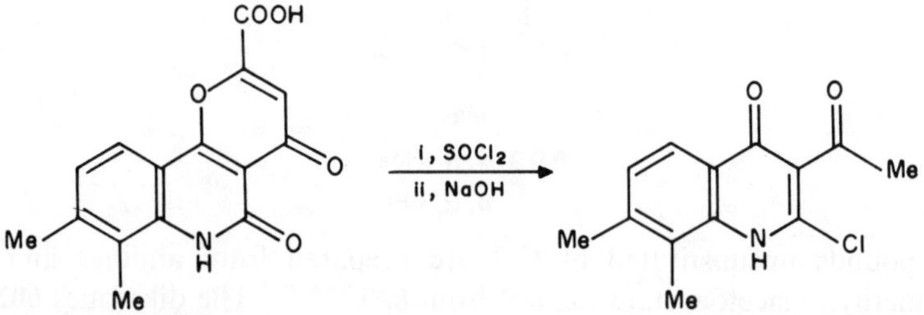

604　　　　　　　　　　　**605**

The 2-aminobenzoxazinone **606** was attacked by the anion from pentane-2,4-dione to give mainly quinazoline-2,4-dione; however, the product contained 25% of the ketone **607**.[999]

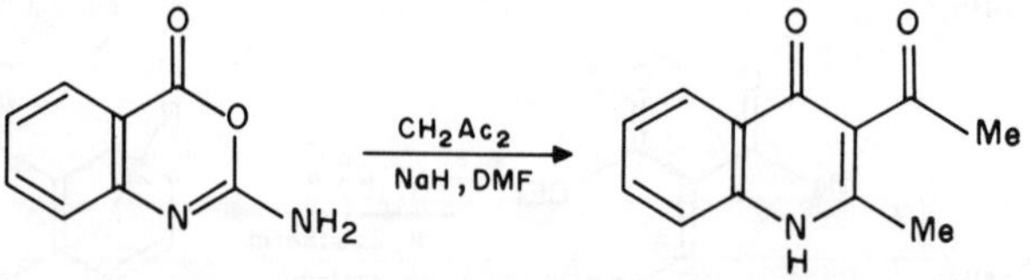

606　　　　　　　　　　**607**

The pyrrolinedione **608** rearranged on pyrolysis to the 4-quinolone **609**. This reaction was thought to go via a ketene intermediate, Scheme 67.[221] The more complicated conversion of diones **610** to ketone **612**, R = Me was also assumed to involve a ketene, **611**.[219] The epoxide **613** similarly rearranged to give the ketone **612**, R = H, Scheme 68.[1000]

Ring opening reactions were used to produce the ketones **614** and **615** from derivatives of melicopine alkaloids (Scheme 69).[1001]

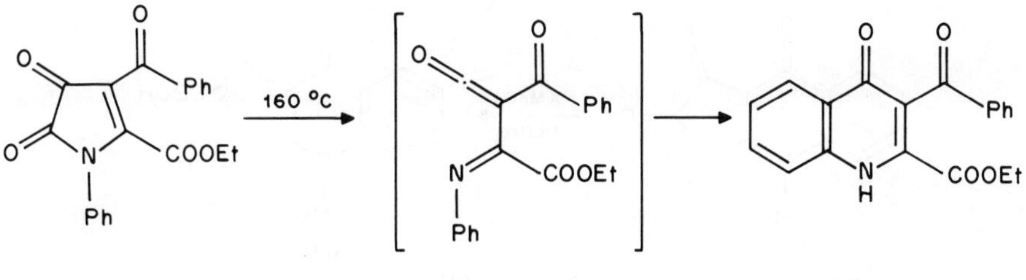

608   **609**

SCHEME 67

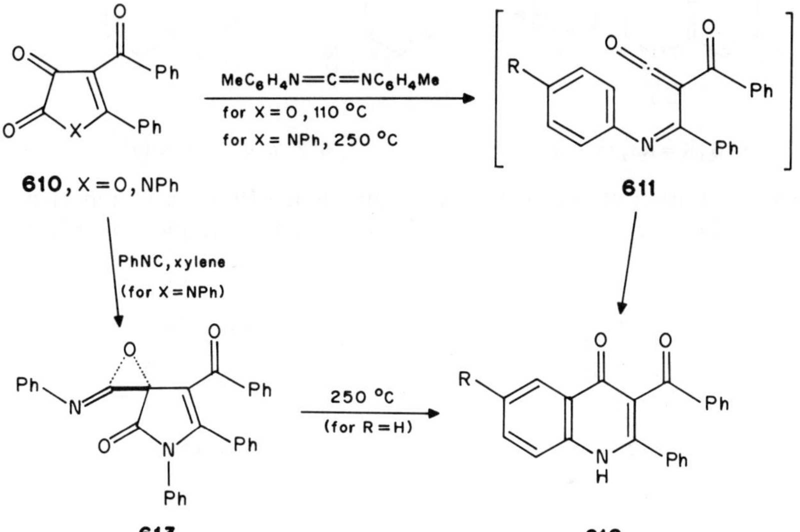

610, X = O, NPh   **611**

PhNC, xylene
(for X = NPh)

613   **612**

SCHEME 68

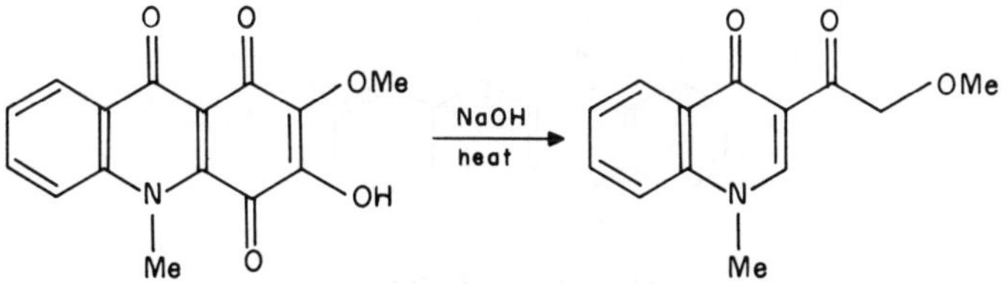

614

Either of the compounds **616** reacted with nitric acid in acetic acid to give a new, high molecular weight product. The authors indicated that formula **617** (drawn by them as the dihydroxy tautomer) agreed with the properties of the product, but were unable to explain how it formed.[1002]

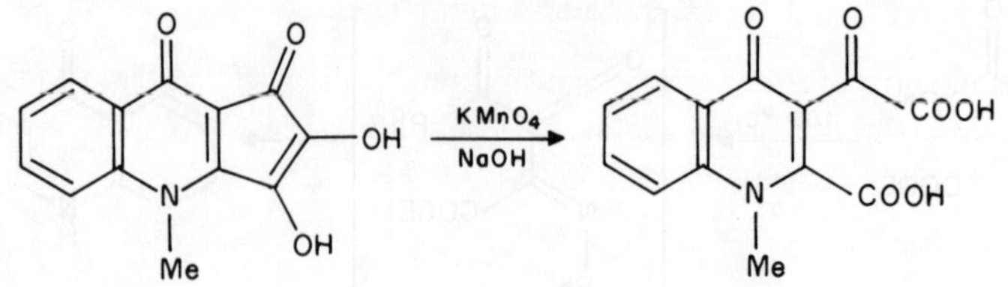

615

SCHEME 69

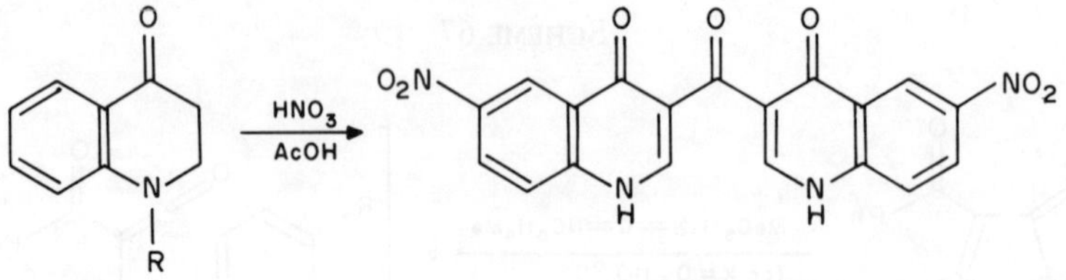

616, R = Ac, Ts　　　　　　　　**617**, m.p. 335 °C (d)

The cyclization of ester **618** to keto-quinolone **619** has been reported. The same product was obtained by hydrolysis of the 4-chloroquinoline **620**.[230]

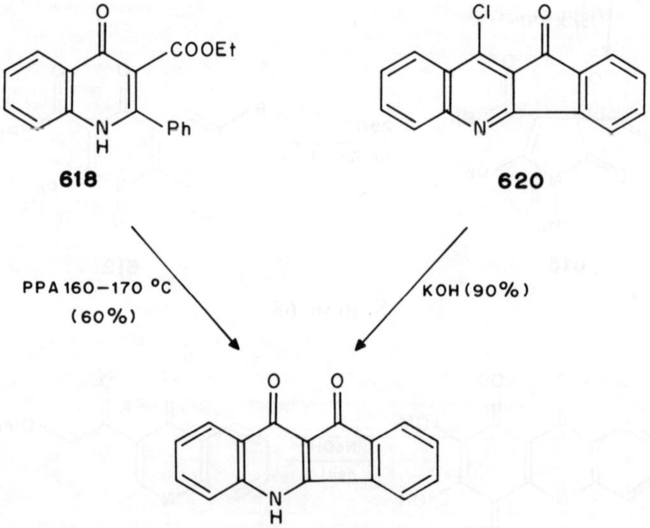

618　　　　　　　　　　　　　　**620**

PPA 160–170 °C　　　　　　　　KOH (90%)
(60%)

619, m.p. > 360 °C

10. 5-Keto-4-quinolone

See Section 12, (p. 433).

11. 6-Keto-4-quinolones

Simple ring closure of the 4-aminoacetophenone derivative **621** gave the ketone **622**.[398] A similar procedure gave the ester **623**, R = Me, which was hydrolyzed with aqueous alkali to the acid **623**, R = H.[418]

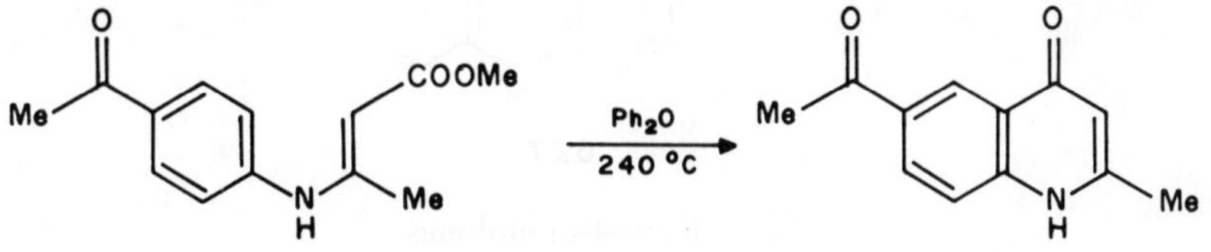

621   **622**

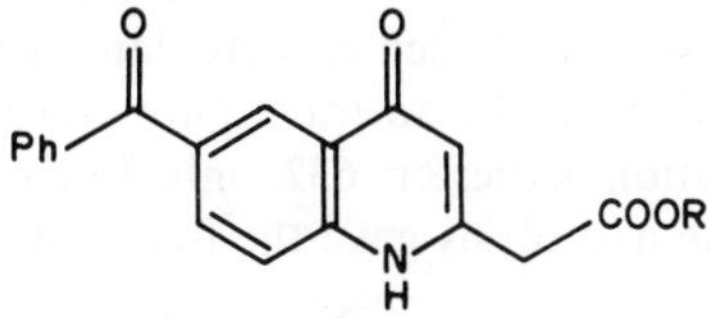

623

The diester **624** cyclized in polyphosphoric acid at 100 °C to the expected ester **625a** and a trace of the decarboxylated product **625b**. Remarkably, there was also a small yield of compound **626**, in which the ring had formed with expulsion of the propyl group.[1003]

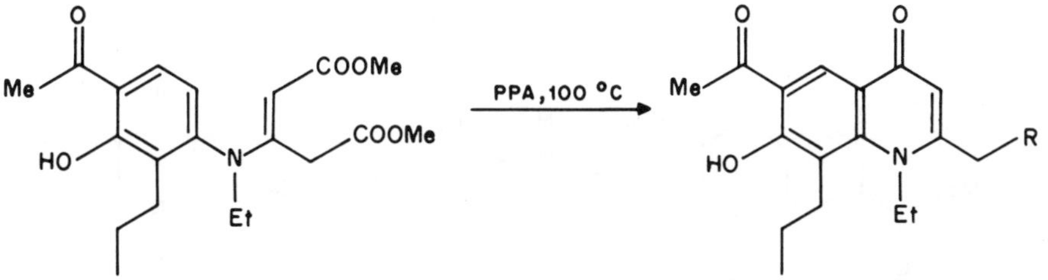

624   **625 a, R = COOMe; b, R = H**

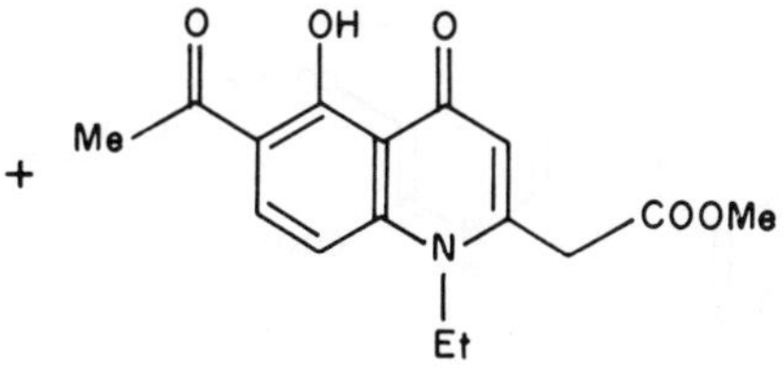

626

Diethyl oxalacetate and 4-aminobenzophenone reacted in chloroform containing hydrogen chloride to give the keto-ester **627**, R = COOEt. Hydrolysis gave the acid **627**, R = COOH, which decarboxylated at 290 °C to compound **627**, R = H.[416] These and other examples are collected in Table 53.

627

12. 7-Keto-4-quinolones

This class of ketone is assembled in Table 54. Ring closure of ester **628** gave the expected mixture (m.p. 189–191 °C) of the 5-, **629**, and 7-, **630**, acetyl-quinolones. After the mixture had been N-methylated and oxidized, the 5-acetyl sulphoxide was removed by h.p.l.c. to leave pure 7-acetyl sulphoxide **631**.[1004] In contrast, from cyclization of diesters **632**, only 7-keto-4-quinolones **633a** and **b** were isolated. Alkaline hydrolysis gave the free acids **633c** and **d**.[1005]

628   **629**

630   **631**

632   **633**

a, R^1, R^2 = Me   b, R^1 = Ph, R^2 = Me

c, R^1 = Me, R^2 = H   d, R^1 = Ph, R^2 = H

Table 53. 6-Keto-4-quinolones

R	Quinoline substituent(s)	Preparation	Yield %	M.p. (°C)	References
Me	None	3-COOH-6-Ac-4-quinolone, Dowtherm, reflux	90	285–286	403, 412
Me		2-COOH-6-Ac-4-quinolone, Ph_2, Ph_2O, 250 °C		284–288	414
Me	1-Me	6-Ac-4-quinolone, Me_2SO_4, NaOH		184	411, 412, 414
Me	1-Me-3-SMe	3-MeS-6-Ac-4-quinolone, MeI, K_2CO_3, DMSO		183–184	1004
Me	1-Me-3-SOMe	1-Me-3-MeS-6-Ac-4-quinolone, MCPBA		254–255	1004
Me	2-Me	4-AcC_6H_4NHC(Me)=CHCOOMe, Ph_2O, 240 °C	82	327–329	398*
Me	1-CH_2Ph-7-OH	2,4-diAc-5-HOC_6H_2NHCHO, NaH; $PhCH_2$Br	90	246–247	404
Me	1-CH_2Ph-7-OCH_2-CH=CH_2	1-$PhCH_2$-6-Ac-7-HO-4-quinolone, $BrCH_2CH=CH_2$, K_2CO_3	87	139–140	404
Me	1-CH_2Ph-7-OH-8-$CH_2CH=CH_2$	Above, $PhNEt_2$, reflux	91	123–124	404
Me	1-CH_2Ph-7-OH-8-Pr	1-$PhCH_2$-6-Ac-7-HO-8-$CH_2CH=CH_2$-4-quinolone, H_2, Pd	84	Oil	404
Me	1-Et	6-Ac-4-quinolone, Et_2SO_4, NaOH		156–158	411, 414
Me	1-Et-2-Me-7-OH-8-Pr	2-Pr-3-HO-4-AcC_6H_2NEtC-(CH_2COOMe)=CHCOOMe, PPA, 100 °C	3[a]	120–123	1003
Me	1-Et-2-CH_2COOMe-5-OH	2-Pr-3-HO-4-AcC_6H_2NEtC-(CH_2COOMe)=CHCOOMe, PPA, 100 °C	3[a]	180–181	1003

Table 53. (*Contd.*)

R	Quinoline substituent(s)	Preparation	Yield %	M.p. (°C)	References
Me	1-Et-2-CH₂COOMe-7-OH-8-Pr	2-Pr-3-HO-4-AcC₆H₂NEtC-(CH₂COOMe)=CHCOOMe, PPA, 100°C	37[a]	124–126	1003
Me	1-Et-3-COOH-7-Cl	3-COOEt-6-Ac-7-Cl-4-quinolone, EtI, K₂CO₃	63	>300	1111
Me	1-Et-3-COOH-7-N-(CH₂CH₂)₂NH	1-Et-3-COOH-6-Ac-7-Cl-4-quinolone, NH(CH₂CH₂)₂NH	94	262–263(d)	1111
Me	1-Et-2-COOMe-7-OH-8-Pr	2-Pr-3-HO-4-AcC₆H₂NEtC-(COOMe)=CHCOOMe, PPA	57	121–123	1112–1114
Me	1-Et-2-COOMe-5-OH-8-Pr	By-product from above	2.6		1112–1114
Me	2-CH₂COOMe-5-OH	3-HO-4-AcC₆H₃NHC(CH₂COOME)=CHCOOMe, C₆H₄CL₂, reflux	12	228–229	1003
Me	2-CH₂COOMe-7-OH-8-Pr	2-Pr-3-HO-4-AcC₆H₂NHC-(CH₂COOMe)=CHCOOMe, PPA, 100°C	33	189–190	1003
Me	1-CH₂CH=CH₂	6-Ac-4-quinolone, CH₂CH=CH₂Br, EtONa		166	411, 414
Me	1-Bu	6-Ac-4-quinolone, BuBr, EtONa		126–128	411, 414
Me	1-CH₂CH₂iPr-2-COOMe-5-OH[b]	3-HO-4-Ac-C₆H₃N(CH₂CH₂iPr)-C(COOMe)=CHCOOMe, PPA	40		407
Me	1-CH₂CH₂iPr-2-COOMe-7-OH[b]		5	108–110	407
Me	7-OH-8-Pr	2-COOMe-6-Ac-7-HO-8-Pr-4-quinolone, NaOH; Ph₂O, reflux	75	250–252	407
Me	2-OEt	4-AcC₆H₄NHC(OEt)=CHCOOEt, Dowtherm, 250°C	65	242–243	410
Me	3-SMe	4-AcC₆H₄NH₂, NaOCH=C(SMe)-COOMe, HCl; Ph₂O, 250°C		265–269	1004
Me	2-COOH	2-COOMe-6-Ac-4-quinolone, aq. NaOH 2-COOMe-6-Ac-4-quinolone, 1m-NaOH	97 75	275–276 266–268	1115 1005

R	Compound	Synthesis	Yield (%)	mp (°C)	References
Me	2-COOH-3-Me	2-COOEt-6-Ac-4-quinolone, 10% NaOH		280(d)	412, 414
		2-COOEt-3-Me-6-Ac-4-quinolone, NaOH		268–269(d)	1116
Me	2-COOH-3-Et	2-COOEt-3-Et-6-Ac-4-quinolone, NaOH		260(d)	1116
Me	2-COOMe	4-AcC$_6$H$_4$NH$_2$, DMAD	41	290	1005, 1115
Me	2-COOMe-7-OH-8-Pr	2-Pr-3-HO-4-AcC$_6$H$_2$NH$_2$, DMAD; Ph$_2$O	74	169–170	406–408
Me	2-COOEt	4-AcC$_6$H$_4$N=C(COOEt)CH$_2$COOEt, Ph$_2$, Ph$_2$O, 250 °C		259	412, 414
Me	2-COOEt-7-OH-8-Pr				406
Me	3-COOH	3-COOEt-6-Ac-4-quinolone, 10% NaOH	100	278	403
Me	3-COOEt	4-AcC$_6$H$_4$NH$_2$, EtOCH=C(COOEt)$_2$, 90 °C; Dowtherm, reflux	70	298–300	403
Me	3-COOEt-7-Cl	3-Cl-4-AcC$_6$H$_3$NH$_2$, EtOCH=C(COOEt)$_2$	81	>300	1111
Me	2-CONHMe-7-OH-8-Pr	2-COOMe-6-Ac-7-HO-8-Pr-4-quinolone, MeNH$_2$			406
Pr	1,7-diMe-3-COOH	1,7-diMe-3-COOEt-6-PrCO-4-quinolone, KOH			415, 1064
Pr	1,7-diMe-3-COOEt	3-Me-4-PrCOC$_6$H$_3$NHMe, DEAD		184–186	415, 1064 / 1064
Pr	1,7,8-triMe-2-COOH	1,7,8-triMe-3-COOEt-6-PrCO-4-quinolone, KOH			415, 1064
Pr	1,7,8-triMe-3-COOH				415, 1064
Pr	1,7,8-triMe-2-COOEt	2,3-diMe-4-PrCOC$_6$H$_2$NMeC(COOEt)=CHCOOEt, PPA			415, 1064
Pr	1-Et-3-COOH-7-Me	1-Et-3-COOEt-6-PrCO-7-Me-4-quinolone, NaOH		208–209	415, 1064
Pr	1-Et-3-COOEt-7-Me	3-COOEt-6-PrCO-7-Me-4-quinolone, NaH, EtI			415, 1064
Pr	3-COOEt-7-Me	3-COOEt-6-PrCO-7-Me-4-quinolone, NaOH		145–146	415, 1064
Pr	1-Pr-3-COOEt-7-Me	3-Me-4-PrCOC$_6$H$_3$NHPr, DEAD		109–111	415, 1064

Table 53. (*Contd.*)

R	Quinoline substituent(s)	Preparation	Yield %	M.p. (°C)	References
Pr	1-CH$_2$CH=CH$_2$-3-COOEt-7-Me	3-Me-4-PrCOC$_6$H$_3$NHCH$_2$CH=CH$_2$, DEAD		137–139	415, 1064
Pr	2-COOH-5-Me	2-COOEt-5-Me-6-PrCO-4-quinolone, NaOH		237–238	415, 1064
Pr	2-COOH-7-Me	2-COOEt-6-PrCO-7-Me-4-quinolone, KOH		255–258	415, 1064
Pr	2-COOH-7,8-diMe	2-COOMe-6-PrCO-7,8-diMe-4-quinolone, KOH		267–269	415, 1064
Pr	2-COOMe-7,8-diMe	2,3-diMe-4-PrCOC$_6$H$_2$NHC(COOMe)=CHCOOMe, PPA		157–159	415, 1064
Pr	2-COOEt-5-Me	3-Me-4-PrCOC$_6$H$_3$NH$_2$, DEAD, 35 °C; Ph$_2$O, 240 °C		173–175	415, 1064
Pr	2-COOEt-7-Me	Chromatographic separation from 5-Me (above)		217–218	415, 1064
Pr	3-COOH-7-Me	3-COOEt-6-PrCO-7-Me-4-quinolone, NaOH		249–252.5	415, 1064
Pr	3-COOEt-7-Me	3-Me-4-PrCOC$_6$H$_3$NH$_2$, EtOCH=C-(COOEt)$_2$, 130 °C; Ph$_2$O, reflux			415, 1064
Pr	3-COOEt-7-OCH$_2$cyclo C$_3$H$_5$	3-cycloC$_3$H$_5$CH$_2$O-4-PrCOC$_6$H$_3$-NHCH=C((COOEt)$_2$, PPA	77.8	298.5–299(d)	1117
Ph	2-CH$_2$COOH	2-CH$_2$COOMe-6-PhCO-4-quinolone, 290 °C	95	244–245	416
Ph		2-CH$_2$COOMe-6-Ph-4-quinolone, KOH, 25 °C		83(d)	418
Ph	2-CH$_2$COOMe	4-PhCOC$_6$H$_4$N=C(CH$_2$COOMe)$_2$, 270–290 °C	50	216–217	418
Ph	2-COOH	2-COOEt-6-PhCO-4-quinolone, aq. NaOH	96	245–246^c	416
		2-COOMe-6-PhCO-4-quinolone, 1m-NaOH	59	>300^c	1005

Ph	2-COOMe	4-PhCOC₆H₄NH₂, DMAD; Ph₂O	35	252–255	1005
Ph	2-COOEt	4-PhCOC₆H₄NH₂, EtOOCCO-CH₂COOEt, HCl	82	225–226	416
CHO		6-Ac-4-quinolone, SeO₂, AcOH		H₂O > 300	414
CHO	1-Me	1-Me-6-Ac-4-quinolone, SeO₂, AcOH			412, 414
CHO	1-Et	1-Et-6-Ac-4-quinolone, SeO₂, AcOH			412, 414
CHO	1-CH₂CH=CH₂	1-CH₂CH=CH₂-6-Ac-4-quinolone, SeO₂, AcOH			412, 414
CHO	1-Bu	1-Bu-6-Ac-4-quinolone, SeO₂, AcOH			412, 414

*This reference includes carbonyl derivatives.
ᵃThese three products came from a single reaction; see text, Section XIV.11.
ᵇThese two compounds were isolated from the same reaction. The 5-hydroxy compound was not characterized.
ᶜInconsistent literature melting points.

Table 54. 7-Keto-4-quinolones

R	Quinoline substituent(s)	Preparation	Yield (%)	M.p. (°C)	References
Me	None	3-COOEt-7-Ac-4-quinolone, NaOH; Dowtherm		280	411
Me	1-Me	7-Ac-4-quinolone, Me$_2$SO$_4$, NaOH		178	411, 414
Me[a]	1-Me-3-SMe	3-MeS-7-Ac-4-quinolone, Me$_2$SO$_4$, NaOH			1004
Me	1-Me-3-SOMe	3-MeS-7-Ac-4-quinolone, MeI, K$_2$CO$_3$; MCPBA		245–246	1004
Me	1-Et-3-COOH-6,8-diF	1-Et-3-COOEt-6,8-diF-7-Ac-4-quinolone, HCl	80	250–251	1118
Me	1-Et-3-COOEt-6-F	3-COOEt-6,8-diF-7-Ac-4-quinolone, EtI, K$_2$CO$_3$		106–108	1118
Me	1-Et-3-COOEt-6,8-diF			129–130	1118
Me	1-CH$_2$CH$_2$F-3-COOH-6,8-diF	1-CH$_2$CH$_2$F-3-COOEt-6,8-diF-7-Ac-4-quinolone, HCl	80	215–216	1118
Me	1-CH$_2$CH$_2$F-3-COOEt-6,8-diF	3-COOEt-6,8-diF-7-Ac-4-quinolone, BrCH$_2$CH$_2$F, NaH	35	155–156	1118
Me[a]	3-SMe	3-AcC$_6$H$_4$NHCH=C(SMe)COOMe, Ph$_2$O, 250 °C			1004
Me	2-COOH	2-COOMe-7-Ac-4-quinolone, aq. NaOH	13	285–288	1005
Me	2-COOMe	3-AcC$_6$H$_4$NHC(COOMe)=CHCOOMe, Ph$_2$O	45	286–287	1005
Me	3-COOH	3-COOEt-7-Ac-4-quinolone, aq. NaOH	92	281–282(d)	1119

		Conditions	%	mp	Ref.
Me	3-COOEt	3-AcC$_6$H$_4$NHCH=C(COOEt)$_2$, Ph$_2$O, 250°C	86	302–303	411, 1119, 1120
Me	3-COOEt-6,8-diF	2,4-diF-3-AcC$_6$H$_2$NHCH=C(COOEt)$_2$, Dowtherm		267–270	1118
CH$_2$Br	1-Et-3-COOEt-6-F	1-Et-3-COOEt-6-F-7-Ac-4-quinolone, KBrO$_3$, 48% HBr	80		1118
CH$_2$Br	1-Et-3-COOH-6,8-diF	1-Et-3-COOH-6,8-diF-7-Ac-4-quinolone, KBrO$_3$, 48% HBr		213–215	1118
CH$_2$Br	1-Et-3-COOEt-6,8-diF				1118
CH$_2$Br	1-CH$_2$CH$_2$F-3-COOH-6,8-diF	1-CH$_2$CH$_2$F-3-COOH-6,8-diF-7-Ac-4-quinolone, KBrO$_3$, 48% HBr		173–175	1118
Me$_2$NCH=CH	1-Et-3-COOEt-6-F	1-Et-3-COOEt-6-F-7-Ac-4-quinolone, (tBuO)$_2$CHNMe$_2$		176–179	1118
Me$_2$NCH=CH	1-Et-3-COOEt-6,8-diF	1-Et-3-COOEt-6,8-diF-7-Ac-4-quinolone, (tBuO)$_2$CHNMe$_2$		175–177	1118
Ph	2-COOH	2-COOMe-7-PhCO-4-quinolone, aq. NaOH	20	269–270	1005
Ph	2-COOMe	3-PhCOC$_6$H$_4$NHC(COOMe)=CHCOOMe, Ph$_2$O	20	263(d)	1005
CHO	1-Me	1-Me-7-Ac-4-quinolone, SeO$_2$, AcOH			414

aObtained as a mixture with the 5-acetyl isomer.

13. 8-Keto-4-quinolones

A similar procedure to that used in the preparation of compounds **633** gave ketones **634** and **635**.[1005] Compound **634**, R = Me was prepared by another group[1006] using the same method, but they reported 49% yield and m.p. 176–177 °C.

	R = Me	46%	m.p. > 250 °C	23%	m.p. 168–171°
	Ph	66	265–268 °C	35	193–195°
	4-MeC$_6$H$_4$	91	245 °C (d)	50	185–187°
	Ph(6-Cl)	32	262–264 °C	2	181–182°

14. 3-Acetyl-2-methyl-7,8-quinolinedione

Oxidation of 7,8-dihydroxy-3-acetyl-2-methylquinoline by chromic acid at room temperature gave the *o*-quinone, m.p. 245 °C(d). The quinone could be reduced back to the dihydroxyquinoline by sulphur dioxide.[74]

XV. Quinolones with Side-chain Ketone Groups

1. 1-Ketoalkyl Quinolones

α-Haloketones and 2-methoxyquinolines gave 1-ketomethyl-2-quinolones **637**, Table 55.

Pyrolysis of acid **638** (from quinoline and chloroacetic acid followed by potassium ferricyanide) gave the ketone **639** in respectable yield.[1123]

Table 55. 1-Ketomethyl-2-quinolones (637)

637

R^1	R^2	Yield (%)	M.p. (°C)	References
Me*	H	30	118–120	1121
Ph	H	38	165.6–169	1122
4-PhC$_6$H$_4$	H	9	180–182	1122
4-BrC$_6$H$_4$	H	17	195–196	1122
Me	Me	28	156–157	1122
Ph	Me	30	179–180.5	1122
4-BrC$_6$H$_4$	Me	19	230.5–233.5	1122

*From MeCOCH$_2$Cl, NaHCO$_3$, EtOH

638　　　　**639**, 43.3%, m.p. 279–280 °C

Methyl isocyanoacetate and 1-acetonylisatoic anhydride in the presence of DBU gave the oxazolo[4,5-c]quinolone **640**, which was hydrolysed to the 1-keto-4-hydroxy-2-quinolone **641a**.[1124] The N-ketomethylanthranilic acids **642**, R^1 = Me, Ph and acetic anhydride gave the oxazolo[3,2-a]quinolones **643a** and **b**, which opened with base to ketones **641a** and **c**. The N-propargyl-quinolone **644** with aqueous sodium hydroxide gave compound **641a**; hydration was accompanied by hydrolysis and decarboxylation. It was suggested that this reaction also went via the oxazoloquinolone **643a**.[1125] Diazomethane converted the phenol **641a** into the ether **641b**.[1126,1127] Compound **643c** with aqueous sodium hydroxide gave compound **641d**, which was methylated (Me$_2$SO$_4$, NaOH) to compound **641e**.[1033] see Scheme 70.

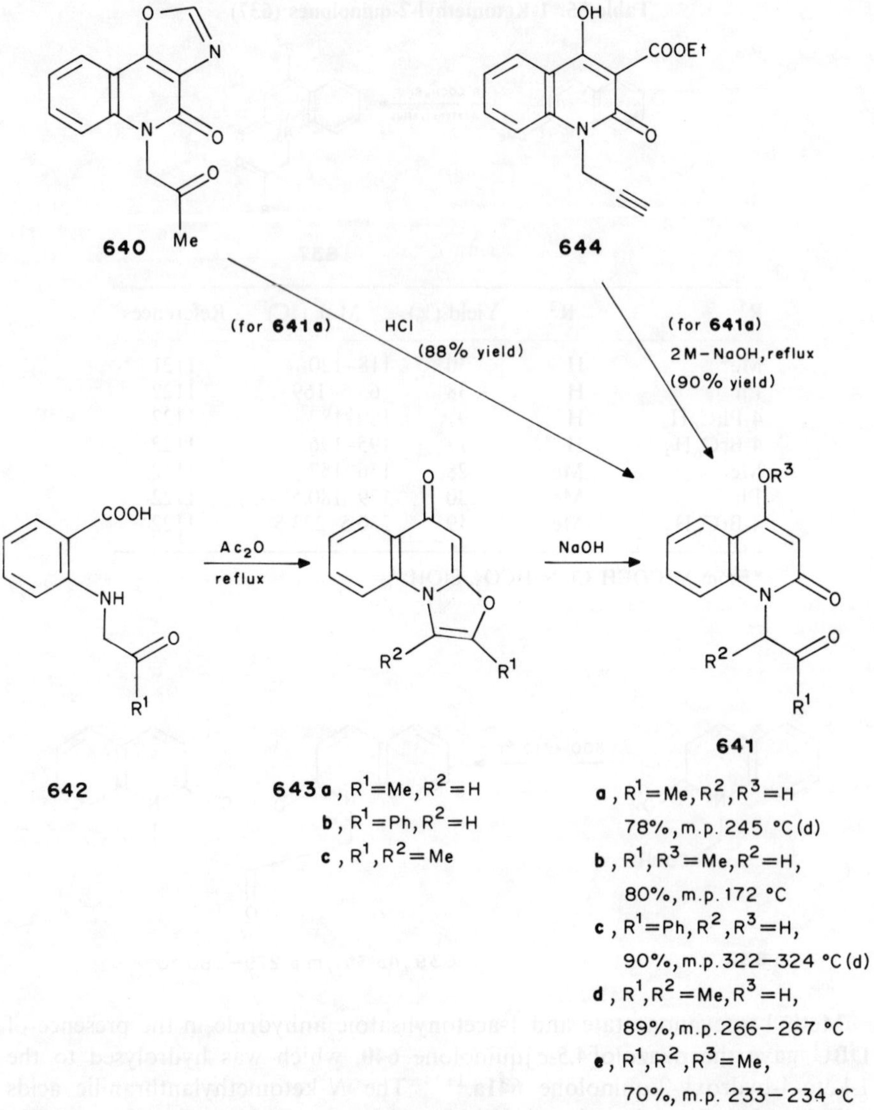

SCHEME 70

In a similar way, the five membered ring of mesoionic compound **645** was readily opened, but a second molecule reacted to give the ketone **647a** (89%). The same product was obtained when the salt **646** was allowed to dissolve in water (90% yield). Diazomethane converted compound **647a** into the ether **647b**.[1032,1033]

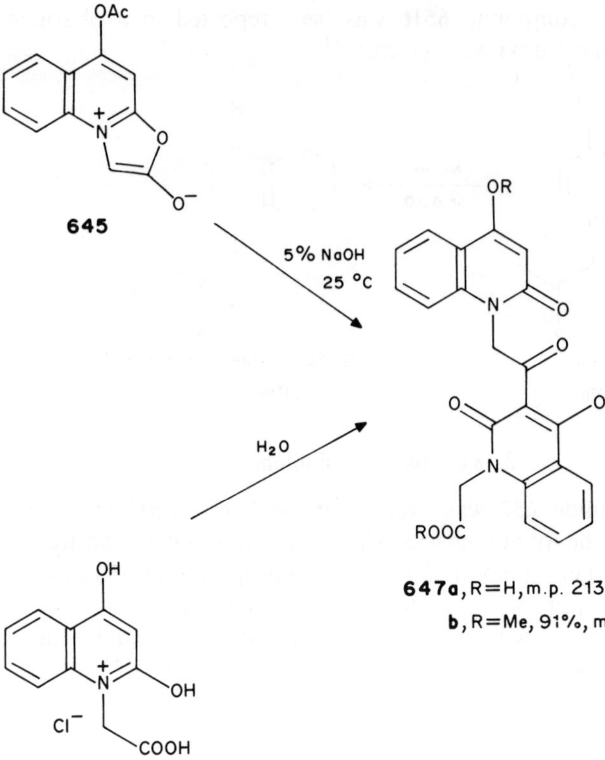

645

5% NaOH
25 °C

H₂O

647a, R=H, m.p. 213 °C(d)

b, R=Me, 91%, m.p. 230–232 °C

646

Methyl vinyl ketone and 2-quinolone in sodium hydroxide gave the adduct **648a**,[1128] while 4-methyl-2-quinolone and 1,3-dichloro-2-butene gave the vinyl chloride **649**, which was hydrolysed to the ketone **648b**.[1129]

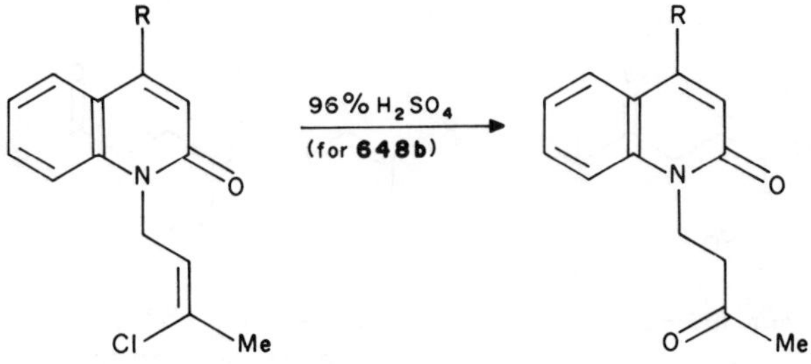

96% H₂SO₄
(for **648b**)

649

648a, R=H, 17%, m.p. 85–88 °C

b, R=Me, 87%

Bromoacetone or phenacyl bromide warmed with 4-aminoquinoline gave the salts **650a** and **650b** respectively.[1130,1131] The former was hydrolysed to the quinolone 651.[1131] Compound 651b was been reported in a pharmacology paper, but no chemical data were given.[1132]

650a, R=Me   651a, R=Me, m.p. 174−177 °C
   b, R=Ph   b, R=Ph

2. Ketoalkyl-2-quinolones

When the keto-anilide 652 was cyclized in acid, the vinyl chloride group hydrolysed to give the quinolone 653, which was 'chlorinated' and treated with thiourea to give the ketoquinoline-2-thione.[740] The quinolone 654 was obtained by hydrolysis of the appropriate 2,4-dichloroquinoline in refluxing acetic acid.[739] Ketones 655 were prepared by aluminium chloride catalysed Friedel–Crafts reactions between the quinolone acid chloride and benzene.[1133]

652   653, 78%

654, 83%, m.p. 199−200 °C   655, R=H, m.p. 237 °C
                                    R=Me, m.p. 178 °C

The pyran ring of the pyranoquinoline 656 opened with base, and the product decarboxylated on neutralization to give the quinolone 657.[742]

The keto-esters 658b and 659 were made from the appropriate 4-methylquinolones and diethyl oxalate with potassium ethoxide[1134] and sodium hydride[1135] respectively. Compound **658b** was hydrolysed (6% H_2SO_4) to give compound **658a**.[1134]

656

KOH, MeOH
25 °C

657, 100%, m.p. 149.5—151 °C

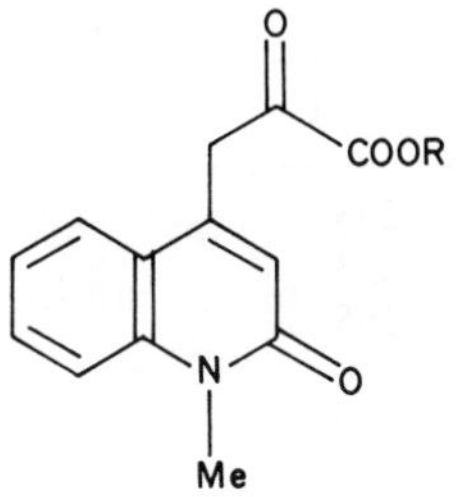

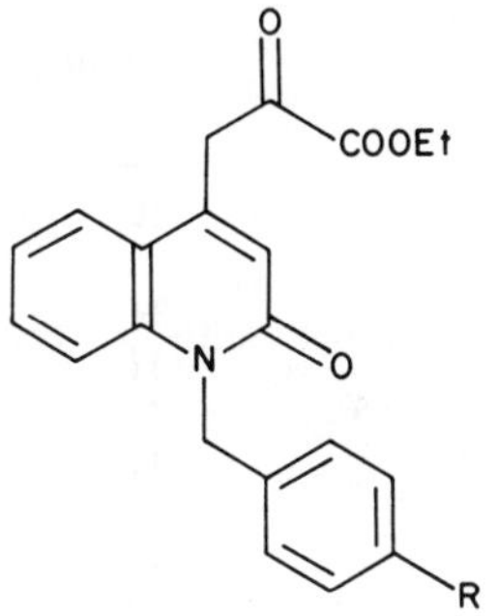

658 a, R=H, 91%, m.p. 231—233 °C(d)
  b, R=Et, 71%, m.p. 181.5—183 °C

659, R=Cl, m.p. 180—182 °C
R=NO₂, Ox m.p. 267 °C

The appropriate 2-methoxyquinoline was hydrolysed with 47% hydrobromic acid to the quinolone **660**, which was reduced to the secondary alcohol without purification.[507,532]

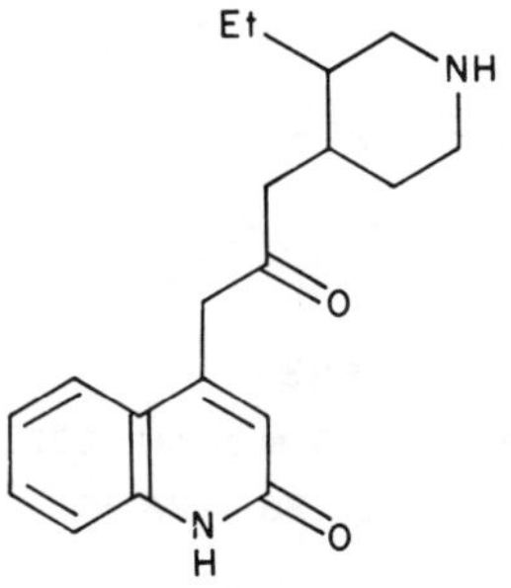

660

The bromomethylquinolones **661** reacted with the anions from aceto-acetanilides (sodium/benzene) to give keto-amides **662**, Table 56. The keto-esters **663** were similarly prepared (ethyl acetoacetate, EtONa).[1136]

A Wittig reaction on 6-formyl-1-methyl-2-quinolone gave the keto-ester **664**.[755]

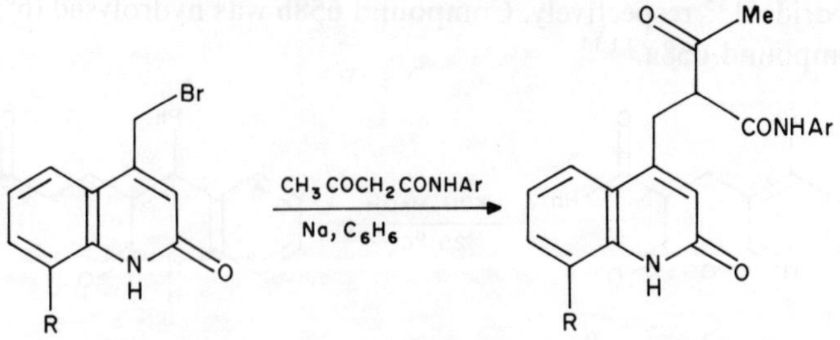

661 → **662**

reagents: CH$_3$COCH$_2$CONHAr, Na, C$_6$H$_6$

Table 56. Acetoacetanilide Derivatives (662)[745]

R	Ar	M.p. (°C)
H	Ph	265
H	3-ClC$_6$H$_4$	233
H	4-ClC$_6$H$_4$	265
H	2-MeOC$_6$H$_4$	178
H	4-MeOC$_6$H$_4$	240
H	1-C$_{10}$H$_7$	280
Me	Ph	242
Me	2-MeC$_6$H$_4$	274

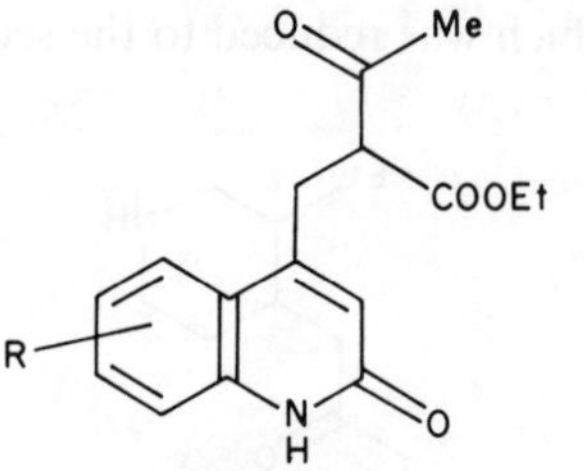

663

R = 6-Me, 60%, m.p. 231–232 °C
R = 7-Cl, m.p. 226–228 °C

3. Ketoalkyl-4-quinolones

The dianion from 1-phenylbutane-1,3-dione reacted with methyl anthranilate to give the trione **665**, which was cyclized to the ketoquinolone **666a**. With 5-chloroisatoic anhydride instead of methyl anthranilate, compound **666b** was obtained.[1137]

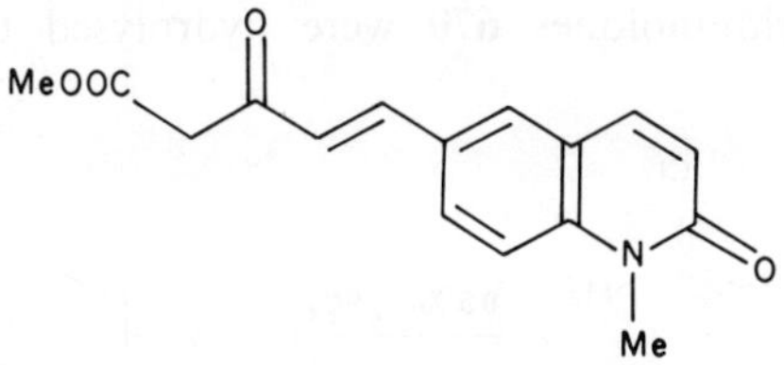

664

665   **666**

a, R = H, 35%, m.p. 237 °C
b, R = Cl, 45%, m.p. 281 °C

A compound claimed to have structure **667** was extracted from *Vepris ampody*. The i.r., u.v., ¹H n.m.r., and mass spectra were consistent with the given structure, but the possibility of a branched side chain should not be overlooked.[1138]

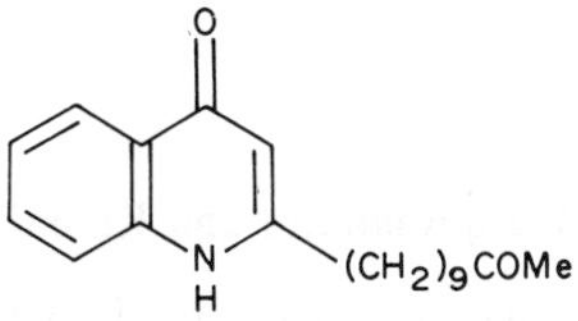

667

Quinoline *N*-oxides and enamines give, generally, 2-ketomethylquinolines, Section V.3. The 4-hydroxy group of compound **668**, however, directed attack to the 3-position to give the ketoquinolone **669**, Scheme 71.[585]

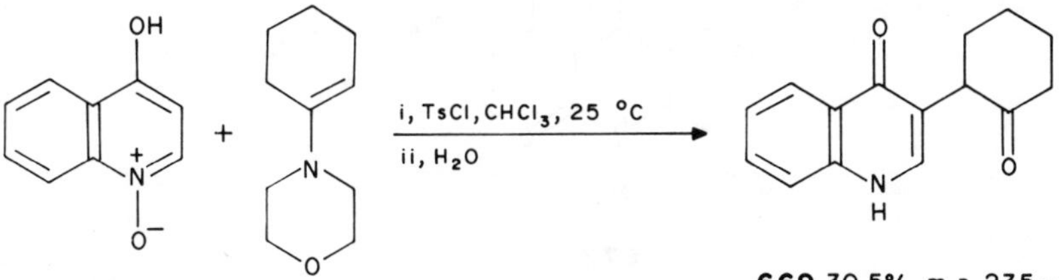

669, 30.5%, m.p. 235—236 °C

668

SCHEME 71

The chlorobutenylquinolones **670** were hydrolysed to ketones **671**; see Table 57.[733-735]

670   **671**

Table 57. 3-(3-Oxobutyl)-4-quinolones
(671)

R	Yield %	M.p. (°C)
H	82.5	243–244
6-Me	80.3	262–263
6-Cl	67	251
6-Br	57.6	255
8-Br	52	214
6-COOH	76.6	272–273
6-NH$_2$		183–186
6-NHAc	67.8	268–270

4. Ketoalkyl-4-hydroxy-2-quinolones and Ketoalkyl-2,4-quinolinediones

Compound **672**, X = CH$_2$ was ozonolized to the ketone **672**, X = O[1139]

Treatment of ketone **673** with excess diazomethane gave a mixture of compounds **674** and **675**, both with extended side chains. However, the more hindered ketone **676** gave the ketone **677** mixed with the simple O-methylated derivative **678**.[630,1037]

672

Reaction between 4-hydroxy-1-methyl-2-quinolone and p-benzoquinone in acetone at room temperature gave the ketone **679**.[1140]

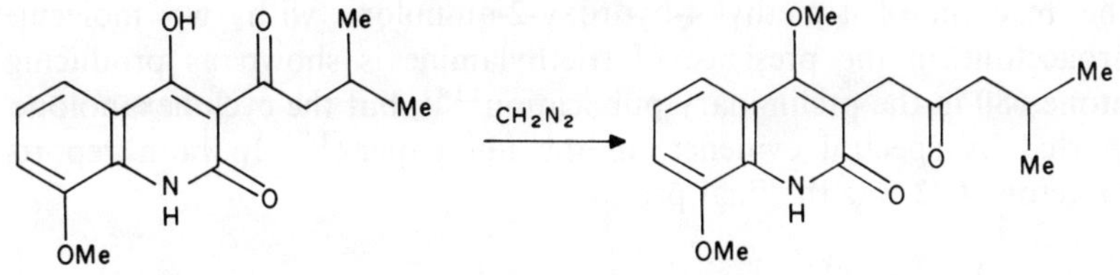

673

CH₂N₂ →

674, 35%, m.p. 144−145 °C

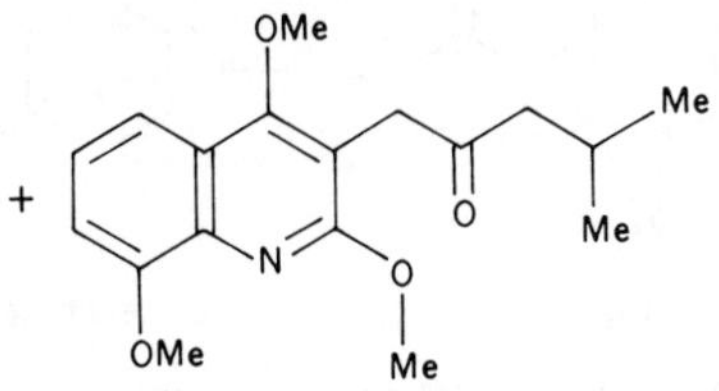

+

675, 1%, m.p. 62−63 °C

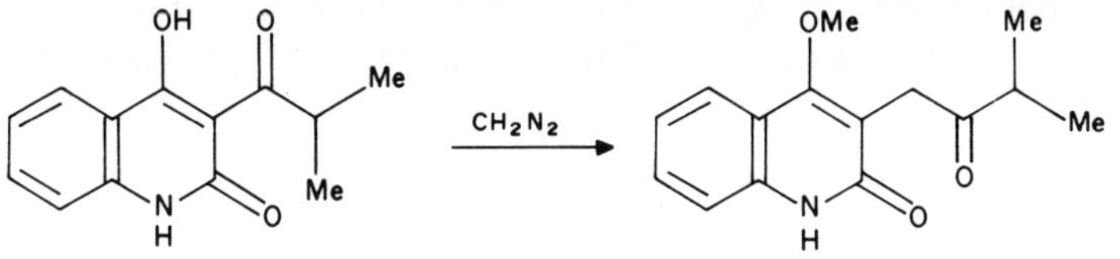

676

CH₂N₂ →

677, 16.2%, m.p. 182−183 °C

+

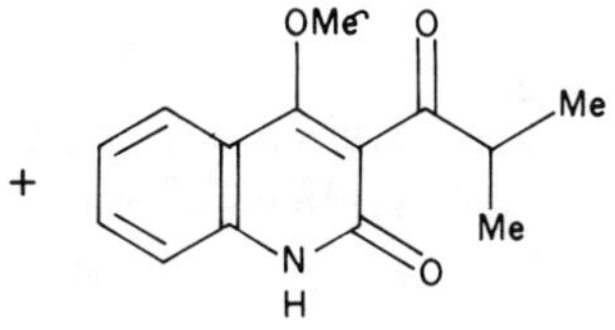

678, 13.8%, m.p. 158−160 °C

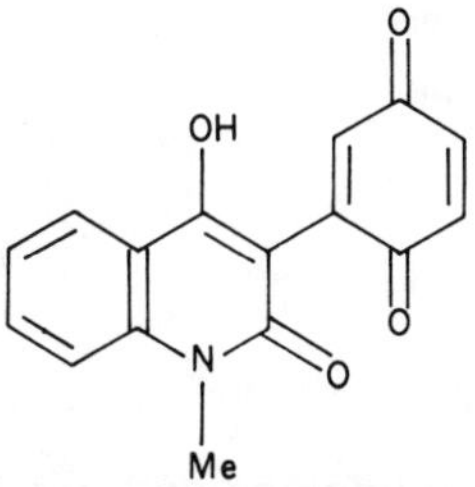

679, 54%, m.p. > 300 °C

The reaction of 1-methyl-4-hydroxy-2-quinolone with two molecules of chloroacetone in the presence of triethylamine is shown as producing the diketone **680** in the preliminary publication[1141], but the cyclohexanolone **681**, supported by spectral evidence, in the full paper.[1142] In both reports the spiro-ketone **682** was the final product.

680   **681**,12%,m.p.231 °C   **682**,85%,m.p.281 °C

Quinoline-2,3,4-trione reacted with acetylacetone, ethyl acetoacetate and dimedone at room temperature without catalyst to give the adducts **683a, b** and **c** respectively.[1143] The same trione and the appropriate Wittig reagent gave compound **684**.[1144] A patent describing oxime derivatives of the trione **685** and the equivalent quinoline-2-thione has recently appeared.[1145]

683

a, R^1=R^2=Me,66%,m.p.178−180 °C
b, R^1=Me,R^2=OEt,75%,m.p.142−144 °C
c, R^1,R^2=CH$_2$CMe$_2$CH$_2$,74%,m.p.158−160 °C

684,90%,m.p.315 °C   **685**

A mixture of 4-methoxy-2-quinolone and hex-3-yne was irradiated to give the adduct **686**. Acid treatment then gave ketone **687**, also available by the same route but in lower yield from 4-acetoxy-2-quinolone. The epoxide **688** was prepared and treated with dilute alkali to give the 3-ketoalkyl quinolone **689**; Scheme 72. A similar sequence produced a mixture of epoxide **690** (53%) and its diastereoisomer **691** (15%); either isomer could be converted into the ketone **692**, Scheme 73.[1146]

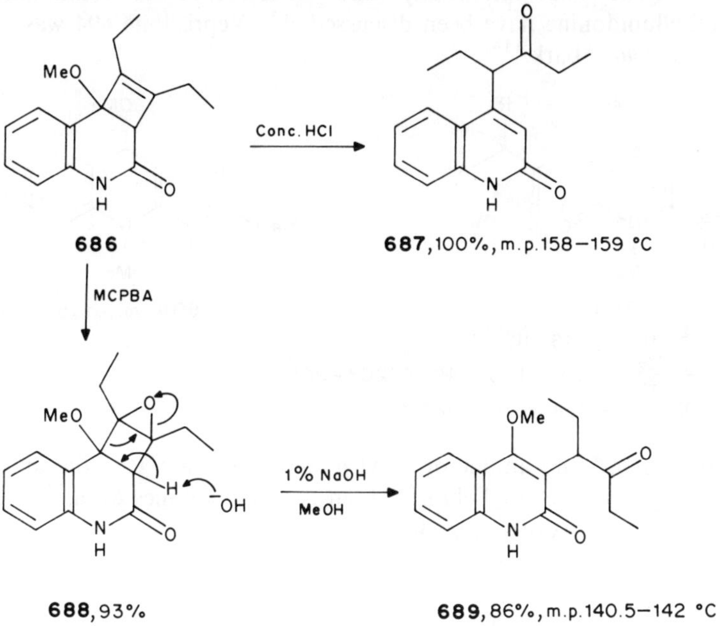
MeO
Conc. HCl
686
687,100%, m.p.158–159 °C
MCPBA
MeO
OMe
1% NaOH
MeOH
688, 93%
689, 86%, m.p.140.5–142 °C

SCHEME 72

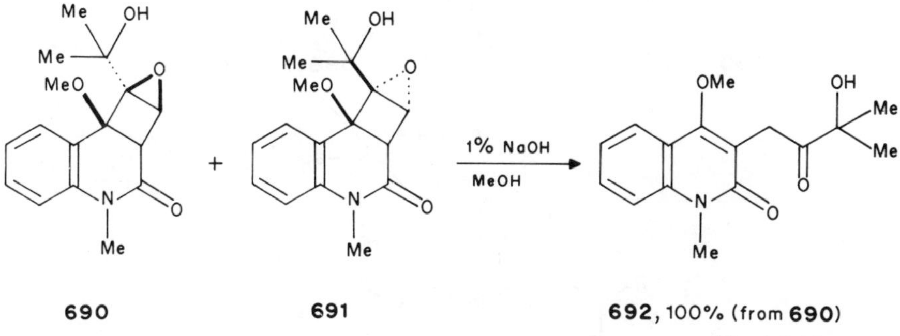
Me OH
Me
MeO
1% NaOH
MeOH
OMe
OH
Me
690
691
692, 100% (from 690)
m.p.112–113 °C

SCHEME 73

The following alkaloids were isolated from *Ptelea trifoliata*: lunidonine **693a**[1147], *N*-demethyllunidonine[1148], hydroxylunidonine **693b**[1147] and 6-

methoxylunidonine **693c**[1147]; the alcohol **693b** was converted into compound
693a by zinc in acetic acid.[1147] However, another group[1149] isolated the alcohol
693b from the same species but found the higher melting point shown in brackets.
Lunidonine has also been prepared by oxidation (CrO₃) of the appropriate
secondary alcohol, lunidine. Both alkaloids have been isolated from *Lunasia
amara*.[1150] The chromatography and fluorescence of lunidonine and
N-demethyllunidonine have been discussed.[1151] Veprisilone **694** was isolated
from *Vepris louisii* bark.[1152]

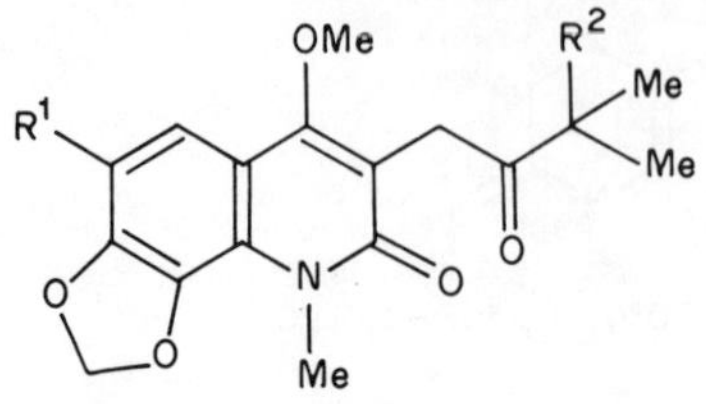

693

a, R¹, R² = H, m.p. 118–119 °C
b, R¹ = H, R² = OH, m.p. 145–149 °C (203–205 °C)
c, R¹ = OMe, R² = H, m.p. 123–125 °C

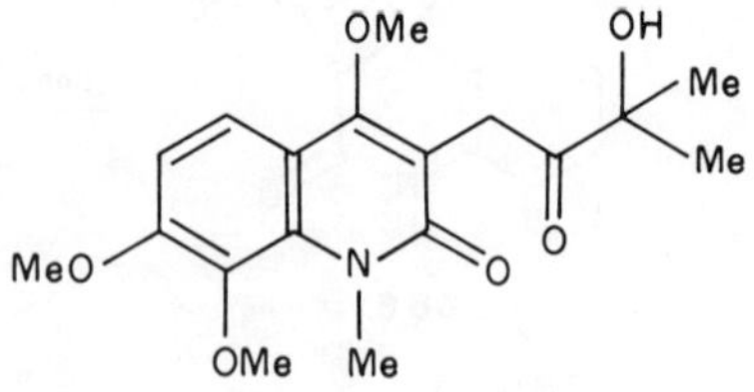

694, m.p. 135–136 °C

The diester **695** and aniline or ethyl *p*-aminobenzoate gave the quinolones
696. Hydrolysis of the vinyl chloride then gave the ketones **697a**,[739] and with
concomitant ester hydrolysis **697b**.[1153]

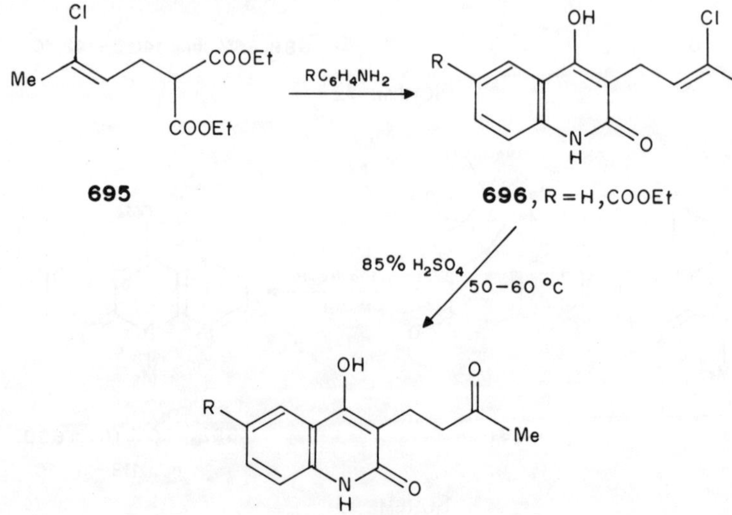

695

696, R = H, COOEt

85% H₂SO₄
50–60 °C

697

a, R = H, 86%, m.p. 182–183 °C
b, R = COOH, 67.3%, m.p. 320 °C

By reaction with a methyl vinyl ketone precursor, 4-hydroxy-2-quinolones **698** were substituted with 3-oxobutyl side chains. The products were shown to exist as the hemi-ketals **700a** and **b** rather than ketones **699a** and **b**.[1154] Compounds **699a–c** (shown in the keto form) were reported by a separate group, who gave the data shown in parentheses for compound **699b**.[1141,1142]

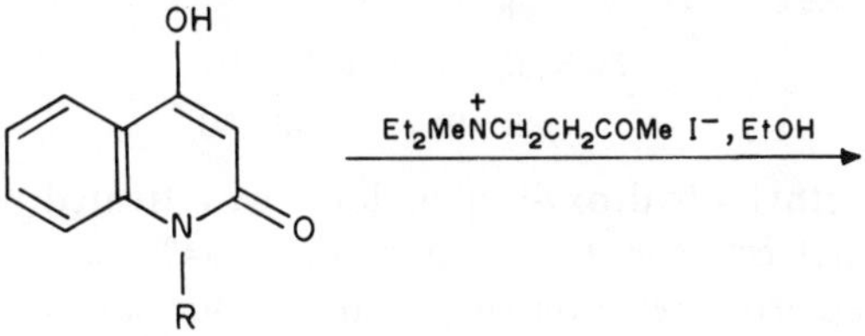

698

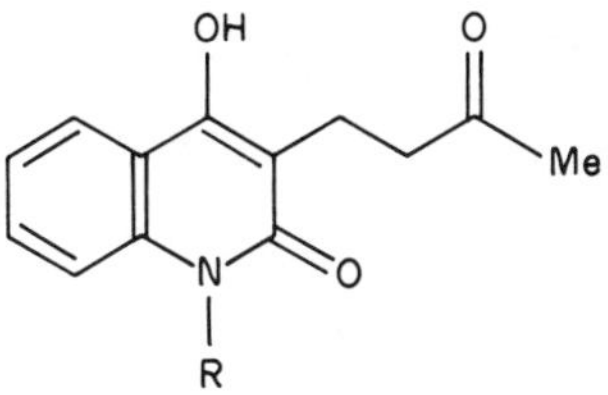

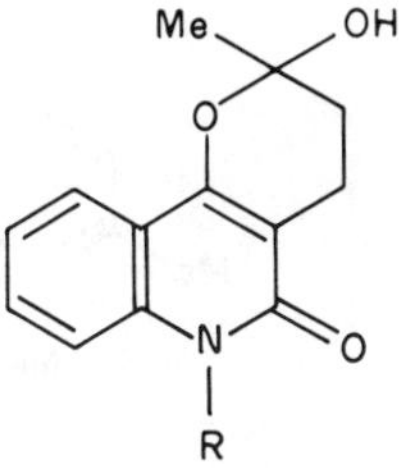

$Et_2MeNCH_2CH_2COMe$ I^-, EtOH

699

700

a, R = H, 73%, m.p. 190–191 °C

b, R = Me, 69%, m.p. 80–81 °C
    (87%, m. p. 131 °C)

c, R = Ph, 75%, m.p. 179 °C

The alkaloid dubinidinone **701** was reduced under Clemmensen conditions to a mixture of ketone **699a** and ketal **702**. The ketal could be hydrolysed (H_2SO_4, dioxane) to the ketone. Diazomethane gave the 4-methoxy derivative of compound **699a** (m.p. 181–182 °C).[1155]

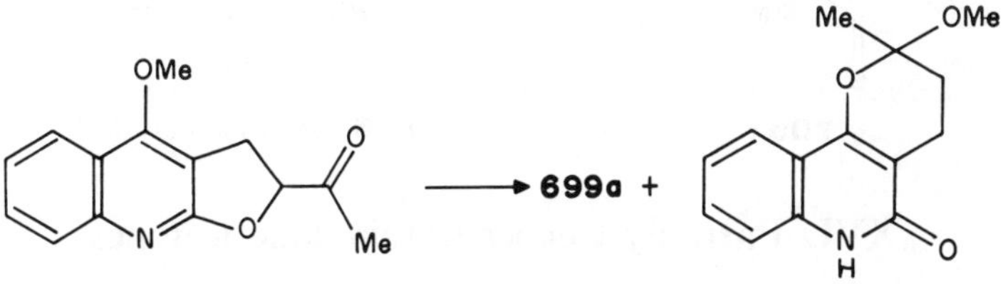

701

→ **699a** +

702

Michael addition of the appropriate 4-hydroxy-2-quinolones to 4-phenylbut-3-en-2-one gave ketones **703**, R = Me, Ph,[1156] while 4-hydroxy-4-phenyl-1-butyne in an acetic/sulphuric acid mixture gave the ketone **703**, R = H.[1157]

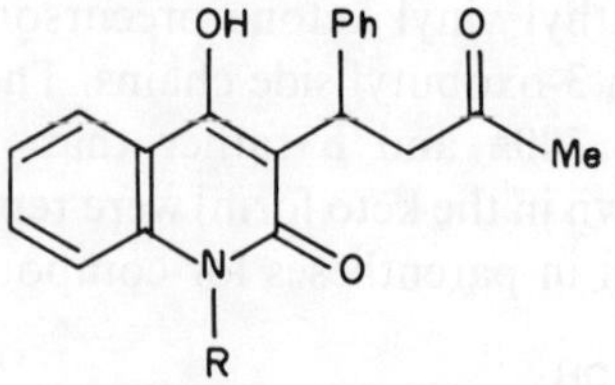

703, R = Me, m.p. 313 °C

R = Ph, m.p. 335 °C

When 7,8-dimethyl-4-hydroxy-2-quinolone was treated with paraformaldehyde and phenylacetylene, the main product (84%) was the methylenebisquinolone derived from reaction only with the formaldehyde. The required ketone **704a** was isolated in low yield and treated with diazomethane to obtain **704b**.[1158]

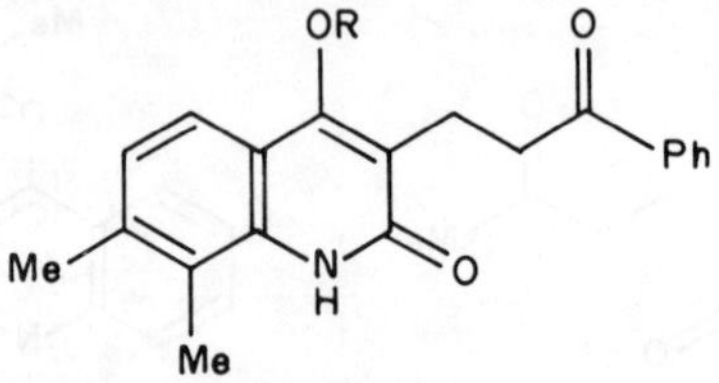

704

a, R = H, 7%, m.p. 250 °C

b, R = Me, 57%, m.p. 187–190 °C

5. Ketoalkyl-5,8-quinolinedione

The quinone **705** underwent Michael addition to methyl vinyl ketone in the presence of pyridine to give the quinone **706**.[1159]

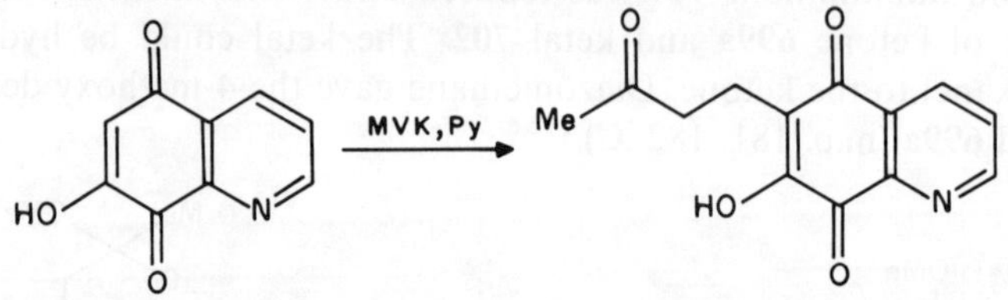

705   **706**, 64%, m.p. 162–163 °C

XVI. Partially Reduced Quinolone Ketones

1. 3,4-Dihydro-3-keto-2-quinolones

Irradiation of the methacrylamide **707** gave the ketone **708**. The suggested mechanism is shown in Scheme 74, with the final step a thermally allowed [1, 5] suprafacial shift of the acetyl group.[1160,1161]

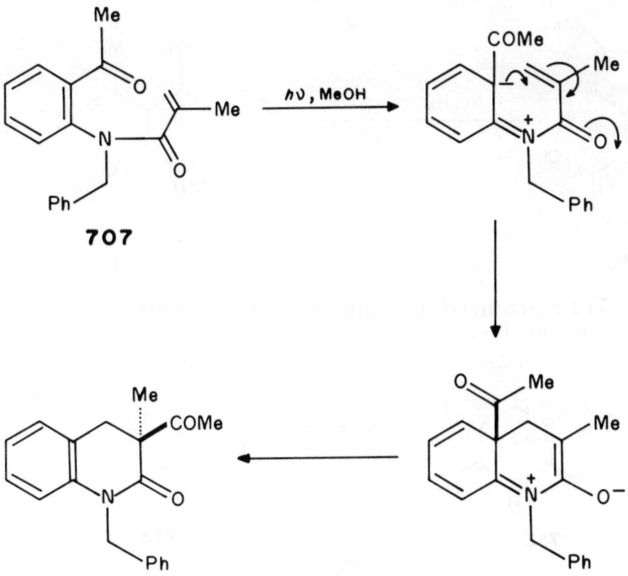

707

708, 25%, m.p. 102–103 °C

SCHEME 74

The keto-esters **709a** and **b** were prepared from the appropriate quinolones and dimethyl oxalate (NaH, DMF) at 110–115 °C; see Chapter 1, Section VII.[1162]

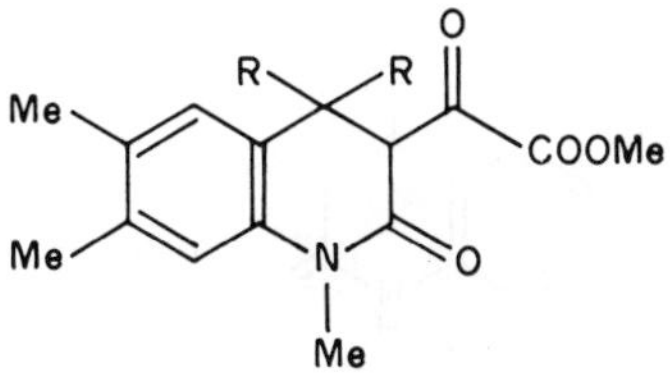

709

a, R = H, 21%, m.p. 252–255 °C

b, R = Me, 22%, oil

The pyrazolone **710** and 2-aminobenzaldehyde reacted at high temperature to give the hydrazone **711**. In other similar reactions water was eliminated to give 3,4-double bonds; see Section XIV.1 and Chapter 1, Section V.1.[957]

2. 3,4-Dihydro-4-keto-2-quinolones

The 3,4-bond of the appropriate 4-ketoquinolone was saturated (zinc/acetic acid) to give compounds **712**. These and similar ketones are in Table 58.

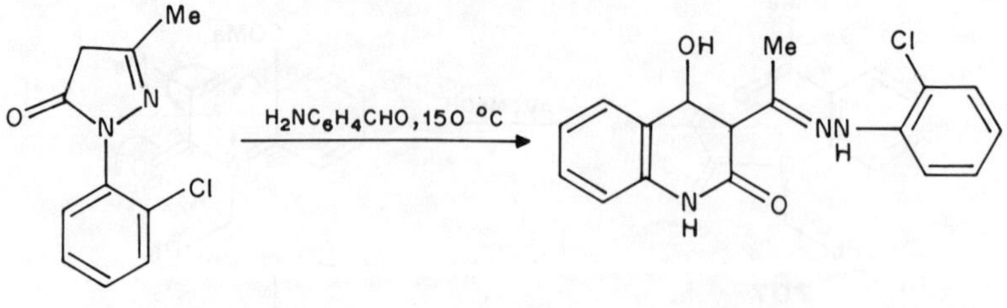

710   **711**

Compounds **712** rearranged in ethanolic acid to the ethyl indolylacetates **713**, Scheme 75.[235,300,301,1163]

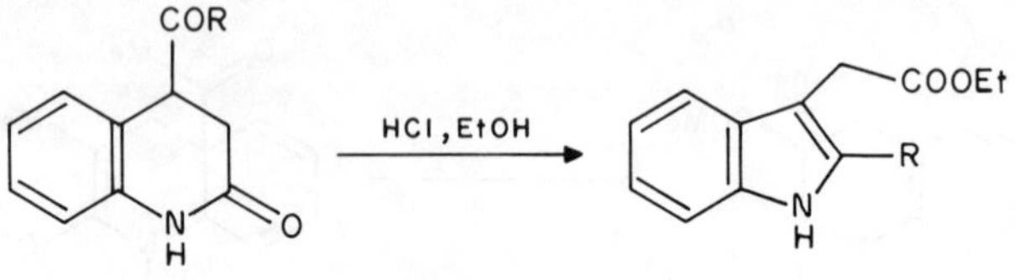

712   **713**

R = H, Me, Ph, 2-Quinolyl

SCHEME 75

Treatment of 4-picolinoyl-2-quinolone with ethylene glycol was performed before hydrogenation (Raney nickel) to ketal **714**.[1164] When the reduction was run at a higher temperature, the pyridine ring was also saturated.[1165]

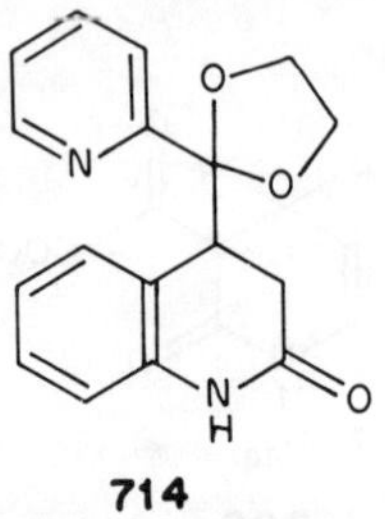

714

The acid **715**, R = OH was, apparently, converted into the ketone **715**, R = Me in 28% yield by warming below 100 °C in an acetic anhydride/pyridine mixture.[1166] Alternatively, the quinolone **715**, with a C-3, C-4 double bond and R = Me, was reduced by zinc in acetic acid to compound **715**, R = Me in 74% yield.[1040]

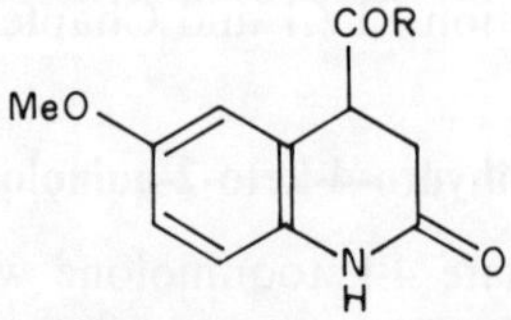

715, R = Me, m.p. 186–188 °C

Table 58. 3,4-Dihydro-4-keto-2-quinolones

R	Quinoline substituent(s)	Preparation	Yield (%)	M.p. (°C)	References
Me	None	4-Ac-2-quinolone, Zn, AcOH or H$_2$, Pd	50	202^a	235, 300
Me	6-Cl	3,4-diH-4-C(OCH$_2$)$_2$Me-6-NH$_2$-2-quinolone, HNO$_2$, HCl		EGAc^b 148–150	284
Me	6-OMe	4-Ac-6-MeO-2-quinolone, Zn, AcOH		189	301
Me	6-NO$_2$	3,4-diH-4-Ac-2-quinolone, KNO$_3$, H$_2$SO$_4$	42.1	255–257(d)	285
Me	6-NH$_2$	4-C(OCH$_2$)$_2$Me-6-NO$_2$-2-quinolone, H$_2$, Ni	82	EGAc^b 210–211	284
Me	1-(COC$_6$H$_4$-4-Cl)-6-OMe	1-(4-ClC$_6$H$_4$CO)-4-Ac-6-MeO-2-quinolone, H$_2$, Pd			1041
Ph		3,4-diH-4-CHOHPh-2-quinolone, CrO$_3$	80	230	235, 300*
Ph	6-Me	4-PhCO-6-Me-2-quinolone, Zn, AcOH	57	211.5–212.5	284
Ph	6-OMe	4-PhCO-6-MeO-2-quinolone, Zn, AcOH	71.2	209–210	285
2-Pyridyl		4-(Py-2-CO)-2-quinolone, H$_2$, Pd		214–216 EGAc^b 253–255	327, 1165
5-Et-2-pyridyl		4-(5-Et-Py-2-CO)-2-quinolone, H$_2$, Ni	29		1163
2-Piperidyl		4-(Py-2-CO)-2-quinolone, (CH$_2$OH)$_2$; H$_2$, Ni		EGAc^b 206–207	1165

*Reference includes carbonyl derivatives.

^{a}Obtained as a 3:2 mixture with the secondary alcohol.

^{b}EGAc = Ethylene glycol acetal.

Either isomer of indolone 716, $R^1 = H$, $R^2 = OH$ or $R^1 = OH$, $R^2 = H$, rearranged to the ketoquinolone 717 in refluxing acetic acid containing 10% ethylenediamine.[1167]

716   717, 30%, m.p. 135–136 °C

The carbazole 718 gave the fused ketone 719 on refluxing with a solution of dimethyl acetylenedicarboxylate.[1168,1169]

718   719, m.p. 222–223 °C

3. Other Keto-3,4-dihydro-2-quinolones

The dihydro-8-hydroxyquinolone 720, R = H gave the normal Friedel–Crafts product 721 on warming in carbon disulphide or nitrobenzene. However, the same reagents at 0 °C gave the ester 720, R = COCH$_2$Cl, which underwent Fries rearrangement to the ketone 721.[1170] Similar Friedel–Crafts reactions are listed in Table 59. The ortho-rearrangement of the ester 722 to 7-acetyl-3,4-dihydro-6-hydroxy-2-quinolone 723 needed more severe conditions.[1077,1171] The hydroxy group could be removed by reduction of its methanesulphonate

720   721

derivative.[1077] The 5-acetyloxy ester similarly rearranged to 6-acetyl-3,4-dihydro-5-hydroxy-2-quinolone. It was reported that 7-acetyloxy-3,4-dihydro-2-quinolone rearranged both to the 6-acetyl-7-hydroxy and to the 8-acetyl-7-hydroxy compounds. It is not clear whether a mixture resulted or one of the reports is in error.[1171]

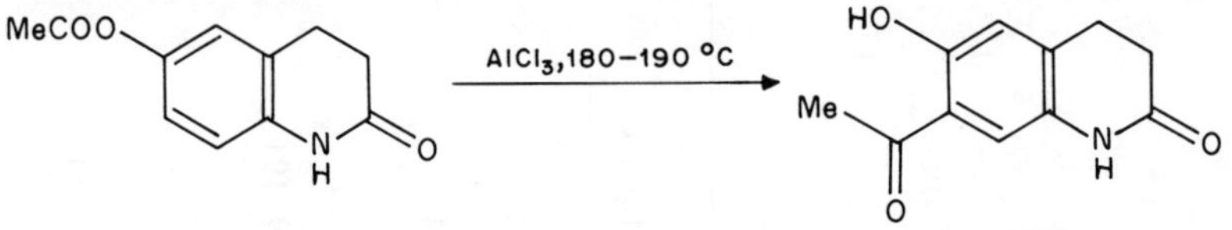

722   **723**

Nitro group reduction converted compound **724** into the 7-ketoquinolone **725**, Table 63, entry 5, which gave the amino-ketone **726**, Table 61 (p. 467), entry 9.[1172]

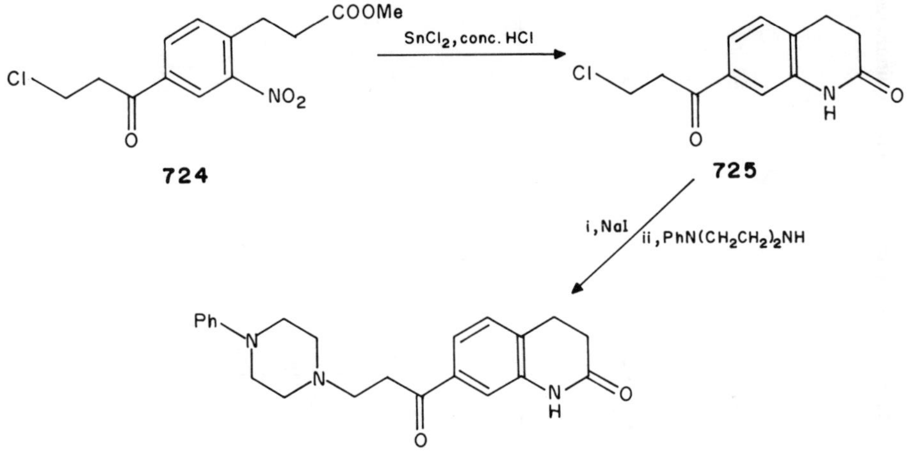

724   **725**

726

4. 3-Keto-5,6,7,8-tetrahydro-2-quinolones

Acetylacetamide and hydroxymethylenecyclohexanone condensed in refluxing ethanol containing triethylamine to give the tetrahydroquinolone **727**.[1173,1174]

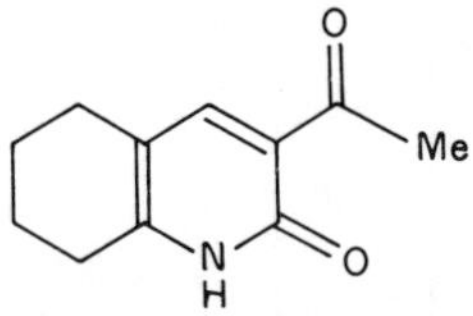

727, 71%, m.p. 217–218 °C

Table 59. 3,4-Dihydro-5-keto-2-quinolones

R	Quinoline substituent(s)	Preparation	Yield (%)	M.p. (°C)	References
Me	8-OH	3,4-diH-5-Ac-8-MeO-2-quinolone, Py·HCl, 200 °C	89	255	1077
Me		Friedel–Crafts reaction			1043, 1044
Me	8-OMe	Friedel–Crafts reaction	96	154	1043
Me	8-OCH₂Ph	3,4-diH-5-Ac-8-HO-2-quinolone, PhCH₂I, Na₂CO₃	77	141	1043
CH₂Cl	1-Me-8-OH	Friedel–Crafts reaction			1050
CH₂Cl	8-OH	Friedel–Crafts reaction	37	189–191	1044, 1170, 1199
		Fries rearrangement			1170
		3,4-diH-5-COCH₂Cl-8-OCOCH₂Cl-2-quinolone, 10% HCl			1170, 1200
CH₂Cl	8-OCOCH₂Cl	Friedel–Crafts reaction		206–207	1050, 1170, 1200
CH₂Cl	8-OMe	Friedel–Crafts reaction	82	187–188	1044, 1050, 1170

R	Substituent	Method	Salt, mp (°C)	IR
CH_2NH_2	8-OH	Method A		1050, 1190, 1200
CH_2NHMe	8-OH	Method A or B	HCl 254–256	1170
CH_2NHMe	8-OMe	Method A or B	HCl 232–234	1170
CH_2NHCH_2Ph	8-OH	Method A		1199, 1200
CH_2NHEt	8-OH	Method A	HCl 258–260	1057, 1170
CH_2NHEt	8-OMe	Method A	HCl 224–227	1057
$CH_2NHCH_2CH_2Ph$	8-OH	Method A		1200
$CH_2NHCH_2CH_2OMe$	8-OMe	Method A		1055
$CH_2NHCH_2CH=CH_2$	8-OH			1062
$CH_2NHCH_2CH=CH_2$	8-OMe			1062
CH_2NHiPr	1-Me-8-OH			1063
CH_2NHiPr	8-OH	Method A or B	HCl 274–275	1170, 1199, 1200
CH_2NHiPr	8-OMe	Method A	HCl 208–209	1050, 1057, 1170
CH_2NHsBu	1-Me-8-OMe	Method A		1057
CH_2NHsBu	8-OH	Method A or B	HCl 269–271	1050, 1170, 1199, 1200
CH_2NHsBu	8-OMe	Method A	HCl 212–214	1170
CH_2NHtBu	8-OH	Method A or B	HCl 253–255	1050, 1170, 1199, 1200
CH_2NHtBu	8-OMe	Method A	HCl 208–210	1057, 1070, 1170
$CH_2NHCMe_2CH_2Ph$	8-OH	Method A		1050, 1200
$CH_2NHcycloC_6H_{11}$	8-OH	Method A		1057
$CH_2NHcycloC_6H_{11}$	8-OMe	Method A		1057
$CH_2N(CH_2)_5$	8-OH	Method A	HCl 239–241	1050, 1170, 1200
$CH_2N(CH_2CH_2)_2O$	8-OMe			1055
Et	8-OMe			1044
CHClMe	8-OH			1050
CH_2CH_2Cl	8-OMe			1077

Table 59. (*Contd.*)

R	Quinoline substituent(s)	Preparation	Yield (%)	M.p. (°C)	References
CHBrMe	8-OMe	Friedel–Crafts reaction	82	154–155	1044, 1201, 1202
$CH_2CH_2N(CH_2CH_2)_2$-$CHPh$[a]	8-OH	Method B		287–291	1077
$CH_2CH_2N(CH_2CH_2)_2$-$CHPh$[a]	8-OSO$_2$Me	Above, MeSO$_2$Cl, KOH		165–167	1077
$CH_2CH_2N(CH_2CH_2)_2$-$CHPh$[a]	8-OMe	Method A			1077
$CH_2CH_2N(CH_2CH_2)_2$-NPh[a]		8-OSO$_2$Me derivative, H$_2$/Pd			1077
$CH_2CH_2N(CH_2CH_2)_2$-NPh[a]	8-OH				1077
$CH_2CH_2N(CH_2CH_2)_2$-NPh[a]	8-OMe				1077
$CH_2CH_2N(CH_2CH_2)_2$-NC_6H_3-2,3-diMe[a]		Mannich reaction	35	231–234	1077
CHMeNHiPr	1-Me-8-OH				1050
CHMeNHiPr	8-OH	Method B		HCl 223–226(d)	1050, 1057, 1201
CHMeNHiPr	8-OMe	Method A		HCl 172–174(d)	1057, 1201
CHMeNHrBu	8-OH	Method B		HBr 224–227(d)	1050, 1057, 1201
CHMeNHrBu	8-OMe	Method A		HCl 207–210(d)	1057, 1201
CHMeNHCH$_2$CH$_2$C$_6$H$_4$-4-OMe	8-OMe	Method A			1060
CHMeN(CH$_2$CH$_2$)$_2$O	8-OH	Method A		235–236(d)	1201
Pr	8-OMe				1044

CHClEt	8-OH	Friedel–Crafts reaction		1050
CHBrEt	8-OH			1052
CHBrEt	8-OMe	Friedel–Crafts reaction	151–152	1044, 1201, 1202
CHEtNHCH$_2$CH$_2$Ph	8-OH			1063
CHEtNHiPr	8-OH	Method B	HBr 165–168	1050, 1057, 1201
CHEtNHiPr	8-OMe	Method A	HCl 204–206(d)	1057, 1201
CHEtNHsBu	8-OH			1050
CHEtNHtBu	8-OH	Method B	HBr 114–116(d)	1057, 1201
CHEtNHtBu	8-OMe	Method A	HCl 260–262(d)	1201
CHEtN(CH$_2$CH$_2$)$_2$O	8-OH	Method A		1200, 1202
CHEtN(CH$_2$CH$_2$)$_2$NH	8-OH	Method A		1202
CHEtN(CH$_2$CH$_2$)$_2$NMe	8-OH	Method A		1202
CHO	8-OH	Method A		1234

[a] About 15 similar amino-ketones are included in the patent.[1077]

Table 60. 3,4-Dihydro-6-keto-2-quinolones

Method A

R	Quinoline substituent(s)	Preparation	M.p. (°C)	References
Me	None	Friedel–Crafts reaction (80% yield)		1023
Me	1,4,4-triMe	Friedel–Crafts reaction (70% yield)	75–77	893
Me	5-OH	3,4-diH-5-AcO-2-quinolone, $AlCl_3$, 180–190 °C (52.5% yield)		1171
Me	5-OH-8-Cl			1204
Me	5-OH-8-Br			1204
Me	7-OH	3,4-diH-7-AcO-2-quinolone, $AlCl_3$, 180–190 °C		1171
Me	8-OMe	3,4-diH-6-NH_2-8-MeO-2-Quinolone, MeCH=NNHCONH_2, Na_2SO_3, AcONa, $CuSO_4$; HCl (37% yield)	150	1073
Me	5-OCH_2CHOHCH_2Cl			1204
Me	5-OCH_2CHOHCH_2Cl-8-Cl			1204
Me	5-OCH_2CHOHCH_2Cl-8-Br			1204
Me	5-OCH_2CHOHCH_2NHCH_2C$H_2$$C_6H_3$-3',4'-diOMe-8-Br			1204

			mp	Ref.
Me	5-OCH$_2$CHOHCH$_2$-NHiPr-8-Br			1204
Me	5-OCH$_2$CHOHCH$_2$-NHtBu			1204
Me	5-OCH$_2$CHOHCH$_2$-NHt-Bu-8-Cl			1204
Me	5-OCH$_2$CHOHCH$_2$-NHtBu-8-Br			1204
Me	5-OCH$_2$CHOHCH$_2$-NHCH$_2$CH$_2$C$_6$H$_3$-3',4'-diOMe-8-Br			1204
CH$_2$Cl	8-Cl	Friedel–Crafts reaction (88% yield)	233–234	1077
CH$_2$Cl	8-Cl	Friedel–Crafts reaction (90% yield)	230–231	1073
CHCl$_2$	1-Me	Friedel–Crafts reaction		1205
CHCl$_2$		Friedel–Crafts reaction		1205
CH$_2$Br		Friedel–Crafts reaction		1077
CH$_2$Br	8-OMe	3,4-diH-6-Ac-2-quinolone, Br$_2$ (76% yield)	206–207	1073
CHBr$_2$		3,4-diH-6-Ac-2-quinolone, Br$_2$, AcOH	168–169	988, 990
CH$_2$NH$_2$		Method A		1079
CH$_2$NH$_2$	1-Me	1-Me-3,4-diH-6-[1,2-C$_6$H$_4$(CO)$_2$NCH$_2$-CO]-2-quinolone, HCl (79% yield)	HCl >300	864
CH$_2$NHMe		Method A		1206
CH$_2$NHiPr	8-OH	Method A or 6-iPrNHCH$_2$CO-8-MeO-2-quinolone, 47% HBr		1207–1209
CH$_2$NHiPr	8-OMe	Method A		1207–1209
CH$_2$NHBu		Method A		1206
CH$_2$NHtBu		Method A		1206
CH$_2$NHtBu	8-OH	Method A or 6-tBuNHCH$_2$CO-8-MeO-2-quinolone, 47% HBr		1207–1209
CH$_2$NHtBu	8-OMe	Method A		1207–1209
CH$_2$NHCH$_2$Ph		Method A		1206, 1210
CH$_2$NMeCH$_2$Ph	8-OMe	Method A		1207–1209
CH$_2$N(CH$_2$)$_5$				1211
CH$_2$N(CH$_2$CH$_2$)$_2$NH		Method A	HCl 265–267	1079

Table 60. (*Contd.*)

R	Quinoline substituent(s)	Preparation	M.p. (°C)	References
$CH_2N(CH_2CH_2)_2NMe$				1211
$CH_2N(CH_2CH_2)_2NCH_2CH_2OPh$[a]				1075, 1078
$CH_2N(CH_2CH_2)_2NAc$[a]		3,4-diH-2-quinolone-6-$COCH_2N(CH_2CH_2)_2NH$, AcCl	HCl 249–252	1075, 1078, 1079
$CH_2N(CH_2CH_2)_2NCOPh$[a]		3,4-diH-2-quinolone-6-$COCH_2N$-$(CH_2CH_2)_2NH$, PhCOCl	207–210	1075, 1078, 1079
$CH_2N(CH_2CH_2)_2N$-COC_6H_4-3-Cl[a]		3,4-diH-2-quinolone-6-$COCH_2N(CH_2CH_2)_2NH$, 3-ClC_6H_4COCl	231–234	1075, 1079
$CH_2N(CH_2CH_2)_2NSO_2$-Me[a]		3,4-diH-2-quinolone-6-$COCH_2N$-$(CH_2CH_2)_2NH$, $MeSO_2Cl$	HCl 191–194	1075, 1078, 1079
$CH_2N(CH_2CH_2)_2NCOOEt$[a]		3,4-diH-2-quinolone-6-$COCH_2N(CH_2CH_2)_2NH$, ClCOOEt	HCl 235–237(d)	1075, 1078, 1079
$CH_2NHCOOEt$	1-Me	1-Me-3,4-diH-6-NH_2CH_2CO-2-quinolone, ClCOOEt, Et_3N (48% yield)		864
$CH_2N(CO)_2$-1,2-C_6H_4	1-Me	Friedel–Crafts reaction (48% yield)		864
Et	1-Me	Friedel–Crafts reaction		1080, 1203
Et	1,4-diMe	Friedel–Crafts reaction		864, 899
Et	1,4,4-triMe	Friedel–Crafts reaction		895
Et	4-Me	Friedel–Crafts reaction		895
Et	4,4-diMe	Friedel–Crafts reaction		895
Et	1-$CH_2CH_2OCH_2Ph$	Friedel–Crafts reaction		895
CH_2CH_2Cl	1-Me	Friedel–Crafts reaction		864
CH_2CH_2Cl	1-CH_2Ph	Friedel–Crafts reaction	121–123	1077, 1212
CH_2CH_2Cl	1-$CH_2CH{=}CH_2$	Friedel–Crafts reaction		1212
CH_2CH_2Cl	1-$CMe{=}CH_2$	Friedel–Crafts reaction		1212
CH_2CH_2Cl		Friedel–Crafts reaction		1212
CHClMe		Friedel–Crafts reaction		1213
CHBrMe		Friedel–Crafts reaction (92% yield)	192–193	1073, 1213

Side chain	Ring substituent	Method/Reaction	mp (°C)	Ref.
CHBrMe	1-Me	Friedel–Crafts reaction (48% yield)	232.5–233.5	1214
CHBrMe	8-Me	Friedel–Crafts reaction (69% yield)	180–182	1073
CHBrMe	8-Cl	Mannich reaction		1073, 1215
$CH_2CH_2N(CH_2)_5$	1,4-diMe	Mannich reaction		1216
$CH_2CH_2N(CH_2CH_2)_2$-NPh[a]		Method A	196–197	1077
$CH_2CH_2NMe_3^+I^-$		Mannich reaction; MeI		899
CH_2CH_2CN		3,4-diH-6-$COCH_2CH_2NMe_3^+$-2-quinolone I^-, KCN		899
C-Me=N-OH	1-CH_2Ph	1-$PhCH_2$-3,4-diH-6-EtCO-2-quinolone, $iPrCH_2CH_2ONO$		864
C-Me=N-OH	1-$CH_2CH_2OCH_2Ph$	1-$PhCH_2OCH_2CH_2$-3,4-diH-6-EtCO-2-quinolone, $iPrCH_2CH_2ONO$ (25% yield)		864
CH_2COOH		3,4-diH-6-Ac-2-quinolone, DBU; CO_2, DMSO		1217
$CHMeNHCH_2Ph$		Method A		1218
$CHMeNHCH_2CH_2C_6H_3$-3,4-diOMe		Method A		1219
CHMeNHCHMePh		Method A		1219
$CHMeNHCHMeCH_2$-OPh		Method A		1218
$CHMeNHCHMeCH_2$-CH_2Ph		Method A		1218
$CHMeNHCMe_2CH_2Ph$		Method A		1219
CHMeNHBu		Method A		1218
$CHMeN(CH_2Ph)_2$		Method A		1219
$CHMeN(CH_2CH_2)_2CH$-CH_2Ph		Method A		1218
$CHMeN(CH_2CH_2)_2NPh$	8-Cl	3,4-diH-6-$PhN(CH_2CH_2)_2NCHMeCO$-2-quinolone, Cl_2, AcOH	158–160	1215
Pr		Friedel–Crafts reaction		1080
$(CH_2)_3Cl$		Friedel–Crafts reaction		898, 1212

Table 60. (*Contd.*)

R	Quinoline substituent(s)	Preparation	M.p. (°C)	References
CHBrEt		Friedel–Crafts reaction		1213
$(CH_2)_3OH$				1077
$(CH_2)_3S$-(1-Me-tetrazol-5-yl)			188–190	898
$(CH_2)_3S$-(1-Me-tetrazol-5-yl)	1-Me		110–113.5	898
$(CH_2)_3N(CH_2CH_2Cl)_2$				1077
$(CH_2)_3N(CH_2CH_2OH)_2$				1077
$(CH_2)_3N(CH_2CH_2)_2$-NPh^a		Friedel–Crafts reaction or Method A	HCl 195–196	1076, 1077, 1212
CH_2CH_2COOH		Friedel–Crafts reaction (succinic anhydride)		1220, 1221
		3,4-diH-6-$COCH_2CH_2CN$-2-quinolone, conc. HCl		899
CH_2CH_2COOH	1,4-diMe	Friedel–Crafts reaction		895
CH_2CH_2COOH	1,4,4-triMe	Friedel–Crafts reaction		895, 896
CH_2CH_2COOH	4-Me	Friedel–Crafts reaction		895
CH_2CH_2COOEt		3,4-diH-6-$COCH_2CH_2COOH$-2-quinolone, $SOCl_2$, EtOH		1221
$CH_2CHCOOiPr$		3,4-diH-6-$COCH_2CH_2COOH$-2-quinolone, $SOCl_2$, iPrOH		1221
CH_2CH_2COOcyclo-C_6H_{11}		3,4-diH-6-$COCH_2CH_2COOH$-2-quinolone, $SOCl_2$, cyclo$C_6H_{11}OH$		1221
CH=CHCOOH		Friedel–Crafts reaction (maleic anhydride)		1220
CH=$CHCOOCH_2Ph$		3,4-diH-2-quinolone-6-COCH=$CHCOOH$, $SOCl_2$; $PhCH_2OH$		1221
$CHMeCH_2NMe_2$	1-Me	Mannich reaction; MeI		1216
$CHMeCH_2N(CH_2)_5$	1-Me	Mannich reaction; MeI		1216
$CHMeCH_2N(CH_2)_5$	4-Me	Mannich reaction; MeI		1216
$CHMeCH_2N(CH_2)_5$	1,4-diMe	Mannich reaction; MeI		1216

Substituent	Ring substituent	m.p.	Method	Ref.
CHMeCH₂N(CH₂CH₂)₂- O			Mannich reaction; MeI	1216
CHMeCH₂N(CH₂CH₂)₂- O	1-Me		Mannich reaction; MeI	1216
CHMeCH₂NMe₃⁺I⁻	1-Me		Mannich reaction; MeI	899
CHMeCH₂NMe₃⁺I⁻	1-Me		Mannich reaction; MeI	899
CHMeCH₂CN			3,4-diH-6-COCHMeCH₂NMe₃⁺-2-quinolone I⁻, KCN	899, 1222
CHMeCH₂CN	1-Me		3,4-diH-6-COCHMeCH₂NMe₃⁺-2-quinolone I⁻, KCN	899
CHEtN(CH₂CH₂)₂NH				1077
CHEtN(CH₂CH₂)₂NC₆H₄-4-NO₂[a]				1077
Bu			Friedel–Crafts reaction	1080
Bu	1-Me		Friedel–Crafts reaction	1080
(CH₂)₃COOH		149–151	Friedel–Crafts reaction (34.6% yield)	1221, 1223
(CH₂)₃COOEt			The carboxylic acid, SOCl₂; EtOH (72% yield)	1223
(CH₂)₃COOCH₂iPr			The carboxylic acid, SOCl₂; iPrCH₂OH	1221
(CH₂)₃COOC₅H₁₁			The carboxylic acid, SOCl₂; C₅H₁₁OH	1221
(CH₂)₃COOCH₂CH₂iPr	1-Me		The carboxylic acid, SOCl₂; iPrCH₂CH₂-OH	1221
(CH₂)₃COOCH₂Ph			The carboxylic acid, SOCl₂; PhCH₂OH	1221
(CH₂)₃CONMecyclo-C₆H₁₁	1-Me			1224
iBu			Friedel–Crafts reaction	1080
CHMeCH₂COOH			3,4-diH-6-COCHMeCH₂CN-2-quinolone, conc. HCl	899
CHMeCH₂COOH	1-Me		1-Me-3,4-diH-6-COCHMeCH₂CN-2-quinolone, conc. HCl	899, 1225
CH₂CH₂COOH	1,4,4-triMe		Friedel–Crafts reaction	895, 896, 1225
CHMeCH₂COOH	4-Me		Friedel–Crafts reaction	895
CHMeCH₂COOH	1-CH₂CH₂NMe₂			1226

Table 60. (*Contd.*)

R	Quinoline substituent(s)	Preparation	M.p. (°C)	References
CHMeCH₂COOH	1-(CH₂CH₂NEt₂)-4,4-diMe			1226
CHMeCH₂COOH	1-[CH₂CH₂N(CH₂-CH₂)₂O]-4,4-diMe			1226
CHMeCH₂COOH	1-[(CH₂)₃NMe₂]-4,4-diMe			1226
CHMeCH₂COOH	1-(CH₂)₃N(CH₂)₄			1226
CHMeCH₂COOH	1-(CH₂)₃N(CH₂CH₂CH₂)₂O			1226
CHMeCH₂COOH	1-(CH₂)₃N(CH₂CH₂CH₂)₂-NMe			1226
CHMeCH₂COOH	8-Cl	Friedel–Crafts reaction		1222
C₅H₁₁		Friedel–Crafts reaction		1080
C₆H₁₃				1080
5-Oxotetrahydrofuran-2-yl	1-Me			1225, 1227
5-Oxotetrahydrofuran-3-yl	1-Me			896
5-Oxotetrahydrofuran-3-yl	1-Et			1225, 1227
4-ClC₆H₄		Friedel–Crafts reaction		1080

ᵃAbout 200 further examples are included in the patents, mainly with extra substitution in the alkyl chain or the phenyl ring. Some are based on piperidine rather than piperazine.

Table 61. 3,4-Dihydro-7-keto-2-quinolones

	R^1	R^2	Preparation	Yield (%)	M.p. (°C)	References
1	H	H	Entry 3, H_2, Pd			1077
2	H	OH	Fries rearrangement			962
3	H	OSO_2Me	Entry 2, MsCl, Py			1077
4	Br	H	Entry 1, Br_2, AcOH		202–203	1077
5	CH_2Cl	H	Entry 7, H_2, Pd			1077, 1172
6	CH_2Cl	OH	Fries rearrangement	60	205–208	1077
7	CH_2Cl	OSO_2Me	Entry 6, MsCl, Py			1077
8	$N(CH_2CH_2)_2CHPh$	H	Entry 4, $PhCH(CH_2CH_2)_2NH$		171–173	1077
9	$CH_2N(CH_2CH_2)_2CHPh$	H	See below			1172
10	$CH_2N(CH_2CH_2)_2CHPh$	OH	Entry 6, $PhCH(CH_2CH_2)_2NH$		240–245	1077
11	$CH_2N(CH_2CH_2)_2CHPh$	OSO_2Me	Entry 7, $PhCH(CH_2CH_2)_2NH$ Entry 10, MsCl, Py			1077

About 15 amino-ketones similar to entries 8 and 9 but derived from different cyclic amines are included in the patent.[1077]

Ketones **728** were made by Grignard reactions on the appropriate 3-cyano-2-quinolones. The methyl ketone **728a** condensed with benzaldehyde to give the enone **729** as shown.[1175]

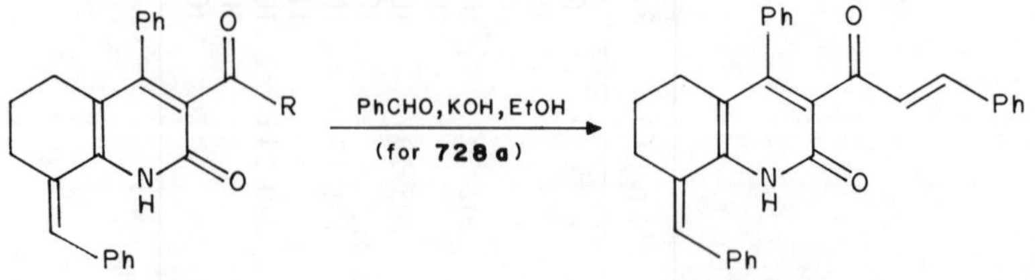

728

a, R = Me, 85%, m.p. 273 °C
b, R = Et, 83%, m.p. 252 °C
c, R = Ph, 90%, m.p. 265 °C

729, 81%, m.p. 273 °C

The isocyanates **730** were prepared from the cyclohexenecarboxylic acids and converted, without isolation, into the quinolones **731**.[1176]

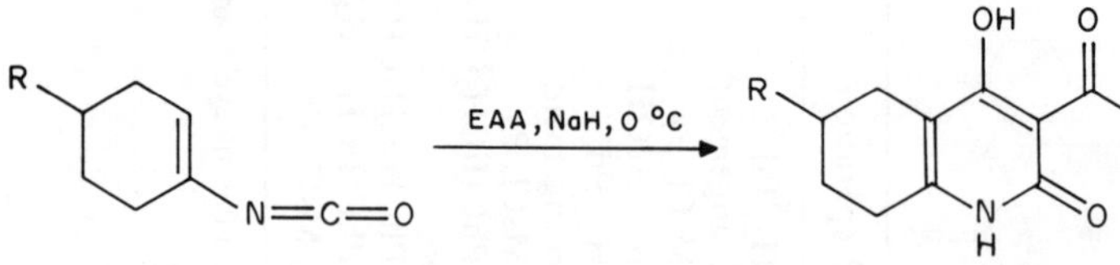

730

731, R = H, 76%, m.p. 241–242 °C
R = C(=CH$_2$)Me, 72%,
m.p. 196.5–197.5 °C

5. 7,8-Dihydro-6-ketoquinoline-2(1H),5(6H)-diones

The 6-ketoquinolinediones **732** were prepared as shown in Scheme 76.[1177]

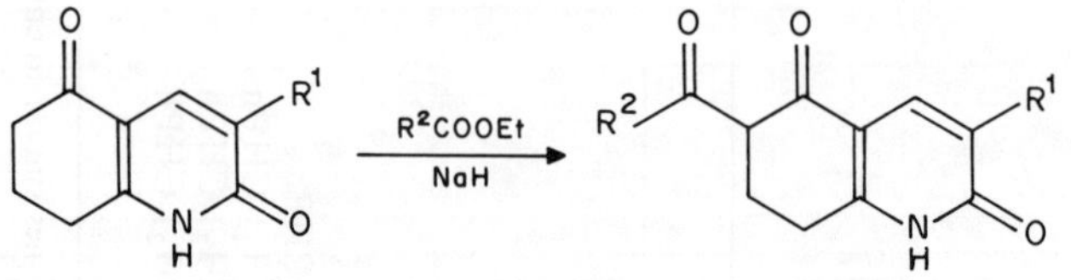

732

for R^1 = CONH$_2$; R^2 = Me, 2,4-diMeOC$_6$H$_3$, 4-Py
for R^1 = CN; R^2 = Me, Et, Ph, 4-MeOC$_6$H$_4$, 2,4-diMeOC$_6$H$_3$,
4-Py, 2-furyl, 2-thienyl

SCHEME 76

6. Keto-2,3-dihydro-4-quinolones

The keto-esters **733a**[1178], **733b**[1179], **733c**[1180] and **733d**[885] were prepared from the appropriate quinolones and oxalic esters under standard conditions.

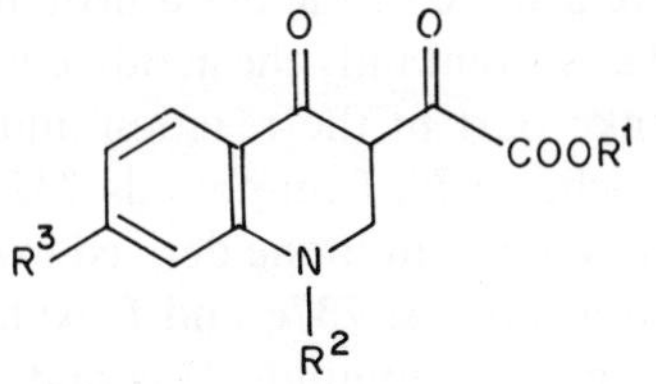
O
O
COOR¹
R³
N
R²

733

a, $R^1 = Me$, $R^2 = Ts$, $R^3 = H$, 94.5%, m.p. 126–127 °C
b, $R^1 = Me$, $R^2 = Ts$, $R^3 = Cl$, 90%, m.p. 137–138 °C
c, $R^1 = Me$, $R^2 = Ts$, $R^3 = OMe$, m.p. 134–136 °C
d, $R^1 = Et$, $R^2 = SO_2Ph$, $R^3 = OMe$, 60%, m.p. 122–123 °C

The reactions of Scheme 77 are self explanatory.[1181]

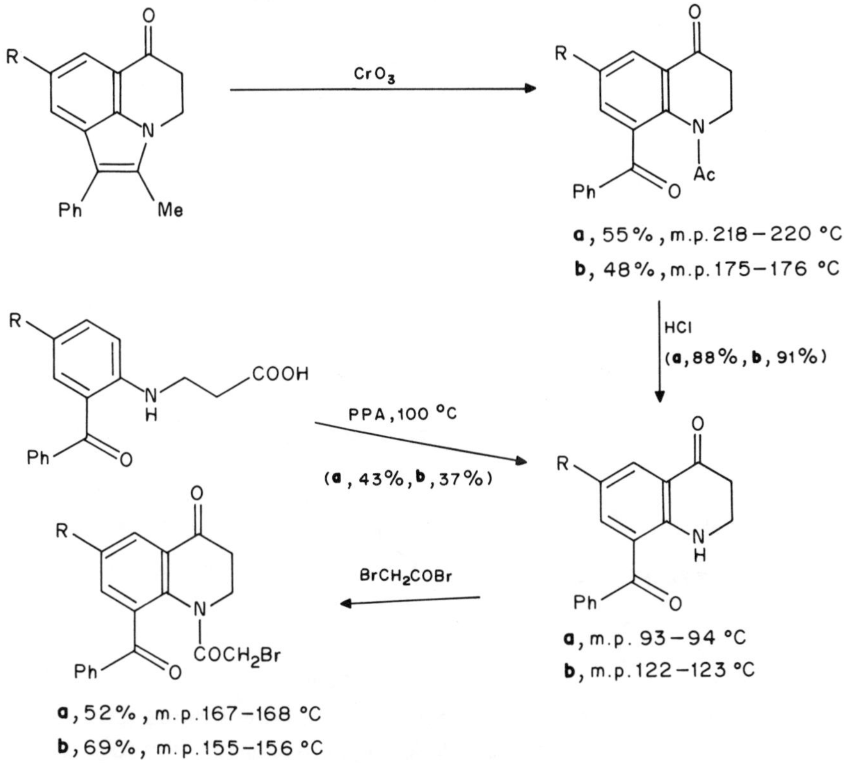
CrO3
R
N
Ph
Me
O
Ac
a, 55%, m.p. 218–220 °C
b, 48%, m.p. 175–176 °C
HCl
(a, 88%, b, 91%)
COOH
N H
PPA, 100 °C
(a, 43%, b, 37%)
BrCH2COBr
COCH2Br
a, m.p. 93–94 °C
b, m.p. 122–123 °C
a, 52%, m.p. 167–168 °C
b, 69%, m.p. 155–156 °C

SCHEME 77 **a**, R=H, **b**, R=Cl

7. Partially Reduced 3-Keto-5-quinolones

The 3-keto-5-quinolones **737** were prepared from enaminone **734** as shown
(Scheme 78); in some examples triethylamine was used as a catalyst. From the
preparation of compound **737c** a trace of the acridinedione was also isolated,
and when acetic acid was used as solvent only the acridinedione was formed.[1182]
Here the diketone failed to take part in the reaction and the product was a
derivative of compound **743**, Scheme 79. Compounds **737a, c** and **f** were also
prepared from dimedone, the acyclic enaminone derived from diketone **735** and
an aldehyde **736**.[1183] For two examples, **737c** and **f** oxidation was shown to
aromatize the pyridine ring to give compounds **738a** and **b**.[1183]

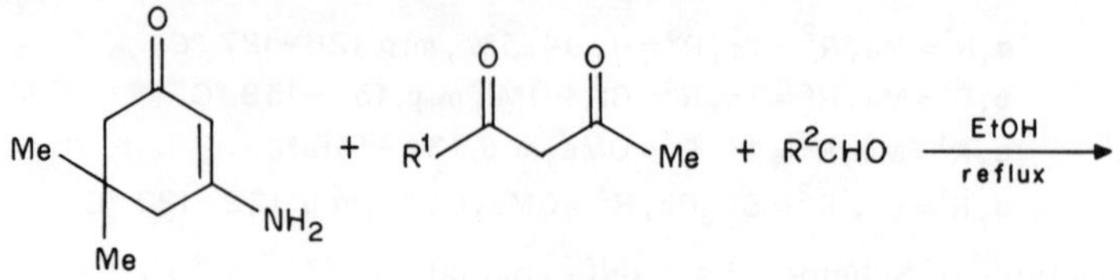

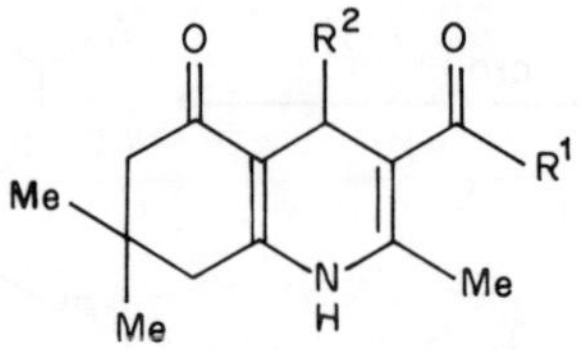

	R^1	R^2	Yield(%)	M.p.(°C)
a,	Me	H	60	205–207
b,	Me	Me	48	169–170
c,	Me	Ph	29	198–200
d,	Ph	H	66	181–183
e,	Ph	Me	52	158–160
f,	Ph	Ph	29	207–209

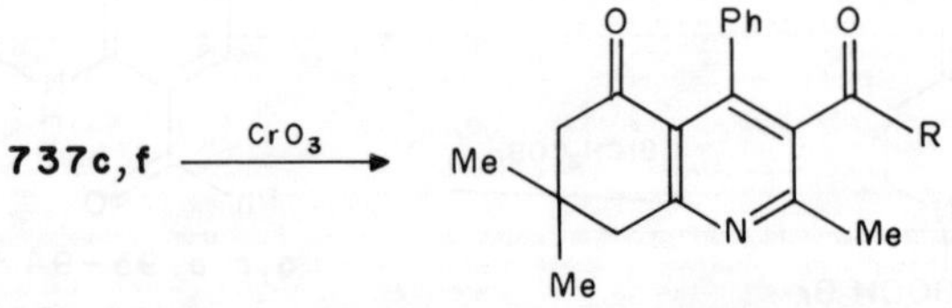

a, R = Me, 51%, m.p. 107–108 °C
b, R = Ph, 76%, m.p. 157–158 °C

SCHEME 78

In a similar way, cyclohexane-1,3-dione, pyridine-4-carboxaldehyde and 4-aminopent-3-en-2-one reacted in refluxing alcohol containing acetic acid to give compound **739**.[1184]

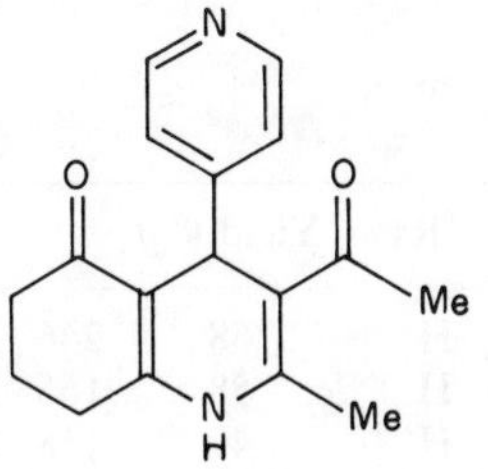

739, 35%, m.p. 257 °C

The enaminone **740** was converted into the methylene-bis-enaminone **741** by aqueous formaldehyde. The use of acidified formaldehyde gave the spiro-ketone **742** rapidly at room temperature. Attempted hydrolysis of dienamine **741**

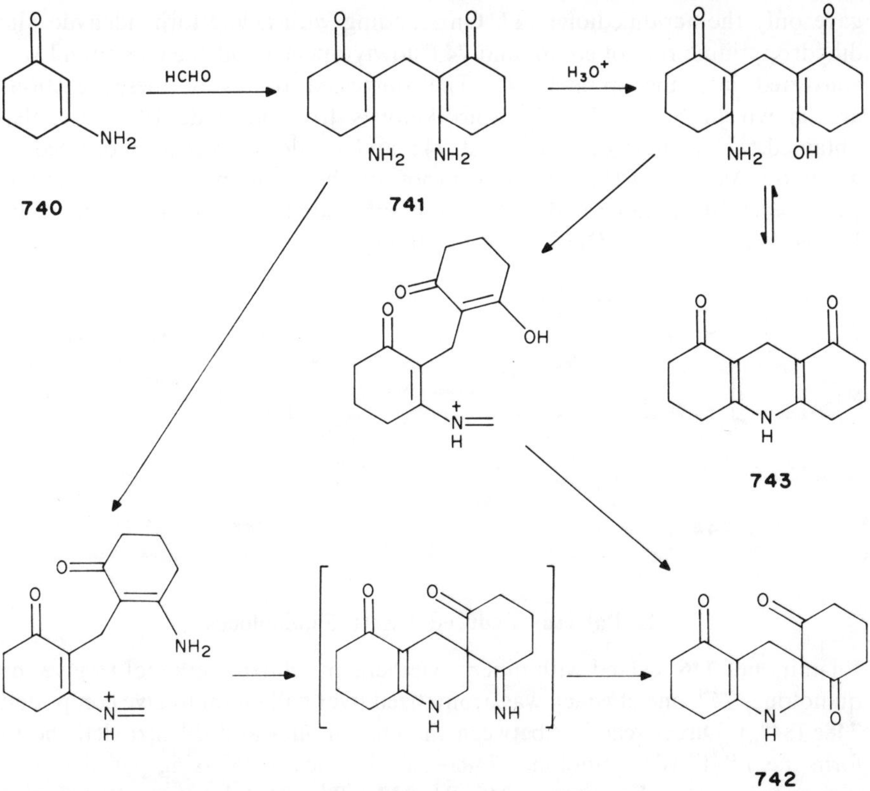

SCHEME 79

Table 62. Quinolone-3-spirocyclohexanediones

R_1	R_2	R_3	Yield (%)	M.p. (°C)
H	H	H	58	246
H	Me	H	58	165–166
H	CH$_2$CH$_2$Ph	H	49	158–159
Me	H	H	79	260–261, HCl 256
Me	H	Me	73	199–200, HCl 252
Me	Me	H	59	182–183, HCl 221–222
Me	Me	Me	90	215

gave only the acridinedione **743**. On standing with acidic formaldehyde, the dihydropyridine ring of compound **743** slowly opened and the compound was converted into the spirane **742**. The suggested routes of these reactions are shown in Scheme 79.[1185] Spiro-ketones from acetaldehyde were also obtained.[1186] Details of compound **742** and its derivatives are collected in Table 62. As part of the structural proof for these spiranes, two compounds **744**, R = H, Me were treated with sodium ethoxide to give the keto-esters **745**, R = H, Me via reverse Dieckmann reactions.[1186]

744   745

8. Partially Reduced 4-Keto-5-quinolones

Enaminones **746** reacted with dibenzoylethene in refluxing ethanol to give the quinolones **747**, one of which was aromatized over palladium to give compound **748c** (84%). Direct reaction between the enaminones and dibenzoylethene to form the 7,8-dihydroquinolones **748a–c** took place in refluxing pyridine with passage of oxygen. Enaminone **746**, R^1 = Me, R^2 = H did not give compound **747**, R^1 = Me, R^2 = H, but rearranged *in situ* to the indole **749**, R^1 = Me, R^2 = H.

Compounds **747** rearranged to the corresponding indoles **749** in refluxing acetic acid.[1187,1188] See Scheme 80.

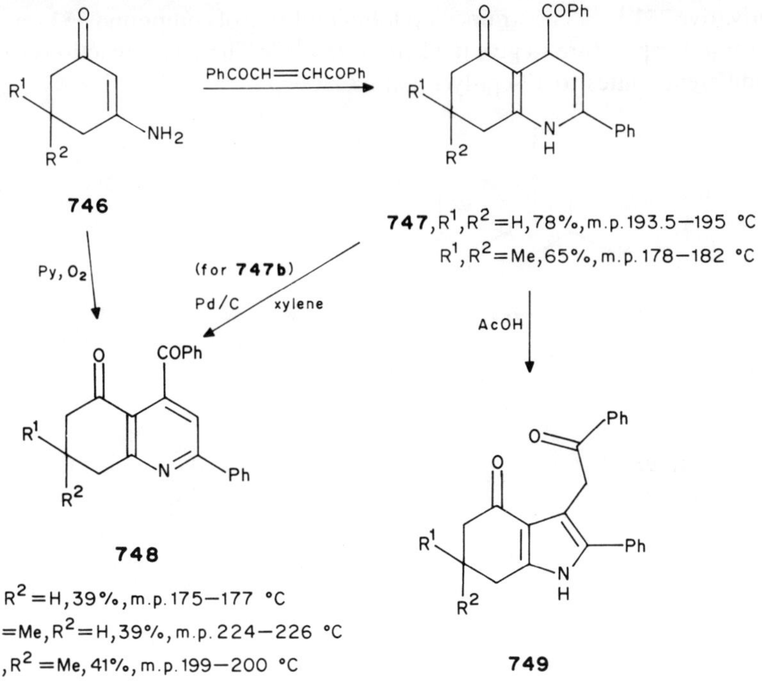
PhCOCH=CHCOPh
746
747, R1,R2=H, 78%, m.p. 193.5–195 °C
R1,R2=Me, 65%, m.p. 178–182 °C
Py, O2
(for 747b)
Pd/C xylene
AcOH
748
a, R1,R2=H, 39%, m.p. 175–177 °C
b, R1=Me, R2=H, 39%, m.p. 224–226 °C
c, R1,R2=Me, 41%, m.p. 199–200 °C
749

SCHEME 80

Note also the preparation of 8-benzoyl-1-methyl-1,2,3,4,7,8-hexahydro-5(6*H*)-quinolone **480** in Section XII.

9. 7-Ketodecahydro-6-quinolone

The decahydroquinolone **750a** was treated with lithium diisopropylamide followed by acetyl chloride to give the ketone **750b**. This was reacted, without purification, with substituted guanidines to give octahydropyrimidino[4,5-*g*]-quinolines.[1189,1190]

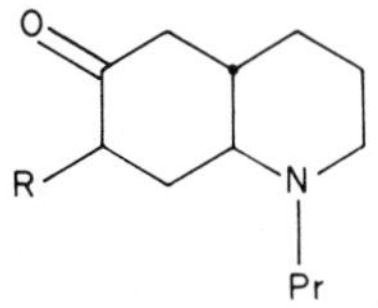

750a, R=H
b, R=Ac

10. Quinolones with Side Chain Ketone Groups

The quinolinedione **751** was alkylated as shown to give the 1-(5-oxohexyl) derivative **752**.[1191] The strained cyclobutanol ring of compound **753** was opened at room temperature to give the ketone **754**.[1192] These two reactions were used in different routes to 12-epi-lycopodine.

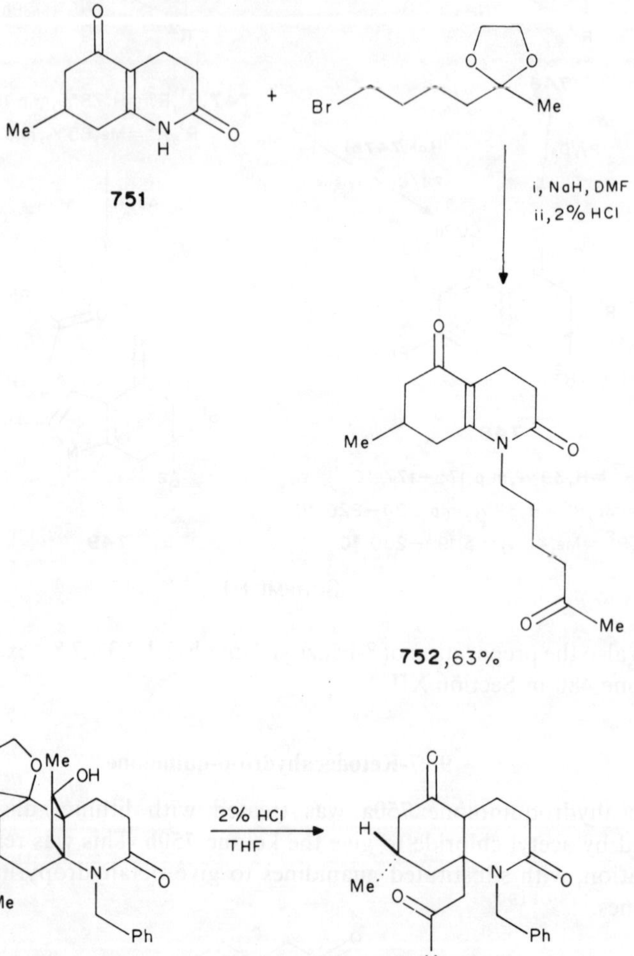

751

i, NaH, DMF
ii, 2% HCl

752, 63%

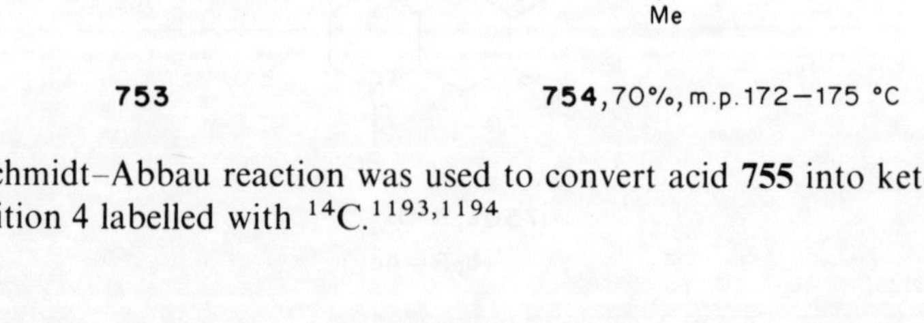

753

2% HCl
THF

754, 70%, m.p. 172–175 °C

The Schmidt–Abbau reaction was used to convert acid **755** into ketone **756** with position 4 labelled with ^{14}C.[1193,1194]

755   **756,** m.p. 111−112.5 °C

Irradiation of a mixture of 1-methyl-2-quinolone and diketene followed by treatment with sodium methoxide gave the keto-ester **757**.[982]

757, 70%, b.p. 195−199 °C/0.001 mm

The dianion from the partially reduced 2-quinolone **758** underwent Michael addition to chalcone to produce ketone **759**.[1195] The dianion from 2-formylcyclohexanone added to chalcone under the same conditions. The crude product was refluxed with cyanoacetamide in acetic acid to give compound **759**.[1196]

758   **759,** 65%, m.p. 261−263 °C

The Mannich bases **761** proved to be alkylating agents for the quinolones **760**. Presumably the ketones **762** were formed by Michael addition to transient vinyl ketones. The products are listed in Table 63.[1197]

Table 63. 1-Ketoethyl-4-quinolones (762)[1197]

R^1	R^2	Yield (%)	M.p. (°C)
Me*	Cl	58	83–83.5
Ph	H	55	84–86
Ph	Cl	68	97–98
$4\text{-PhC}_6\text{H}_4$	H	53	101.5–105.5
$4\text{-PhC}_6\text{H}_4$	Cl	61	111.5–112.5
$4\text{-FC}_6\text{H}_4$	H	64	111–112
$4\text{-FC}_6\text{H}_4$	Cl	71	132–133
$4\text{-MeOC}_6\text{H}_4$	H	61	127.5–129.5
$4\text{-MeOC}_6\text{H}_4$	Cl	80	110.5–111.5
$4\text{-NO}_2\text{C}_6\text{H}_4$	H	76	147–150
$4\text{-NO}_2\text{C}_6\text{H}_4$	Cl	88	148–151.5
2-Thienyl	H	64	95.5–97
2-Thienyl	Cl	75	132–132.5

*Prepared from 3-buten-2-one.

2,3-Dihydro-3-formyl-3-(3-oxobutyl)-4-quinolones are mentioned in Chapter 1, Section VII, Scheme 18.

The pyran derivatives 763 reacted with primary amines in a ring opening–ring

	R^1	R^2	R^3	Yield%	M.p. °C
a,	Me	Me	Me	66	236–238(d)
b,	Me	Me	Et	52	253–255
c,	—(CH$_2$)$_5$—		Et	93	236

Table 64. 3,4-Dihydro-8-keto-2-quinolones

(Reaction scheme: R-substituted 3,4-dihydro-8-keto-2-quinolone with Hal, reacting with R^1R^2NH to give the R^1/R^2 amino derivative.)

R	Quinoline substituent(s)	Preparation	References
Me	5-OH		1204
Me	5-OH-6-Br		1204
Me	5-OCH$_2$CHOHCH$_2$NHiPr-6-Br		1204
Me	5-OCH$_2$CHOHCH$_2$NHCHMePh-6-Br		1204
Me	5-OCH$_2$CHOHCH$_2$NHCH$_2$CH$_2$Ph-6-Br		1204
Me	5-OCH$_2$CHOHCH$_2$NHCH$_2$CH$_2$(C$_6$H$_4$-4-OMe)		1204
Me	7-OH[a]	3,4-diH-7-AcO-2-quinolone, AlCl$_3$, 180–190 °C	1171
CH$_2$Cl	5-OCOCH$_2$Cl	Friedel–Crafts reaction	1228, 1229
CH$_2$Cl	5-OMe	Friedel–Crafts reaction (41% yield)	1228–1230
CH$_2$NHMe	5-OH		1231
CH$_2$NHEt	5-OH	Method A	1230, 1232
CH$_2$NHiPr	5-OH	Method A	1232
CH$_2$NHiPr	5-OMe	Method A	1230, 1232
CH$_2$NHtBu	5-OH		1233
CH$_2$NHtBu	5-OMe	Method A	1230
CH$_2$NH(CH$_2$)$_6$Cl	5-OMe		1231
CH$_2$NHcycloC$_6$H$_{11}$	5-OMe	Method A	1229, 1230
CHBrEt	5-OCOCHBrEt	Friedel–Crafts reaction	1228, 1230
CHEtNHMe	5-OH	Method A	1230
CHEtNHiPr	5-OH	Method A	1230
CHBrPr	5-OCOCHBrPr	Friedel–Crafts reaction	1232
CHPrNHiPr	5-OH	Method A	1230, 1232

[a] Comparison of *Chemical Abstracts*, **85**, 32874 and 32875 (1976) suggests that this compound should be 7-acetyl-3,4-dihydro-8-hydroxy-2-quinolone. According to **85**, 32874 the 7-acetyloxy compound undergoes Fries rearrangement to give both a 6-acetyl and an 8-acetyl derivative.

closure sequence to give the ketones **764**. One example was also prepared from the benzopyran **765** as shown (Scheme 81).[1198]

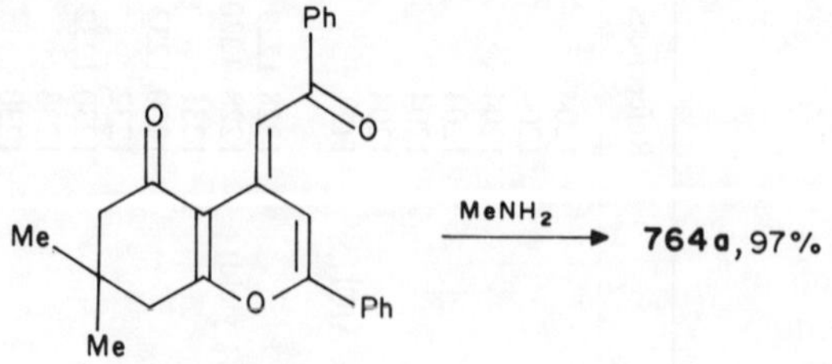

765

SCHEME 81

XVII. Appendix

Abbreviations used in Text and Tables

AA	Pentane-2,4-dione
DBA	Dibenzoylacetylene
DBU	Diazabicycloundecene
DCCI	Dicyclohexylcarbodiimide
DCM	Dichloromethane
DDQ	Dichlorodicyano-*p*-benzoquinone
DEAD	Diethyl acetylenedicarboxylate
DEM	Diethyl malonate
DIBAH	Diisobutyl aluminium hydride
DMA	Dimethylacetamide
DMAD	Dimethyl acetylenedicarboxylate
DMAP	Dimethylaminopyridine
DME	Dimethoxyethane
DMF	Dimethylformamide
DMSO	Dimethyl sulphoxide
DNP	2,4-Dinitrophenylhydrazone
EAA	Ethyl acetoacetate
Hy	Hydrazone
LAH	Lithium aluminium hydride
LDA	Lithium diisopropylamide
MAA	Methyl acetoacetate
MCPBA	Metachloroperbenzoic acid
MVK	Methyl vinyl ketone
NBS	*N*-Bromosuccinimide
NPH	4-Nitrophenylhydrazone

Ox	Oxime
PH	Phenylhydrazone
Pic.	Picrate
PPA	Polyphosphoric acid
Prl	Picrolonate
Py	Pyridine
PyClCrO$_3$	Pyridinium chlorochromate
Q	Quinoline
SC	Semicarbazone
SDA	Sodium diisopropylamide
TFA	Trifluoroacetic acid
TFAA	Trifluoroacetic anhydride
THQ	1,2,3,4-Tetrahydroquinoline
Ts	p-CH$_3$C$_6$H$_4$SO$_2$
TSC	Thiosemicarbazone

XVIII.   References

1. Y. Yamamoto and A. Yanagi, *Heterocycles*, **19**, 41 (1982).
2. Y. Yamamoto and A. Yanagi, *Chem. Pharm. Bull.*, **30**, 2003 (1982).
3. A. R. Patel, C. J. Ohnmacht, D. P. Clifford, A. S. Crosby, and R. E. Lutz, *J. Med. Chem.*, **14**, 198 (1971).
4. H. Andrews, S. Skidmore, and H. Suschitzky, *J. Chem. Soc.*, **1962**, 2370.
5. F. H. Case and G. Maerker, *J. Amer. Chem. Soc.*, **75**, 4920 (1953).
6. E. Lorz and R. Baltzly, *J. Amer. Chem. Soc.*, **70**, 1904 (1948).
7. S. Pavlov and V. Arsenijevic, *Glas. Hem. Drus. Beograd*, **32**, 469 (1967); *Chem. Abstr.*, **71**, 49734 (1969).
8. B. R. Brown and D. L. Hammick, *J. Chem. Soc.*, **1949**, 173.
9. B. R. Brown, *J. Chem. Soc.*, **1949**, 2577.
10. P. Dyson and D. L. Hammick, *J. Chem. Soc.*, **1937**, 1724.
11. T. Cohen, R. W. Berninger, and J. T. Wood, *J. Org. Chem.*, **43**, 837 (1978).
12. G. Karlivan and R. Valters, *Khim. Geterotsikl. Soedin.*, **1984**, 1231; *Chem. Heterocycl. Compd. (Engl. Transl.)*, **20**, 1007 (1984); *Chem. Abstr.*, **101**, 230319 (1984).
13. S. V. Tsukerman, K.-S. Chang, Yu. S. Rozym, and V. F. Lavrushin, *Zh. Prikl. Spektrosk.*, **5**, 489 (1966); *Chem. Abstr.*, **66**, 60444 (1967).
14. A. Citterio, A. Gentile, M. Serravalle, L. Tinucci, and E. Vismara, *J. Chem. Res. Synop.*, **1982**, 272; *Miniprint*, p., 2801.
15. T. Caronna, G. Fronza, F. Minisci, O. Porta, and G. P. Gardini, *J. Chem. Soc., Perkin Trans.* **II**, **1972**, 1477.
16. T. Caronna, G. Fronza, F. Minisci, and O. Porta, *J. Chem. Soc., Perkin Trans II*, **1972**, 2035.
17. A. J. Saggiomo, S. Kano, T. Kikuchi, K. Okubo, and M. Shinbo, *J. Med. Chem.*, **15**, 989 (1972).
18. G. Kempter and G. Möbius, *J. Prakt. Chem.*, **34**, 298 (1966).
19. M. J. Haddadin, N. C. Chelhot, and M. Pieridou, *J. Org. Chem.*, **39**, 3278 (1974).
20. M. S. Sinsky and R. G. Bass, *J. Heterocycl. Chem.*, **21** 759 (1984).

21. K. T. Potts and A. J. Elliott, *J. Org. Chem.*, **38**, 1769 (1973).
22. D. W. Bayne, A. J. Nicol, and G. Tennant, *J. Chem. Soc., Chem. Commun.*, **1975**, 782.
23. F. M. Abdel-Galil, B. Y. Riad, S. M. Sherif, and M. H. Elnagdi, *Chem. Lett.*, **1982**, 1123.
24. G. Casnati, R. Cavalleri, F. Piozzi, and A. Quilico, *Gazz. Chim. Ital.*, **92**, 105 (1962); *Chem. Abstr.*, **58**, 3504 (1963).
25. V. Boekelheide and J. Weinstock, *J. Amer. Chem. Soc.*, **74**, 660 (1952).
26. B. C. Uff, R. S. Budhram, M. F. Consterdine, J. K. Hicks, B. P. Slingsby, and J. A. Pemblington, *J. Chem. Soc., Perkin Trans I*, **1977**, 2018.
27. H. V. Kamath, K. S. Nargund, and S. N. Kulkarni, *Indian J. Chem., Sect. B*, **16B**, 903 (1978).
28. O. Tsuge and H. Watanabe, *Heterocycles*, **4**, 1905 (1976).
29. P. F. Devitt, A. Timoney, and M. A. Vickars, *J. Org. Chem.*, **26**, 4941 (1961).
30. D. J. Donnelly, J. A. Donnelly, and E. M. Philbin, *Proc. R. Ir. Acad., Sect. B*, **73B**, 129 (1973).
31. D. J. Donnelly, R. Geoghegan, C. O'Brien, E. M. Philbin, and T. S. Wheeler, *J. Med. Chem.*, **8**, 872 (1965).
32. M. Cushman and J. Mathew, *J. Org. Chem.*, **46**, 4921 (1981).
33. I. Ono and N. Hata, *Bull. Chem. Soc. Japan*, **56**, 3667 (1983).
34. E. G. Podrebarac and W. E. McEwen, *J. Org. Chem.*, **26**. 1165 (1961).
35. G. Scheibe and G. Schmidt, *Ber.*, **55B**, 3157 (1922).
36. G. Scheibe and H. J. Friedrich, *Chem. Ber.*, **94**, 1336 (1961).
37. K. Funakoshi, Y. Kuchino, H. Shigyo, H. Sonoda, and M. Hamana, *Chem. Pharm. Bull.*, **24**, 2356 (1976).
38. E. Ochiai and M. Takahashi, *Itsuu Kenkyusho Nempo*, **1968**, (15), 1; *Chem. Abstr.*, **71**, 101753 (1969).
39. E. Besthorn and J. Ibele, *Ber.*, **37**, 1236 (1904).
40. E. Besthorn, *Ber.*, **41**, 2001 (1908).
41. A. A. Ardakani, N. Malcki, and M. R. Saadein, *J. Org. Chem.*, **43**, 4128 (1978).
42. M. Los and W. H. Stafford, *J. Chem. Soc.*, **1959**, 1680.
43. L. E. Kholdov, E. M. Peresleni, I. F. Tishchenkova, N. P. Lostyuchenko, and V. G. Yashchunskii, *Dokl. Akad. Nauk SSSR*, **195**, 868 (1970); *Chem. Abstr.*, **74**, 132700 (1971).
44. G. B. Bachman and R. M. Schisla, *J. Org. Chem.*, **22**, 1302 (1957).
45. M. Watanabe, M. Kodera, T. Kinoshita, and S. Furukawa, *Chem. Pharm. Bull.*, **23**, 2598 (1975).
46. E. C. Taylor, M. L. Chittenden, and S. F. Martin, *Heterocycles*, **1**, 59 (1973).
47. T. Endo, S. Saeki, and M. Hamana, *Heterocycles*, **3**, 19 (1975).
48. S. Ikegami and Y. Hatano, *Ger. Offen.* 2 634 987 (1977); *Chem. Abstr.*, **87**, 54530 (1977); *U.S. Pat.* 4 200 751 (1980).
49. Y. Mizuno and J. Kobayashi, *J. Chem. Soc., Chem. Commun.*, **1975**, 308.
50. W. Borsche and W. Ried, *Justus Liebigs Ann. Chem.*, **554**, 269 (1943).
51. W. Borsche and W. Noll, *Justus Liebigs Ann. Chem.*, **532**, 127 (1937).
52. W. Borsche and F. Sinn, *Justus Liebigs Ann. Chem.*, **532**, 146 (1937).
53. R. E. Pratt, W. J. Welstead, Jr., and R. E. Lutz, *J. Heterocycl. Chem.*, **7**, 1051 (1970).
54. W. R. Ashcroft, L. Dalton, M. G. Beal, G. L. Humphrey, and J. A. Joule, *J. Chem. Soc., Perkin Trans I*, **1983**, 2409.
55. V. Ehrig, H. S. Bien, E. Klanke, and D. I. Schuetze, *Ger. Offen.* 2 730 061 (1979); *Chem. Abstr.*, **90**, 152 027 (1979).
56. J. Eliasberg and P. Friedländer, *Ber.*, **25**, 1752 (1892).

57. T. Pavolini, F. Gambarin, and E. Giusto, *Ann. Chim. (Rome)*, **43**, 242 (1953) *Chem. Abstr.*, **48**, 11419 (1954).
58. G. Stefanović, M. Pavićić-Wass, L. Lorenc, and M. L. Mihailović, *Tetrahedron*, **6**, 97 (1959).
59. C. J. Ohnmacht, Jr., F. Davis, and R. E. Lutz, *J. Med. Chem.*, **14**, 17 (1971).
60. B. K. Nandi, *J. Indian Chem. Soc.*, **17**, 285 (1940).
61. A. Ide, K. Matsumori, and H. Watanabe, *Nippon Kagaku Kaishi*, **91**, 578 (1970); *Chem. Abstr.*, **73**, 130860 (1970).
62. R. E. Fuson and J. J. Miller, *J. Amer. Chem. Soc.*, **79**, 3477 (1957).
63. R. Robinson and A. Zaki, *J. Chem. Soc.*, **1927**, 2488.
64. M. J. Haddadin, G. E. Zahr, T. N. Rawdah, N. C. Chelhot, and C. H. Issidorioles, *Tetrahedron*, **30**, 659 (1974).
65. T. Kurihara, H. Sano, and H. Hirano, *Chem. Pharm. Bull.*, **23**, 1155 (1975).
66. H. Bretschneider, K. Hohenlohe-Oehringen, and A. Rhomberg, *U.S. Pat.* 3 311 632 (1967).
67. F. Hoffmann-La Roche & Co. A.-G., *Neth. Appl.* 6 401 199 (1964); *Chem. Abstr.*, **62**, 10421 (1965).
68. K. Hohenlohe-Oehringen, A. Rhomberg, and H. Bretschneider, *Monatsh. Chem.*, **97**, 135 (1966).
69. O. S. Wolfbeis, *Synthesis*, **1977**, 723.
70. J. F. Biellmann and M. P. Goeldner, *Tetrahedron*, **27**, 1789 (1971).
71. N. H. Cromwell and D. J. Pokorny, *J. Heterocycl. Chem.*, **12**, 533 (1975).
72. B. B. Neelima and A. P. Bhaduri, *Z. Naturforsch.*, **40B**, 990 (1985).
73. N. H. Cromwell and J. C. David, *J. Amer. Chem. Soc.*, **82**, 1138 (1960).
74. W. Ried, A. Berg, and G. Schmidt, *Chem. Ber.*, **85**, 204 (1952).
75. G. Kollenz, E. Ziegler, H. Igel, and C. Labes, *Chem. Ber.*, **109**, 2503 (1976).
76. G. Kollenz, G. Penn, G. Dolenz, Y. Akcamur, K. Peters, E.-M. Peters, and H. G. von Schuering, *Chem. Ber.*, **117**, 1299 (1984).
77. B. Bhat and A. P. Bhaduri, *Synthesis*, **1984**, 673.
78. K. Tobinaga, M. Mukaino, T. Inazu, and T. Yoshino, *Bull. Chem. Soc. Japan*, **45**, 3485 (1972).
79. E. Uhlemann, W. Mickler, E. Ludwig, and G. Klose, *J. Prakt. Chem.*, **323**, 521 (1981).
80. S. A. Osadchii, V. A. Tatarinov, L. Y. Marochkina, N. A. Pankrushina, and L. M. Gindin, *Izv. Sib. Otd. Akad. Nauk SSSR, Ser. Khim. Nauk*, **1976**, (9), 59; *Chem. Abstr.*, **85**, 142961 (1976).
81. S. A. Osadchii, N. A. Pankrushina, I. N. Gribanova, and A. V. Nikolaev, *Izv. Sib. Otd. Akad. Nauk SSSR, Ser. Khim Nauk*, **1976** (6), 48; *Chem. Abstr.*, **86**, 121136 (1977).
82. G. Manecke and H. P. Aurich, *Makromol. Chem.*, **133**, 83 (1970); *Chem. Abstr.*, **72**, 132474 (1970).
83. W. Borsche and H. Groth, *Justus Liebigs Ann. Chem.*, **549**, 238 (1941).
84. H. Matsumoto, M. Nagata, J. Kawahira and T. Yamazaki, *Yakugaku Zasshi*, **88**, 1412 (1968); *Chem. Abstr.*, **70**, 77747 (1969).
85. P. K. Sharma and R. N. Khanna, *Monatsh. Chem.*, **116**, 353 (1985).
86. O. Dieb and O. Staehlin, *Ber.*, **35**, 3275 (1902).
87. R. Gopalchari and M. L. Dhar, *J. Sci. Ind. Res.*, **21B**, 266 (1962); *Chem. Abstr.*, **57**, 11160 (1962).
88. Farbenfabriken Bayer A.-G., *Brit. Pat.* 735 547 (1955); *Chem. Abstr.*, **50**, 4238 (1956); *U.S. Pat.* 2 770 619 (1956); *Chem. Abstr.*, **51**, 36736 (1957).
89. K. Matsumura and C. Sone, *J. Amer. Chem. Soc.*, **53**, 1490 (1931).

90. K. Matsumura, H. Okuma, M. Onishi, and A. Ikemura, *Bull. Chem. Soc. Japan*, **43**, 3282 (1970).
91. H. Möhrle and R. Schaltenbrand, *Pharmazie*, **40**, 307 (1985).
92. H. Möhrle and R. Schaltenbrand, *Pharmazie*, 40, 387 (1985).
93. H. Möhrle and R. Schaltenbrand, *Pharmazie*, **40**, 767 (1985).
94. R. A. Glenn and J. R. Bailey, *J. Amer. Chem. Soc.*, **63**, 641 (1941).
95. J. V. Dejardin and C. L. Lapiere, *Pharm. Acta Helv.*, **45**, 198 (1970); *Chem. Abstr.*, **72**, 100458 (1970).
96. H. Kühnis and H. de Diesbach, *Helv. Chim. Acta*, **41**, 894 (1958).
97. J. M. Maheshvari and K. A. Thaker, *J. Inst. Chem. (India)*, **55**, 189 (1983); *Chem. Abstr.*, **100**, 156473 (1984).
98. A. M. van Leusen and J. W. Terpstra, *Tetrahedron Lett.*, **22**, 5097 (1981).
99. F. H. Case and J. A. Brennan, *J. Amer. Chem. Soc.*, **81**, 6297 (1959).
100. M. Iklobdzija, T. Kovac, O. Christofis, R. Taso, and V. Sunjic, *Kem. Ind.*, **27**, 453 (1978); *Chem. Abstr.*, **90**, 186909 (1979).
101. W. J. Suggs, *J. Amer. Chem. Soc.*, **100**, 640 (1978).
102. E. V. Brown and M. G. Frazer, *J. Heterocycl. Chem.*, **6**, 567 (1969).
103. T. Kametani, K. Ogasawara, A. Kozuka, and M. Shio, *Yakugaku Zasshi*, **87**, 254 (1967); *Chem. Abstr.*, **67**, 54316 (1967).
104. A. Kaufmann, P. Dandliker, and H. Burkhardt, *Ber.*, **46**, 2929 (1913).
105. T. Nakashima, *Yakugaku Zasshi*, **77**, 1298 (1957); *Chem. Abstr.*, **52**, 6345 (1958).
106. K. N. Campbell, C. H. Helbing, and J. F. Kerwin, *J. Amer. Chem. Soc.*, **68**, 1840 (1946).
107. J. D. Kendall, G. F. Duffin, and J. Voltz, *Brit. Pat.* 885 520 (1961); *Chem. Abstr.*, **57**, 74286 (1962).
108. T. Sakamoto, Y. Kondo, M. Shiraiwa, and H. Yamanaka, *Synthesis*, **1984**, 245.
109. L. Capuano, *Chem. Ber.*, **92**, 2670 (1959).
110. A. Arnoldi, E. Betto, G. Farina, A. Formigoni, R. Galli, and A. Griffini, *Pestic. Sci.*, **13**, 670 (1982), *Chem. Abstr.*, **99**, 153745 (1983).
111. T. Yamazaki, K. Matoba, and S. Imoto, *Heterocycles*, **4**, 713 (1976).
112. H. A. El-Sayad and R. L. McKee, *Org. Prep. Proced. Int.*, **10**, 85 (1978).
113. A. Burger and L. R. Modlin, Jr., *J. Amer. Chem. Soc.*, **62**, 1079 (1940).
114. A. Citterio, A. Gentile, and F. Minisci, *Tetrahedron Lett.*, **23**, 5587 (1982).
115. E. Hayashi and N. Shimada, *Yakugaku Zasshi*, **97**, 641 (1977); *Chem. Abstr.*, **88**, 22570 (1978).
116. H. C. Richards, *U.S. Pat.* 3 821 228 (1974); *Chem. Abstr.*, **81**, 120498 (1974); *U.S. Pat.* 3 929 784 (1975); *Chem. Abstr.*, **85**, 32 869 (1976).
117. A. L. Gershuns, A. N. Brizitskaya, and P. Ya. Pustavar, *Khim. Geterotsikl. Soedin.*, **1973**, 1536; *Chem. Heterocycl. Compd. (Engl. Transl.)*, **9**, 1391 (1973); *Chem. Abstr.*, **80**, 70773 (1974).
118. R. R. G. Haber and E. Schoenberger, *S. African Pat.* 67 03,320 (1968); *Chem. Abstr.*, **71**, 91337 (1969).
119. R. R. G. Haber and E. Schoenberger, *Israel Pat.* 27993 (1972); *Chem. Abstr.*, **78**, 147821 (1973).
120. M. S. Ouali, M. Vaultier, and R. Carrie, *Tetrahedron*, **36**, 1821 (1980).
121. A. Boehringer, *Brit. Pat.* 311387 (1928); *Chem. Abstr.*, **24**, 919 (1930).
122. G. Scheuing and L. Winterhalder, *Ger. Pat.* 594849 (1934); *Chem. Abstr.*, **28**, 4541 (1934).
123. J. M. Smith, Jr., *U.S. Pat.* 2 482 521 (1949); *Chem. Abstr.*, **44**, 2570 (1950).
124. J. M. Smith, Jr., H. W. Stewart, B. Roth, and E. H. Northey, *J. Amer. Chem. Soc.*, **70**, 3997 (1948).

125. C. A. Buehler and J. O. Harris, *J. Amer. Chem. Soc.*, **72**, 5015 (1950).
126. A. Walker, W. E. Baldwin, H. I. Thayer, and B. B. Corson, *J. Org. Chem.*, **16**, 1805 (1951).
127. M. Carissimi, P. G. De Meglio, F. Ravenna and G. Riva, *Farmaco, Ed. Sci.*, **24**, 478 (1969); *Chem. Abstr.*, **71**, 124175 (1969).
128. A. Markovac, C. L. Stevens, and A. B. Ash, *J. Med. Chem.*, **15**, 490 (1972).
129. A. Jordaan, V. P. Joynt, and R. R. Arndt, *J. Chem. Soc.*, **1965**, 3001.
130. Ciba Ltd., *Belg. Pat.* 633575 (1963); *Chem. Abstr.*, **61**, 7152 (1964).
131. B. Estert and E. Endres, *Justus Liebigs Ann. Chem.*, **734**, 56 (1970).
132. F. Ozawa, N. Kawasaki, T. Yamamoto, and A. Yamamoto, *Chem. Lett.*, **1985**, 567.
133. C. Seelkopf, *Rev. Fac. Farm., Univ. Los Andes*, **15**, 157 (1974); *Chem. Abstr.*, **83**, 78998 (1975).
134. C. R. Warner, E. J. Walsh, and R. F. Smith, *J. Chem. Soc.*, **1962**, 1232.
135. M.-C. Dubroeucq, C. G. A. Gueremy, R. G. LeFur, and J. Mizoule, *Eur. Pat. Appl.* 42781 (1981); *Chem. Abstr.*, **96**, 162553 (1982).
136. G. R. Clemo and M. Hoggarth, *J. Chem. Soc.*, **1954**, 95.
137. S. V. Tsukerman, K.-S. Ch'ang, and V. F. Lavrushin, *Zh. Obshch. Khim.*, **34**, 2881 (1964); *J. Gen. Chem. USSR (Engl. Transl.)* **34**, 2914 (1964); *Chem. Abstr.*, **62**, 520 (1965).
138. M. Sato, H. Tagawa, A. Kossasayama, F. Uchimaru, H. Kojina, T. Yamasaki, and T. Sakurai, *Chem. Pharm. Bull.*, **26**, 3296 (1978).
139. E. E. Glover and G. Jones, *J. Chem. Soc.*, **1958**, 3021.
140. K. B. Prasad and S. C. Shaw, *Indian J. Chem.*, **12**, 344 (1974).
141. K. B. Prasad and G. A. Swan, *J. Chem. Soc.*, **1958**, 2024.
142. H. Michel and F. Zymalkowski, *Arch. Pharm. (Weinheim)*, **307**, 689 (1974).
143. S. Hauptmann and K. Hirshberg, *J. Prakt. Chem.*, **34**, 272 (1966).
144. T. Caronna, A. Citterio, and M. Bellatti, *J. Chem. Soc., Chem. Commun.*, **1976**, 987.
145. A. Arnoldi, M. Bellatti, T. Caronna, A. Citterio, F. Minisci, O. Porta, and G. Sesana, *Gazz. Chim. Ital.*, **107**, 491 (1977).
146. S. V. Tsukerman, V. P. Maslennikova, and V. F. Lavrushin, *Opt. Spektrosk.*, **23**, 396 (1967); *Chem. Abstr.*, **68**, 73776 (1968).
147. S. Pavlov and V. Arsenijevic, *Croat. Chem. Acta*, **41**, 251 (1969); *Chem. Abstr.*, **72**, 66754 (1970).
148. T. L. Hough and G. Jones, *J. Chem. Soc. (C)*, **1967**, 1112.
149. G. Jones and J. Wood, *Tetrahedron*, **21**, 2529 (1965).
150. D. L. Hammick and W. P. Dickinson, *J. Chem. Soc.*, **1929**, 214.
151. R. A. Seibert, T. R. Norton, A. A. Benson, and F. W. Bergstrom, *J. Amer. Chem. Soc.*, **68**, 2721 (1946).
152. M. Hamana, H. Noda, and J. Uchida, *Yakugaku Zasshi*, **90**, 1001 (1970); *Chem. Abstr.*, **73**, 130920 (1970).
153. F. Piozzi and E. Cavalieri, *Gazz. Chim. Ital.*, **92**, 454 (1962); *Chem. Abstr.*, **57**, 13722 (1962).
154. Z. Chen, Y. Wu, M. Yin, and M. Jiang, *Gaodeng Xuexiao Huaxue Xuebao*, **5**, 67 (1984); *Chem. Abstr.*, **100**, 208971 (1984).
155. P. Nickel, R. Zimmerman, L. Preissinger, and E. Fink, *Arzneim-Forsch.*, **28**, 723 (1978); *Chem. Abstr.*, **89**, 99743 (1978).
156. F. D. Popp and J. M. Wefer, *J. Heterocycl. Chem.*, **4**, 183 (1967).
157. F. D. Popp and J. Kaut, *J. Heterocycl. Chem.*, **22**, 869 (1985).
158. K. Funakoshi, M. Mizuoka, K. Wada, S. Saeki, and M. Hamana, *Chem. Pharm. Bull.*, **32**, 3886 (1984).
159. R. F. Knott, J. Ciric, and J. G. Breckenridge, *Can. J. Chem.*, **31**, 615 (1953).

160. E. A. Fehnel, *J. Heterocycl. Chem.*, **4**, 565 (1967).
161. N. C. Rose and W. E. McEwen, *J. Org. Chem.*, **23**, 337 (1958).
162. C. Kaiser, *U.S. Pat.* 3 691 172 (1972); *Chem. Abstr.*, **77**, 151996 (1972).
163. C. Kaiser, *U.S. Pat.* 3 773 954 (1973); *Chem. Abstr.*, **80**, 47866 (1974).
164. R. Valters, L. Karole, G. Karlivans, and A. Bace, *Latv. PSR Zinat. Akad. Vestis, Kim. Ser.*, **1984**, 500; *Chem. Abstr.*, **102**, 6160 (1985).
165. W. J. Becker, S. Farber, and T. E. Hoover, *U.S. Pat.* 4 275 206 (1981); *Chem. Abstr.*, **96**, 21285 (1982).
166. G. Karlivan and R. Valters, *Zh. Org. Khim.*, **20**, 665 (1984); *J. Org. Chem. USSR (Engl. Transl.)*, **1984**, 605; *Chem. Abstr.*, **101**, 23294 (1984).
167. P.-L. Chien, D. J. McCaustland, W. H. Burton, and C. C. Cheng, *J. Med. Chem.*, **15**, 28 (1972).
168. J. Klosa, *Arch. Pharm. Ber. Dtsch. Pharm. Ges.*, **289**, 104 (1956).
169. M. Hamana and M. Yamazaki, *Chem. Pharm. Bull.*, **11**, 415 (1963).
170. T. Kinoshita, Y. Sasada, M. Watanabe, and S. Furukawa, *Chem. Pharm. Bull.*, **28**, 2892 (1980).
171. H. Reihlen and W. Hühn, *Justus Liebigs Ann. Chem.*, **489**, 42 (1931).
172. R. G. Jones, Q. F. Soper, O. K. Behrens, and J. W. Corse, *J. Amer. Chem. Soc.*, **70**, 2843 (1948).
173. B. K. Nandi, *Proc. Indian Acad. Sci.*, **12A**, 1 (1940); *Chem. Abstr.*, **34**, 7918 (1940).
174. R. B. Woodward and E. C. Kornfeld, *J. Amer. Chem. Soc.*, **70**, 2508 (1948).
175. P. Bamberg and B. Johansson, *Acta Chem. Scand.*, **22**, 2422 (1968).
176. U. Haug and H. Fürst, *Chem. Ber.*, **93**, 593 (1960).
177. G. Koller, H. Ruppersberg, and E. Strang, *Monatsh. Chem.*, **52**, 59 (1929).
178. B. R. Brown, D. L. Hammick, and R. Robinson, *J. Chem. Soc.*, **1950**, 780.
179. O. Stark, *Ber.*, **40**, 3425 (1907); *J. Chem. Soc. Abstr.*, **92**, 973 (1907).
180. S. Yamada and I. Chibata, *Pharm. Bull. (Japan)*, **3**, 21 (1955).
181. M. Wilk, H. Schwab and J. Rochlitz, *Justus Liebigs Ann. Chem.*, **698**, 149 (1966).
182. T. Higashino, Y. Nagano, and E. Hayashi, *Chem Pharm. Bull.*, **21**, 1943 (1973).
183. Etablissements Chin-Byla, *Fr. Demande* 2 134 169 (1973); *Chem. Abstr.*, **79**, 42371 (1973).
184. M. A. Haq, M. L. Kapur and J. N. Rây, *J. Chem. Soc.*, **1933**, 1087.
185. F. Gaedcke and F. Zymolkowski, *Arch. Pharm. (Weinheim)*, **312**, 767 (1979).
186. E. A. Fehnel and D. E. Cohn, *J. Org. Chem.*, **31**, 3852 (1966).
187. Y. Toi, K. Isagawa, and Y. Fushizaki, *Nippon Kagaku Zasshi*, **90**, 81 (1969); *Chem. Abstr.*, **70**, 106347 (1969).
188. A. Walser, T. Flynn, and R. I. Fryer, *J. Heterocycl. Chem.*, **12**, 737 (1975).
189. G. Dolschall and K. Lempert, *Tetrahedron*, **30**, 3997 (1974).
190. W. Ried and H. Schiller, *Chem. Ber.*, **85**, 216 (1952).
191. I. P. Sword, *J. Chem Soc. (C)*, **1970**, 1916.
192. H. Schaefer, K. Sattler, and K. Gerwald, *J. Prakt. Chem.*, **321**, 695 (1979).
193. T. Kurichara, T. Michida, and H. Hirano, *Chem. Pharm Bull.*, **23**, 2451 (1975).
194. P. C. Anderson and B. Staskun, *J. Org. Chem.*, **30**, 3033 (1965).
195. G. Singh and G. V. Nair, *J. Amer. Chem. Soc.*, **78**, 6105 (1956).
196. B. P. Bangdiwala and C. M. Desai, *J. Indian Chem. Soc.*, **31**, 553 (1954).
197. T. B. Desai and R. C. Shah, *J. Indian Chem. Soc.*, **26**, 121 (1949).
198. U. Usgaonkar and G. V. Jadhav, *J. Indian Chem Soc.*, **40**, 75 (1963).
199. B. Mills and K. Schofield, *J. Chem. Soc.*, **1956**, 4213.
200. A. Walser, G. Zenchoff, and R. I. Fryer, *J. Heterocycl. Chem.*, **13**, 131 (1976).
201. R. Hayes and O. Meth-Cohn, *Tetrahedron Lett.*, **23**, 1613 (1982).
202. B. B. Neelima and A. P. Bhaduri, *J. Heterocycl. Chem.*, **21**, 1469 (1984).

203. R. H. Baker, J. G. Van Oot, S. W. Tinsley, Jr., D. Butler, and B. Riegel, *J. Amer. Chem. Soc.*, **71**, 3060 (1949).
204. K. Elliott and E. Tittensor, *J. Chem. Soc.*, **1959**, 484.
205. J. Szmuszkovicz, L. Baczynskyj, C. C. Chidester and D. J. Dunchamp, *J. Org. Chem.*, **41**, 1743 (1976).
206. E. A. Harrison, Jr., and K. C. Rice, *J. Org. Chem.*, **44**, 2977 (1979).
207. W. A. Jacobs and M. Heidelberger, *J. Biol. Chem.*, **21**, 455 (1915).
208. K. Posselt and K. Thiele, *U.S. Pat.* 3 715 369 (1973); *Chem. Abstr.*, **78**, 136051 (1973); U.S. Pat, 3 859 305 (1975); *Chem. Abstr.*, **83**, 43189 (1975).
209. G. Koller and H. Ruppersberg, *Monatsh. Chem.* **58**, 238 (1931).
210. G. A. Sennikov and L. I. Dranik, *Farm. Zh. (Kiev)*, **27**, 68 (1972); *Chem. Abstr.*, **77**, 58873 (1972).
211. M. A. Haq, J. N. Rây and M. Tuffail-Malkana, *J. Chem. Soc.*, **1934**, 1326.
212. K. Shimizu, I. Sakamoto, and S. Fukishima, *Yakugaku Zasshi*, **87**, 672 (1967); *Chem. Abstr.*, **68**, 12876 (1968).
213. A. Stener, *Gazz. Chim. Ital.*, **90**, 1365 (1960).
214. V. S. Ruskikh, Yu. P. Dormidontov, and I. L. Lapkin, *Khimiya Elementoorganicheskikh Soedinenii. II, IV, V, VI Grupp Periodich Sistemy*, Perm, 1979, 58.
215. T. Kurihara, K. Nasu, and K. Mihara, *J. Heterocycl. Chem.*, **20**, 289 (1983).
216. J. P. Wibaut and L. G. Heeringa, *Recl. Trav. Chim. Pays-Bas*, **74**, 1003 (1955).
217. O. Stark and F. Hoffmann, *Ber.*, **42**, 715 (1909).
218. K. R. Huffman, M. Loy, and E. F. Ullman, *J. Amer. Chem. Soc.*, **87**, 5417 (1965); **88**, 601 (1966).
219. G. Kollenz, H. Igel, and E. Ziegler, *Monatsh. Chem.*, **103**, 450 (1972).
220. C. D. Nenitzescu, E. Cioránescu, and L. Birladeanu, *Comun. Acad. Repub. Pop. Rom.* **8**, 775 (1958); *Chem. Abstr.*, **53**, 18003 (1959).
221. Y. Tsuda, Y. Horiguchi and T. Sano, *Heterocycles*, **4**, 1237 (1976).
222. H. Schafer and K. Gewald, *Monatsh. Chem.*, **109**, 527 (1978).
223. G. M. Badger and R. Pettit, *J. Chem. Soc.*, **1953**, 2774.
224. C. Musante and S. Fatutta, *Gazz. Chim. Ital.*, **88**, 879 (1958).
225. E. Noelting and H. Blum, *Ber.*, **34**, 2467 (1901).
226. N. S. Prostakov, V. G. Pleshakov, M. Zainul Abedin, I. R. Kordova, V. F. Zakharov, and V. P. Zvolinskii, *Zh. Org. Khim*, **18**, 640 (1982); *J. Org. Chem. USSR*, *(Engl. Transl.)* **18**, 556 (1982).
227. N. H. Cromwell and R. A. Mitsche, *J. Org. Chem.*, **26**, 3812 (1961).
228. E. Noelting and A. Herzbaum, *Ber.*, **44**, 2585 (1911).
229. M. Bala, M. Michankow, E. Chojnacka-Wojcik, B. Wiczynska, and E. Tatarczynska, *Pol. J. Pharmacol. Pharm.*, **35**(6), 523 (1983).
230. M. Bala, *Pol. J. Chem.*, **55**, 121 (1981); *Chem. Abstr.*, **95**, 168957 (1981).
231. M. Bala, *Pol. Pat.* 116879 (1983); *Chem. Abstr.*, **99**, 22335 (1983).
232. T. Sakamoto, Y. Kondo, M. Shiraiwa, and H. Yamanaka, *Synthesis*, **1984**, 245.
233. M. Ihara, K. Noguchi, K. Fukumoto, and T. Kametani, *Heterocycles*, **20**, 421 (1983).
234. A. Kaufmann, H. Peyer, and M. Kunkler, *Ber.*, **45**, 3090 (1912).
235. E. Ochiai, M. Takahashi, Y. Tamai, and H. Kataoka, *Chem. Pharm. Bull.*, **11**, 137 (1963).
236. E. E. Van Tamelen and V. B. Haarstad, *Tetrahedron Lett.*, **1961**, (12), 390.
237. W. A. Skinner, H. T. Crawford, H. Tong, D. Skidmore, and H. I. Maibach, *J. Pharm. Sci.*, **65**, 1404 (1976).
238. P. Blumbergs, M-S. Ao, M. P. LaMontagne, A. Markovac, J. Novotny, C. H. Collins, and F. W. Starks, *J. Med. Chem.*, **18**, 1122 (1975).

239. R. Leardini, G. F. Pedulli, A. Tundo, and G. Zanardi, *J. Chem. Soc., Chem. Commun.*, **1984**, 1320.
240. K. Micscher, *U.S. Pat.* 1 434 306 (1922); *Chem. Abstr.*, **17**, 402 (1923).
241. C. M. Atkinson and A. R. Mattocks, *J. Chem. Soc.*, **1957**, 3722.
242. S.A.I.C.B., *Swiss Pat.* 98482; 98712; 98713; *Chem. Abstr.*, **18**, 2174 (1924).
243. R. B. Turner and A. C. Cope, *J. Amer. Chem. Soc.*, **68**, 2214 (1946).
244. D. W. Boykin, A. R. Patel, R. E. Lutz, and A. Burger, *J. Heterocycl. Chem.*, **4**, 459 (1967).
245. G. A. Epling, N. K. N. Ayengar, and E. F. McCarthy, *Tetrahedron Lett.*, **1977**, 517.
246. M. L. Belli, S. Fatutta, M. Forchiassin, G. Illuminati, P. Linda, G. Marino, and E. Zinato, *Ric. Sci., Parte 2: Sez. A*, **3**, 530 (1963); *Chem. Abstr.*, **59**, 8743 (1963).
247. R. E. Lutz, J. F. Codington, and N. H. Leake, *J. Amer. Chem. Soc.*, **69**, 1260 (1947).
248. S. J. Lewis, M. S. Mirrlees, and P. J. Taylor, *Quant. Struct.-Act. Relat. Pharmacol., Chem. Biol.*, **2**, 100 (1983).
249. J. B. Wommack, Jr., T. G. Barbee, Jr., K. N. Subbaswami, and D. E. Pearson, *J. Med. Chem.*, **14**, 1218 (1971).
250. M. Bodanszky, J. Fried, J. T. Sheehan, N. J. Williams, J. Alicino, A. I. Cohen, B. T. Keeler, and C. A. Birkhimer, *J. Amer. Chem. Soc.*, **86**, 2478 (1964).
251. C. N. C. Drey, G. W. Kenner, H. D. Law, R. C. Sheppard, M. Bodanszky, J. Fried, N. J. Williams, and J. T. Sheehan, *J. Amer. Chem. Soc.*, **83**, 3906 (1961).
252. A. Kaufmann, *U.S. Pat.* 1 145 487 (1915); *Chem. Abstr.*, **9**, 2423 (1915).
253. P. Rabe, W. Schuler, G. Suszka, and E. Federer, *Chem. Ber.*, **81**, 139 (1948).
254. H. Wojahn, *Arch. Pharm. Ber. Dtsch. Pharm. Ges.* **274**, 83 (1936).
255. A. Kaufmann, M. Kunkler, and H. Peyer, *Ber.*, **46**, 57 (1913).
256. I. Muramatsu, E. Hikawa, A. Hagitani, and N. Miyairi, *J. Antibiot.*, **25**, 537 (1972).
257. M. S. Puar, A. K. Ganguly, A. Afonso, R. Brambilla, P. Mangiaracina, O. Sarre, and R.D. MacFarlane, *J. Amer. Chem. Soc.*, **103**, 5231 (1981).
258. S. Winstein, T. L. Jacobs, G. B. Linden, D. Seymour, E. F. Levy, B. F. Day, J. H. Robson, R. B. Henderson, and W. H. Florsheim, *J. Amer. Chem. Soc.*, **68**, 1831 (1946).
259. R. B. Turner, J. Mills, and A. C. Cope, *J. Amer. Chem. Soc.*, **68**, 2220 (1946).
260. P. Rabe, *Ber.*, **45**, 2163 (1912).
261. P. Rabe and R. Pasternack, *Ber.*, **46**, 1026 (1913).
262. K. Feist, W. Awe, M. Kuklinski and W. Völksen, *Arch. Pharm Ber. Dtsch. Pharm. Ges.*, **276**, 271 (1938).
263. L. E. Hewitt, D. R. Wade, J. E. Sinsheimer, J. Wang, J. C. Drach, and J. H. Burckhalter, *J. Med. Chem.*, **21**, 1339 (1978).
264. R. E. Olsen, *U. S. Pat.* 3 714 168 (1973); *Chem. Abstr.*, **78**, 124463 (1973).
265. J. Schaefer, K. S. Kulkarni, R. Costin, J. Higgins, and L. Honig, *J. Heterocycl. Chem.*, **7**, 607 (1970).
266. A. Kaufmann, *Ber.*, **46**, 1823 (1913).
267. M. Lazarevic and K. Colanceska-Radjenovic, *Arh. Farm.*, **32**, 125 (1982); *Chem. Abstr.*, **98**, 143250 (1983).
268. Ciba Ltd., *Belg. Pat.* 612971 (1962); *Chem. Abstr.*, **58**, 3438 (1963).
269. C. M. Atkinson and B. N. Biddle, *J. Chem. Soc (C)*, **1966**, 2053.
270. A. M. Fahmy, M. Z. A. Badr, Y. S. Mohamed, and F. F. Abdel-Latif, *J. Heterocycl. Chem.*, **21**, 1233 (1984).
271. V. Prelog, R. Seiwerth, V. Hahn, and E. Cerkovinkov, *Ber.*, **72**, 1325 (1939).
272. C. G. A. Gueremy, M. A. P. Mestre, and C. L. A. Renault, *Eur. Pat. Appl.* 53964 (1982).
273. H. Weidel, *Monatsh. Chem.*, **17**, 401 (1896).

274. M. Otani, H. Otaki, and K. Kawamura, *Japan Pat.* **72**, 37185 (1972); *Chem. Abstr.*, **78**, 16056 (1973).

275. M. Otani, H. Otaki, and K. Kawamura, *Japan Pat.* **72**, 37186 (1972); *Chem. Abstr.*, **78**, 16055 (1973).

276. E. R. Buchman, H. Sargent, T. C. Meyers, and D. R. Howton, *J. Amer. Chem. Soc.*, **68**, 2710 (1946).

277. H. Sargent, *J. Amer. Chem. Soc.*, **68**, 2688 (1946).

278. J. C. Shivers and E. R. Hauser, *J. Amer. Chem. Soc.*, **70**, 437 (1948).

279. A. K. Chizhov, *Zh. Obshch. Khim.*, **31**, 3469 (1961); *J. Gen. Chem. USSR (Engl. Transl.*), **31**, 3232 (1961); *Chem. Abstr.*, **57**, 2183 (1962).

280. P. Remfry and H. Decker, *Ber.*, **41**, 1007 (1908); *J. Chem. Soc. Abstr.*, **94**, 364 (1908).

281. H. de Diesbach, A. Pugin, F. Morard, W. Nowaczinski, and J. Dessibourg, *Helv. Chim. Acta* **35**, 2322 (1952).

282. K. Feist, W. Awe, and M. Kuklinski, *Arch. Pharm. Ber Dtsch. Pharm. Ges.*, **276**, 420 (1938).

283. L. I. Mastafanova, L. F. Linberg, T. Ya. Filipenko, and L. N. Yakhontov, *Khim. Geterotsikl. Soedin.*, **1978**, 368; *Chem. Heterocycl. Compd. (Engl. Transl.*), **14**, 301 (1978); *Chem. Abstr.*, **89**, 43065 (1978).

284. M. Takahashi and R. Tanabe, *Itsuu Kenkyusho Nempo*, **1968**, (15), 41; *Chem. Abstr.*, **72**, 90182 (1970).

285. E. Ochiai and M. Takahashi, *Chem. Pharm. Bull.*, **14**, 1272 (1966).

286. P. Karrer, *Ber.*, **50**, 1499 (1917).

287. P.-L. Chien and C. C. Cheng, *J. Med. Chem.*, **19**, 170 (1976).

288. H. King and T. S. Work, *J. Chem. Soc.*, **1940**, 1307.

289. K. N. Campbell and J. F. Kerwin, *J. Amer. Chem. Soc.*, **68**, 1837 (1946).

290. C. F. Koelsch, *J. Amer. Chem. Soc.*, **65**, 2460 (1943).

291. P. Rabe and R. Pasternack, *Ber.*, **46**, 1032 (1913).

292. I. W. Elliott, *J. Org. Chem.*, **25**, 1256 (1960).

293. R. Rabe, R. Pasternack, and K. Kindler, *Ber.*, **50**, 144 (1917).

294. E. Ferber and H. Bendix, *Ber.*, **72**, 839 (1939).

295. H. King and J. Wright, *Proc. R. Soc. (London), Ser. B*, **135**, 271 (1948).

296. J. C. Shivers, M. L. Dillon and C. R. Hauser, *J. Amer. Chem. Soc.*, **69**, 119 (1947).

297. E. Schwenk and D. Papa, *J. Org. Chem.*, **11**, 798 (1946).

298. Ges. f. Chem. Ind. Basel, *D. R. Pat.* 462, 136 (1928).

299. A. J. Saggiomo, K. Kato, and T. Kaiya, *J. Med. Chem.*, **11**, 277 (1968).

300. E. Ochiai and Y. Tamai, *Itsuu Kenkyusho Nempo*, **1963**, (13), 5, 13; *Chem. Abstr.*, **60**, 488 (1964).

301. M. Takahashi, *Itsuu Kenkyusho Nempo*, **1965**, (14), 23; *Chem. Abstr.*, **67**, 64157 (1967).

302. E. Rabald and H. Dietrich, *Ger. Pat.* 858249 (1952); *Chem. Abstr.*, **52**, 5484 (1958).

303. J. B. Wommach, Jr. and D. E. Pearson, *J. Med. Chem.*, **13**, 383 (1970).

304. L. Ruzicka, C. F. Seidel, and F. Liebel, *Helv. Chim. Acta*, **7**, 995 (1924).

305. A. E. Senear, H. Sargent, J. F. Mead, and J. B. Koepfli, *J. Amer. Chem. Soc.*, **68**, 2695 (1946).

306. E. R. Buchman, H. Sargent, T. C. Myers, and J. A. Seneker, *J. Amer. Chem. Soc.*, **68**, 2692 (1946).

307. J. F. Mead, A. E. Senear, and J. B. Koepfli, *J. Amer. Chem. Soc.*, **68**, 2708 (1946).

308. F. I. Carroll and R. E. Olsen, *U.S. Pat. Appl.* 684040 (1976); *Chem. Abstr.*, **88**, 22664 (1978).

309. M. M. Rapport, A. E. Senear, J. F. Mead, and J. B. Koepfli, *J. Amer. Chem. Soc.*, **68**, 2697 (1946).

310. E. R. Buchman and D. R. Howton, *J. Amer. Chem. Soc.*, **68**, 2718 (1946).
311. R. F. Brown, T. L. Jacobs, S. Winstein, and J. A. Terek, *J. Amer. Chem. Soc.*, **68**, 2705 (1946).
312. S. Winstein, T. L. Jacobs, E. F. Levy, D. Seymour, G. B. Linden, and R. B. Henderson, *J. Amer. Chem. Soc.*, **68**, 2714 (1946).
313. A. A. Champseix, G. R. LeFur, and C. L. A. Renault, *Eur. Pat. Appl.* 31753 (1981); *Chem. Abstr.*, **96**, 122650 (1982).
314. L. Ruzicka, *Helv. Chim. Acta*, **4**, 486 (1921).
315. R. E. Lutz, P. S. Bailey, M. T. Clark, J. F. Codington, A. J. Deinet, J. A. Freek, G. H. Harnest, N. H. Leake, T. A. Martin, R. J. Rowlett, Jr., J. M. Salsburg, N. H. Shearer, Jr., J. D. Smith, and J. W. Wilson, III, *J. Amer. Chem. Soc.*, **68**, 1813 (1946).
316. R. B. Fugitt and R. M. Roberts, *J. Med. Chem.*, **16**, 875 (1973).
317. I. M. Roushdi and A. I. El-Sebai, *Egypt. Pharm Bull.*, **44**, 425 (1962); *Chem. Abstr.*, **64**, 11170 (1966).
318. R. C. Elderfield, M. Israel, J. H. Ross, and J. A. Waters, *J. Org. Chem.*, **26** 2827 (1961).
319. B. Zorn and A. Mankel, *Hoppe-Syler's Z. Physiol. Chem.*, **296**, 239 (1954); *Chem. Abstr.*, **49**, 9649 (1955).
320. H. R. Munson, Jr., R. E. Johnson, J. M. Sanders, C. J. Ohnmacht, and R. E. Lutz, *J. Med. Chem.*, **18**, 1232 (1975).
321. E. R. Atkinson and A. J. Puttick, *J. Med. Chem.*, **11**, 1223 (1968).
322. J. S. Gillespie, Jr., R. J. Rowlett, Jr., and R. E. Davis, *J. Med. Chem.*, **11**, 425 (1968).
323. E. R. Atkinson and A. J. Puttick, *J. Med. Chem.*, **13**, 537 (1970).
324. H. Gilman, L. Tolman, and S. P. Massie, Jr., *J. Amer. Chem. Soc.*, **68**, 2399 (1946).
325. C. R. Wetzel, J. R. Shanklin, Jr., and R. E. Lutz, *J. Med. Chem.*, **16**, 528 (1973).
326. S. A. I. C. B., *Swiss Pat.* 92001, 92607, 92819; *Chem. Abstr.*, **17**, 2118 (1923).
327. M. Takahashi, *Itsuu Kenkyusho Nempo*, **1963**, (13), 25; *Chem. Abstr.*, **60**, 1694 (1964).
328. M. Takahashi and R. Tanabe, *Chem. Pharm. Bull.*, **15**, 793 (1967).
329. M. Takahashi, *Chem. Pharm. Bull.*, **14**, 1144 (1966).
330. R. M. Pinder and A. Burger, *J. Med. Chem.* **11**, 267 (1968).
331. C. J. Ohnmacht, A. R. Patel, and R. E. Lutz, *J. Med. Chem.*, **14**, 926 (1971).
332. E. Hickmann and H. G. Oeser, *Eur. Pat. Appl.* 49776 (1982); *Chem. Abstr.*, **97**, 127517 (1982).
333. E. Hickmann, H. G. Oeser, and L. Moebius, *Ger. Offen.* 2 940 443 (1981); *Chem. Abstr.*, **95** 62019 (1981).
334. W. H. Yanko and G. F. Deebel, *J. Labelled Compd. Radiopharm.*, **17**, 431 (1980); *Chem. Abstr.*, **93**, 186125 (1980).
335. J. Novotny, C. H. Collins, and F. W. Starks, *J. Pharm. Sci.*, **63**, 1264 (1974).
336. D. W. Boykin, Jr., A. R. Patel, and R. E. Lutz, *J. Med. Chem.*, **11**, 273 (1968).
337. E. Hodel and H. Gysin, *Ger. Pat.* 961668 (1957); *Chem. Abstr.*, **54**, 14563 (1960).
338. E. Hodel, *Fr. Pat.* 1 450 570 (1966); *Chem. Abstr.*, **69**, 10380 (1968).
339. W. H. Edgerton and J. H. Burkhalter, *J. Amer. Chem. Soc.*, **74**, 5209 (1952).
340. G. Madgeney and J. Pechmeze, *Fr. Pat.* 1 172 432 (1959); *Chem. Abstr.*, **54**, 19722 (1960).
341. J. R. Geigy A.-G., *Brit. Pat.* 1 003 478 (1965); *Chem. Abstr.*, **64**, 12651 (1966).
342. H. J. Sohn and J. C. Kim, *Nonmunjip – Sanop Kwahak Kisul Yonguso (Inha Taehakkyo)*, **13**, 249 (1985); *Chem. Abstr.*, **104**, 90977 (1986).
343. J. R. Geigy A.-G., *Brit. Pat.* 1 003 478 (1965); *Chem. Abstr.*, **64**, 12651 (1966).
344. I. C. Popoff and C. B. Thanawalla, *J. Med. Chem.*, **13**, 1002 (1970).
345. A. Makriyannis, J. S. Frazee, and J. W. Wilson, *J. Med. Chem.*, **16**, 118 (1973).
346. E. Sturm, *Ger. Offen.* 2 523 093 (1975); *Chem. Abstr.*, **84**, 150522 (1976).

347. G. Hamprecht, J. Markert, W. Spiegler, W. Richarz, H. Graf, E. Ammermann, and E. H. Pommer, *Ger. Offen.* 3225169 (1984); *Chem. Abstr.*, **100**, 209833 (1984); *Eur. Pat. Appl.* 98486.

348. K. Matsumura and M. Ito, *J. Org. Chem.*, **25**, 853 (1960).

349. K. Matsumura and C. Sone, *J. Amer. Chem. Soc.*, **53**, 1406 (1931).

350. G. Madgeney and J. Pechmeze, *Fr. Pat.* 1172432 (1959); *Chem. Abstr.*, **54**, 19722 (1960).

351. P. Umapathy, S. N. Bhide, K. D. Ghuge, and D. N. Sen, *J. Indian Chem. Soc.*, **58**, 33 (1981).

352. G. M. Badger and A. G. Moritz, *J. Chem. Soc.*, **1958**, 3437.

353. V. M. Thakor and R. C. Shah, *J. Indian Chem. Soc.*, **31**, 597 (1954).

354. J. Debat, *Brit. Pat.* 1100870 (1968); *Chem. Abstr.*, **69**, 27270 (1968).

355. K. W. Rosenmund and G. Karst, *Arch. Pharm. Ber. Dtsch. Pharm. Ges.*, **279**, 154 (1941).

356. K. Matsumura, *J. Amer. Chem. Soc.*, **52**, 4433 (1930).

357. G. V. Villeval'd, N. N. Anshits, S. A. Osadchii, I. N. Gribanova and A. V. Nikolaev, *U.S.S.R. Pat.* 579271 (1977); *Chem. Abstr.*, **88**, 37637 (1978).

358. F. Sztaricskai, I. Miskolczi, E. R. Farkas, and R. Bognár, *Acta Chim. Acad. Sci. Hung.*, **94**, 169 (1977); *Chem. Abstr.*, **88**, 170068 (1978).

359. T. Iwakuma, A. Tsunashima, K. Ikezawa, and O. Takaiti, *Eur. Pat. Appl.* 147719 (1985); *Chem. Abstr.*, **104**, 88453 (1986).

360. E. Hodel, *Swiss Pat.* 408007 (1966); *Chem. Abstr.*, **67**, 21845 (1967).

361. Y. Gutter, *Plant Dis. Rep.*, **53**, 474 (1969); *Chem. Abstr.*, **71**, 69595 (1969); C. Van Assche, E. Van Wambeke, and A. Vanachter, *Parasitica*, **31**, 123 (1975); *Chem. Abstr.*, **85**, 88377 (1976).

362. J. Drabek, A. A. Oswald, and P. L. Valint, *Ger. Offen.* 2304128; *Chem. Abstr.* **79**, 115293 (1973).

363. Y. Kawasaki, T. Takahashi, and A. Fujioka, *J. Organomet. Chem.*, **131**, 239 (1977).

364. L. F. Sytsma and R. J. Kline, *J. Organomet. Chem.*, **54**, 15 (1973).

365. G. Haas, K. A. Jaeggi, A. Rossi, and A. Sele, *U.S. Pat.* 4297358 (1980); *Eur. Pat. Appl.* 13666 (1980); *Chem. Abstr.*, **94**, 15704 (1981).

366. A. M. Clark, S. L. Evans, C. D. Hufford, and J. D. McChesney, *J. Nat. Prod.*, **45**, 574 (1982).

367. J. J. Johnson, D. L. Worth, N. L. Colbry, and L. M. Werbel, *J. Heterocycl. Chem.*, **21**, 1093 (1984).

368. J. D. McChesney and R. C. Gupta, *Pharm. Res.*, **1985**, 46; *Chem. Abstr.*, **103**, 6202 (1985).

369. A. Burger, K. C. Bass, Jr., and J. M. Fredricksen, *J. Org. Chem.*, **9**, 373 (1944).

370. S. Fränkel and O. Grauer, *Ber.*, **46**, 2551 (1913).

371. S. Yoshizaki, S. Tamada, and K. Nakagawa, *Japan Kokai*, 7783,380 (1977); *Chem. Abstr.*, **87**, 201339 (1977).

372. P. Da Re, P. Valenti, G. Geccarelli, and G. Caceia, *Ann. Chim. (Rome)*, **60**, 215 (1970); *Chem. Abstr.*, **73**, 25338 (1970).

373. K. Matsumura, *J. Amer. Chem. Soc.*, **57**, 124 (1935).

374. Societa Farmaceutici Italia, *Italian Pat.* 600483 (1959); *Chem. Abstr.*, **58**, 9092 (1963).

375. H. K. Pujari and M. K. Rout, *J. Indian Chem. Soc.*, **32**, 431 (1955).

376. K. Matsumura, M. Ito, and S. T. Lee, *J. Org. Chem.*, **25**, 854 (1960).

377. E. Schraufstätter and M. Bock, *Ger. Pat.* 1039518 (1958); *Chem. Abstr.*, **54**, 22687 (1960).

378. E. Uhlemann and W. Weber, *Z. Chem.*, **23**, 334 (1983).

379. H. R. Henze, *U.S. Pat.* 2 526 232 (1950); *Chem. Abstr.*, **45**, 2974 (1951).
380. W. H. Linnell and S. V. Vora, *J. Pharm. Pharmacol.*, **3**, 670 (1951).
381. N. K. Chawla and M. M. Jones, *Inorg. Chem.*, **3**, 1549 (1964).
382. L. F. Fieser and E. B. Hershberg, *J. Amer. Chem. Soc.*, **62**, 1640 (1940).
383. G. M. Pandit and K. S. Nargund, *J. Univ. Bombay, Phys. Sci.*, 1960, **29** (Pt. 3 & 5, No. 48–49), 71; *Chem. Abstr.*, **58**, 3388 (1963).
384. M. Ito and K. Matsumura, *J. Org. Chem.*, **25**, 856 (1960).
385. E. Hodel and H. Gysin, *U.S. Pat.* 2 875 126 (1959).
386. T. Nógrádi, L. Vargha, G. Ivánovics, and I. Koczka, *Acta Chim. Acad. Sci. Hung.*, **6**, 287 (1955); *Chem. Abstr.*, **51**, 5077 (1957).
387. Farbenfabriken Bayer A.-G., *Brit. Pat.* 708013 (1954); *Chem. Abstr.*, **50**, 1086 (1956).
388. S. S. Misra and B. Nath, *Indian J. Appl. Chem.*, **35**, 95 (1972); *Chem. Abstr.*, **81**, 151957.
389. S. C. Kushwaha, Dinkar and J. B. Lal, *Indian J. Chem.*, **5**, 82 (1967).
390. S. S. Misra, *Monatsh. Chem.*, **104**, 11 (1973).
391. S. S. Misra, *J. Chin. Chem. Soc. (Taipei)*, **24**, 45 (1977); *Chem. Abstr.*, **87**, 134300 (1977).
392. Unicler S. A., *Fr. Demande* 2 387 956 (1978); *Chem. Abstr.*, **92**, 58620 (1980).
393. J. R. E. Hoover and B. Black, *J. Med. Chem.*, **6**, 628 (1963).
394. S. G. Waley, *J. Chem. Soc.*, **1948**, 2008.
395. M. Ferles, P. Jančar and O. Kocián, *Collect. Czech. Chem. Commun.*, **44**, 2672 (1979).
396. B. P. Lugovkin, *Zh. Obsch. Khim.*, **25**, 392 (1955); *J. Gen. Chem. U.S.S.R. (Engl. Transl.)*, **25**, 371 (1955); *Chem. Abstr.*, **50**, 2594 (1956).
397. L. Berend and E. Thomas, *Ber.*, **25**, 2548 (1892).
398. C. E. Kaslow, J. D. Genzer, and J. G. Goodspeed, *Proc. Indiana Acad. Sci.*, **59**, 134 (1949); *Chem. Abstr.*, **45**, 8534 (1951).
399. A. A. Tolmachev, D. I. Sheiko, and Yu. L. Slominskii, *Khim. Geterotsikl. Soedin.*, **1984**, 943; *Chem. Abstr.*, **101**, 112414 (1984).
400. T. P. Dabhi, *Ph. D. Thesis*, Bhavnagar University, Bhavnagar, India.
401. D. J. Bhatt, G. C. Kamdar, and A. R. Parikh, *J. Indian Chem. Soc.*, **61**, 816 (1984).
402. W. Borsche, *Ber.*, **41**, 3884 (1908).
403. B. Riegel, G. R. Lappin, B. H. Adelson, R. I. Jackson, C. J. Albisetti, R. M. Dodson, and R. H. Baker, *J. Amer. Chem. Soc.*, **68**, 1264 (1946).
404. I. D. Dicker, K. J. Gould, and J. L. Suschitzky, *Eur. Pat. Appl.* 150996 (1985); *Chem. Abstr.*, **104**, 68836 (1986).
405. S. Oguchi, *Bull. Chem. Soc. Japan*, **47**, 1291 (1974).
406 D. Cox, H. Cairns, N. Chadwick, and J. L. Suschitzky, *Brit. Pat.* 2 035 312 (1980); *Ger. Offen.* 2 943 658 (1980); *Chem. Abstr.*, **94**, 103331 (1981).
407. H. Cairns, D. Cox, K. J. Gould, A. H. Ingall, and J. L. Suschitzky, *J. Med. Chem.*, **28**, 1832 (1985).
408. D. W. Payling and J. L. Suschitzky, *Eur. Pat. Appl.* 37187 (1981); *Chem. Abstr.*, **96**, 122775 (1982) *U.S. Pat.* 4 356 181.
409. J. R. E. Hoover, *U.S. Pat.* 3 014 036 (1961); *Chem. Abstr.*, **56**, 15491 (1962).
410. J. G. Atkinson and B. K. Wasson, *U.S. Pat.* 4 348 398 (1982); *Chem. Abstr.*, **98**, 16594 (1983).
411. M. S. Chodnekar, A. F. Crowther, W. Hepworth, R. Howe, B. J. McLoughlin, A. Mitchell, B. S. Rao, R. P. Slatcher, L. H. Smith, and M. A. Stevens, *J. Med. Chem.*, **15**, 49 (1972).
412. W. Hepworth, A. Mitchell, M. S. Chodnekar, and R. Howe, *French Pat.* M 3697 (1965); *Chem. Abstr.*, **67**, 82119 (1967).

413. G. A. Melentleva and S. I. Kanevskaya, *Zh. Obshch. Khim.*, **23**, 310 (1953); *Chem. Abstr.*, **48**, 2712 (1954).
414. Imperial Chemical Industries Ltd., *Belg. Pat.* 633973 (1963); *Chem. Abstr.*, **61**, 672 (1964); *Brit. Pat.* 1013224 (1965).
415. R. Göschke, P. G. Ferrini, and A. Sallmann, *Eur. Pat. Appl.* 62001 (1982); *Chem. Abstr.*, **98**, 160599 (1983).
416. C. E. Kaslow and E. Arnoff, *J. Org. Chem.*, **19**, 857 (1954).
417. E. Hinz, *Justus Liebigs Ann. Chem.*, **242**, 321 (1887).
418. C. E. Kaslow and S. J. Nix, *J. Org. Chem.*, **16**, 895 (1951).
419. E. Noelting and C. Schwartz, *Ber.*, **24**, 1606 (1891)
420. E. Hodel, *U.S. Pat.* 3113135 (1963); *Chem. Abstr.*, **60**, 5469 (1964).
421. D. R. Patel, C. S. Choxi, and S. R. Patel, *Can. J. Chem.*, **47**, 105 (1969).
422. K. Schofield and R. S. Theobald, *J. Chem. Soc.*, **1950**, 395.
423. H. Hennig, R. Spitzner, and E. Uhlemann, *Chem. Ber.*, **99**, 3806 (1966).
424. D. W. Ockenden and K. Schofield, *J. Chem. Soc.*, **1953**, 3440.
425. L. P. Kulev and L. I. Aristov, *Izv, Tomsk. Politekh. Inst.*, **102**, 29 (1959); *Chem. Abstr.*, **59**, 6852 (1963).
426. K. N. Campbell, J. F. Kerwin, R. A. LaForge, and B. K. Campbell, *J. Amer. Chem. Soc.*, **68**, 1844 (1946).
427. N. V. Smolentseva, I. A. Red'kin, and A. I. Tochilkin, *Khim. Geterotsikl. Soedin.*, **1975**, 514; *Chem. Heterocycl. Compd. (Engl. Transl.)* **11**, 452 (1975); *Chem. Abstr.*, **83**, 79050 (1975).
428. R. A. Glenn and J. R. Bailey, *J. Amer. Chem. Soc.*, **63**, 639 (1941).
429. A. McConbrey and W. Webster, *J. Chem. Soc.*, **1948**, 97.
430. J. W. Suggs and C.-H. Jun, *J. Amer. Chem. Soc.*, **106**, 3054 (1984).
431. J. W. Suggs and C.-H. Jun, *J. Chem. Soc., Chem. Commun.*, **1985**, 92.
432. G. Stefanovica, L. Lorenc, and M. Miljovica, *Glas. Hem. Drus. Beograd*, **27**, 13 (1962); *Chem. Abstr.*, **59**, 9978 (1963).
433. Z. J. Allan, *Chem. Listy*, **46**, 224 (1952); *Chem. Abstr.*, **47**, 87326 (1953).
434. R. C. Schnur, *Fr. Demande* 2487350 (1982); *Chem. Abstr.*, **97**, 23775 (1982); *U.S, Pat.* 4332952 (1982).
435. J. W. Suggs, M. J. Wovkulich, and S. D. Cox, *Organometallics*, **4**, 1101 (1985).
436. J. Howitz and O. Köpke, *Justus Liebigs Ann. Chem.*, **396**, 38 (1913).
437. J. W. Suggs and S. D. Cox, *J. Organomet. Chem.*, **221**, 199 (1981).
438. D. Nasipuri and S. K. Konar, *J. Chem. Soc., Perkin Trans II*, **1979**, 269.
439. K. S. Topchiev and A. F. Bekhli, *Zh. Obshch. Khim.*, **18**, 1710 (1948); *Chem. Abstr.*, **43**, 2623 (1949).
440. E. R. Buchman and H. Sargent, *J. Amer. Chem. Soc.*, **68**, 2720 (1946).
441. R. J. Lee and L. M. R. Murphy, *U.S. Pat.* 4390437 (1983); *Chem. Abstr.*, **99**, 107994 (1983).
442. R. Geigy and W. Königs, *Ber.*, **18**, 2400 (1885); *J. Chem. Soc. Abstr.*, **48**, 1236 (1885).
443. H. P. Härter and S. Liisberg, *Acta Chem. Scand.*, **22**, 3332 (1968).
444. F. D. Popp, W. R. Schleigh, and L. E. Katz, *J. Chem. Soc. (C)*, **1968**, 2253.
445. T. Ukai, S. Kanahara, and S. Kanetomo, *Yakugaku Zasshi*, **74**, 43 (1954); *Chem. Abstr.*, **49**, 1723 (1955).
446. F. Raschig, *Brit. Pat.* 772520 (1957); *Chem. Abstr.*, **52**, 449 (1958).
447. D. Oda, *Mem. Def. Acad., Math., Phys., Chem. Eng. Japan*, **4**, 355 (1965); *Chem. Abstr.*, **63**, 13205f (1965).
448. D. Oda, *Bull. Chem. Soc. Japan*, **34**, 557 (1961).
449. C. A. Buehler and S. P. Edwards, *J. Amer. Chem. Soc.*, **74**, 977 (1952).
450. E. W. Gill and E. D. Morgan, *Nature*, **183**, 248 (1959).

451. B. R. Brown and D. L. Hammick, *Nature*, **164**, 831 (1949).
452. D. R. Davies and H. M. Powell, *Nature*, **168**, 386 (1951).
453. B. R. Brown and D. L. Hammick, *J. Chem. Soc.*, **1950**, 628.
454. T. Ishiguro and I. Utsumi, *Yakugaku Zasshi*, **72**, 865 (1952); *Chem. Abstr.*, **47**, 6416 (1953).
455. B. P. Lugovkin, *Zh. Obshch. Khim.*, **23**, 1696 (1953); *Chem. Abstr.*, **49**, 326 (1955).
456. U. Jacoby and F. Zymalkowski, *Arch. Pharm. (Weinheim)*, **304**, 271 (1971).
457. A. P. Phillips, *J. Amer. Chem. Soc.*, **68**, 2568 (1946).
458. H. Albers, *Ger. Pat.* 880444 (1953); *Chem. Abstr.*, **52**, 11127 (1958).
459. M. Weissenfels and J. Punkt, *Tetrahedron*, **34**, 311 (1978).
459a. M. Weissenfels, B. Ulrici and S. Kaubisch, *Z. Chem.*, **18**, 138 (1978); *Chem. Abstr.*, **89**, 24108 (1979).
460. R. B. Turner and R. B. Woodward, in R. H. F. Manske and H. L. Holmes (Eds.), *The Alkaloids*, Vol. 3, p. 1, Academic Press, New York, 1953.
461. M. R. Uskoković and G. Grethe, in R. H. F. Manske (Ed.), *The Alkaloids*, Vol. 14, p. 181, Academic Press, New York, 1973.
462. M. R. Uskoković and G. Grethe, in K. Wiesner (Ed.), *Organic Chemistry*, Series Two, Vol. 9, *Alkaloids*, p. 91, Butterworth, London, 1976.
463. E. W. Warnhoff and P. Reynolds-Warnhoff, *J. Org. Chem.*, **28**, 1431 (1963).
464. J. P. Gignier and J. Bournelly, *U.S. Pat.* 4 175 192 (1979); *Chem. Abstr.*, **92**, 147017 (1980).
465. F. X. Jarreau and J. J. Koenig, *Ger. Offen.* 2 652 685 (1977); *Chem. Abstr.*, **87**, 136113 (1977).
466. S. A. Devinter, *Fr. Demande* 2 394 545 (1979); *Chem. Abstr.*, **92**, 42213 (1980).
467. J. J. Koenig, J. DeRostolan, J. C. Bourbier, and F. X. Jarreau, *Tetrahedron Lett.*, **1978**, 2779.
468. Ref. 460, p. 15.
469. G. G. Lyle and W. Gaffield, *Tetrahedron*, **23** 51 (1967).
470. J. Gutzwiller and M. R. Uskoković, *J. Amer. Chem. Soc.*, **100**, 576 (1978).
471. P. Rabe, E. Kuliga, O. Marshall, W. Naumann, and W. F. Russell, *Justus Liebigs Ann. Chem.*, **373**, 85 (1910).
472. R. B. Woodward, N. L. Wendler, and F. J. Brutschy, *J. Amer. Chem. Soc.*, **67**, 1425 (1945).
473. D. V. McCalley, *J. Chromatogr.*, **260**, 184 (1983).
474. Th. Mulder-Krieger, R. Verpoorte, A. de Water, M. van Gessel, B. C. J. A. van Oeveren, and A. B. Svendsen, *Planta Med.* **46**, 19 (1982).
475. A. R. Battersby and R. J. Parry, *J. Chem. Soc., Chem. Commun.*, **1971**, 31.
476. J. Staunton, in J. E. Saxton (Ed.), *The Alkaloids*, Vol. 2, p. 4 Specialist Periodical Reports, The Chemical Society, London, 1972.
477. A. G. Renfrew and L. H. Cretcher, *J. Amer. Chem. Soc.*, **57**, 738 (1935).
478. W. Von Miller and G. Rhode, *Ber.*, **28**, 1056 (1895); *J. Chem. Soc. Abstr.*, **68**, 434 (1895).
479. P. Rabe and A. McMillan, *Ber.*, **43**, 3308 (1910).
480. A. Machado, *Rev. Soc. Bras. Quim.*, **8**, 59 (1939); *Chem. Abstr.*, **34**, 3752^3 (1940).
481. *The Merck Index*, 10th edn. Rahway, NJ., 1983, No. 9808.
482. C. Guérémy, A. Uzan, and J. C. Tamen, *Arzneim.-Forsch.*, **22**, 1336 (1972).
483. W. Von Miller and G. Rohde, *Ber.*, **33**, 3214 (1900).
484. P. Rabe, *Justus Liebigs Ann. Chem.*, **350**, 180 (1906).
485. J. Suszko and K. Golankiewicz, *Poznan. Tow. Przyj. Nauk, Wydz. Mat.-Przyr., Pr. Kom. Mat.-Przyr.*, **10**, 3, 15 (1962); *Chem. Abstr.*, **61**, 10727 (1964).
486. Pharmindustrie, *Japan Kokai Tokkyo Koho*, 79 135 775 (1979); *Chem. Abstr.*, **92**, 94254 (1980).

487. R. Tixier, *U.S. Pat.* 3 828 048 (1974); *Chem. Abstr.*, **83**, 10588 (1975).
488. Ref. 460, p. 19.
489. T. Dubas, A. Konopincki, and J. Suszko, *Rocz. Chem.*, **13**, 464 (1933).
490. ACF Chemifarma N. V., *Neth. Appl.* 79 08,030 (1981); *Chem. Abstr.*, **95**, 187095 (1981).
491. J. Gutzwiller and M. R. Uskoković, unpublished results quoted in R. H. F. Manske (Ed.), *The Alkaloids*, Vol. 14, p. 185, Academic Press, New York, 1973.
492. J. Gutzwiller and M. Uskoković, *Helv. Chim. Acta*, **56**, 1494 (1973).
493. J. Gutzwiller, C. Reese and M. R. Uskoković, unpublished results quoted in Ref. 491, p. 205.
494. M. R. Uskoković, D. L. Pruess, C. W. Despreaux, S. Shivey, G. Pizzolato, and J. Gutzwiller, *Helv. Chim. Acta*, **56**, 2834 (1973).
495. D. H. R. Barton, D. J. Lester, W. B. Motherwell, and M. T. B. Papoula, *J. Chem. Soc., Chem. Commun.*, **1980**, 246.
496. D. H. R. Barton, R. Bengelmans, and R. N. Young, *Nouv. J. Chim.*, **2**, 363 (1978).
497. M. V. Rubtsov, *J. Gen. Chem. USSR (Engl. Transl.)*, **9**, 1493 (1939); Chem. Abstr., **34**, 2850 (1940).
498. M. V. Rubtsov, *J. Gen. Chem. USSR (Engl. Transl.)*, **13**, 593 (1943); *Chem. Abstr.*, **39**, 705 (1945).
499. P. Rabe, K. Kindler, and O. Wagner, *Ber.*, **55B**, 532 (1922).
500. P.-C. Wirth, *Brit. Pat.* 1 441 244 (1976); *Fr. Demande* 2 206 944 (1974); *Chem. Abstr.*, **82**, 72805 (1975).
501. A. A. Champseix, C. G. Gueremy, and G. R. Le Fur, *U.S. Pat.* 4 237 139 (1980); *Chem. Abstr.*, **94**, 156763 (1981).
502. A. A. Champseix, C. A. Gueremy, and G. R. Le Fur, *Ger. Offen.* 2 727 144 (1977); *Chem. Abstr.*, **88**, 121003 (1978).
503. M. Kleiman and S. Weinhouse, *J. Org. Chem.*, **10**, 562 (1945).
504. B. Prelog, R. Seiwerth, S. Heimbach-Juhász, and P. Stern, *Ber.*, **74**, 647 (1941).
505. H. B. Trijzelaar, R. De Bode, and H. B. A. Welle, *Eur. Pat. Appl.* 59 009 (1982); *Chem. Abstr.*, **98**, 126449 (1983).
506. H. B. Trijzelaar, R. De Bode, and H. B. A. Welle, *Eur. Pat. Appl.* 35 819 (1981); *Chem. Abstr.*, **96**, 104575 (1982).
507. H. B. Trijzelaar, R. De Bode, and H. B. A. Welle, *Eur. Pat. Appl.* 35 820 (1981); *Chem. Abstr.*, **96**, 218090 (1982).
508. J. Gutzwiller and M. R. Uskoković, *Ger. Offen.* 1 933 599 (1970); *Chem. Abstr.*, **72**, 90696 (1970); *U.S. Pat.* 3 753 992 (1973); 3 864 461 (1975).
509. P. C. Wirth, *Belg. Pat.* 797852 (1973); *Chem. Abstr.*, **80**, 124762 (1974).
510. C. Kan-Fan, M.-H. Brillanceau, J. Pusset, G. Chauviere, and H.-P. Husson, *Phytochemistry*, **24**, 2773 (1985).
511. H. B. Trijzelaar, R. De Bode, and H. B. A. Welle, *Eur. Pat. Appl.* 30044 (1981); *Chem. Abstr.*, **96**, 123074 (1982).
512. P. Rabe, K. Dussel and R. Teske-Guttmann, *Justus Liebigs Ann. Chem.*, **561**, 159 (1949).
513. M. Proštenik and V. Prelog, *Helv. Chim. Acta*, **26**, 1965 (1943).
514. R. B. Woodward and W. E. Doering, *J. Amer. Chem. Soc.*, **66**, 849 (1944).
515. R. B. Woodward and W. E. Doering, *J. Amer. Chem. Soc.*, **67**, 860 (1945).
516. Ref. 491, p. 193.
517. G. Grethe, H. L. Lee, T. Mitt, and M. R. Uskoković, *Helv. Chim. Acta*, **56**, 1485 (1973).
518. A. J. Quevauviller, J. Soleıl, G. Sarrazin, and G. Bellot, *Ann. Pharm. Fr.*, **24**, 39 (1966).
519. G. Engler, H. Strub, J. P. Fleury, and H. Fritz, Helv. Chim. Acta, **68**, 789 (1985).
520. B. R. Carpenter and E. E. Turner, *J. Chem. Soc.*, **1934**, 869.

521. J. A. Hannart, A. J. Quevauviller and G. Sarrazin, *Ger. Offen.* 2 315 148 (1973); *Chem. Abstr.*, **80**, 14856 (1974).
522. R. Tixier, *Ger. Offen.* 2 151 054 (1972); *Chem. Abstr.*, **77**, 34750 (1972).
523. P. Rabe, *Ber.*, **44**, 2088 (1911).
524. P. Rabe and K. Kindler, *Ber.*, **51**, 446 (1918).
525. A. J. Quevauviller and J. A. Hannart, *Belg. Pat.* 772399 (1972); *Chem. Abstr.*, **77**, 126453 (1972).
526. G. Grethe, J. Gutzwiller, H. L. Lee, and M. R. Uskoković, *Helv. Chim. Acta*, **55**, 1044 (1972).
527. P. Rabe and B. Böttcher, *Ber.*, **50**, 127 (1917).
528. A. Kaufmann and P. Haensler, *Ber.*, **50**, 702 (1917).
529. R. Ludwiczakówna, *Rocz. Chem.*, **22**, 138 (1948); *Chem. Abstr.*, **44**, 1519 (1950).
530. P. Rabe, *Ber.*, **55B**, 522 (1922).
531. M. Gates, B. Sugavanam, and W. L. Schreiber, *J. Amer. Chem. Soc.*, **92**, 205 (1970).
532. H. B. Trijzelaar, R. De Bode, and H. B. A. Welle, *Eur. Pat. Appl.* 35821 (1981); *Chem. Abstr.*, **96**, 69227 (1982).
533. F. Hoffmann-La Roche and Co. A.-G., *Brit. Pat.* 1 441 519 (1976); equivalent to: G. Grethe and M. Uskoković, *Ger. Offen.* 2 458 146 (1975); *Chem. Abstr.*, **83**, 114718 (1975); *U.S. Pat.* 3 953 453 (1976); *Chem. Abstr.*, **85**, 108882 (1976).
534. J. A. W. Gutzwiller and M. R. Uskoković, *U.S. Pat.* 3 864 347 (1975).
535. J. A. W. Gutzwiller and M. R. Uskoković, *U.S. Pat.* 3 929 795 (1975); *Chem. Abstr.*, **84**, 135912 (1976); *U.S. Pat.* 3 772 302 (1973); *Chem. Abstr.*, **80**, 37350; *U.S. Pat.* 3 914 235 (1975).
536. J. A. W. Gutzwiller and M. R. Uskoković, *J. Amer. Chem. Soc.*, **92**, 204 (1970).
537. J. A. W. Gutzwiller and M. R. Uskoković, *Ger. Offen.* 1 933 600 (1970); *Chem. Abstr.*, **72**, 90698 (1970).
538. Hoffmann-La Roche Inc., *Austrian Pat.* 323 337 (1975); *Chem. Abstr.*, **84**, 59231 (1976).
539. V. Prelog, P. Stern, R. Seiwerth, and S. Heimbach-Juhász, *Naturwissenschaften*, **28**, 750 (1940).
540. P. Rabe and G. Hagen, *Ber.*, **74**, 636 (1941).
541. A. Kaufmann and M. Huber, *Ber.*, **46**, 2913 (1913).
542. P. Rabe, W. Naumann, and E. Kuliga, *Justus Liebigs Ann. Chem.*, **364**, 330 (1909); *J. Chem. Soc. Abstr.*, **46**, 252 (1909).
543. G. Grethe and M. R. Uskoković, *U.S. Pat.* 3 931 192 (1976); *Chem. Abstr.*, **84**, 135914 (1976); *U.S. Pat.* 3 931 193 (1976); *Chem. Abstr.*, **84**, 135913 (1976); *U.S. Pat.* 3 907 806 (1975); 3 917 612 (1975).
544. G. Grethe and M. R. Uskoković, *U.S. Pat.* 3 823 146 (1974); 3 898 237 (1975); 3 898 238 (1975); 3 899 498 (1975).
545. E. Ochiai, H. Kataoka, T. Dodo, and M. Takahashi, *Itsuu Kenkyusho Nempo*, **1962**, (12), 11; *Chem. Abstr.*, **59**, 14040 (1963).
546. G. Grethe and M. R. Uskoković, *Ger. Offen.* 2 111 938 (1971); *Chem. Abstr.*, **76**, 25468 (1972).
547. G. Grethe, H. L. Lee, T. Mitt, and M. R. Uskoković, *J. Amer. Chem. Soc.*, **100**, 589 (1978).
548. G. R. Pettit and S. K. Gupta, *J. Chem. Soc.* (*C*), **1968**, 1208.
549. P. Rabe, *Ber.*, **40**, 3655 (1907).
550. J. Russo, M. E. Russo, R. A. Smith, and L. K. Pershing, *J. Clin. Pharmacol.*, **22**, 264 (1982).
551. R. Roussel, M. Oteyza de Guerrero, P. Spegt, and J. C. Galin, *J. Heterocycl. Chem.*, **19**, 785 (1982).

552. M. Yamazaki, K. Noda, and M. Hamana, *Chem. Pharm. Bull.*, **18**, 908 (1970).
553. A. R. E. Carey, G. Fukata, R. A. M. O'Ferrall, and M. G. Murphy, *J. Chem. Soc., Perkin Trans. II*, **1985**, 1711.
554. G. Fukata, C. O'Brien, and R. A. M. O'Ferrall, *J. Chem. Soc., Perkin Trans. II*, **1979**, 792.
555. J. E. Douglass and H. D. Fortner, *J. Heterocycl. Chem.*, **10**, 115 (1973).
556. R. Mondelli and L. Merlini, *Tetrahedron*, **22**, 3253 (1966).
557. J. D. Baty, G. Jones, and C. Moore, *J. Org. Chem.*, **34**, 3295 (1969).
558. A. M. Stock, W. E. Donahue, and E. D. Amstutz, *J. Org. Chem.*, **23**, 1840 (1958).
559. B. Golankiewicz and K. Golankiewicz, *Bull. Acad. Pol. Sci., Ser. Sci. Chem.*, **14**, 199 (1966); *Chem. Abstr.*, **65**, 10466 (1966).
560. P. L. Compagnon, F. Gasquez, O. Compagnon, and T. Kimny, *Bull. Soc. Chim. Belg.*, **91**, 931 (1982).
561. M. Hamana and H. Noda, *Chem. Pharm. Bull.*, **14**, 762 (1966).
562. J. V. Greenhill, H. Loghmani-Khouzani, and D. J. Maitland, *Tetrahedron*, **44**, 3319 (1988).
563. B. M. Baum and R. Levine, *J. Heterocycl. Chem.*, **3**, 272 (1966).
564. J. V. Hay, T. Hudlicky, and J. F. Wolfe, *J. Amer. Chem. Soc.*, **97**, 374 (1975).
565. J. V. Hay and J. F. Wolfe, *J. Amer. Chem. Soc.*, **97**, 3702 (1975).
566. M. P. Moon and J. F. Wolfe, *J. Org. Chem.*, **44**, 4081 (1979).
567. G. Coudert, G. Guillaumet, L. Lalloz, and P. Canbere, *Synthesis*, **1976**, 764.
568. E. Hayashi and T. Saito, *Yakugaku Zasshi*, **89**, 74 (1969); *Chem. Abstr.*, **70**, 96580 (1969).
569. S. Kanno, M. Shiraiwa, and H. Yamanaka, *Chem. Pharm. Bull.*, **29**, 3554 (1981).
570. J. M. Smith, Jr., *U.S. Pat.* 2 414 398 (1945); *Chem. Abstr.*, **41**, 5904 (1947).
571. M. Hamana and M. Yamazaki, *Chem. Pharm. Bull.*, **11**, 411 (1963).
572. K. Funakoshi, H. Sonoda, Y. Sonoda, and M. Hamana, *Chem. Pharm. Bull.*, **26**, 3504 (1978).
573. M. Yamazaki, K. Noda, and M. Hamana, *Chem. Pharm. Bull.*, **18**, 901 (1970).
574. M. Yamazaki, K. Noda, N. Honijo, and M. Hamana, *Chem. Pharm. Bull.*, **21**, 712 (1973).
575. T. Okamoto and H. Takayama, *Chem. Pharm. Bull.*, **11**, 514 (1963).
576. V. M. Dziomko, I. A. Krasavin, and Yu. P. Radin, *Metody Poluch. Khim. Reakt. Prep.*, **1969**, (20), 164; *Chem. Abstr.*, **76**, 113034 (1972).
577. M. Hamana, K. Funakoshi, H. Shigyo, and Y. Kuchino, *Chem. Pharm. Bull.*, **23**, 346 (1975).
578. M. Hamana and H. Noda, *Chem. Pharm. Bull.*, **18**, 26 (1970).
579. E. Fischer and H. Kuźel, *Ber.*, **16**, 163 (1883).
580. E. Fischer and H. Kuźel, *Ber.*, **16**, 164 (1883); *J. Chem. Soc. Abstr.*, **44**, 588 (1883).
581. T. Kinoshita, T. Onoue, M. Watanabe, and S. Furakawa, *Chem. Pharm. Bull.*, **28**, 795 (1980).
582. F. Kröhnke and I. Vogt, *Justus Liebigs Ann. Chem.*, **600**, 228 (1956).
583. K. Thomae, *Brit. Pat.* 1 068 698 (1967); *Chem. Abstr.*, **68**, 21858 (1968).
584. M. Hamana and H. Noda, *Chem. Pharm. Bull.*, **13**, 912 (1965).
585. M. Hamana and H. Noda, *Chem. Pharm. Bull.*, **15**, 474 (1967).
586. P. Bruni and A. Strocchi, *Ann. Chim. (Rome)*, **56**, 767 (1966); *Chem. Abstr.*, **65**, 13649 (1966).
587. H. Stetter and J. Krasselt, *J. Heterocycl. Chem.*, **14**, 573 (1977).
588. P. Canonne, G. Lemay, and R. A. Abramovitch, *Heterocycles*, **9**, 1217 (1978).
589. R. A. Abramovitch, G. Grins, R. B. Rogers, J. L. Atwood, M. D. Williams, and S. Crider, *J. Org. Chem.*, **37**, 3383 (1972).

590. R.A. Abramovitch and I. Shinkai, *J. Chem. Soc., Chem. Commun.*, **1973**, 569.
591. R. A. Abramovitch, G. Grins, R. B. Rogers and I. Shinkai, *J. Amer. Chem. Soc.*, **98**, 5671 (1976).
592. A. Eibner and K. Hofmann, *Ber.*, **37**, 3011 (1904).
593. A. Eibner, *Ber.*, **39**, 2202 (1906); *J. Chem. Soc. Abstr.*, **90**, 588 (1906).
594. S. Furukawa, T. Kinoshita, and M. Watanabe, *Yakugaku Zasshi*, **93**, 1064 (1973); *Chem. Abstr.*, **79**, 105189 (1973).
595. I. F. Tishchenkova, L. E. Kholodor, and V. G. Yashunskii, *Khim.-Farm., Zh.*, **2**, 7 (1968); *Chem. Abstr.*, **70**, 87522 (1969); *Khim. Geterotsikl. Soedin.* **1971**, 87; *Chem. Heterocycl. Compd. (Engl. Transl.)*, **1971**, 82; *Chem. Abstr.*, **75**, 35656 (1971).
596. H. Yamanaka, M. Komatsu, S. Ogawa, and S. Konno, *Chem. Pharm. Bull.*, **27**, 806 (1979).
597. L. P. Zalukaievs and N. P. L'vova, *Zh. Obshch. Khim.*, **33**, 3056 (1963); *J. Gen. Chem. USSR (Engl. Transl.)* **33**, 2981 (1963); *Chem. Abstr.*, **60**, 4107 (1964).
598. L. P. Zalukaievs and N. P. L'vova, *Khim. Geterotsikl. Soedin.*, **1965**, 872; *Chem. Abstr.*, **64**, 12639 (1966).
599. V. Carelli, M. Cardellini, and F. Liberatore, *Ann. Chim. (Rome)*, **49**, 709 (1959); *Chem. Abstr.*, **53**, 21954 (1959).
600. W. Borsche and R. Manteuffel, *Justus Liebigs Ann. Chem.*, **526**, 22 (1936); *Chem. Abstr.*, **31**, 405 (1937).
601. W. Borsche, W. Doeller, and M. Wagner-Roemmich, *Ber.*, **76**, 1099 (1943).
602. F. Al-Tai, G. Y. Sarkis, and F. A. Al-Najjar, *Bull. Coll. Sci., Univ. Baghdad*, **10**, 93 (1967); *Chem. Abstr.*, **72**, 43386 (1970).
603. N. J. Leonard and J. H. Boyer, *J. Amer. Chem. Soc.*, **72**, 2980 (1950).
604. O. Compagnon and P. L. Compagnon, *Bull. Soc. Chim. Fr.*, **50**, 2596 (1974).
605. M. Regitz and A. Liedhegener, *Chem. Ber.*, **99**, 2918 (1966).
606. F. H. Case and A. A. Schilt, *J. Heterocycl. Chem.*, **16**, 1135 (1979).
607. M. Augustin and W. Doelling, *J. Prakt. Chem.*, **324**, 322 (1982).
608. M. Augustin and W. Doelling, *Ger. Pat. (East)* 158103 (1982); *Chem. Abstr.*, **99**, 22330 (1983).
609. A. Eibner and O. Lange, *Justus Liebigs Ann. Chem.*, **315**, 303 (1901).
610. R. Kulm, *Naturwissenschaften*, **20**, 618 (1932).
611. J. Ogilvie, *U.S. Pat.*, 1 963 374 (1934).
612. Y. Yamazaki, T. Ishii, and K. Shiro, *Yûki Gôsei Kagaku Kyokaishi*, **17**, 229 (1959); *Chem. Abstr.*, **53**, 14105 (1959).
613. Colour Index, Vol. 4, Society of Dyers and Colourists, Bradford, 1971.
614. R. Kulm and F. Bär, *Justus Liebigs Ann. Chem.*, **516**, 155 (1935); *Chem. Abstr.*, **29**, 3677 (1935).
615. E. Jacobsen and K. L. Reimer, *Ber.*, **16**, 1082 (1883).
616. D. G. Manly, A. Richardson, A. M. Stock, C. H. Tilford, and E. D. Amstutz, *J. Org. Chem.*, **23**, 373 (1958).
617. B. K. Manukian, P. Niklaus, and H. Ehrsam, *Helv. Chim. Acta*, **52**, 1259 (1969).
618. T. Kitao, M. Koga, T. Harada, S. Nagahama, and K. Shimada, *Japan Kokai* 76 102,023 (1976); *Chem. Abstr.*, **86**, 44766 (1977).
619. T. Kitao, M. Koga, T. Harada, S. Nagahama, and K. Shimada, *Japan Kokai* 76 102,024 (1976); *Chem. Abstr.*, **86**, 44765 (1977).
620. T. Kitao, M. Matsuoka, and H. Oda, *Japan Kokai* 77 10,329 (1977); *Chem. Abstr.*, **86**, 141629 (1977).
621. T. Kitao, M. Matsuoka, and H. Oda, *Japan Kokai* 77 10,330 (1977); *Chem. Abstr.*, **86**, 141628 (1977).

622. T. Kitao, M. Matsuoka, and H. Oda, *Japan Kokai* 77 10,341 (1977); *Chem. Abstr.*, **86**, 141631 (1977).
623. T. Kitao, M. Matsuoka, and H. Oda, *Japan Kokai* 77 10,342 (1977); *Chem. Abstr.*, **86**, 141630 (1977).
624. W. I. Awad and O. M. Aly, *J. Org. Chem.*, **25**, 1872 (1960).
625. H. Kazoka and I. Meirovics, *Latv. PSR Zinat. Akad. Vestis, Kim. Ser.*, **1984**, 496; *Chem. Abstr.*, **102**, 62054 (1985).
626. A. Eibner, *Ber.*, **37**, 3605 (1904); *J. Chem. Soc. Abstr.*, **86**, 1049 (1904).
627. A. Eibner and M. Löbering, *Ber.*, **39**, 2215 (1906); *J. Chem. Soc. Abstr.*, **90**, 606 (1906).
628. K. Shimada, T. Harada, and M. Koga, *Ger. Offen.* 2 706 872 (1977); *Chem. Abstr.*, **90**, 205785 (1979).
629. S. Imahori, M. Kaneko, and H. Ono, *Japan Kokai* 76 50,330 (1976); *Chem. Abstr.*, **86**, 44764 (1977).
630. J. W. Huffman and J. H. Cecil, *J. Org. Chem.*, **34**, 2183 (1969).
631. N. N. Goldberg and R. Levine, *J. Amer. Chem. Soc.*, **74**, 5217 (1952).
632. N. N. Goldberg, L. B. Barkley, and R. Levine, *J. Amer. Chem. Soc.*, **73**, 4301 (1951).
633. J. F. Wolfe and T. P. Murray, *J. Org. Chem.*, **36**, 354 (1971).
634. W. L. Ruigh, *U.S. Pat.* 2 351 069 (1944); *Chem. Abstr.*, **38**, 5370 (1944).
635. M. J. Weiss and C. R. Hauser, *J. Amer. Chem. Soc.*, **71**, 2023 (1949).
636. R. M. Acheson and G. Procter, *J. Chem. Soc., Perkin Trans I*, **1979**, 2171.
637. H. Gnichtel and B. Möller, *Liebigs Ann. Chem.*, **1981**, 1751.
638. T. Kato, H. Yamanaka, T. Sakamoto, and T. Shiraishi, *Chem. Pharm. Bull.*, **22**, 1206 (1974).
639. W. Borsche and F. Sinn, *Justus Liebigs Ann. Chem.*, **538**, 283 (1939).
640. J. F. O'Leary, D. E. Leary and I. H. Slater, *Proc. Soc. Exp. Biol. Med.*, **76**, 738 (1951); *Chem. Abstr.*, **45**, 7244e (1951).
641. M. Iwao and T. Kuraishi, *J. Heterocycl. Chem.*, **15**, 1425 (1978).
642. A. Pollak, B. Stanovnik, M. Tišler and J. Venetič-Fortuna, *Monatsh. Chem.*, **106**, 473 (1975).
643. F. S. Babichev and Yu. M. Volovenko, *Ukr. Khim. Zh. (Russ. Ed.)*, **43**, 41 (1977); *Chem. Abstr.*, **86**, 171330 (1977).
644. C. A. Blank, *Diss. Abstr.*, **14**, 2190 (1954); *Chem. Abstr.*, **49**, 3820 (1955).
645. F. S. Babichev, Yu. M. Volovenko, and A. A. Oleinik, *Khim. Geterotsikl. Soedin.*, **1977**, 1515; *Chem. Heterocycl. Compd. (Engl. Transl.)*, **1977**, 1211; *Chem. Abstr.*, **88**, 62276 (1978).
646. E.-S. A. Ibrahim, I. Chaaban and S. M. El-Khawass, *Pharmazie*, **32**, 155 (1977); *Chem. Abstr.*, **87**, 68114 (1977).
647. R. Beugelmans and G. Roussi, *Tetrahedron, Suppl.*, **1981**, 393.
648. R. F. Meyer and C. D. Stratton, *U.S. Pat.* 3 794 650 (1974); *Chem. Abstr.*, **80**, 108404 (1974).
649. R. F. Meyer, C. D. Stratton, S. G. Hastings, and R. M. Corey, *J. Med. Chem.*, **16**, 1113 (1973).
650. F. S. Babichev and Yu. M. Volovenko, *Chem. Heterocycl. Compd. (Engl. Transl.)*, **1975**, 881; *Khim. Geterotsikl. Soedin.*, **1975**, 1005; *Chem. Abstr.*, **83**, 193051 (1975).
651. Yu. M. Volovenko, F. S. Babichev, and Yu. M. Pustovit, *USSR Pat.* 1 027 166 (1983); *Chem. Abstr.*, **99**, 212524 (1983).
652. P. L. Compagnon, M. Miocque, and J. A. Gautier, *Bull. Soc. Chim. Fr.*, **44**, 4127 (1968).
653. O. Compagnon and P. L. Compagnon, *Bull. Soc. Chim. Fr.*, **49**, 3385 (1973).

654. R. Roussel, M. O. de Guerrero, and J. C. Galin, *Macromolecules*, **19**, 291 (1986).
655. P. L. Compagnon, M. Miocque, and J. A. Gautier, *Bull. Soc. Chim. Fr.*, **44**, 4132 (1968).
656. P. L. Compagnon, M. Miocque, and J. A. Gautier, *Bull. Soc. Chim. Fr.*, **44**, 4136 (1968).
657. T. A. Crabb and J. S. Mitchell, *J. Chem. Soc., Perkin Trans. II*, **1977**, 1592.
658. P. Bruni and G. Guerra, *Ann. Chim. (Rome)*, **57**, 688 (1967); *Chem. Abstr.*, **67**, 116779 (1967).
659. H. Noda, T. Yamamori, M. Yoshida, and M. Hamana, *Heterocycles*, **4**, 453 (1976).
660. M. Nakanishi and M. Yatabe, *Japan Kokai* 75 24,278; *Chem. Abstr.*, **86**, 35758 (1977).
661. J. F. Wolfe, D. E. Portlock, and D. J. Feuerbach, *J. Org. Chem.*, **39**, 2006 (1974).
662. E. H. Sund and W. D. Lowe, *J. Chem. Eng. Data*, **28**, 137 (1983).
663. Ciba Ltd., *Brit. Pat.* 1 102 428 (1968); *Neth. Appl.* 6 513 784 (1966); *Chem. Abstr.*, **65**, 15342 (1966).
664. F. N. Stepanov and S. L. Davydova, *Zh. Obshch. Khim.*, **28**, 891 (1958); *J. Gen. Chem.*, USSR (Eng. Transl.) **28**, 864 (1958); *Chem. Abstr.*, **52**, 17243 (1958).
665. F. W. Bergstrom and A. Moffat, *J. Amer. Chem. Soc.*, **59**, 1494 (1937).
666. P. E. Wright and W. E. McEwen, *J. Amer. Chem. Soc.*, **76**, 4540 (1954).
667. C. R. Hauser and J. G. Murray, *J. Amer. Chem. Soc.*, **77**, 3858 (1955).
668. J. Kacens and O. Neilands, *Latv. PSR Zinat. Akad. Vestis, Kim. Ser.*, **1970**, 606; *Chem. Abstr.*, **74**, 53473 (1971).
669. American Cyanamid Co., *Brit. Pat.* 611991 (1948); *Chem. Abstr.*, **43**, 3045 (1949).
670. J. M. Smith, Jr., *U.S. Pat.* 2 442 865 (1948); *Chem. Abstr.*, **42**, 7796 (1948).
671. I. P. Kupchevskaya and V. P. Khilya, *Dopov. Akad. Nauk Ukr. RSR, Ser. B*, **1981**, 46; *Chem. Abstr.*, **96**, 6615 (1982).
672. V. G. Pivovarenko, V. P. Khilya, and F. S. Babichev, *Dopov. Akad. Nauk Ukr. RSR, Ser. B*, **1985**, 56; *Chem. Abstr.*, **103**, 141792 (1985).
673. V. G. Pivovarenko and V. P. Khilya, *Dopov. Akad. Nauk Ukr. RSR, Ser. B*, **1985**, 44; *Chem. Abstr.*, **104**, 129748 (1986).
674. L. P. Zalukaievs and E. V. Vanag, *Zh. Obshch. Khim.*, **25**, 2639 (1956); *Chem. Abstr.*, **51**, 5073 (1957).
675. W. Wislicenus, *Ber.*, **30**, 1479 (1897).
676. W. Wislicenus and E. Kleisinger, *Ber.*, **42**, 1140 (1909).
677. W. Borsche and O. Vorbach, *Justus Liebigs Ann. Chem.*, **537**, 22 (1938).
678. Y. Ishiguro, K. Funakoshi, S. Saeki, M. Hamana, and I. Ueda, *Heterocycles*, **14**, 179 (1980).
679. K. Golankiewicz, *Bull. Acad. Pol. Sci., Ser. Sci. Chim.*, **9**, 547 (1961); *Chem. Abstr.*, **60**, 4107 (1964).
680. M. M. Yousif, S. Saeki, and M. Hamana, *Chem. Pharm. Bull.*, **30**, 2326 (1982).
681. F. Kehrer, P. Niklaus, and B. K. Mannukian, *Helv. Chim. Acta*, **50**, 2200 (1967).
682. L. P. Zalukaievs, *Zh. Obshch. Khim.*, **29**, 1637 (1959); *Chem. Abstr.*, **54**, 8835 (1960).
683. E. Jacobsen and C. L. Reimer, *Ber.*, **16**, 2602 (1883); *J. Chem. Soc. Abstr.*, **46**, 335 (1884).
684. W. Koenigs and J. U. Nef, *Ber.*, **19**, 2427 (1886).
685. J. Kacens, A. Cebure, and O. Nielands, *Latv. PSR Zinat. Akad. Vestis, Kim. Ser.*, **1973**, 100; *Chem. Abstr.*, **78**, 124420 (1973).
686. Mitsui Toatsu Chemicals Inc., *Japan Kokai Tokkyo Koho* 83 59,971 (1983); *Chem. Abstr.*, **99**, 175605 (1983).
687. M. Groll, V. Hederich, and H. S. Bien, *Ger. Offen.* 2 225 960 (1973); *Chem. Abstr.*, **80**, 122407 (1974).

688. M. Groll, V. Hederich, and H. S. Bien, U.S. Pat. 3 932 419 (1976).
689. K. Watanabe and H. Harada, Japan Kokai 76 92,371 (1976); Chem. Abstr., 86, 74331 (1977).
690. Mitsubishi Chemical Industries Ltd., Japan Kokai Tokkyo Koho 84 128,373 (1984); Chem. Abstr., 102, 24505 (1984).
691. L. Capuano and F. Jamaigne, Chem. Ber., 96, 798 (1963).
692. S. Blechert, R. Gericke, and E. Winterfeldt, Chem. Ber., 106, 355 (1973).
693. S. Blechert, R. Gericke, and E. Winterfeldt, Ger. Offen. 2 301 401 (1974); Chem. Abstr., 81, 120598 (1974).
694. W. J. Donnelly and M. F. Grundon, J. Chem. Soc., Perkin Trans. I, 1972, 2116.
695. R. M. Bowman, G. A. Gray, and M. F. Grundon, J. Chem. Soc., Perkin Trans. I, 1973, 1051.
696. M. M. Baradarani and J. A. Joule, J. Chem. Soc., Perkin Trans. I, 1980, 72.
697. F. Kröhnke and I. Vogt, Chem. Ber., 90, 2227 (1957).
698. D. Monti, P. Gramatica, G. Speranza, and P. Manitto, Tetrahedron Lett., 24, 417 (1983).
699. D. Monti, P. Gramatica, and P. Manitto, Farmaco, Ed. Sci., 36, 412 (1981).
700. D. R. Carver, T. D. Greenwood, J. S. Hubbard, A. P. Komin, Y. P. Sachdeva, and J. F. Wolfe, J. Org. Chem., 48, 1180 (1983).
701. H. Yamanaka, H. Abe, and T. Sakamoto, Chem. Pharm. Bull., 25, 3334 (1977).
702. Z. S. Ariyan and H. Suschitzky, J. Chem. Soc., 1961, 2242.
703. W. Borsche and L. Butschli, Justus Liebigs Ann. Chem., 529, 266 (1937).
704. N. S. Lee, C. K. Lim, T. S. Cho, C. H. Won, D. W. Moon, I. S. Park, M. G. Kim, Y. S. Min, J. S. Chung, Y. B. Cho, and Y. S. No, Yakhak Hoe Chi, 18, 59 (1974); Chem. Abstr., 82, 139878 (1975).
705. V. P. Khilya, V. G. Pivovarenko, A. S. Ogorodniichuk, and F. S. Babichev, Dopov. Akad. Nauk Ukr. RSR, Ser. B, 1985, 55; Chem. Abstr., 103, 105233 (1985).
706. I. P. Kupchevskaya and V. P. Khilya, Dopov. Akad. Nauk Ukr. RSR, Ser. B, 1978, 234; Chem. Abstr., 89, 146734 (1978).
707. I. P. Kupchevskaya and V. P. Khilya, Dopov. Akad. Nauk Ukr. RSR, Ser. B, 1979, 117; Chem. Abstr., 90, 186749 (1979).
708. B. R. Brown, D. L. Hammick, and B. H. Thewlis, J. Chem. Soc., 1951, 1145.
709. I. Yanagisawa, M. Ohta, and T. Takagi, Eur. Pat. Appl. 86647 (1983); Chem. Abstr., 100, 6543 (1984).
710. J. A. Gautier, M. Miocque, and N. M. Hung, Bull. Soc. Chim. Fr., 1961, 2092.
711. W. J. Irwin and D. G. Wibberley, J. Chem. Soc., Perkin Trans. I, 1974, 250.
712. T. Melton, J. Taylor, and D. G. Wibberley, J. Chem. Soc., Chem. Commun., 1965, 151.
713. J. Hurst, T. Melton, and D. G. Wibberley, J. Chem. Soc., 1965, 2948.
714. Z. S. Ariyan, B. Mooney, and H. I. Stonehill, J. Chem. Soc., 1962, 2239.
715. S. M. McElvain and H. G. Johnson, J. Amer. Chem. Soc., 63, 2213 (1941).
716. F. Bell, J. Chem. Soc., 1953, 348.
717. J. Klosa, Arch. Pharm. Ber. Dtsch. Pharm. Ges., 289, 177 (1956).
718. R. Palland, T. Chaudron, B. Marcot, and P. Delest, Chim. Ind. (Paris), 89, 283 (1963); Chem. Abstr., 59, 6363 (1963).
719. C. E. Kwartler and H. G. Lindwall, J. Amer. Chem. Soc., 59, 524 (1937).
720. Z. S. Ariyan and B. Mooney, J. Chem. Soc., 1962, 1519.
721. S. V. Tsukerman, Kuo-Seng Ch'ang, and V. F. Lavrushin, Zh. Obshch. Khim., 34, 832 (1964); J. Gen. Chem. USSR (Engl. Transl.), 34, 827 (1964); Chem. Abstr., 60, 15826 (1964).
722. A. Corvaisier, Bull. Soc. Chim. Fr., 1962, 528.

723. V. M. Zubarovskii and Yu. L. Briks, *Khim. Geterotsikl. Soedin.*, **1982**, 644; *Chem. Heterocycl. Compd. (Engl. Transl.)*, **1982**, 485; *Chem. Abstr.*, **97**, 72281 (1982).
724. V. M. Zubarovskii and Yu. L. Briks, *Ukr. Khim. Zh. (Russ. Ed.)*, **48**, 761 (1982); *Chem. Abstr.*, **97**, 144812 (1982).
725. J. Ebersberg, R. Haller, and K. W. Merz, *Naturwissenschaften*, **52**, 514 (1965).
726. R. Haller and J. Ebersberg, *Arch. Pharm. Ber. Dtsch. Pharm. Ges.*, **302**, 932 (1969); *Chem. Abstr.*, **72**, 90215 (1970).
727. Kalle A.-G., *Brit. Pat.* 943266 (1963); *Chem. Abstr.*, **60**, 14696 (1964).
728. F. Piozzi and C. Fuganti, *Ann. Chim. (Rome)*, **56**, 1248 (1966); *Chem. Abstr.*, **67**, 73473 (1967).
729. H. Gilman and L. F. Cason, *J. Amer. Chem. Soc.*, **72**, 3469 (1950).
730. F. Zymalkowski, *Arch. Pharm. Ber. Dtsch. Pharm. Ges.*, **288**, 162 (1955).
731. S. O. Korotkov, V. D. Orlov and V. F. Lavrushin, *Vestn. Khar'k. Univ.*, **1972**, (84), 72; *Chem. Abstr.*, **78**, 147695 (1973).
732. G. J. Bird, G. J. Farquharson, and K. G. Watson, *Eur. Pat. Appl.* 124992 (1984); *Chem. Abstr.*, **103**, 5933 (1985).
733. L. V. Gyul'budagyan, N. A. Margaryan, and V. G. Durgaryan, *Khim. Geterotsikl. Soedin.*, **1971**, 1681; *Chem. Heterocycl. Compd. (Engl. Transl.)*, **1971**, 1562; *Chem. Abstr.*, **76**, 153534 (1972).
734. L. V. Gyul'budagyan and V. G. Durgaryan, *Khim. Geterotsikl. Soedin.*, **1972**, 534; *Chem. Heterocycl. Compd. (Engl. Transl.)*, **1972**, 487; *Chem. Abstr.*, **77**, 61851 (1972).
735. L. V. Gyul'budagyan and G. A. Gevorkyan, *Uch. Zap., Erevan. Gos. Univ.*, **1969**, (2), 70; *Chem. Abstr.*, **73**, 25272 (1970).
736. L. V. Gyul'budagyan and V. G. Durgaryan, *Arm. Khim. Zh.*, **24**, 937 (1971); *Chem. Abstr.*, **77**, 19553 (1972).
737. L. V. Gyul'budagyan, Van Ngok Quong, and V. G. Durgaryan, *Arm. Khim. Zh.*, **29**, 629 (1976); *Chem. Abstr.*, **86**, 55262 (1977).
738. L. V. Gyul'budagyan, Van Ngoc Quong, and R. S. Asriyan, *Arm. Khim. Zh.*, **30**, 493 (1977); *Chem. Abstr.*, **87**, 167919 (1977).
739. L. V. Gyul'budagyan and V. G. Durgaryan, *Khim. Geterotsikl. Soedin.*, **1973**, 836; *Chem. Heterocycl. Compd. (Engl. Transl.)*, **1973**, 769; **1973**, 836; *Chem. Abstr.*, **79**, 105107 (1973).
740. L. V. Gyul'budagyan and I. L. Aleksanyan, *Arm. Khim. Zh.*, **36**, 540 (1983); *Chem. Abstr.*, **100**, 34441 (1984).
741. K. K. Hsu, S. W. Sun, and Y. H. Chen, *J. Chin. Chem. Soc. (Taipei)*, **29**, 29 (1982); *Chem. Abstr.*, **96**, 181118 (1982).
742. W. Zankowska-Jasińska and A. Kolasa, *Rocz. Chem.*, **50**, 625 (1976).
743. T. V. Smirnova and V. N. Marshalkin, *Deposited Doc.*, VINITI, 5134-81 (1981); *Chem. Abstr.*, **98**, 88936 (1983).
744. M. Keil, R. Becker, N. Goetz, D. Jahn, W. Speigler, and B. Wuerzer, *Eur. Pat. Appl.* 107156 (1984); *Chem. Abstr.*, **101**, 191714 (1984).
745. P. V. Thakore and K. K. Trivedi, *J. Indian Chem. Soc.*, **54**, 1204 (1977).
746. K. W. Merz and R. Haller, *Pharm. Acta Helv.*, **38**, 442 (1963); *Chem. Abstr.*, **60**, 15824 (1964).
747. H. Meyer, F. Bossert, W. Vater, and K. Stoepel, *Ger. Offen.* 2 117 571 (1972); *Chem. Abstr.*, **78**, 16038 (1973).
748. D. W. Ockenden and K. Schofield, *J. Chem. Soc.*, **1953**, 1915.
749. N. S. Kozlov, L. K. Slobodchikova, and G. S. Shmanai, *Dokl. Akad. Nauk SSSR*, **187**, 803 (1969).
750. G. R. Clemo and E. Hoggarth, *J. Chem. Soc.*, **1939**, 1241.
751. C. E. Kaslow and J. M. Schlatter, *J. Amer. Chem. Soc.*, **77**, 1054 (1955).

752. J. A. Gainor and S. M. Weinreb, *J. Org. Chem.*, **46**, 4317 (1981).
753. J. A. Gainor and S. M. Weinreb, *J. Org. Chem.*, **47**, 2833 (1982).
754. B. P. Lugovkin, *Zh. Obshch. Khim.*, **28**, 1007 (1958); *J. Gen. Chem. USSR (Engl. Transl.)*, **28**, 980 (1958); *Chem. Abstr.*, **52**, 17273 (1958).
755. H. Narita, Y. Konishi, J. Nitta, S. Misumi, and I. Kitayama, *Brit. Pat.* 2 130 580 (1984); *Chem. Abstr.*, **101**, 171101 (1984).
756. P. Ruggli and E. Girod, *Helv. Chim. Acta*, **27**, 1464 (1944).
757. M. Von Strandtmann, M. P. Cohen, C. Puchalski, and J. Shavel, Jr., *J. Org. Chem.*, **33**, 4306 (1968).
758. P. Ruggli and F. Brandt, *Helv. Chim. Acta*, **27**, 274 (1944).
759. J. Krepelka, D. Vlckova, and J. Benes, *Czech. Pat.* 218532 (1985); *Chem. Abstr.*, **104**, 19721 (1986).
760. S. V. Tsukerman, V. M. Nikitchenko, V. P. Maslennikova, V. E. Bondarenko, and V. F. Lavrushin, *Khim. Geterotsikl. Soedin.*, **1968**, 1093; *Chem. Heterocycl. Compd. (Engl. Transl.)*, **1968**, 794; *Chem. Abstr.*, **70**, 87653 (1969).
761. S. V. Tsukerman, L. N. Thiem, V. M. Nikitchenko and V. F. Lavrushin, *Khim. Geterotsikl. Soedin.*, **1968**, 980; *Chem. Heterocycl. Compd. (Engl. Transl.)*, **1968**, 708; *Chem. Abstr.*, **70**, 57727 (1969).
762. G. R. Clemo and R. Howe, *J. Chem. Soc.*, **1955**, 3552.
763. A. G. Anderson, *Eur. Pat. Appl.* 39025 (1981); *Chem. Abstr.*, **96**, 77549 (1982); *U.S. Pat.* 4 268 667 (1981); *Chem. Abstr.*, **95**, 33394 (1981).
764. N. S. Kozlov, G. S. Shmanai, and K. N. Gusak, *Dokl. Akad. Nauk BSSR*, **29**, 141 (1985); *Chem. Abstr.*, **102**, 220716 (1985).
765. J. Market, H. Hagen, R.-D. Kohler and B. Würzer, *Ger. Offen.* 3 229 175 (1984); *Chem. Abstr.*, **101**, 23360 (1984).
766. S. V. Tsukerman, V. D. Orlov, and V. F. Lavrushin, *Khim. Str., Svoistva Reakt. Org. Soedin.*, **1969**, 63; *Chem. Abstr.*, **72**, 99698 (1970).
767. S. V. Tsukerman, V. P. Izvekov and V. F. Lavrushin, *Khim. Geterotsikl. Soedin.*, **1968**, 823; *Chem. Abstr.*, **71**, 2798 (1969).
768. S. V. Tsukerman, V. D. Orlov, and V. F. Lavrushin, *Khim. Geterotsikl. Soedin.*, **1969**, 67; *Chem. Abstr.*, **70**, 114400 (1969).
769. S. V. Tsukerman, V. P. Izvekov, Yu. S. Rozym, and V. F. Lavrushin, *Khim. Geterotsikl. Soedin.*, **1968**, 1011; *Chem. Abstr.*, **70**, 72307 (1969).
770. S. V. Tsukerman, K.-S. Ch'ang, and V. F. Lavrushin, *Khim. Geterotsikl. Soedin., Akad. Nauk. Latv. SSSR*, **1965**, 386; *Chem. Abstr.*, **63**, 14650 (1965).
771. S. V. Tsukerman, C. K'uo-Sheng, Yu. S. Rozym, and V. F. Lavrushin, *Zh. Prikl. Spektrosk.*, **5**, 489 (1966); *Chem. Abstr.*, **66**, 60444 (1967).
772. S. V. Tsukerman, K.-S. Ch'ang, and V. F. Lavrushin, *Zh. Fiz. Khim.*, **40**, 160 (1966); *Russ. J. Phys. Chem. (Engl. Transl.)*, **40**, 80 (1966); *Chem. Abstr.*, **64**, 14070 (1966).
773. K.-K. Hsu and S.-F. Chang, *T'ai-wan K'o Hsueh*, **29**, 51 (1975); *Chem. Abstr.*, **84**, 164725 (1976).
774. K.-K. Hsu, S.-F. Chang, and S.-W. Sun, *T'ai-wan K'o Hsueh*, **34**, 104 (1980); *Chem. Abstr.*, **94**, 208655 (1981).
775. R. Kuhn and H. R. Hensel, *Chem. Ber.*, **86**, 1333 (1953).
776. K.-K. Hsu and S.-F. Chang, *T'ai-wan K'o Hsueh*, **31**, 130 (1977); *Chem. Abstr.*, **88**, 169921 (1978).
777. K.-K. Hsu and T. S. Wu, *J. Chin. Chem. Soc. (Taipei)*, **26**, 17 (1979); *Chem. Abstr.*, **91**, 91475 (1979).
778. L. Szucs, J. Durinda, L. Krasnec, and J. Heger, *Chem. Zvesti*, **20**, 817 (1966); *Chem. Abstr.*, **67**, 90619 (1967).
779. J. Boichard, J. P. Monin, and J. Tironflet, *Bull. Soc. Chim. Fr.*, **1963**, 851.
780. F. Zymalkowski and P. Tinapp, *Justus Liebigs Ann. Chem.*, **699**, 98 (1966).

781. G. Kobayashi, Y. Matsuda, R. Natsuki, and H. Yamaguchi, *Yakugaku Zasshi*, **91**, 934 (1971); *Chem. Abstr.*, **75**, 151643 (1971).
782. T. Shono, S. Kodama, and R. Oda, *Kogyo Kagaku Zasshi*, **58**, 917 (1955); *Chem. Abstr.*, **50**, 13017 (1956).
783. R. Oda, K. Teramura, S. Tanimoto, M. Nomura, H. Suda, and K. Matsuda, *Bull. Inst. Chem. Res., Kyoto Univ.*, **33**, 117 (1955); *Chem. Abstr.*, **51**, 11355 (1957).
784. V. Boekelheide and G. Marinetti, *J. Amer. Chem. Soc.*, **73**, 4015 (1951).
785. M. J. Weiss and C. R. Hauser, *J. Amer. Chem. Soc.*, **71**, 2026 (1949).
786. L. P. Zalukaievs and E. Vanags, *Zh. Obshch. Khim.*, **30**, 145 (1960); *Chem. Abstr.*, **54**, 24746 (1960).
787. J. A. Gautier, M. Miocque, and N. M. Hùng, *Compt. Rend.*, **253**, 128 (1961).
788. E. M. Kaiser, G. J. Barthing, W. R. Thomas, S. V. Nichols, and D. R. Nash, *J. Org. Chem.*, **38**, 71 (1973).
789. F. Kröhnke and I. Vogt, *Justus Liebigs Ann. Chem.*, **600**, 211 (1956).
790. E. A. Chairka, G. I. Matyushcheva, and L. M. Yagupol'skii, *Zh. Org. Khim.*, **18**, 186 (1982); *J. Org. Chem. USSR (Engl. Transl.)*, **18**, 165 (1982); *Chem. Abstr.*, **96**, 201246 (1982).
791. G. H. Alt, *J. Org. Chem.*, **31**, 2384 (1966).
792. E. Vongerichten and W. Rotta, *Ber.*, **44**, 1419 (1911).
793. N. A. Damir and N. N. Sveshuikov, *Zh. Vses. Khim. O-va.*, **10**, 592 (1965); *Chem. Abstr.*, **64**, 3730 (1966).
794. H. Tomisawa, T. Tanbara, H. Kato, H. Hongo, and R. Fujita, *Heterocycles*, **15**, 277 (1981).
795. E. D. Sych and L. T. Gorb, *Khim. Geterotsikl. Soedin.*, **1980**, 312; *Chem. Heterocycl. Compd. (Engl. Transl.)*, **1980**, 227; *Chem. Abstr.*, **93**, 114378 (1980).
796. E. B. Knott, *J. Chem. Soc.*, **1955**, 916.
797. J. D. Mee, *J. Amer. Chem. Soc.*, **96**, 4712 (1974).
798. J. D. Mee, *U.S. Pat.* 4 026 884 (1977); *Chem. Abstr.*, **87**, 69754 (1977).
799. J. D. Mee, *U.S. Pat.* 4 025 349 (1977); *Chem. Abstr.*, **87**, 119270 (1977).
800. A. K. Sheinkman, T. S. Chinilenko and T. M. Baranova, *Khim. Geterotsikl. Soedin.*, **1982**, 849; *Chem. Heterocycl. Compd. (Engl. Transl.)*, **1982**, 648; *Chem. Abstr.*, **97**, 127471 (1982).
801. A. K. Sheinkman, T. S. Chinilenko and T. M. Baranova, *Khim. Geterotsikl. Soedin.*, **1983**, 961; *Chem. Heterocycl. Compd. (Engl. Transl.)*, **1983**, 733; *Chem. Abstr.*, **99**, 175559 (1983).
802. B. N. Dashkevich, I. V. Smolanka, and Yu. Yu. Tsmur, *Nauchn. Zap. Uzhgorod. Gos. Univ.*, **22**, 81 (1957); *Chem. Abstr.*, **54**, 14100 (1960).
803. Yu. L. Slominskii and I. D. Radchenko, *Ukr. Zhim. Zh. (Russ. Ed.)*, **40**, 546 (1974); *Chem. Abstr.*, **81**, 91410 (1974).
804. H. Zenno, *Yakugaku Zasshi*, **72**, 1633 (1952); *Chem. Abstr.*, **47**, 4611 (1953).
805. B. N. Baijal and B. Hirsch, *Ger. Pat. (East)*, 68583 (1969); *Chem. Abstr.*, **72**, 134136 (1970).
806. A. I. Kiprianov and G. G. Dyadyusha, *Zh. Obshch. Khim.*, **29**, 1708 (1959); *Chem. Abstr.*, **54**, 8817 (1960).
807. T. Takahashi and K. Satake, *Yakugaku Zasshi*, **75**, 20 (1955); *Chem. Abstr.*, **50**, 1004 (1956).
808. A. I. Tolmachev, *Zh. Obshch. Khim.*, **32**, 3745 (1962); *Chem. Abstr.*, **58**, 12530 (1963).
809. A. Ya. Il'chenko, L. I. Trushanina, M. M. Kul'chitskii, V. I. Krokhtyak, and L. M. Yagupol'skii, *Zh. Org. Khim.*, **20**, 434 (1984); *J. Org. Chem. USSR (Engl. Transl.)*, **20**, 390 (1984); *Chem. Abstr.*, **100**, 211626 (1984).
810. G. Reichardt, *Justus Liebigs Ann. Chem.*, **715**, 74 (1968); *Tetrahedron Lett.*, **1965**, 429.

811. D. W. Heseltine and L. L. Lincon, *U.S. Pat.* 3 125 448 (1964); *Chem. Abstr.,* **61**, 5829 (1964).
812. A. van Dormael and T. Ghys, *Bull. Soc. Chim. Belg.,* **57**, 24 (1948).
813. L. G. S. Brooker and F. L. White, *U.S. Pat.* 2 341 357 (1944); *Chem. Abstr.,* **38**, 4453 (1944).
814. A. Ya. Il'chenko, L. I. Trushanina and L. M. Yagupol'skii, *Zh. Org. Khim.,* **8**, 1779 (1972); *J. Org. Chem. USSR (Engl. Transl.),* **20**, 1768 (1972); *Chem. Abstr.,* **78**, 45023 (1973).
815. A. S. Bailey, J. H. Ellis, J. M. Harvey, A. N. Hilton, and J. M. Peach, *J. Chem. Soc., Perkin Trans. I,* **1983**, 795.
816. M. Coenen, *Ger. Pat.* 823, 599 (1951); *Chem. Abstr.,* **49**, 2517 (1955).
817. J. D. Kendall and D. J. Fry, *Brit. Pat.* 577 259 (1946); *Chem. Abstr.,* **42**, 5362 (1948).
818. B. R. Baker and F. J. McEvoy, *J. Org. Chem.,* **20**, 118 (1955).
819. T. V. Stupnikova, Z. M. Skorobogatova, A. V. Sheinkman, and V. A. Golubev, *Dopov. Akad. Nauk Ukr. RSR, Ser. B,* **1978**, 144; *Chem. Abstr.,* **89**, 24105 (1978).
820. M. Farcasan, R. Chira, Gh. Ciurdara, and V. I. Denes, *Rev. Roum. Chim.,* **18**, 2103 (1973); *Chem. Abstr.,* **80**, 95812 (1974).
821. J. R. Wade, *Eur. Pat. Appl.* 135348 (1985); *Chem. Abstr.,* **103**, 45823 (1985).
822. J. R. Wade, R. M. Potts, and J. M. Pratt, *Eur. Pat. Appl.* 125140 (1984); *Chem. Abstr.,* **102**, 70278 (1985); *Eur. Pat. Appl.* 125875 (1984); *Chem. Abstr.,* **102**, 140898 (1985).
823. V. I. Dénes and G. Ciurdaru, *J. Chem. Soc., Chem. Commun.,* **1970**, 510.
824. J. D. Kendall and G. F. Duffin, *Brit. Pat.* 835275 (1960); *Chem. Abstr.,* **55**, 187 (1961).
825. B. A. Lea and R. W. Burrows, *U.S. Pat.* 4 283 487 (1981); *Chem. Abstr.,* **95**, 195181 (1981).
826. A. K. Sheinkman, A. N. Prilepskaya, A. O. Ginzburg, A. K. Tokarev, and A. A. Deikalo, *Khim. Geterotsikl. Soedin.,* **1971**, 421; *Chem. Abstr.,* **76**, 14294 (1972).
827. R. M. Acheson and J. M. Woollard, *J. Chem. Soc., Perkin Trans. I,* **1975**, 446.
828. C. A. Goffe, P. W. Jenkins, R. E. Bernard, D. M. Sturmer, and D. W. Heseltine, *Res. Discl.,* **147**, 24 (1976); *Chem. Abstr.,* **86**, 24432 (1977).
829. C. A. Goffe, D. W. Heseltine, and R. E. Bernard, *U.S. Pat.* 4 081 279 (1978); *Chem. Abstr.,* **89**, 51431 (1978).
830. D. W. Heseltine, R. E. Bernard, and C. A. Goffe, *Belg. Pat.* 854990 (1977); *Chem. Abstr.,* **89**, 68591 (1978).
831. J. Meisenheimer, L. Angermann, O. Finn, and E. Vieweg, *Ber.,* **57**, 1744 (1924).
832. J. Hill, M. M. Zakira, and D. Mumford, *J. Chem. Soc., Perkin Trans. I,* **1983**, 2455.
833. C. T. Bahner, W. K. Easley, and E. Stephen, *J. Amer. Chem. Soc.,* **74**, 4198 (1952).
834. V. Grinsteins and J. Veinbergs, *Latv. PSR Zinat. Akad. Vestis, Kim. Ser.,* **1974**, 364; *Chem. Abstr.,* **81**, 105207 (1974).
835. L. J. Sargent and L. Small, *J. Org. Chem.,* **13**, 608 (1948).
836. D. Nightingale, H. E. Ungnade, and H. E. French, *J. Amer. Chem. Soc.,* **67**, 1262 (1945).
837. E. L. May and E. Mosettig, *J. Org. Chem.,* **11**, 1 (1946).
838. V. V. Klyushin, V. A. Portnyagina, and V. Ya. Pochinok, *Ukr. Khim. Zh. (Russ. Ed.),* **50**, 1291 (1984); *Chem. Abstr.,* **103**, 37163 (1985).
839. R. E. Lutz, R. K. Allison, G. Ashburn, P. S. Bailey, M. T. Clark, J. F. Codington, A. J. Deinet, J. A. Freek, R. H. Jordan, N. H. Leake, T. A. Martin, K. C. Nicodemus, R. J. Rowlett, Jr., N. H. Shearer, Jr., J. D. Smith, and J. W. Wilson, III, *J. Org. Chem.,* **12**, 617 (1947).
840. N. H. Cromwell and D. J. Cram, *J. Amer. Chem. Soc.,* **65**, 301 (1943).

841. N. H. Cromwell and I. H. Witt, *J. Amer. Chem. Soc.*, **65**, 308 (1943).
842. N. H. Cromwell, J. A. Caughlan, and G. F. Gilbert, *J. Amer. Chem. Soc.*, **66**, 401 (1944).
843. N. H. Cromwell, C. E. Harris, and D. J. Cram, *J. Amer. Chem. Soc.*, **66**, 134 (1944).
844. L. Lafon, *Fr. Demande* 2 534 914 (1984); *Chem. Abstr.*, **101**, 171122 (1984).
845. N. N. Sveshnikov and N. S. Stakovskaya, *USSR Pat.*, 176922 (1965); *Chem. Abstr.*, **64**, 12552 (1966).
846. S. A. Kheifets and N. N. Sveshnikov, *Dokl. Akad. Nauk. SSSR*, **163**, 1177 (1965); *Chem. Abstr.*, **64**, 843 (1966).
847. Y. Sato, H. Kojima, and H. Shirai, *Tetrahedron*, **30**, 2695 (1974).
848. I. Ugi and E. Böttner, *Justus Liebigs Ann. Chem.*, **670**, 74 (1963).
849. A. Einhorn, *Ber.*, **19**, 904 (1886); *J. Chem. Soc. Abstr.*, **50**, 721 (1886).
850. O. Kocian and M. Ferles, *Collect. Czech. Chem. Commun.*, **44**, 1167 (1979).
851. K. Eiter and E. Mrazek, *Monatsh. Chem.*, **83**, 1491 (1952).
852. R. L. Cobb and W. E. McEwen, *J. Amer. Chem. Soc.*, **77**, 5042 (1955).
853. W. G. Dauben and C. W. Vaughan, Jr., *J. Amer. Chem. Soc.*, **75**, 4651 (1953).
854. M. Nógrádi, B. Vermess, A. Gergely, and G. Toth, *Pharmazie*, **37**, 126 (1982).
855. T. E. Whiteley and S. M. Krueger, *Res. Discl.*, **127**, 15 (1974); *Chem. Abstr.*, **83**, 186261 (1975).
856. P. W. Jenkins, D. M. Sturmer, and C. A. Goffe, *Belg. Pat.* 854992 (1977); *Chem. Abstr.*, **89**, 120856 (1978); *U.S. Pat.* 4 095 981 (1978).
857. V. Lucchini, M. Prato, U. Quintily, and G. Scorrano, *J. Chem. Soc., Chem. Commun.*, **1984**, 48.
858. G. P. Zecchini and M. P. Paradisi, *J. Heterocycl. Chem.*, **16**, 1589 (1979).
859. D. A. Schwartz and P. Yates, *Can. J. Chem.*, **61**, 1126 (1983).
860. F. Sannicolo, *Tetrahedron Lett.*, **25**, 3101 (1984).
861. R. R. Arndt, S. H. Eggers, and A. Jordaan, *Tetrahedron*, **25**, 2767 (1969).
862. N. Hughes and G. H. Keats, *Ger. Offen.* 2 243 436 (1973); *Chem. Abstr.*, **78**, 136112 (1973).
863. N. Hughes and G. H. Keats, *Brit. Pat.* 1 398 075 (1975); *Chem. Abstr.*, **83**, 113933 (1975).
864. T. Teraji, Y. Shiokawa, K. Okumura, and Y. Sato, *Eur. Pat. Appl.* 122494 (1984); *Chem. Abstr.*, **102**, 149290 (1985).
865. W. W. Hoffman and R. A. Kraska, *Eur. Pat. Appl.* 130795 (1985); *Chem. Abstr.*, **103**, 22476 (1985).
866. M. Kondo, H. Iwasaki, K. Yasui, and M. Miyake, *Ger. Offen.* 2 709 580 (1977); *Chem. Abstr.*, **88**, 22393 (1978); *U.S. Pat.* 4 323 700 (1982).
867. C. C. Barker and G. Hallas, *J. Chem. Soc. (B)*, **1969**, 1068.
868. A. G. Anderson and T. E. Dueber, *Eur. Pat. Appl.* 127762 (1984); *Chem. Abstr.*, **102**, 123114 (1985).
869. Y. Kumagae and Y. Iwasaki, *Eur. Pat. Appl.* 80814 (1983); *Chem. Abstr.*, **99**, 185058 (1983).
870. T. Kappe, G. Baxevanidis, and E. Ziegler, *Monatsh. Chem.*, **101**, 932 (1970).
871. J. B. Hester, Jr., A. D. Rudzik, and W. Veldkamp, *J. Med. Chem.*, **13**, 827 (1970).
872. J. B. Hester, Jr., *U.S. Pat.* 3 793 328 (1974); *Chem. Abstr.*, **80**, 95763 (1974).
873. A. H. Robins Co. Inc., *Brit. Pat.* 1 290 277 (1972); *Chem. Abstr.*, **76**, 59669 (1972).
874. J. B. Hester, Jr., *U.S. Pat.* 3 714 149 (1973); *Chem. Abstr.*, **78**, 111380 (1973).
875. F. Kunckell and E. Vollhase, *Chem. Zentralbl.*, **78**, (2), 660 (1910).
876. F. Kunckell, *Ber. Dtsch. Pharm. Ges.*, **20**, 277 (1910); *Chem. Abstr.*, **5**, 872 (1911).
877. F. Kunckell and E. Vollhase, *Ber.*, **42**, 3196 (1909).
878. H. Ahlbrecht and F. Kröhnke, *Justus Liebigs Ann. Chem.*, **717**, 96 (1968).
879. K. Akiba, T. Kobayashi, and Y. Yamamoto, *Heterocycles*, **22**, 1519 (1984).

880. J. Metzger, H. Larive, R. Dennilauler, R. Baralle, and C. Gaurat, *Bull. Soc. Chim. Fr.*, **43**, 30 (1967).
881. O. Kocian and M. Ferles, *Collect. Czech. Chem. Commun.*, **46**, 503 (1981).
882. T. Severin, D. Bätz, and H. Lerche, *Chem. Ber.*, **101**, 2731 (1968).
883. R. V. Heinzelman, *U.S. Pat.* 2 773 069 (1956); *Chem. Abstr.*, **52**, 11964 (1958).
884. H. von Dobeneck and W. Goltzsche, *Chem. Ber.*, **95**, 1484 (1962).
885. S. V. Kessar, A. K. Lumb, and R. K. Mayor, *Indian J. Chem.*, **5**, 18 (1967).
886. J. F. Eggler, M. R. Johnson, and L. S. Melvin, Jr., *U.S. Pat.* 4 486 428 (1984); *Chem. Abstr.*, **103**, 178170 (1985).
887. F. G. Webster, M. T. Regan, and L. J. Rossi, *Eur. Pat. Appl.* 0599 (1979); *U.S. Appl.* 818697 (1977); *Chem. Abstr.*, **92**, 102298 (1980).
888. F. G. Webster, M. T. Regan, and L. J. Rossi, *U.S. Pat.* 4 272 595 (1981); *Chem. Abstr.*, **95**, 71047 (1981).
889. O. M. Radul, S. M. Bukhanyuk, and M. A. Rekhter, *Khim. Geterotsikl. Soedin.*, **1985**, 1131; *Chem. Heterocycl. Compd. (Engl. Transl.)*, **1985**, 948; *Chem. Abstr.*, **104**, 5753 (1986).
890. I. Ugi, F. Beck, and E. Boettner, *Belg. Pat.* 628827 (1963); *Chem. Abstr.*, **60**, 11993 (1964).
891. Farbenfabriken Bayer A.-G., *Neth. Appl.* 289529 (1965); *Chem. Abstr.*, **63**, 9925 (1965).
892. T. Kurihara, Y. Ohshita, and Y. Sakamoto, *Heterocycles*, **6**, 123 (1977).
893. M. Klaus and M. Loeliger, *Ger. Offen.* 3 316 932 (1983); *Brit. Pat.* 2 119 801 (1983); *Chem. Abstr.*, **100**, 51468 (1984).
894. S. Cacchi and G. Palmieri, *Tetrahedron*, **39**, 3373 (1983).
895. Yoshitomi Pharmaceutical Industries Ltd., *Japan Kokai Tokkyo Koho* 80 66,560 (1980); *Chem. Abstr.*, **94**, 139634 (1981).
896. Yoshitomi Pharmaceutical Industries Ltd., *Neth. Appl.* 78 10,396 (1980); *Chem. Abstr.*, **93**, 239442 (1980).
897. T. Teraji, Y. Shiokawa, K. Okumura, and Y. Sato, *Eur. Pat. Appl.* 77983 (1983).
898. M. Uchida, T. Nishi and K. Nakagawa, *Eur. Pat. Appl.* 35228 (1981); *Chem. Abstr.*, **96**, 6738 (1982).
899. T. Nakao, *Japan Kokai Tokkyo Koho* 79 141 785 (1979); *Chem. Abstr.*, **92**, 181031 (1980).
900. M. Ozutsumi, Y. Miyazawa, K. Motohasi, and T. Watanabe, *Ger. Offen* 2 412 509 (1974); *Chem. Abstr.*, **82**, 113184 (1975).
901. Y. Miyazawa, M. Ozutsumi, and S. Ogawa, *Ger. Offen.* 2 435 408 (1975); *Chem. Abstr.*, **83**, 44730 (1975); *U.S. Pat.*, 4 002 631.
902. R. L. Stuz, C. A. Reynolds, and W. E. McEwen, *J. Org. Chem.*, **26**, 1685 (1961).
903. A. K. Sheinkman, A. K. Tokarer, S. G. Potashnikova, A. A. Deikalo, A. P. Kucherenko, and S. N. Baranov, *Khim. Geterotsikl. Soedin.*, **1971**, 643; *Chem. Heterocycl. Compd. (Engl. Transl.)*, **1971**, 602; *Chem. Abstr.*, **76**, 59409 (1972).
904. G. F. Palladino and W. E. McEwen, *J. Org. Chem.*, **43**, 2420 (1978).
905. R. Ya. Omel'chenko, V. A. Sapronov, A. K. Sheinkman, and G. A. Blokh, *Kauch. Rezina*, **1974**, 24; *Chem. Abstr.*, **82**, 32079 (1975).
906. E. Reimann and D. Voss, *Arch. Pharm (Weinheim)*, **310**, 2 (1977).
907. A. R. Katritzky, L. Ürögdi, and R. C. Patel, *J. Chem. Soc., Perkin Trans. I*, **1982**, 1349.
908. A. R. Katritzky, R. T. Langthorne, H. A. Muathin, and R. C. Patel, *J. Chem. Soc., Perkin Trans. I*, **1983**, 2601.
909. D. E. Beattie, R. Crossley, A. C. W. Curran, D. G. Hill, and A. E. Lawrence, *J. Med. Chem.*, **20**, 718 (1977).
910. D. E. Ames and B. T. Warren, *J. Chem. Soc.*, **1964**, 5518.

911. K. Winterfeld and K. Nonn, *Chem. Ber.*, **100**, 2274 (1967).
912. H. W. Smith, *Eur. Pat. Appl.*, 161867 (1985); *Chem. Abstr.*, **104**, 224839 (1986).
913. U. Basu, *Justus Liebigs Ann. Chem.*, **514**, 292 (1934).
914. G. Bouchon, K. H. Spohn, and E. Breitmaier, *Chem. Ber.*, **106**, 1736 (1973).
915. M. Ohashi, H. Kamachi, H. Kakisawa, and G. Stork, *J. Amer. Chem. Soc.*, **89**, 5460 (1967).
916. G. Stork, M. Ohashi, H. Kamachi, and H. Kakisawa, *J. Org. Chem.*, **36**, 2784 (1971).
917. U. Basu, *Justus Liebigs Ann. Chem.*, **530**, 131 (1937).
918. W. Wunderlich, *J. Prakt. Chem.*, **274**(2), 302 (1955).
919. U. Basu, *Justus Liebigs Ann. Chem.*, **512**, 131 (1934).
920. B. M. Gutsnlyak and P. D. Romanko, *Khim. Geterotsikl. Soedin.*, **1972**, 359; *Chem. Heterocycl. Compd. (Engl. Transl.)*, **1972**, 326; *Chem. Abstr.*, **77**, 48174 (1972).
921. M. N. Tilichenko, G. V. Parel, and A. D. Chumak, *Khim. Geterotsikl. Soedin.*, **1977**, 1356; *Chem. Heterocycl. Compd. (Engl. Transl.)*, **1977**, 1086; *Chem. Abstr.*, **88**, 62320 (1978).
922. G. Stork and J. E. McMurry, *J. Amer. Chem. Soc.*, **89**, 5463 (1967).
923. J. E. McMurry, *Org. Synth.*, **53**, 70 (1973).
924. V. G. Gramik, A. M. Zhidkova, R. G. Glushkov, A. B. Grigorev, M. K. Polievktvov, T. F. Vlasova, and O. S. Anisimova, *Khim. Geterotsikl Soedin.*, **1977**, 1348; *Chem. Heterocycl. Compd. (Engl. Transl.)*, **1977**, 1079; *Chem. Abstr.*, **88**, 50631 (1978).
925. D. Schumann and A. Naumann, *Liebigs Ann. Chem.*, **1984**, 1519.
926. C. H. Heathcock and E. F. Kleinman, *J. Amer. Chem. Soc.*, **103**, 222 (1981).
927. C. H. Heathcock, E. F. Kleinman, and E. S. Binkley, *J. Amer. Chem. Soc.*, **104**, 1054 (1982).
928. H. Meyer, F. Bossert, and H. Horstmann, *Liebigs Ann. Chem*, **1977**, 1888.
929. A. A. Akhrem, L. I. Ukhova, and N. F. Uskova, *Izv. Akad. Nauk SSSR, Ser. Khim.*, **1970**, 2305; *Bull. Acad. Sci. USSR, Div. Chem. Sci.*, *(Engl. Transl.)* **1970**, 2167; *Chem. Abstr.*, **75**, 5100 (1971).
930. A. A. Akhrem, L. I. Ukhova, and N. F. Uskova, *Izv. Akad. Nauk. SSSR, Otd. Khim. Nauk*, **1962**, 304; *Bull. Acad. Sci. USSR, Div. Chem. Sci. (Engl. Transl.)*, **1962**, 281; *Chem. Abstr.*, **57**, 11163 (1962).
931. B. B. Kuz'mitskii, T. N. Ignat'eva, L. I. Ukhova, G. P. Kukso, S. S. Prikhozhii, and Yu. D. Uksusov, *Vestsi Akad. Navuk BSSR, Ser. Khim. Navuk*, **1985**, 76; *Chem. Abstr.*, **104**, 28372 (1986).
932. K. D. Pralier, A. A. Andrusenko, T. T. Omarov, D. V. Sokolov, L. I. Ukhova and A. A. Akhrem, *Izv. Akad. Nauk Kaz. SSR. Ser. Khim.*, **17**, 49 (1967); *Chem. Abstr.*, **69**, 35251 (1968).
933. L. I. Ukhova, G. P. Kukso, V. A. Tokareva, and V. G. Zaikin, *Khim. Geterotsikl. Soedin.*, **1982**, 72; *Chem. Heterocycl. Compd. (Engl. Transl.)*, **1982**, 62; *Chem. Abstr.*, **96**, 162520 (1982).
934. A. A. Akhrem, L. I. Ukhova, and G. V. Bludova, *Vestsi Akad. Navuk BSSR, Ser. Khim. Navuk*, **1979**, 84; *Chem. Abstr.*, **92**, 76248 (1980).
935. A. A. Akhrem, L. I. Ukhova, and A. N. Sergeeva, *Vestsi Akad. Navuk BSSR, Ser. Khim. Navuk*, **1975**, (6), 66; *Chem. Abstr.*, **84**, 58473 (1976).
936. A. A. Akhrem, L. I. Ukhova, and A. N. Sergeeva, *Vestsi Akad. Navuk BSSR, Ser. Khim. Navuk*, **1975**, 75; *Chem. Abstr.*, **84**, 58474 (1976).
937. A. A. Akhrem, L. I. Ukhova, N. F. Uskova and T. E. Prokof'ev, *Khim. Geterotsikl. Soedin.*, **1977**, 796; *Chem. Heterocycl. Compd. (Engl. Transl.)*, **1977**, 647; *Chem. Abstr.*, **88**, 62272 (1978).
938. V. S. Bogdanov, L. I. Ukhova, N. F. Uskova, A. P. Marochkin, A. M. Moiseenkov,

and A. A. Akhrem, *Izv. Akad. Nauk SSSR, Ser. Khim.*, **1971**, 2349; *Bull. Acad. Sci. USSR, Div. Chem. Sci.* (*Engl. Transl.*), **1971**, 2235; *Chem. Abstr.*, **76**, 58448 (1972).

939. V. S. Bogdanov, L. I. Ukhova, N. F. Uskova, A. N. Sergeeva, A. S. Marochkin, A. M. Moiseenkov, and A. A. Akhrem, *Khim. Geterotsikl. Soedin.*, **1973**, 1078, *Chem. Heterocycl. Compd.* (*Engl. Transl.*), **1973**, 999; *Chem. Abstr.*, **79**, 125684 (1973).

940. A. A. Akhrem, L. I. Ukhova, A. P. Marochkin, B. B. Kuz'mitskii, and S. S. Prikhozhii, *USSR Pat.* 601937 (1980); *Chem. Abstr.*, **93**, 46441 (1980).

941. A. A. Akhrem, L. I. Ukhova, A. P. Marochkin, and V. G. Zaikin, *Khim. Geterotsikl. Soedin.*, **1980**, 1079, *Chem. Heterocycl. Compd.* (*Engl. Transl.*), **1980**, 837; *Chem. Abstr.*, **94**, 47184 (1981).

942. J. Szychowski and D. B. MacLean, *Can. J. Chem.*, **57**, 1631 (1979).

943. E. L. May and E. Mossetig, *J. Org. Chem.*, **11**, 10 (1946).

944. M. Prost, M. Urbain, A. Schumer, V. Van Cromphant, C. Houben, C. Van Meerbeeck, A. Christiaens, M. Colot, and R. Charlier, *Eur. J. Med. Chem. – Chim. Ther.*, **11**, 337 (1976).

945. M. Prost and M. Urbain, *Ger. Offen.* 2 257 397 (1973); *Chem. Abstr.*, **79**, 42372 (1973); *U.S. Pat.* 3 882 129 (1975).

946. M. Prost, M. Urbain, C. Houben, A. Schumer, V. Van Cromphant, C. Van Meerbeeck, W. Verstraeten, M. Colot, and M. De Claviere, *Ann. Pharm. Fr.*, **43**, 139 (1985); *Chem. Abstr.*, **104**, 224 (1986).

947. M. Prost, M. Urbain, A. Schumer, C. Houben, C. Van Meerbeeck, M. Colot, C. Tornay, and R. Charlier, *Eur. J. Med. Chem. – Chim. Ther.*, **10**, 470 (1975).

948. U. P. Arya, *S. African Pat.* 68 00,955 (1968); *Chem. Abstr.*, **70**, 57831 (1969); *U.S. Pat.*, 3 778 444 (1973).

949. M. Prost, V. Van Cromphant, W. Verstraeten, M. Dirks, C. Tornay, M. Colot, and M. De Claviere, *Eur. J. Med. Chem. – Chim. Ther.*, **15**, 215 (1980).

950. M. Prost, *Ger. Offen.* 2 656 678 (1977); *Chem. Abstr.*, **87**, 102193 (1977); *U.S. Pat.* 4 173 636 (1979).

951. T. Ibuka, Y. Inubushi, I. Saji, K. Tanaka, and N. Masaki. *Tetrahedron Lett.*, **1975**, 323.

952. T. Ibuka, N. Masaki, I. Saji, K. Tanaka, and Y. Inubushi, *Chem. Pharm. Bull.*, **23**, 2779 (1975).

953. T. Ibuka, N. G. Chu, and F. Yoneda, *J. Chem. Soc., Chem. Commun.*, **1984**, 597.

954. L. E. Overman and C. Fukaya, *J. Amer. Chem. Soc.*, **102**, 1454 (1980).

955 P. Friedländer and C. F. Gohring, *Ber.*, **16**, 1833 (1883); *J. Chem. Soc. Abstr.*, **44**, 1148 (1883).

956. W. Borsche and A. Herbert, *Justus Liebigs Ann. Chem.*, **546**, 293 (1941).

957. D. Tomasik, P. Tomasik, and R. A. Abramovitch, *J. Heterocycl. Chem.*, **20**, 1539 (1983).

958. O. M. Radul, S. M. Bukhanyuk, M. A. Prekhter, G. I. Zhungietu, and I. P. Ivanova, *Khim. Geterotsikl. Soedin.*, **1982**, 1427; *Chem. Heterocycl. Compd.* (*Engl. Transl.*), **1982**, 1113; *Chem. Abstr.*, **98**, 71891 (1983).

959. G. Roma, A. Ermili, and A. Balbi, *J. Heterocycl. Chem.*, **12**, 1103, (1975).

960. G. M. Coppola, A. D. Kahle, and M. J. Shapiro, *Org. Magn. Reson.*, **17**, 242 (1981).

961. V. G. Zaikin, Z. S. Ziyavidinova, and N. S. Vul'fson, *Khim. Geterotsikl. Soedin.*, **1974**, 1516; *Chem. Heterocycl. Compd.* (*Engl. Transl.*), **1974**, 1332; *Chem. Abstr.*, **82**, 57022 (1975).

962. C. M. Foltz and Y. Kondo, *Tetrahedron Lett.*, **1970**, 3163.

963. Fuji Kogaku Kogyo Co. Ltd., *Japan Kokai Tokkyo Koho* 59 16,878 (1984); *Chem. Abstr.*, **101**, 23362 (1984).

964. K. Tomita, *Yakugaku Zasshi.*, **71**, 1100 (1951); *Chem. Abstr.*, **46**, 5044 (1952).

965. H. J. Sturm and H. Goerth, *Fr. Demande* 2 009 119 (1970); *Chem. Abstr.*, **73**, 45369 (1970); *U.S. Pat.* 3 652 571 (1972).
966. R. N. Lacey, *J. Chem. Soc.*, **1954**, 850.
967. Y. Morinaka, K. Takahashi, S. Hata, and S. Yamada, *Eur. J. Med. Chem. – Chim. Ther.*, **16**, 251 (1981).
968. H. Enomoto, T. Nomura, Y. Aoyagi, S. Chokai, T. Kono, M. Murase, K. Inoue, and M. Adachi, *Eur. Pat. Appl.* 101330; *Japan Kokai Tokkyo Koho* 83 225 065 (1983); *Chem. Abstr.*, **101**, 72625 (1984).
969. P. Venturella and A. Bellino, *J. Heterocycl. Chem.*, **12**, 669 (1975).
970. G. M. Coppola and G. E. Hardtmann, *J. Heterocycl. Chem.*, **16**, 1605 (1979).
971. W. R. Vaughan, *J. Amer. Chem. Soc.*, **68**, 324 (1946).
972. R. E. Bowman, A. Campbell, and E. M. Tanner, *J. Chem. Soc.*, **1959**, 444.
973. P. Venturella, A. Bellino, and F. Piozzi, *Gazz. Chim. Ital.*, **104**, 297 (1974).
974. N. S. Vul'fson and R. B. Zhurin, *Zh. Obshch. Khim.*, **32**, 991 (1962); *Chem. Abstr.*, **58**, 1448 (1963).
975. K. Faber and T. Kappe, *J. Heterocycl. Chem.*, **21**, 1881 (1984).
976. G. J. B. Cajipe, G. Landen, B. Semler, and H. W. Moore, *J. Org. Chem.*, **40**, 3874 (1975).
977. K. Ogura, H. Sasaki, and S. Seto, *Bull. Chem. Soc. Japan,* **38**, 306 (1965).
978. H. J. Sturm and H. Goerth, *Fr. Demande* 2 009 125 (1970); *Chem. Abstr.*, **73**, 45370 (1970); *U.S. Pat.* 3 682 928 (1972).
979. W. Stadlbauer and T. Kappe, *Z. Naturforsch.*, **36B**, 739 (1981).
980. T. Kappe, W. Gober, M. Hariri, and W. Stadlbauer, *Chem. Ber.*, **112**, 1585 (1979).
981. B. Eistert and G. Borggrefe, *Justus Liebigs Ann. Chem.*, **718**, 142 (1968).
982. T. Chiba, M. Okada and T. Kato, *J. Heterocycl. Chem.*, **19**, 1521 (1982).
983. G. Jones, *J. Chem. Soc. (C)*, **1967**, 1808.
984. C. Kaneko, H. Fuji, S. Kawai, K. Hashiba, Y. Karasawa, M. Wakai, R. Hayashi, and M. Somei, *Chem. Pharm. Bull.*, **30**, 74 (1982).
985. Y. K. Sawa and H. Matsumura, *Tetrahedron* **26**, 2923 (1970).
986. M. Ishikawa, *Chem. Pharm. Bull.*, **6**, 67 (1958).
987. E. Ziegler and H. Junek, *Monatsh. Chem.*, **89**, 323 (1958).
988. Otsuka Pharmaceutical Co. Ltd., *Japan Kokai Tokkyo Koho* 58 148 861 (1983); *Chem. Abstr.*, **100**, 51465 (1984).
989. M. Uchida, M. Komatsu, and K. Nakagawa, *Ger. Pat.* 3 324 034 (1984); *Chem. Abstr.*, **101**, 54936 (1984).
990. Y. H. Yang, M. Tominaga, H. Ogawa, and K. Nakagawa, *Brit. Pat.* 2 094 789 (1982); *Chem. Abstr.*, **98**, 34510 (1983).
991. W. E. Noland and R. F. Modler, *J. Amer. Chem. Soc.*, **86**, 2086 (1964).
992. C. C. Price and B. H. Velzen, *J. Org. Chem.*, **12**, 386 (1947).
993. M. S. Mayadeo and S. H. Hussain, *Indian J. Chem.*, **23A**, 599 (1984).
994. S. A. Kulkarni, V. M. Thakor, and R. C. Shah, *J. Indian Chem. Soc.*, **28**, 688 (1951); *Chem. Abstr.*, **47**, 2175 (1953).
995. G. Singh and G. V. Nair, *J. Sci. Ind. Res. (India)*, **13B**, 79 (1954); *Chem. Abstr.*, **49**, 1724 (1955).
996. G. Singh and G. V. Nair, *Chem. Ind. (London)*, **1954**, 253.
997. F. K. Hess and P. B. Stewart, *Ger. Offen.* 2 446 497 (1976); *Chem. Abstr.*, **85**, 32979 (1976).
998. T. S. Gore, R. J. Kamat, and S. V. Sunthankar, *Indian J. Chem.*, **3**, 90 (1965).
999. D. J. Le Count, *J. Chem. Soc., Perkin Trans. I*, **1983**, 813.
1000. E. Zeigler, G. Kollenz, and W. Ott, *Justus Liebigs Ann. Chem.*, **1976**, 2071.
1001. W. D. Crow and J. R. Price, *Aust. J. Sci. Res.*, **2A**, 282 (1949).

1002. G. Bendz, C. C. J. Culvenor, L. J. Goldsworthy, K. S. Kirby, and R. Robinson, *J. Chem. Soc.*, **1950**, 1130.
1003. N. S. Nixon, F. Scheinmann, and J. L. Suschitzky, *J. Chem. Res. Synop.*, **1984**, 380; *Miniprint*, p. 3401.
1004. R. V. Davies, J. Fraser, K. J. Nichol, R. Parkinson, M. F. Sim, and D. B. Yates, *Ger. Offen.* 3 011 994 (1980); *Chem. Abstr.*, **94**, 83968 (1981); *U.S. Pat.* 4 302 460 (1981).
1005. E. H. Erickson, L. R. Lappi, T. K. Rice, K. F. Swingle, and M. Van Winkle, *J. Med. Chem.*, **21**, 984 (1978).
1006. E. C. Taylor and N. D. Heindel, *J. Org. Chem.*, **32**, 1666 (1967).
1007. A. Musierowicz, S. Niementowski, and Z. Tomasik, *Rocz. Chem.*, **8**, 325 (1928); *Chem. Abstr.*, **23**, 1900 (1929).
1008. E. Ziegler, T. Kappe, and H. G. Foraita, *Monatsh. Chem.*, **97**, 409 (1966).
1009. E. Wohnlich, *Arch. Pharm. (Weinheim)*, **251**, 526 (1913).
1010. Y. Sato, *Sankyo Kenkyusho Nempo*, **15**, 51 (1963); *Chem. Abstr.*, **60**, 11979 (1964).
1011. J. Tröger and C. Cohaus, *J. Prakt. Chem.*, **117**, 97 (1927).
1012. J. Tröger and E. Dunker, *J. Prakt. Chem.*, **111**, 207 (1925).
1013. P. Ruggli and H. Frey, *Helv. Chim. Acta* **22**, 1413 (1939).
1014. H.-J. Gais, K. Hafner, and M. Neuenschwander, *Helv. Chim. Acta*, **52**, 2641 (1969).
1015. H. Gourlaouen and P. Pastour, *Compt. Rend.*, **256**, 3319 (1963).
1016. H. Tomisawa, M. Watanabe, R. Fujita, and H. Hongo, *Chem. Pharm. Bull.*, **18**, 919 (1970).
1017. H. Tomisawa, R. Fujita, H. Hongo, and K. Kato, *Chem. Pharm. Bull.*, **23**, 592 (1975).
1018. W. F. Smith, Jr., K. L. Eddy, and D. P. Harnish, *Ger. Pat.* 2 422 087 (1974); *Chem. Abstr.*, **84**, 52076 (1976).
1019. H. Tomisawa, R. Fujita, H. Hongo, and H. Kato, *Chem. Pharm. Bull.*, **22**, 2091 (1974).
1020. K. R. Huffman, M. Burger, W. A. Henderson, Jr., M. Loy, and E. F. Ullman, *J. Org. Chem.*, **34**, 2407 (1969).
1021. J. Tröger and J. Bohnekamp, *J. Prakt. Chem.*, **117**, 161 (1927).
1022. P. Ruggli, P. Hindermann, and H. Frey, *Helv. Chim. Acta*, **21**, 1066 (1938).
1023. T. Sakan, K. Kusada, and T. Miwa, *Bull. Chem Soc. Japan*, **42**, 2952 (1969).
1024. E. Ziegler, H. Junek, and A. Metallidis, *Monatsh. Chem.*, **101**, 92 (1970).
1025. E. Ziegler and H. Junek, *Monatsh. Chem.*, **90**, 762 (1959).
1026. N. S. Vul'fson and R. B. Zhurin, *Zh. Vses. Khim. O-va.*, **5**, 352 (1960); *Chem. Abstr.*, **54**, 24733 (1960).
1027. N. S. Vul'fson and R. B. Zhurin, *Zh. Obshch. Khim.*, **29**, 3677 (1959); *J. Gen. Chem. USSR (Engl. Transl.)*, **29**, 3635 (1959); *Chem. Abstr.*, **54**, 19676 (1960).
1028. R. F. C. Brown, G. K. Hughes, and E. Ritchie, *Aust. J. Chem.*, **9**, 277 (1956).
1029. S. K. Phadtare, S. K. Kamat, and G. T. Panse, *Indian J. Chem.*, **22B**, 496 (1983).
1030. A. A. Sayed, S. M. Sami, E. Elfayoumi, and E. A. Mohamed, *Acta Chim. Acad. Sci. Hung.*, **94**, 131 (1977); *Chem. Abstr.*, **88**, 152388 (1978).
1031. M. Mazzei, G. Roma, A. Ermili, and C. Cacciatore, *Farmaco, Ed. Sci.*, **34**, 469 (1979).
1032. E. Tighineanu, F. Chiraleu, and D. Raileanu, *Tetrahedron Lett.*, **1978**, 1887.
1033. E. Tighineanu, F. Chiraleu, and D. Raileanu, *Tetrahedron*, **36**, 1385 (1980).
1034. A. A. Sayed, S. M. Sami, A. Elfayoumi, and E. C. Mohamed, *Egypt. J. Chem.*, **19**, 811 (1976); *Chem. Abstr.*, **91**, 193222 (1979).
1035. H. H. Zoorob, E. M. Afsah, and W. S. Hamama, *Pharmazie*, **40**, 246 (1985).
1036. J. W. Huffman and L. E. Browder, *J. Org. Chem.*, **29**, 2598 (1964).
1037. J. W. Huffman, S. P. Garg, and J. H. Cecil, *J. Org. Chem.*, **31**, 1276 (1966).
1038. J. Moszew and A. Kolasa, *Zesz. Nauk. Univ. Jagiellon., Pr. Chem.*, **1969**, (14), 83; *Chem. Abstr.*, **72**, 31569 (1970).

510 Quinoline Ketones

1039. J. Moszew and S. Kusmierczyk, *Zesz. Nauk. Univ. Jagiellon.*, *Pr. Chem.*, **1968**, 31; *Chem. Abstr.*, **72**, 31585 (1970).
1040. H. Tani, G. Otani, and T. Mizutani, *Japan Pat.* 24782 (1965); *Chem. Abstr.*, **64**, 8155 (1966).
1041. J. M. Chemerda and M. Sletzinger, *Fr. Demande* 2 000 747 (1969); *Chem. Abstr.*, **72**, 66815 (1970).
1042. Y. S. Mohammed, A. E. M. N. Gohar, F. F. Abdel-Latif, and M. Z. A. Badr, *Pharmazie*, **40**, 312 (1985).
1043. N. Amlaiky, G. Leclerc, N. Decker, and J. Schwartz, *Eur. J. Med. Chem. – Chim. Ther.*, **19**, 341 (1984).
1044. K. Nakagawa, S. Yoshiyaki, K. Tanimura, and S. Tamada, *Japan Kokai* 76 136,681 (1976); *Chem. Abstr.*, **87**, 39301 (1977).
1045. K. Nakagawa, S. Yoshizaki, K. Tanimura, and S. Tamada, *Japan Kokai* 76 136,682 (1976); *Chem. Abstr.*, **87**, 39302 (1977).
1046. K. Nakagawa, S. Yoshizaki, K. Tanimura, and S. Tamada, *U.S. Pat.* 4 022 784 (1977).
1047. S. Yoshizaki, K. Tanimura, S. Tamada, Y. Yabuuchi, and K. Nakagawa, *J. Med. Chem.*, **19**, 1138 (1976).
1048. K. Nakagawa, S. Yoshizaki, K. Tanimura, and S. Tamada, *Japan Kokai* 76 143,677 (1976); *Chem. Abstr.*, **87**, 68180 (1977).
1049. K. Nakagawa, S. Yoshizaki, K. Tanimura, and S. Tamada, *U.S. Pat.* 4 022 776 (1977).
1050. K. Nakagawa, S. Yoshizaki, K. Tanimura, and S. Tamada, *Japan Kokai* 76 128,982 (1976); *Chem. Abstr.*, **87**, 53105 (1977).
1051. K. Nakagawa, S. Yoshizaki, K. Tanimura, and S. Tamada, *Japan Kokai* 75 111,081 (1975); *Chem. Abstr.*, **84**, 164633 (1976).
1052. K. Nakagawa, S. Yoshizaki, K. Tanimura, and S. Tamada, *Belg. Pat.* 823841 (1975); *Chem. Abstr.*, **85**, 123 768 (1976).
1053. K. Nakagawa, S. Yoshizaki, K. Tanimura, and S. Tamada, *Japan Kokai* 76 141,879 (1976); *Chem. Abstr.*, **87**, 23077 (1977).
1054. K. Nakagawa, S. Yoshizaki, K. Tanimura, and S. Tamada, *Japan Kokai* 76 143,678 (1976); *Chem. Abstr.*, **87**, 53101 (1977).
1055. Otsuka Pharmaceutical Co. Ltd., *Japan Kokai Tokkyo Koho* 80 85,520 (1980); *Chem. Abstr.*, **94**, 20391 (1981).
1056. K. Nakagawa, S. Yoshizaki, K. Tanimura, and S. Tamada, *Japan Kokai* 76 149,283 (1976); *Chem. Abstr.*, **87**, 53103 (1977).
1057. S. Yoshizaki, T. Tamada, K. Tanimura, and K. Nakagawa, *Japan Kokai* 77 00,282 (1977); *Chem. Abstr.*, **87**, 68182 (1977).
1058. S. Yoshizaki, E. Yo. K. Nakagawa, and Y. Tamura, *Chem. Pharm. Bull.*, **28**, 3699 (1980).
1059. K. Nakagawa, S. Yoshizaki, K. Tanimura, and S. Tamada, *Japan Kokai* 76 149,284 (1976); *Chem. Abstr.*, **87**, 68183 (1977).
1060. S. Yoshizaki, S. Tamada, and K. Nakagawa, *Japan Kokai* 77 83,379 (1977); *Chem. Abstr.*, **88**, 6752 (1978).
1061. S. Yoshizaki, K. Tanimura, S. Tamada, and K. Nakagawa, *Japan Kokai* 78 09,777 (1978); *Chem. Abstr.*, **89**, 129422 (1978).
1062. S. Yoshizaki, S. Tamada, N. Yo, and K. Nakagawa, *Japan Kokai Tokkyo Koho* 78 112,883 (1978); *Chem. Abstr.*, **90**, 103853 (1979).
1063. S. Yoshizaki, S. Tamada, K. Tanimura, and K. Nakagawa, *Japan Kokai* 77 00,283 (1977); *Chem. Abstr.*, **87**, 53102 (1977).
1064. R. Goschke, P. G. Ferrini, and A. Sallmann, *Can. Pat.* 1 192 554 (1985); *Chem. Abstr.*, **104**, 168378 (1986); *U.S. Pat.* 4 476 132 (1984).

1065. K. Nakagawa, W. Yoshizaki, K. Tanimura, and S. Tamada *Japan Kokai* 76 136,680 (1976); *Chem. Abstr.*, **87**, 39300 (1977).
1066. K. Nakagawa, S. Yoshizaki, K. Tanimura, and S. Tamada, *U.S. Pat.* 4 026 897 (1977).
1067. K. Nakagawa, S. Yoshizaki, and S. Tamada, *Japan Kokai* 77 19,676 (1977); *Chem. Abstr.*, **87**, 117843 (1977).
1068. S. Yoshizaki, E. Yo, and K. Nakagawa, *Chem. Pharm. Bull.*, **28**, 3439 (1980).
1069. K. Nakagawa, S. Yoshizaki, K. Tanimura, and S. Tamada, *Japan Kokai* 77 03,076 (1977); *Chem. Abstr.*, **87**, 53111 (1977).
1070. Otsuka Pharmaceutical Co. Ltd., *Japan Kokai Tokkyo Koho* 57 154,129 (1982); *Chem. Abstr.*, **98**, 78131 (1983).
1071. T. Nakao, T. Saito, O. Yaoku, and T. Tawara, *Japan Kokai Tokkyo Koho* 60 237,072 (1985); *Chem. Abstr.*, **104**, 148762 (1986).
1072. K. Gu, L. Qian, and R. Ji, *Yaoxue Xuebao*, **20**, 277 (1985); *Chem. Abstr.*, **104**, 50765 (1986).
1073. T. Michiaki, Y. H. Yang, N. Kazuyuki, and O. Hidenori, *Eur. Pat. Appl.* 52016 (1982); *Chem. Abstr.*, **98**, 16680 (1983).
1074. Otsuka Pharmaceutical Co. Ltd., *Japan Kokai Tokkyo Koho* 58 144,371 (1983); *Chem. Abstr.*, **100**, 34414 (1984).
1075. Otsuka Pharmaceutical Co. Ltd., *Belg. Pat.* 894105 (1983); *Chem. Abstr.*, **99**, 5649 (1983).
1076. Otsuka Pharmaceutical Co. Ltd., *Japan Kokai Tokkyo Koho* 57 142,972 (1982); *Chem. Abstr.*, **98**, 143285 (1983).
1077. K. Banno, T. Fujioka, M. Osaki, and K. Nakagawa, *Fr. Demande* 2 477 542 (1981); *Chem. Abstr.*, **96**, 85434 (1982); *Brit. Pat.*, 2 071 09A (1981).
1078. Otsuka Pharmaceutical Co. Ltd., *Japan Kokai Tokkyo Koho* 59 13 722 (1984); *Chem. Abstr.*, **101**, 23497 (1984).
1079. M. Tominaga, Y. Yang, H. Ogawa, and K. Nakagawa, *U.S. Pat.* 4 514 401 (1985).
1080. T. Shimizu and Y. Tamura, *Japan Kokai* 77 73,867 (1977); *Chem. Abstr.*, **87**, 167900 (1977).
1081. Z. Meszaras, I. Hermez, E. Somfai, A. Horvath, and L. Vasvari, *Brit. Pat.* 1 505 427 (1978); *Chem. Abstr.*, **89**, 109138 (1978).
1082. R. K. Mapara and C. M. Desai, *J. Indian Chem. Soc.*, **31**, 951 (1954).
1083. F. Kröhnke, H. Dickhäuser, and I. Vogt, *Justus Leibigs Ann. Chem.*, **644**, 93 (1961).
1084. N. Barton, A. F. Crowther, W. Hepworth, D. N. Richardson, and G. W. Driver, *Brit. Pat.* 830832 (1960); *Chem. Abstr.*, **55**, 7442 (1961).
1085. R. R. Lorenz and W. H. Thielking, *U.S. Pat.* 4 107 167 (1978); *Chem. Abstr.*, **90**, 72073 (1979).
1086. J. D. Loudon and G. Tennant, *J. Chem. Soc.*, **1962**, 3092.
1087. J. D. Loudon and I. Wellings, *J. Chem. Soc.*, **1960**, 3470.
1088. P. G. Ferrini, G. Haas, and A. Rossi, *Ger. Offen.* 2 421 121 (1975); *Chem. Abstr.*, **82**, 156126 (1975).
1089. B. Staskun, *J. Org. Chem.*, **26**, 2791 (1961).
1090. B. Staskun, *J. Chem. Soc. (C)*, **1966**, 2311.
1091. S. A. Kulkarni and R. C. Shah, *J. Indian Chem. Soc.*, **27**, 111 (1950); *Chem. Abstr.*, **45**, 8522 (1951).
1092. S. L. Kishinchandani, B. S. R. Murty, and M. L. Khorana, *Indian J. Pharm.*, **32**, 29 (1970); *Chem. Abstr.*, **73** 45306 (1970).
1093. R. K. Mapara and C. M. Desai, *Sci. Cult.*, **19**, 105 (1953); *Chem. Abstr.*, **48**, 13689 (1954).
1094. H. R. Snyder and R. E. Jones, *J. Amer. Chem. Soc.*, **68**, 1253 (1946).

1095. S. V. Sunthankar, J. K. Mehra, and R. H. Nayak, *J. Sci. Ind. Res. (India)*, **20B**, 455 (1961); *Chem. Abstr.*, **56**, 5941 (1962).
1096. A. Vaidyanathan and S. V. Sunthankar, *Indian J. Technol.*, **2**, 53 (1964); *Chem. Abstr.*, **60** 16016 (1964).
1097. O. S. Wolfbeis and H. Junek, *Monatsh. Chem.*, **110**, 1387 (1979).
1098. M. Von Strandtmann, J. Shavel, Jr., and S. Klutchko, *U.S. Pat.* 3801 644 (1974); *Chem. Abstr.*, **81**, 3595 (1974).
1099. M. Von Strandtmann, J. Shavel, Jr., S. Klutchko, and M. Cohen, *U.S. Pat* 3 892 739 (1975); *Chem. Abstr.*, **84**, 43632 (1976).
1100. M. Von Strandtmann, J. Shavel, Jr., S. Klutchko, and M. Cohen, *U.S. Pat.* 3 880 861 (1975); *Chem. Abstr.*, **83**, 97057 (1975).
1101. M. Von Strandtmann, S. Klutchko, M. Cohen, and J. Shavel, Jr., *J. Heterocycl. Chem.*, **9**, 171 (1972).
1102. Z. Meszaros, I. Hermecz, A. Horvath, and A. Vasvari, *Hung. Teljes* 13 272 (1977); *Chem. Abstr.*, **87**, 167 895 (1977).
1103. K. Elliott and E. Tittensor, *J. Chem. Soc.*, **1961**, 2796.
1104. K. H. Huffman, E. F. Ullman, and M. Loy, *Belg. Pat.* 665812 (1965); *Chem. Abstr.*, **65** 2233 (1966).
1105. F. Serratosa, *An. R. Soc. Esp. Fis. Quim.* **51B**, 619 (1955); *Chem. Abstr.*, **50**, 11341 (1956).
1106. S. A. Kulkarni, V. M. Thaker, and R. C. Shah, *Current Sci. (India)*, **18**, 444 (1949) *Chem. Abstr.*, **44**, 4910 (1950).
1107. H. Briehl, A. Lukosch, and C. Wentrup, *J. Org. Chem.*, **49**, 2772 (1984).
1108. G. Kollenz, E. Ziegler, W. Ott, and G. Kriwetz, *Z. Naturforsch.*, **32B**, 701 (1977).
1109. G. Kollenz, G. Penn, W. Ott, K. Peters, E.-M. Peters, and H. G. von Schnering, *Chem. Ber.*, **117**, 1310 (1984).
1110. P. Gompper and R. Kunz, *Chem. Ber.*, **98**, 1391 (1965).
1111. H. Koga, A. Itoh, S. Murayama, S. Suzue, and T. Irikura, *J. Med. Chem.*, **23**, 1358 (1980).
1112. H. Cairns and D. Cox, *Brit. UK Pat. Appl.* 2022 078 (1979); *Chem. Abstr.*, **93**, 95259 (1980).
1113. H. Cairns and D. Cox, *Ger. Offen.* 2 819 215 (1978); *Chem. Abstr.*, **90**, 72164 (1979).
1114. H. Cairns and D. Cox, *U.S. Pat.* 4 474 787 (1984); *Chem. Abstr.*, **102**, 95624 (1985).
1115. K.-C. Lu, W.-C. Liu, and L.-C. Lee, *T'ai-wan Yao Hseuh Tsa Chih*, **31**, 80 (1979); *Chem. Abstr.*, **94**, 174832 (1981).
1116. D. P. Evans, D. J. Gilman, D. M. O'Mant, and D. S. Thompson, *Ger. Offen.* 2 130 408 (1971); *Chem. Abstr.*, **76** 113093 (1972).
1117. R. H. Mizzoni, F. Goble, and G. De Stevens, *J. Med. Chem.*, **13**, 870 (1970).
1118. T. P. Culbertson, T. F. Mich, J. M. Domagala, and J. B. Nichols, *Eur. Pat. Appl.* 153163 (1985); *Chem. Abstr.*, **104**, 34013 (1986).
1119. K. J. Shah and E. A. Coats, *J. Med. Chem.*, **20**, 1001 (1977).
1120. L. B. Kier and L. H. Hall, *J. Pharm. Sci.*, **72**, 1170 (1983).
1121. J. E. Clifton, I. Collins, P. Hallett, D. Hartley, L. H. C. Lunts, and P. D. Wicks, *J. Med. Chem.*, **25**, 670 (1982).
1122. C. K. Bradsher and M. F. Zinn, *J. Heterocycl. Chem.*, **4**, 66 (1967).
1123. M. Mottier, *Helv. Chim. Acta*, **58**, 337 (1975).
1124. M. Suzuki, K. Matsumoto, M. Miyoshi, N. Yoneda, and R. Ishida, *Chem. Pharm. Bull.*, **25**, 2602 (1977).
1125. G. M. Coppola, G. E. Hardtmann, and O. R. Pfister, *J. Org. Chem.*, **41**, 825 (1976).
1126. S. I. Savlyalov, N. N. Makhova, N. I. Aronova, and I. F. Mustafaeva, *Izv. Akad. Nauk SSSR, Ser. Khim.*, **1971**, 2826; *Bull. Acad. Sci. USSR, Div. Chem. Sci. (Engl. Transl.)*, **20**, 2692 (1971); *Chem. Abstr.*, **77**, 5302 (1972).

1127. R. W. Franck and J. M. Gilligan, *J. Org. Chem.*, **36**, 222 (1971).
1128. S. Kamiya and K. Koshinuma, *Chem. Pharm. Bull.*, **15**, 1985 (1967).
1129. L. V. Gyul'budagyan and I. L. Aleksanyan, *Arm. Khim. Zh.*, **33**, 338 (1980); *Chem. Abstr.*, **93** 204425 (1980).
1130. B. Avramova, L. Zhelyazkov, L. Daleva and D. Stefanova, *Khim. Prir. Soedin.*, **6**, 98 (1970); *Chem. Nat. Compd. (Engl. Transl.)*, **6**, 92 (1970); *Chem. Abstr.*, **73**, 45287 (1970).
1131. B. Avramova, L. Zhelyazkov, L. Daleva, and A. Dimova, *Farmatsiya (Sofia)*, **19**, 22 (1969); *Chem. Abstr.*, **72**, 31566 (1970).
1132. L. Daleva, M. Nikolova and St. Vankov, *Tr. Nauchnoizsled Khim.-Farm. Inst.*, **7** 217 (1972); *Chem. Abstr.*, **79**, 27212 (1973).
1133. R. J. Chudgar and K. N. Trivedi, *J. Indian Chem. Soc.*, **49**, 41 (1972).
1134. C. E. Kashov and D. J. Cook, *J. Amer. Chem. Soc.*, **67**, 1969 (1945).
1135. D. R. Brittain, E. D. Brown, W. Hepworth, and G. J. Stacey, *Brit. Pat.* 1 502 312; *U.S. Pat.* 4 066 651 (1978); *Ger. Offen.* 2 611 824 (1976); *Chem. Abstr.*, **86**, 16558 (1977); *Austrian Pat.* 347956 (1979); *Chem. Abstr.*, **91**, 56836 (1979).
1136. M. T. El-Wassimi, M. Kamel, and G. El-Sarraf, *Pharmazie*, **38**, 201 (1983).
1137. J. Brown, K. L. Sides, T. D. Fulmer, and C. F. Beam, *J. Heterocycl. Chem.*, **16**, 1669 (1979).
1138. C. Kan-Tan, B. C. Das, P. Boiteau, and P. Potier, *Phytochemistry*, **9**, 1283 (1970).
1139. M. F. Grundon and K. J. James, Unpublished work, reported in R. H. F. Manske and R. Rodrigo (Eds.), *The Alkaloids*, Vol. 17, p. 172, Academic Press, New York, 1979.
1140. K. A. Khan and A. Shoeb, *Indian J. Chem.*, **24B**, 62 (1985).
1141. C. Fournier and J. Decombe, *Compt. Rend.*, **265C**, 1169 (1967).
1142. C. Fournier and J. Decombe, *Bull. Soc. Chim. Fr.*, **1968**, 364.
1143. A. Schönberg and E. Singer, *Chem. Ber.*, **103**, 3871 (1970).
1144. M. M. Sidky, L. S. Boulos, and E. M. Yakout, *Z. Naturforsch.*, **40B**, 839 (1985).
1145. D. Jahn, R. Becker, M. Keil, W. Spiegler, and B. Würzer, *Ger. Offen.* 3 310 418 (1984); *Chem. Abstr.*, **102**, 78716 (1985).
1146. T. Naito and C. Kaneko, *Chem. Pharm. Bull.*, **31**, 366 (1983).
1147. L. A. Mitscher, M. S. Bathala, G. W. Clark, and J. L. Beal, *Lloydia*, **38**, 117 (1975).
1148. K. Szendrei, M. Petz, I. Novak, J. Reisch, H. E. Bailey and V. L. Bailey, *Herba Hung.*, **13**, 49 (1974); *Chem. Abstr.*, **83**, 40169 (1975).
1149. J. Reisch, J. Körösi, K. Szendrei, I. Novak, and E. Minker, *Phytochemistry*, **14**, 2722 (1975).
1150. A. Rüegger and D. Stanffacher, *Helv. Chim. Acta*, **46**, 2329 (1963).
1151. Zs. Rozsa, K. Szendrei, I. Novak, E. Minker, M. Koltai, and J. Reisch, *J. Chromatogr.*, **100**, 218 (1974).
1152. J. F. Ayafor, B. L. Sondengam, and B. T. Ngadjui, *Phytochemistry*, **21**, 955 (1982).
1153. L. V. Gyul'budagyan and E. T. Grigoryan, *Arm. Khim. Zh.*, **22**, 936 (1969); *Chem. Abstr.*, **72**, 43388 (1970).
1154. J. L. Asherson and D. W. Young, *J. Chem. Soc., Perkin Trans. I*, **1980**, 512.
1155. A. Bessonova, Ya. V. Rashkes, and S. Yu. Yunesov, *Khim. Prir. Soedin.*, **1976**, 217; *Chem. Nat. Compd. (Engl. Transl.)* **12**, 193 (1976); *Chem. Abstr.*, **85**, 177707 (1976).
1156. M. Boetius, E. Carstens, and C. Meyer, *Ger. Pat. (East)*, 13870; *Chem. Abstr.*, **53**, 11415 (1959).
1157. W. Spitzner, *Ger. Pat.* 1 253 711 (1967); *Chem. Abstr.*, **68** 105019 (1968).
1158. K. E. Schulte, V. von Weisenborn, and G. L. Tittel, *Chem. Ber.*, **103**, 1250 (1970).
1159. R. Cassis, R. Tapia, and J. A. Valderrama, *J. Heterocycl. Chem.*, **21**, 865 (1984).
1160. I. Ninomiya, T. Kiguchi, S. Yamauchi, and T. Naito, *J. Chem. Soc., Perkin Trans. I*, **1980**, 197.

1161. I. Ninomiya, T. Naito, and T. Kiguchi, *Japan Kokai* 75 69,085 (1975); *Chem. Abstr.*, **84**, 43879 (1976).

1162. E. J. Reist, H. P. Hamlow, I. G. Junga, R. M. Silverstein, and B. R. Baker, *J. Org. Chem.*, **25**, 1368 (1960).

1163. M. Takahashi, *Itsuu Kenkyusho Nempo*, **1971**, 65; *Chem. Abstr.*, **77**, 61854 (1972).

1164. E. Ochiai and T. Dodo, *Itsuu Kenkyusho Nempo*, **1967**, (14) 33; *Chem. Abstr.*, **67**, 64159 (1967).

1165. E. Ochiai and M. Takahashi, *Itsuu Kenkyusho Nempo*, **1965**, (14), 5; *Chem. Abstr.*, **67**, 64158 (1967).

1166. H. Tani, M. Otani, T. Mitzutani, and K. Mashimo, *Japan Pat.* 69 05,227 (1969); *Chem. Abstr.*, **70**, 115018 (1969).

1167. G. I. Zhungietu, L. A. Vlad, and F. N Chukhrii, *Khim. Geterotsikl. Soedin.*, **1976**, 797; *Chem. Heterocycl. Compd. (Engl. Transl.)* **1976** 665; *Chem. Abstr.*, **85**, 123886 (1976).

1168. R. M. Acheson, M. C. K. Choi, and R. M. Letcher, *J. Chem. Soc., Perkin Trans. I*, **1981**, 3141.

1169. P. J. Abbott, R. M. Acheson, G. Procter, and D. J. Watkin, *Acta Crystallogr., Sect. B.*, **B34**, 2165 (1978).

1170. K. Nakagawa, S. Yoshizaki, K. Tanimura, and S. Tamada, *U.S. Pat.* 3 975 391 (1976).

1171. T. Shimizu and Y. Tamura, *Japan Kokai* 76 16,678 (1976); *Chem. Abstr.*, **85**, 32874 (1976).

1172. Otsuka Pharmaceutical Co. Ltd., *Japan Kokai Tokkyo Koho* 58 43,952 (1983); *Chem. Abstr.*, **99**, 158466 (1983).

1173. T. Kato and M. Noda, *Chem. Pharm. Bull.*, **24** 303 (1976).

1174. T. Chiba, H. F. Fukumi, M. Noda, and T. Kato, *Hukusokan Kagaku Toronkai Koen Yoshishu*, 8th, 1975, p. 79; *Chem. Abstr.*, **85** 5461 (1976).

1175. A. Sammour, A. R. A. Raouf, A. H. Mustafa, and M. M. Habashy, *Egypt. J. Chem.*, **19**, 27 (1976): *Chem. Abstr.*, **92**, 6382 (1980).

1176. J. H. Rigby and F. J. Burkhardt, *J. Org. Chem.*, **51**, 1374 (1986).

1177. H. Juraszyk, H. J. Enenkel, K. O. Minck, H. J. Shliep, and J. Piulats, *Eur. Pat. Appl.* 154190 (1985); *Chem. Abstr.*, **104**, 224890 (1986).

1178. T. Moriwake and H. Namba, *J. Med. Chem.*, **11**, 636 (1968).

1179. M. McHugh and G. R. Proctor, *J. Chem. Res., Synop.*, **1984**, 246; *Miniprint*, p. 2230.

1180. W. N. Speckamp and H. O. Huisman, *Recl. Trav. Chim. Pays-Bas*, **85**, 681 (1966).

1181. F. Gatta, M. Tomassetti, V. Zaccari, and R. Landi Vittory, *Eur. J. Med. Chem. – Chim. Ther.*, **9**, 133 (1974).

1182. A. Dreimane, E. E. Grinshtein, and E. I. Stankevic, *Khim. Geterotsikl. Soedin.*, **1980**, 791; *Chem. Heterocycl. Compd. (Engl. Transl.)*, **1980**, 612; *Chem. Abstr.*, **93**, 186126 (1980).

1183. E. Grinshtein, E. I. Stankevic, and G. Dubur, *Khim. Geterotsikl. Soedin.*, **1967**, 1118; *Chem. Heterocycl. Compd. (Engl. Transl.)*, **1967** 866; *Chem. Abstr.*, **69**, 77095 (1968).

1184. F. Bossert and W. Vater, *Ger. Offen.* 2 003 148 (1971); *Chem. Abstr.*, **75**, 98459 (1971).

1185. I. Chaaban, J. V. Greenhill, and P. Akhtar, *J. Chem. Soc., Perkin Trans. I*, 1979, 1593.

1186. J. V. Greenhill, *J. Chem. Soc. (C)*, **1971**, 2699.

1187. Y. Tamura, T. Sakaguchi, T. Kawasaki, and Y. Kita, *Heterocycles*, **2**, 645 (1974).

1188. Y. Tamura, T. Sakaguchi, T. Kawasaki, and Y. Kita, *Chem. Pharm. Bull.*, **24**, 1160 (1976).

1189. C. L. Nichols, E. C. Kornfeld, M. M. Foreman, J. M. Schaus, D. L. Huser, R. N.

Booker, and D. T. Wong, *Eur. Pat. Appl.* 139393 (1985); *Chem. Abstr.*, **103**, 196115 (1985).

1190. C. L. Nichols and E. C. Kornfeld, *U.S. Pat.* 4 501 890 (1985); *Chem. Abstr.*, **103**, 123525 (1985).
1191. K. Wiesner, V. Musil, and K. J. Wiesner, *Tetrahedron Lett.*, **1968**, 5643.
1192. H. Dugas, M. E. Hazenburg, Z. Valenta, and K. Wiesner, *Tetrahedron Lett.* **1967**, (49), 4931.
1193. H. Inouye, S. Ueda, K. Inoue, T. Hayashi, and T. Hibi, *Tetrahedron Lett.*, **1975**, 2395.
1194. H. Inouye, S. Ueda, K. Inoue, T. Hayashi, and T. Hibi, *Chem. Pharm. Bull.*, **23**, 2523 (1975).
1195. S. Boatman, T. M. Harris, and C. R. Hauser, *J. Amer. Chem. Soc.*, **87**, 5198 (1965).
1196. T. M. Harris, S. Boatman, and C. R. Hauser, *J. Amer. Chem. Soc.*, **87**, 3186 (1965).
1197. G. Bradley, J. Clark, and W. Kernick, *J. Chem. Soc., Perkin Trans. I*, **1972**, 2019.
1198. B. Eistert, A. Schmitt, and T. J. Arackal, *Chem. Ber.*, **109**, 1549 (1976).
1199. K. Nakagawa, S. Yoshizaki, K. Tanimura, and S. Tamada, *Ger. Offen.* 2 461 604 (1975); *Chem. Abstr.*, **83**, 164017 (1975).
1200. K. Nakagawa, S. Yoshizaki, K. Tanimura, and S. Tamada, *Ger. Offen* 2 461 861 (1975); *Chem. Abstr.*, **83**, 193105 (1975).
1201. K. Nakagawa, S. Yoshizaki, K. Tanimura, and S. Tamada, U.S. Pat. 3 994 901 (1976).
1202. K. Nakagawa, S. Yoshizaki, K. Tanimura, and S. Tamada, *U.S. Pat.* 4 145 542 (1979).
1203. T. Shimizu and Y. Tamura, *Japan Kokai* 76 115,480 (1976); *Chem. Abstr.*, **86**, 121190 (1977).
1204. Y. Oshiro and K. Nakagawa, *Japan Kokai* 78 28,180 (1978); *Chem. Abstr.*, **89**, 24179 (1978).
1205. K. Nakagawa, T. Sato, T. Nishi, Y. Oshiro, and K. Yamamoto, *Japan Kokai* 77 73,874 (1977); *Chem. Abstr.*, **87**, 167896 (1977).
1206. K. Nakagawa, T. Nishi, and Y. Oshiro, *Japan Kokai*, 76 68,574 (1976); *Chem. Abstr.*, **86**, 72463 (1977).
1207. Otsuka Pharmaceutical Co. Ltd., *Japan Kokai Tokkyo Koho* 83 126,868 (1983); *Chem. Abstr.*, **100**, 6353 (1984).
1208. Otsuka Pharmaceutical Co. Ltd., *Japan Kokai Tokkyo Koho* 83 135,865 (1983); *Chem. Abstr.*, **100**, 6354 (1984).
1209. K. Sakano and K. Nakagawa, *Japan Kokai Tokkyo Koho*, 79 44,678 (1979); *Chem. Abstr.*, **91**, 107911 (1979).
1210. K. Nakagawa, T. Nishi, and Y. Oshiro, *Japan Kokai* 76 68,575 (1976); *Chem. Abstr.* **86**, 89632 (1977).
1211. K. Nakagawa, T. Nishi, and Y. Oshiro, *Japan Kokai* 76 125,291 (1976); *Chem. Abstr.*, **87**, 53098 (1977).
1212. Otsuka Chemical Co. Ltd., *Japan Kokai Tokkyo Koho* 81 156,263 (1981); *Chem. Abstr.*, **96**, 104110 (1982).
1213. K. Nakagawa, T. Nishi, and Y. Oshiro, *Japan Kokai* 76 118,770 (1976); *Chem. Abstr.*, **86**, 171277 (1977).
1214. Otsuka Pharmaceutical Co. Ltd., *Japan Kokai Tokkyo Koho* 83 96,022 (1983); *Chem. Abstr.*, **99**, 93757 (1983).
1215. K. Sakano, T. Fujioka, and K. Nakagawa, *Japan Kokai Tokkyo Koho* 79 16,478 (1979); *Chem. Abstr.*, **91**, 5123 (1979).
1216. Yoshitomi Pharmaceutical Industries Ltd., *Japan Kokai Tokkyo Koho* 80 83,749 (1980); *Chem. Abstr.*, **94**, 30584 (1981).

1217. K. Nakagawa, T. Sato, T. Nishi, Y. Oshiro, and K. Yamamoto, *Japan Kokai* 77 73,878 (1977); *Chem. Abstr.*, 87, 167904 (1977).

1218. K. Nakagawa, T. Nishi, and Y. Oshiro, *Japan Kokai* 76 118,771 (1976); *Chem. Abstr.*, **86**, 171276 (1977).

1219. T. Nishi, Y. Oshiro, and K. Nakagawa, *Japan Kokai* 77 118,474 (1977); *Chem. Abstr.*, **88**, 62306 (1978).

1220. K. Nakagawa, T. Sato, T. Nishi, Y. Oshiro, and K. Yamamoto, *Japan Kokai* 77 73,866 (1977); *Chem. Abstr.*, **87**, 167899 (1977).

1221. K. Nakagawa, T. Sato, T. Nishi, Y. Oshiro, and K. Yamamoto, *Japan Kokai* 77 73,879 (1977); *Chem. Abstr.*, **87** 167902 (1977).

1222. Yoshitomi Pharmaceutical Industries Ltd., *Japan Kokai Tokkyo Koho* 80 141,483 (1980); *Chem. Abstr.*, **94**, 121578 (1981).

1223. T. Nishi, K. Yamamoto, T. Shimizu, T. Kanbe, Y. Kimura, and K. Nakagawa, *Chem. Pharm. Bull.*, **31**, 798 (1983).

1224. M. Uchida, T. Nishi, and K. Nakagawa, *Japan Kokai Tokkyo Koho* 80 35,018 (1980); *Chem. Abstr.*, **93**, 26297 (1980).

1225. T. Nakao, S. Setoguchi, and O. Yaoka, *U.S. Pat.* 4 258 185 (1981); *Belg. Pat.* 871310 (1979); *Chem. Abstr.*, **91**, 74639 (1979).

1226. Yoshitomi Pharmaceutical Industries Ltd., *Japan Kokai Tokkyo Koho* 80 83,781 (1980); *Chem. Abstr.*, **94**, 47349 (1981).

1227. Yoshitomi Pharmaceutical Industries Ltd., *Japan Kokai Tokkyo Koho* 80 53,284 (1980); *Chem. Abstr.*, **93**, 220765 (1980).

1228. K. Nakagawa, Y. Oshiro, and T. Nishi, *Japan Kokai* 76 110,574 (1976); *Chem. Abstr.*, **86** 155526 (1977).

1229. K. Nakagawa, Y. Oshiro, and T. Nishi, *Japan Kokai* 78 23,305 (1978); *Chem. Abstr.*, **89**, 215234 (1978).

1230. K. Nakagawa, Y. Oshiro, and T. Nishi, *Japan Kokai* 76 110,575 (1976); *Chem. Abstr.*, **86**, 171274 (1977).

1231. K. Nakagawa, Y. Oshiro, and T. Nishi, *Japan Kokai* 78 23,306 (1978); *Chem. Abstr.*, **89**, 215235 (1978).

1232. K. Nakagawa, Y. Oshiro, and T. Nishi, *Japan Tokkyo Koho* 78 37,867 (1978); *Chem. Abstr.*, **90**, 152020 (1979).

1233. K. Nakagawa, Y. Oshiro, and T. Nishi, *Japan Kokai* 76 110,576 (1976); *Chem. Abstr.*, **86**, 189737 (1977).

1234. K. Nakagawa, S. Yoshizaki, K. Tanimura, and S. Tamada, *Japan Kokai* 51 141,880 (1976); *Chem. Abstr.*, **87**, 5825 (1977).

Author Index

This author index is intended to enable the reader to locate an author's name and work with the aid of the reference numbers in the text. Page numbers appear first and are followed by the reference numbers in parentheses.

Subject Index

Compounds which are in the tables, but not in the text do not appear in the subject index, although table headings are included. Most compounds are under their systematic names, listed in the style of Chemical Abstracts Chemical Substance Index.